Comprehensive

DISCRETE MATHEMATICS

B.A./B.Sc. III Year

Salient features of the present edition :

- ★ It has **detailed theory** supplemented with well explained **examples.**
- ★ It has **adequate number of unsolved problems** of all types in exercises.
- ★ It has **Working rules for solving problems** before exercises.
- ★ It has **Hints of tricky problems** after relevant exercises.
- ★ It has **questions** appeared in the latest university examinations.

Other books by the same author :

1. **Comprehensive Mathematics XI** (*CBSE New Syllabus*)
2. **Comprehensive Mathematics XI** (*Haryana Board*)
3. **Comprehensive Mathematics XII** (*CBSE New Syllabus and Haryana Board*)
4. **Comprehensive IIT-JEE Screening Model Test Papers**
5. **Comprehensive Objective Mathematics** (*For Engg. Entrance Tests*)
6. **Comprehensive Mathematics** (*For AIEEE*)
7. **Comprehensive Applied Mathematics** (*For Diploma Students*)
8. **Comprehensive Mathematics** (*For Indian Civil Services Exam.*)
9. **Comprehensive Algebra and Trigonometry** (*For B.A./B.Sc. I*)
10. **Comprehensive Calculus** (*For B.A./B.Sc. I*)
11. **Comprehensive Differential Equations and Calculus of Variations** (*For B.A./B.Sc. II*)
12. **Comprehensive Differential Equations and Differential Geometry** (*For B.A./B.Sc. II*)
13. **Comprehensive Abstract Algebra** (*For B.A./B.Sc. III*)
14. **Comprehensive Business Mathematics** (*For B.Com. I and B.T.M.*)
15. **Comprehensive Business Statistics** (*For B.Com. II, B.I.M. and B.B.A.*)
16. **An Introduction to Computers**

Comprehensive

DISCRETE MATHEMATICS

(B.A./B.Sc. IIIrd Year)

Paper III

(Strictly as per MDU and KU Haryana, and other Indian Universities following U.G.C. Syllabus)

(Also for B.E./B.Tech. (Computer Science), B.C.A., M.C.A., M.Sc. (Computer Science))

By

PARMANAND GUPTA
M.Sc., M.Phil. (Maths)
M.A.(Economics)
Head, Deptt. of Mathematics
Indira Gandhi National College, Ladwa
Kurukshetra University, Kurukshetra
Haryana

LAXMI PUBLICATIONS (P) LTD
An ISO 9001:2015 Company
BENGALURU • CHENNAI • GUWAHATI • HYDERABAD
KOCHI • KOLKATA • LUCKNOW • RANCHI
NEW DELHI

Printed and bound in India

Second Edition : 2008, Reprint : 2011, New Edition : 2025

ISBN : 978-81-3180-263-2

Branches

✆	Bengaluru	080-26 75 69 30
✆	Chennai	044-24 34 47 26
✆	Guwahati	0361-254 36 69
✆	Hyderabad	040-27 55 53 83
✆	Kochi	0484-405 13 03
✆	Kolkata	033-40 04 77 79
✆	Lucknow	0522-430 36 13
✆	Ranchi	0651-224 24 64

PUBLISHED IN INDIA BY

LAXMI PUBLICATIONS (P) LTD

An ISO 9001:2015 Company
113, GOLDEN HOUSE, GURUDWARA ROAD, DARYAGANJ,
NEW DELHI - 110002, INDIA

Corporate Office : B-33, Sector-5, Noida,
Gautambuddha Nagar (U.P.) - 201 301, INDIA
Telephone : +91-120-44 46 900
www.laxmipublications.com
info@laxmipublications.com

PREFACE

It is a great pleasure for me to write the preface of this book. I express my heartfelt thanks to my worthy colleagues for conveying their suggestions during the writing work of this book.

The topics covered in the subject **Discrete Mathematics** are peculiar in nature. No abstract mathematical background is required for understanding the topics in Discrete mathematics. Knowledge relating to computers is also not required. Fundamentals of mathematics studied upto 10th class are sufficient to study this subject.

The following are some of the salient features of the present edition :

★ **The present edition is in strict confirmity with the U.G.C. Model Curriculum for the new session.**

★ **A new feature "Working Rules for Solving Problems" has been introduced. This would help the students to tackle various types of problems more confidently.**

★ **Questions appeared in latest examinations have also been included in the book.**

★ **The related, solved and unsolved questions have been grouped together to facilitate their location. This would help the readers to enjoy solving all problems relating to a particular topic, at a stretch.**

★ **The problems, which are comparatively harder, are provided with suitable hints. The author has not desired to dishearten the students by incorporating very tedious problems, which are not expected from the students of average calibre.**

I firmly believe that there is always scope for improvement. Suggestions and comments from worthy teachers would be gratefully received and duly acknowledged.

—Author

SYLLABUS

BM 304—DISCRETE MATHEMATICS

(FOR B.A./B.Sc. III students, Paper III(ii))

SECTION I (Three Questions)

Sets: Countability of sets, Cardinality of sets, Principle of inclusion and exclusion, Partially ordered (p.o.) sets, Dual p.o. sets, Totally ordered sets, Chains, Antichains, Isomorphism of ordered sets, Hasse diagram, Pigeonhole principle and its applications.

Discrete probability : Mutually exclusive and exhaustive system of events, Axiom of calculus of probability, Law of addition of probability, Conditional probability, Multiplication law of probability, Independent events, Bayes theorem, Probability distribution, Binomial distribution.

SECTION II (Two Questions)

Recurrence relations : Explicit formula for a sequence, Fibonacci sequence, Solution of recurrence relations using backtracking, Linear recurrence relations with constant coefficients, Homogeneous solutions, Particular solution and total solution of linear recurrence relations, Discrete numeric functions, Generating functions, Convolution of numeric functions, Solution of recurrence relations by the method of generating functions.

Propositions: Definition and types, Basic logical operations, Truth tables, DeMorgan's and commutative laws.

SECTION III (Three Questions)

Lattices : Definition, Examples and simple properties of lattices, Lattices as algebraic system, Equivalence of two definitions, Definition and examples of distributive and complimented lattices.

Boolean algebra : Definition, Examples and basic properties of Boolean algebras, Switching algebras, Boolean expressions, Fundamental product, Sum-of-product forms, Logic gates and circuits, OR gate, AND gate, NOT gate, Simple logic circuits, Boolean functions and their truth tables.

SECTION IV (Two Questions)

Graphs : Definition and examples, Multigraphs, Euler's first theorem of graph theory, Complete graph, Regular graph, Bipartite graph, Subgraph, Paths and circuits, Eulerian paths and circuits, Weighted graphs, Shortest paths, Travelling salesman problem, Definition and examples of planar graphs. Trees : Definition and simple examples of trees.

Finite state machines : Simple examples of finite state machines, Equivalence of finite state machines.

Note : The paper-setter is requested to set ten questions in all. The candidate is required to attempt five questions, selecting at least one question from each section.

CONTENTS

Dedicated to my teachers whose teachings have made me able to grasp some of the fundamentals of Mathematics

SECTION I

1

Principle of Inclusion and Exclusion

1.1. INTRODUCTION

We know that the concept of set is indispensible in almost every branch of mathematics. A set having finitely many elements is called a *finite set* and otherwise it is called an *infinite set*. In this chapter, we shall concentrate only on finite sets and study to find the number of elements in finite sets by using the *principle of inclusion and exclusion*.

The number of elements in a finite set A is denoted by $n(A)$ or by $|A|$.

$n(A)$ is also called the *cardinality* of the set A.

For example $n(\phi) = 0$, $n(A) = 4$, where $A = \{2, 3, 7, 8\}$.

Theorem 1. If A and B be two finite disjoint sets, then prove that

$$\mathbf{n(A \cup B) = n(A) + n(B).}$$

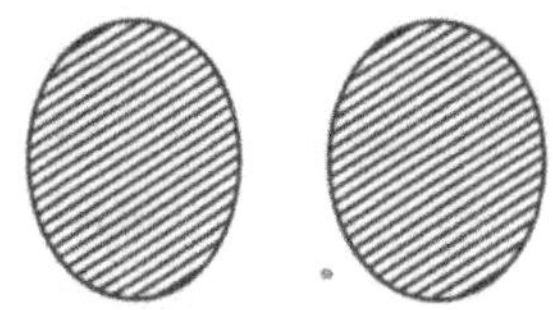

Proof. $A \cup B$ contains all the elements of A and these are $n(A)$ in number. The only other elements of $A \cup B$ are those elements of B which are not in A. Since A and B are disjoint, no element of B is in A, so there are $n(B)$ elements which are in B and not in A.

$$\therefore \quad \mathbf{n(A \cup B) = n(A) + n(B).}$$

Corollary. If A and B are finite sets, then prove that

$$\mathbf{n(A - B) = n(A) - n(A \cap B).}$$

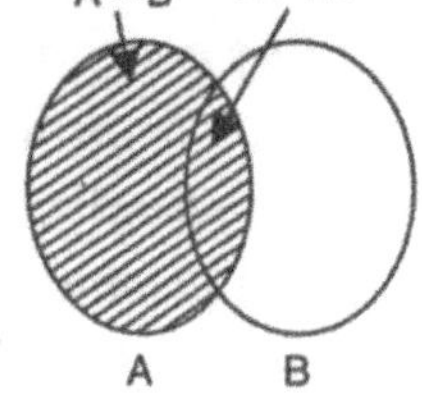

Proof. We have $A = (A - B) \cup (A \cap B)$ and the sets $A - B$ and $A \cap B$ are disjoint sets.

$\therefore$ By **Theorem 1,**

$$n(A) = n(A - B) + n(A \cap B)$$

$$\Rightarrow \quad \mathbf{n(A - B) = n(A) - n(A \cap B).}$$

1.2. PRINCIPLE OF INCLUSION AND EXCLUSION FOR TWO FINITE SETS

Statement. If A and B be two finite sets then prove that

$$\mathbf{n(A \cup B) = n(A) + n(B) - n(A \cap B).}$$

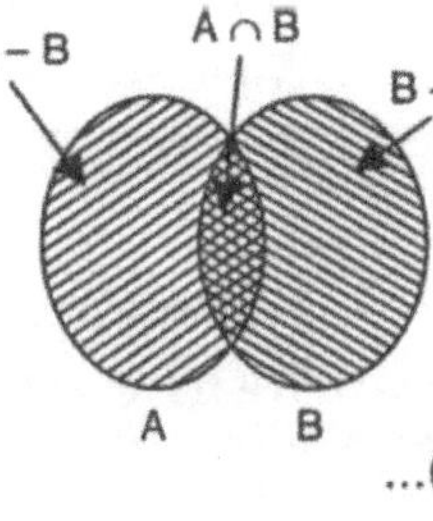

Proof. The sets $A - B$, $A \cap B$ and $B - A$ are mutually disjoint and their union is $A \cup B$.

$$\therefore \quad n(A \cup B) = n(A - B) + n(A \cap B) + n(B - A) \quad ...(1)$$

The sets $A - B$ and $A \cap B$ are disjoint and their union is A.

$$\therefore \quad n(A) = n(A - B) + n(A \cap B)$$

$$\Rightarrow \quad n(A - B) = n(A) - n(A \cap B) \quad ...(2)$$

The sets $A \cap B$ and $B - A$ are disjoint and their union is B.

$\therefore \quad n(B) = n(A \cap B) + n(B - A)$

$\Rightarrow \quad n(B - A) = n(B) - n(A \cap B)$ ...(3)

Putting the values of $n(A - B)$ and $n(B - A)$ from (2) and (3) in (1), we get

$$n(A \cup B) = (n(A) - n(A \cap B)) + n(A \cap B) + (n(B) - n(A \cap B))$$

$\Rightarrow \quad \mathbf{n(A \cup B) = n(A) + n(B) - n(A \cap B).}$

Example 1. *Let A and B be two finite sets such that $n(A - B) = 15$, $n(A \cup B) = 90$, $n(A \cap B) = 30$. Find $n(B)$.*

Sol. We have $n(A \cup B) = n(A - B) + n(A \cap B) + n(B - A)$

$\Rightarrow \quad 90 = 15 + 30 + n(B - A) \qquad \therefore \quad n(B - A) = 90 - 45 = 45$

Now $\quad n(B) = n(A \cap B) + n(B - A) \qquad \therefore \quad n(B) = 30 + 45 = \mathbf{75}.$

Example 2. *Out of 500 car owners investigated, 400 owned cars A, 200 owned cars B and 50 owned both A and B cars. Is this data correct ?*

Sol. Let $\quad X$ = set of all car owners

S = set of all owners of car A

T = set of all owners of car B.

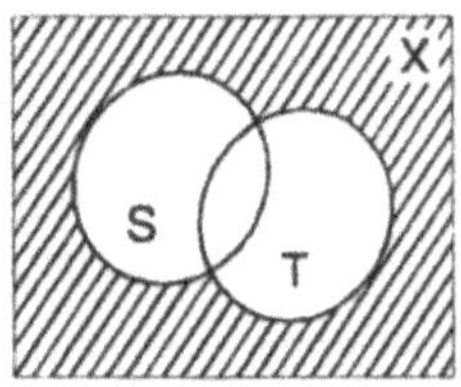

$\therefore \quad n(X) = 500, n(S) = 400, n(T) = 200, \; n(S \cap T) = 50$

We have $\quad n(S \cup T) = n(S) + n(T) - n(S \cap T).$

$\Rightarrow \quad n(S \cup T) = 400 + 200 - 50 = 550$

Also $\quad S \cup T \subseteq X$

$\Rightarrow \quad n(S \cup T) \le n(X) \;\Rightarrow\; 550 \le 500$, which is impossible.

$\therefore$ Given data is not correct.

Example 3. *There are 200 individuals with a skin disorder, 120 had been exposed to the chemical C_1, 50 to chemical C_2, and 30 to both the chemicals C_1 and C_2. Find the number of individuals exposed to :*

(*i*) *chemical C_1 but not chemical C_2*

(*ii*) *chemical C_2 but not chemical C_1*

(*iii*) *chemical C_1 or chemical C_2.*

Sol. Let $\quad X$ = set of individuals with skin disorder

A = set of individuals exposed to chemical C_1

B = set of individuals exposed to chemical C_2.

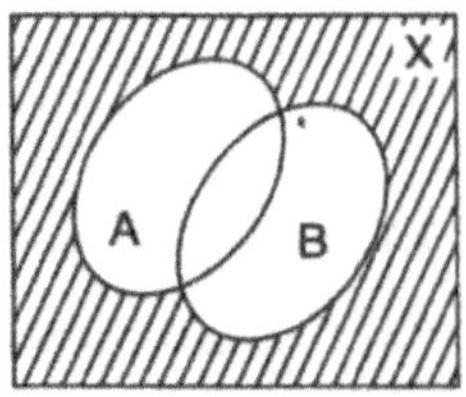

$\therefore \quad n(X) = 200, n(A) = 120, n(B) = 50, n(A \cap B) = 30$

(*i*) Required number $= n(A - B)$

We have $\quad n(A) = n(A - B) + n(A \cap B)$

$\therefore \quad 120 = n(A - B) + 30$

$\therefore \quad n(A - B) = 120 - 30 = \mathbf{90}.$

(*ii*) Required number $= n(B - A)$

We have $\quad n(B) = n(B - A) + n(B \cap A)$

$\therefore \quad 50 = n(B - A) + 30$

$\therefore \quad n(B - A) = 50 - 30 = \mathbf{20}.$

(*iii*) Required number = $n(A \cup B)$

We have $n(A \cup B) = n(A) + n(B) - n(A \cap B)$

$\therefore$ $n(A \cup B) = 120 + 50 - 30 =$ **140.**

Example 4. *In a survey of 600 students in a school, 150 students were found to be drinking tea and 225 drinking coffee, 100 were drinking both tea and coffee. Find how many students were drinking neither tea nor coffee.*

Sol. Let X = set of all students

T = set of students drinking tea

C = set of students drinking coffee

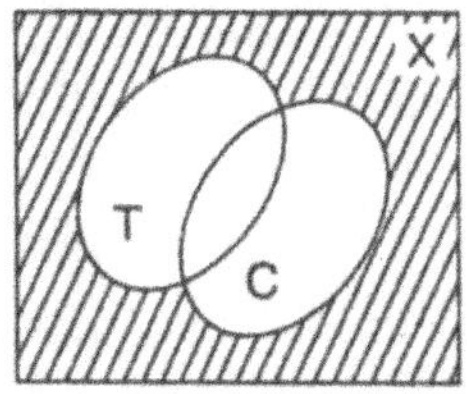

$\therefore$ $n(X) = 600, n(T) = 150, n(C) = 225, n(T \cap C) = 100$

We have $n(T \cup C) = n(T) + n(C) - n(T \cap C)$

$\therefore$ $n(T \cup C) = 150 + 225 - 100 = 275$

$\therefore$ $n(T' \cap C') = n((T \cup C)')$

$= n(X) - n(T \cup C)$

$= 600 - 275 = 325$

$\therefore$ No. of students drinking neither tea nor coffee = **325.**

Example 5. *A market research group conducted a survey of 1000 consumers and reported that 720 consumers liked product A and 450 consumers liked product B. What is the least number that must have liked both products ?*

Sol. Let X = set of all consumers

P = set of consumers liking A

Q = set of consumers liking B

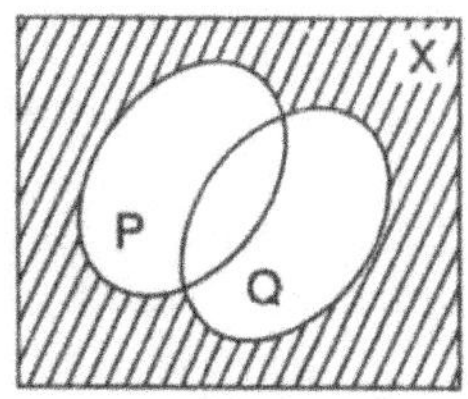

$\therefore$ $n(X) = 1000, n(P) = 720, n(Q) = 450$

We have $n(P \cup Q) = n(P) + n(Q) - n(P \cap Q)$

$\therefore$ $n(P \cup Q) = 720 + 450 - n(P \cap Q)$

$\Rightarrow$ $n(P \cup Q) = 1170 - n(P \cap Q)$

$\Rightarrow$ $n(P \cap Q) = 1170 - n(P \cup Q)$ ...(1)

$\therefore$ $n(P \cap Q)$ is least when $n(P \cup Q)$ is maximum.

Maximum value of $n(P \cup Q) = n(X)$ *i.e.*, 1000

$\therefore$ (1) $\Rightarrow$ Least value of $n(P \cap Q) = 1170 - 1000 =$ **170.**

Example 6. *There are 20 students in a Chemistry class and 30 students in a Physics class. Find the number of students which are either in Physics class or Chemistry class in the following cases :*

(*i*) *two classes meet at the same hour*

(*ii*) *the two classes meet at different hours and ten students are enrolled in both the courses.*

Sol. Let C = set of students in Chemistry class

P = set of students in Physics class

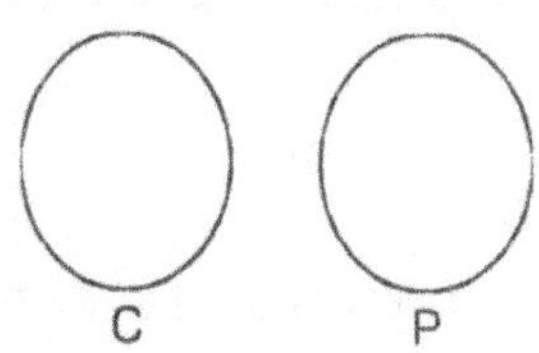

$\therefore$ $n(C) = 20, n(P) = 30$

(*i*) Since the classes meet at the same hour, there cannot be any student enrolled in both the courses.

$\therefore$ $n(C \cap P) = \phi$

$\therefore$ $n(C \cup P) = n(C) + n(P) = 20 + 30 = \mathbf{50}.$

(*ii*) We have $n(C \cap P) = 10$

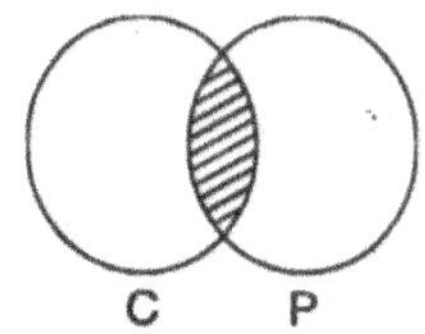

Since the classes meet at different hours, these 10 students would be able to attend both the courses.

$\therefore$ $n(C \cup P) = n(C) + n(P) - n(C \cap P)$

$= 20 + 30 - 10 = \mathbf{40}.$

Example 7. *A town has total population 25,000 out of which 13,000 read 'The Hindustan Times' and 10,500 read 'The Indian Express' and 2,500 read both papers. Find the percentage of population who read neither of these newspapers.*

Sol. Let X = set of persons in the town

A = set of persons who read 'The Hindustan Times'

B = set of persons who read 'The Indian Express'

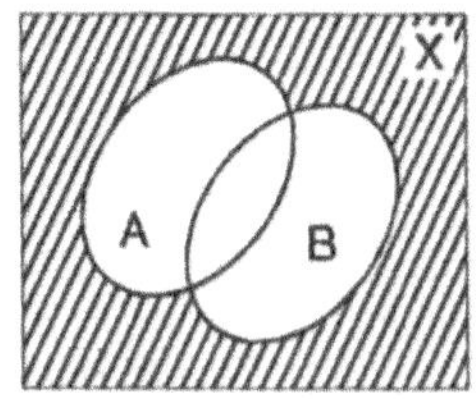

$\therefore$ $n(X) = 25{,}000,\ n(A) = 13{,}000,$

$n(B) = 10{,}500,\ n(A \cap B) = 2{,}500$

We have $n(A \cup B) = n(A) + n(B) - n(A \cap B)$

$= 13{,}000 + 10{,}500 - 2{,}500 = 21{,}000$

Number of persons not reading any of these newspapers is $n(X) - n(A \cup B) = n(A \cup B)'$ and the corresponding set is shown by shaded lines in the diagram.

$$n((A \cup B)') = 25{,}000 - 21{,}000 = 4{,}000$$

$\therefore$ % of persons not reading any of these newspapers $= \dfrac{4{,}000}{25{,}000} \times 100 = \mathbf{16\%}.$

Example 8. *A survey shows that 63% of the Americans like cheese whereas 76% like apples. If x% of the Americans like both cheese and apples, find the value of x.*

Sol. Let A = set of Americans liking cheese

B = set of Americans liking apples

$\therefore$ By the given data : $n(A) = 63\%,\ n(B) = 76\%,\ n(A \cap B) = x\%.$

Let total number of Americans be 100.

$\therefore$ $n(A) = 63,\ n(B) = 76,\ n(A \cap B) = x$

We have $n(A \cup B) \le 100$

$\Rightarrow$ $n(A) + n(B) - n(A \cap B) \le 100$

$\Rightarrow$ $63 + 76 - x \le 100 \Rightarrow x \ge 39$ ...(1)

Also, $A \cap B \subseteq A$ implies $n(A \cap B) \le n(A)$

$\Rightarrow$ $x \le 63$ ...(2)

Combining (1) and (2), we get $\mathbf{39 \le x \le 63}.$

EXERCISE 1.1

1. Let A and B be two finite disjoint sets such that $n(A \cup B) = 475$ and $n(A) = 435$. Find $n(B)$.
2. Let A and B be two finite sets such that $n(A \cap B) = 12$, $n(A - B) = 24$, $n(B - A) = 45$. Find $n(A)$ and $n(B)$.

3. Let A and B be two finite sets such that $n(A) = 115$, $n(B) = 326$ and $n(A - B) = 47$. Find $n(A \cup B)$ and $n(A \cap B)$.
4. There are 210 members in a club, 100 of them drink tea and 65 drink tea but not coffee.
   (*i*) How many drink coffee ? (*ii*) How many drink coffee but not tea ?
5. For a certain test, a candidate could offer English or Hindi or both subjects. Total number of candidates was 500, of whom 350 appeared in English and 90 in both subjects.
   (*i*) How many appeared in English only ? (*ii*) How many appeared in Hindi only ?
   (*iii*) How many appeared in Hindi ?
6. In a group of 400 people, 250 can speak English only and 70 can speak Hindi only.
   (*i*) How many can speak English ? (*ii*) How many can speak Hindi ?
   (*iii*) How many can speak both English and Hindi ?
7. If A and B be two sets containing 12 and 27 distinct elements respectively, what can be the minimum number of elements in $A \cup B$? Find also the maximum number of elements in $A \cup B$.
8. In a survey of 400 students in a school, 100 were listed as drinking apple juice, 150 as drinking orange juice and 75 were listed as drinking both apple as well as orange juice. Find how many students were drinking neither apple juice nor orange juice.
9. Among the positive integers less than or equal to 100, let P be the set of even integers and let Q be the set of integers divisible by 3. Determine how many integers are even or divisible by 3.
10. Among the positive integers less than or equal to 500, let P be the set of odd integers and let Q be the set of integers divisible by 5. Determine how many integers are odd or divisible by 5.

Answers

1.	40	2. 36, 57	3. 373, 68	4. (*i*) 145	(*ii*) 110	
5.	(*i*) 260	(*ii*) 150	(*iii*) 240	6. (*i*) 330	(*ii*) 150	(*iii*) 80
7.	27, 39	8. 225	9. 67	10. 300.		

1.3. PRINCIPLE OF INCLUSION AND EXCLUSION FOR THREE FINITE SETS

Statement. If A, B and C be three finite sets then prove that

$$\mathbf{n(A \cup B \cup C) = n(A) + n(B) + n(C) - n(A \cap B) - n(B \cap C) - n(A \cap C) + n(A \cap B \cap C).}$$

Proof. Considering $B \cup C$ as one set, we have

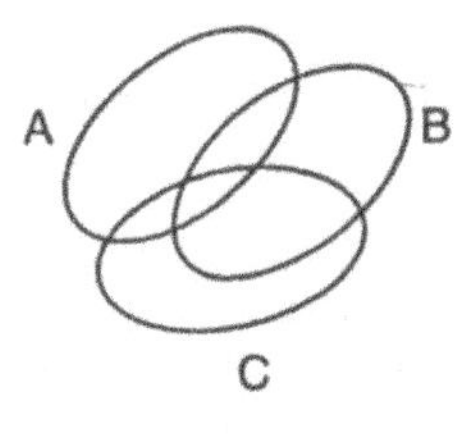

$$\begin{aligned} n(A \cup B \cup C) &= n(A \cup (B \cup C)) \\ &= n(A) + n(B \cup C) - n(A \cap (B \cup C)) \\ &= n(A) + [n(B) + n(C) - n(B \cap C)] \\ &\qquad - n((A \cap B) \cup (A \cap C)) \\ &= n(A) + n(B) + n(C) - n(B \cap C) \\ &\qquad - [n(A \cap B) + n(A \cap C) - n((A \cap B) \cap (A \cap C)) \\ &= n(A) + n(B) + n(C) - n(B \cap C) - n(A \cap B) \\ &\qquad - n(A \cap C) + n(A \cap B \cap C) \end{aligned}$$

$$\therefore \quad \mathbf{n(A \cup B \cup C) = n(A) + n(B) + n(C) - n(A \cap B) - n(B \cap C) - n(A \cap C) + n(A \cap B \cap C).}$$

1.4. PRINCIPLE OF INCLUSION AND EXCLUSION IN GENERAL

Statement. If $A_1, A_2, ..., A_n$ be n finite sets, then

$$n(A_1 \cup A_2 \cup ...... \cup A_n)$$

$$= \sum_{1 \le i \le n} n(A_i) - \sum_{1 \le i < j \le n} n(A_i \cap A_j) + \sum_{1 \le i < j < k \le n} n(A_i \cap A_j \cap A_k) + ...$$

$$+ (-1)^{n-1}\, n(A_1 \cap A_2 \cap ...... \cap A_n).$$

Remark. This result can be proved by using the '*principle of mathematical induction*'.

Example 1. *Of the number of three athletic teams in a school, 21 are in the basketball team, 26 in hockey team and 29 in the football team. 14 play hockey and basketball, 15 play hockey and football, 12 play football and basketball and 8 play all the games. How many members are there in all ?*

Sol. Let B = set of members of the basketball team

H = set of members in the hockey team

and F = set of members in the football team.

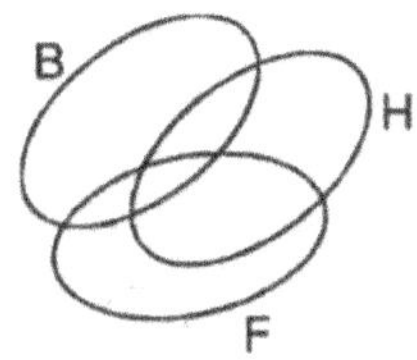

$\therefore$ $n(B) = 21, n(H) = 26, n(F) = 29, n(H \cap B) = 14, n(H \cap F) = 15,$

$n(F \cap B) = 12$ and $n(B \cap H \cap F) = 8$

$\therefore$ Total number of members $= n(B \cup H \cup F)$

$$= n(B) + n(H) + n(F) - n(B \cap H) - n(H \cap F) - n(B \cap F) + n(B \cap H \cap F)$$

$$= 21 + 26 + 29 - 14 - 15 - 12 + 8 = \mathbf{43}.$$

Example 2. *Each student in a class of 40, studies at least one of the subjects English, Mathematics and Economics. 16 study English, 22 Economics and 26 Mathematics, 5 study English and Economics, 14 Mathematics and Economics and 2 English, Economics and Mathematics. Find the number of students who study :*

(*i*) *English and Mathematics* (*ii*) *English, Mathematics but not Economics.*

Sol. Let A, B and C be the sets of students studying English, Economics and Mathematics respectively.

$\therefore$ $n(A \cup B \cup C) = 40,\ n(A) = 16,\ n(B) = 22,\ n(C) = 26,\ n(A \cap B) = 5,$

$n(C \cap B) = 14$ and $n(A \cap B \cap C) = 2$

(*i*) We have $n(A \cup B \cup C) = n(A) + n(B) + n(C) - n(A \cap B) - n(B \cap C)$

$- n(C \cap A) + n(A \cap B \cap C)$

$\therefore$ $40 = 16 + 22 + 26 - 5 - 14 - n(C \cap A) + 2$

$\Rightarrow$ $n(C \cap A) = 66 - 19 - 40 = 7$

$\therefore$ No. of students who study English and Mathematics $= n(A \cap C) = n(C \cap A) = \mathbf{7}.$

(*ii*) No. of students who study English, Mathematics but not Economics

$$= n((A \cap C) - B) = n(A \cap C) - n((A \cap C) \cap B)$$

$$= n(A \cap C) - n(A \cap B \cap C) = 7 - 2 = \mathbf{5}.$$

Example 3. *In a class of 60 students, 23 play Hockey, 15 play Basketball and 20 play Cricket. 7 play Hockey and Basketball, 5 play Cricket and Basketball, 4 play Hockey and Cricket and 15 students do not play any of these games.*

(*i*) *How many play Hockey, Basketball and Cricket ?*

(*ii*) *How many play Hockey but not Cricket ?*

(*iii*) *How many play Hockey and Cricket but not Basketball ?*

Sol. Let H, B and C be the sets of students playing Hockey, Basketball and Cricket respectively.

$\therefore$ $n(H) = 23$, $n(B) = 15$, $n(C) = 20$, $n(H \cap B) = 7$, $n(C \cap B) = 5$, $n(H \cap C) = 4$, $n(H' \cap B' \cap C') = 15$

(*i*) $n(H' \cap B' \cap C') = n((H \cup B \cup C)') = 60 - n(H \cup B \cup C)$

$\therefore$ $60 - n(H \cup B \cup C) = 15$ or $n(H \cup B \cup C) = 60 - 15 = 45$

Also, $n(H \cup B \cup C) = n(H) + n(B) + n(C) - n(H \cap B) - n(B \cap C) - n(H \cap C) + n(H \cap B \cap C)$

$\Rightarrow$ $45 = 23 + 15 + 20 - 7 - 5 - 4 + n(H \cap B \cap C)$

$\therefore$ $n(H \cap B \cap C) = 45 - 42 = 3$

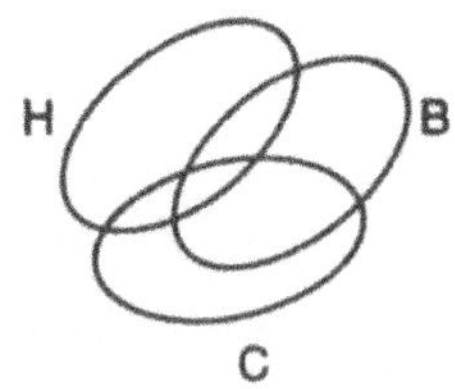

$\therefore$ No. of students playing Hockey, Basketball and Cricket

$= n(H \cap B \cap C) = \mathbf{3}$.

(*ii*) No. of students playing Hockey but not Cricket

$= n(H - C) = n(H) - n(H \cap C) = 23 - 4 = \mathbf{19}$.

(*iii*) No. of students playing Hockey and Cricket but not Basketball

$= n((H \cap C) - B) = n(H \cap C) - n((H \cap C) \cap B)$

$= 4 - 3 = \mathbf{1}$.

Example 4. *In a town of 10,000 families, it was found that 40% families buy newspaper A, 20% families buy newspaper B and 10% families buy newspaper C. 5% families buy A and B, 3% buy B and C and 4% buy A and C. If 2% families buy all the newspapers, find the number of families which buy (i) A only (ii) B only (iii) none of A, B and C.*

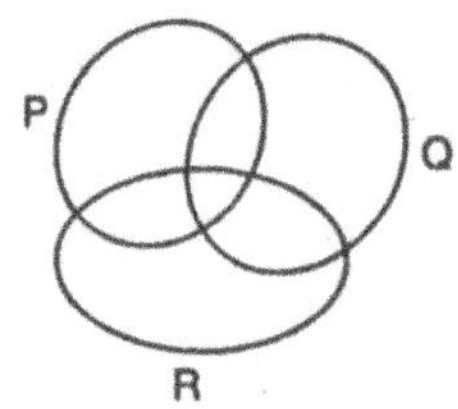

Sol. Let P = set of families buying A,

Q = set of families buying B

and R = set of families buying C.

$\therefore$ $n(P) = 40\%$ of $10,000 = 4,000$ $\quad n(P \cap Q) = 5\%$ of $10,000 = 500$

$n(Q) = 20\%$ of $10,000 = 2,000$ $\quad n(Q \cap R) = 3\%$ of $10,000 = 300$

$n(R) = 10\%$ of $10,000 = 1,000$ $\quad n(P \cap R) = 4\%$ of $10,000 = 400$

$n(P \cap Q \cap R) = 2\%$ of $10,000 = 200$

(*i*) No. of families buying A only

$= n(P - (Q \cup R)) = n(P) - n(P \cap (Q \cup R))$

$= n(P) - n((P \cap Q) \cup (P \cap R))$

$= n(P) - [n(P \cap Q) + n(P \cap R) - n((P \cap Q) \cap (P \cap R)]$

$= n(P) - n(P \cap Q) - n(P \cap R) + n(P \cap Q \cap R)$

$= 4,000 - 500 - 400 + 200 = \mathbf{3,300}$

(*ii*) No. of families buying B only

$= n(Q - (P \cup R)) = n(Q) - P(Q \cap (P \cup R))$

$= n(Q) - n((Q \cap P) \cup (Q \cap R))$

$= n(Q) - [n(Q \cap P) + n(Q \cap R) - n((Q \cap P) \cap (Q \cap R))]$

$= n(Q) - n(Q \cap P) - n(Q \cap R) + n(P \cap Q \cap R)$

$= 2,000 - 500 - 300 + 200 = \mathbf{1,400}$

(*iii*) No. of families buying none of A, B and C

$= n((P \cup Q \cup R)') = 10,000 - n(P \cup Q \cup R)$

$= 10,000 - [n(P) + n(Q) + n(R) - n(P \cap Q) - n(Q \cap R) - n(P \cap R) + n(P \cap Q \cap R)]$

$= 10,000 - [4,000 + 2,000 + 1,000 - 500 - 300 - 400 + 200] = \mathbf{4,000}$.

Example 5. *In a survey of 25 students, it was found that 15 had taken Mathematics, 12 had taken Physics and 11 had taken Chemistry, 5 had taken Mathematics and Chemistry, 9 had taken Mathematics and Physics, 4 had taken Physics and Chemistry and 3 had taken all three subjects. Find the number of students that had (i) only Chemistry (ii) only Mathematics (iii) only Physics (iv) Physics and Chemistry but not Mathematics (v) Mathematics and Physics but not Chemistry (vi) only one of the subjects (vii) at least one of three subjects (viii) none of three subjects.*

Sol. Let M = set of students who have taken Mathematics

P = set of students who have taken Physics

and C = set of students who have taken Chemistry.

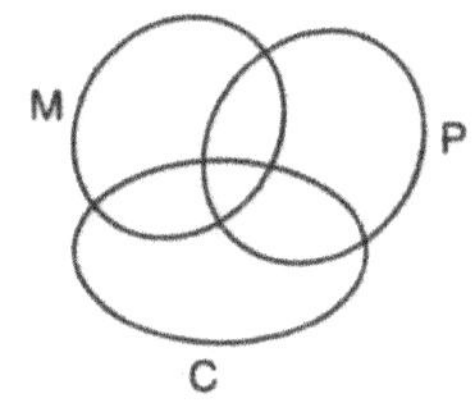

$\therefore \quad n(M) = 15, n(P) = 12, n(C) = 11,$

$n(M \cap C) = 5, n(M \cap P) = 9, n(P \cap C) = 4,$

$n(M \cap P \cap C) = 3$

(*i*) No. of students taking Chemistry only

$$= n(C - (M \cup P)) = n(C) - n(C \cap (M \cup P))$$
$$= n(C) - n((C \cap M) \cup (C \cap P))$$
$$= n(C) - [n(C \cap M) + n(C \cap P) - n((C \cap M) \cap (C \cap P))]$$
$$= n(C) - n(C \cap M) - n(C \cap P) + n(C \cap M \cap P)$$
$$= 11 - 5 - 4 + 3 = \mathbf{5}.$$

(*ii*) No. of students taking Mathematics only

$$= n(M - (P \cup C)) = n(M) - n(M \cap (P \cup C))$$
$$= n(M) - n((M \cap P) \cup (M \cap C))$$
$$= n(M) - [n(M \cap P) + n(M \cap C) - n((M \cap P) \cap (M \cap C))]$$
$$= n(M) - n(M \cap P) - n(M \cap C) + n(M \cap P \cap C)$$
$$= 15 - 9 - 5 + 3 = \mathbf{4}.$$

(*iii*) No. of students taking Physics only

$$= n(P - (M \cup C)) = n(P) - n(P \cap (M \cup C))$$
$$= n(P) - n((P \cap M) \cup (P \cap C))$$
$$= n(P) - [n(P \cap M) + n(P \cap C) - n((P \cap M) \cap (P \cap C))]$$
$$= n(P) - n(P \cap M) - n(P \cap C) + n(P \cap M \cap C)$$
$$= 12 - 9 - 4 + 3 = \mathbf{2}.$$

(*iv*) No. of students taking Physics and Chemistry but not Mathematics

$$= n((P \cap C) - M) = n(P \cap C) - n(P \cap C \cap M) = 4 - 3 = \mathbf{1}.$$

(*v*) No of students taking Mathematics and Physics but not Chemistry

$$= n((M \cap P) - C) = n(M \cap P) - n(M \cap P \cap C) = 9 - 3 = \mathbf{6}.$$

(*vi*) No. of students taking only one subject

$$= n((\text{only } M) \cup (\text{only } P) \cup (\text{only } C))$$
$$= n(\text{only } M) + n(\text{only } P) + n(\text{only } C) = 4 + 2 + 5 = \mathbf{11}.$$

(*vii*) No. of students taking at least one of the subjects

$$= n(M \cup P \cup C) = n(M) + n(P) + n(C) - n(M \cap P)$$
$$- n(P \cap C) - n(M \cap C) + n(M \cap P \cap C)$$
$$= 15 + 12 + 11 - 9 - 4 - 5 + 3 = \mathbf{23}.$$

(*viii*) No. of students taking none of the subjects

$$= 25 - n(M \cup P \cup C) = 25 - 23 = \mathbf{2}.$$

Example 6. *A college awarded 38 medals in Football, 15 in Basketball and 20 in Cricket. If these medals went to a total of 58 men and only three men got medals in all the three sports, how many received medals in exactly two of the three sports ?*

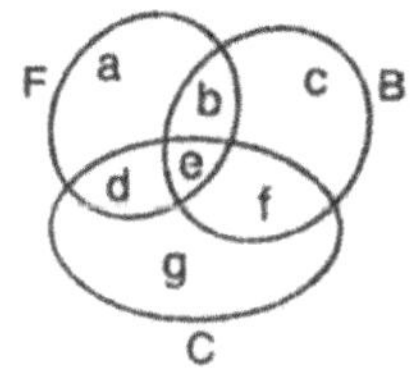

Sol. Let F = set of men receiving medal in Football,
B = set of men receiving medal in Basketball
and C = set of men receiving medal in Cricket.

In the adjoining Venn diagram, let a, b, c, d, e, f, g denote the number of elements in the respective region.

$$\therefore \quad n(F) = a + b + d + e = 38 \quad \ldots(1)$$

$$n(B) = b + c + e + f = 15 \quad \ldots(2)$$

$$n(C) = d + e + f + g = 20 \quad \ldots(3)$$

$$n(F \cup B \cup C) = a + b + c + d + e + f + g = 58 \quad \ldots(4)$$

$$n(F \cap B \cap C) = e = 3 \quad \ldots(5)$$

We have

$$n(F \cup B \cup C) = n(F) + n(B) + n(C) - n(F \cap B) - n(B \cap C) - n(C \cap F) + n(F \cap B \cap C)$$

$$\therefore \quad 58 = 38 + 15 + 20 - n(F \cap B) - n(B \cap C) - n(C \cap F) + 3$$

$$\Rightarrow \quad n(F \cap B) + n(B \cap C) + n(C \cap F) = 18$$

$$\Rightarrow \quad (b + e) + (e + f) + (d + e) = 18$$

$$\Rightarrow \quad (b + d + f) + 3e = 18$$

$$\Rightarrow \quad b + d + f = 18 - 3(3) = 9$$

$\therefore$ No. of men who received medals in exactly two sports = **9.**

Alternative method

No. of men who received medals in exactly two sports

$$= n((F \cap B) - C) + n((B \cap C) - F) + n((F \cap C) - B)$$

$$= [n(F \cap B) - n((F \cap B) \cap C)] + [n(B \cap C) - n((B \cap C) \cap F)] + [n(F \cap C) - n((F \cap C) \cap B)]$$

$$= n(F \cap B) + n(B \cap C) + n(F \cap C) - 3n(F \cap B \cap C)$$

$$= n(F \cap B) + n(B \cap C) + n(F \cap C) - 3(3)$$

$$= n(F \cap B) + n(B \cap C) + n(F \cap C) - 9 \quad \ldots(1)$$

Also, $n(F \cup B \cup C) = n(F) + n(B) + n(C) - n(F \cap B) - n(B \cap C) - n(F \cap C) + n(F \cap B \cap C)$

$$\Rightarrow \quad 58 = 38 + 15 + 20 - n(F \cap B) - n(B \cap C) - n(F \cap C) + 3$$

$$\Rightarrow \quad n(F \cap B) + n(B \cap C) + n(F \cap C) = 76 - 58 = 18$$

$\therefore$ Required number of men = 18 – 9 = **9.** (Using (1))

Example 7. *Among the first 500 positive integers :*

(*i*) *Determine the integers which are neither divisible by 2 nor by 3 nor by 5.*

(*ii*) *Determine the integers which are exactly divisible by one of them.*

Sol. Let A, B, C be the sets of positive integers less than 500 and divisible by 2, divisible by 3, divisible by 5 respectively.

$\therefore$ $A = \{2, 4, 6, \ldots\ldots, 500\}$

$B = \{3, 6, 9, \ldots\ldots, 498\}$

$C = \{5, 10, 15, \ldots\ldots, 500\}$

Also $A \cap B = \{6, 12, \ldots\ldots, 498\}$

$B \cap C = \{15, 30, \ldots\ldots, 495\}$

$A \cap C = \{10, 20, \ldots\ldots, 500\}$

$A \cap B \cap C = \{30, 60, \ldots\ldots, 480\}$

$\therefore$ $n(A) = 250, n(B) = 166, n(C) = 100, n(A \cap B) = 83,$

$n(B \cap C) = 33, n(A \cap C) = 50, n(A \cap B \cap C) = 16.$

(*i*) $n(A \cup B \cup C) = n(A) + n(B) + n(C) - n(A \cap B) - n(B \cap C) - n(A \cap C) + n(A \cap B \cap C)$

$= 250 + 166 + 100 - 83 - 33 - 50 + 16 = 366$

No. of positive integers less than 500 and neither divisible by 2 nor by 3 nor by 5

$= 500 - n(A \cup B \cup C) = 500 - 366 = \mathbf{134}.$

(*ii*) $n\,(\text{only } A) = n(A - (B \cup C)) = n(A) - n(A \cap (B \cup C))$

$= n(A) - n((A \cap B) \cup (A \cap C))$

$= n(A) - [n(A \cap B) + n(A \cap C) - n((A \cap B) \cap (A \cap C))]$

$= n(A) - n(A \cap B) - n(A \cap C) + n(A \cap B \cap C)$

$= 250 - 83 - 50 + 16 = 133$

Similarly, $n(\text{only } B) = 66$ and $n\,(\text{only } C) = 33$

$\therefore$ No. of positive integers less than 500 and exactly divisible by one of 2, 3, 5

$= n(\text{only } A) + n(\text{only } B) + n(\text{only } C) = 133 + 66 + 33 = \mathbf{232}.$

EXERCISE 1.2

1. In a group of 100 persons, liking at least one of tea, coffee, ice-cream, 49 like tea, 41 like coffee, 62 like ice-cream. 25 persons like tea and coffee both, 18 persons like coffee and ice-cream both and 17 persons like both tea and ice-cream. Find the number of persons who like all the three items.
2. Out of 1020 boys in a school, 406 play cricket, 324 play hockey and 250 play football. 80 boys play cricket and hockey, 64 play hockey and football, 92 play football and cricket while 30 play all the three games. How many boys play none of the games ?
3. A class has 175 students. The following is the description showing the number of students studying one or more of the following subjects in this class :

   Mathematics 100, Physics 70, Chemistry 46, Mathematics and Physics 30, Mathematics and Chemistry 28, Physics and Chemistry 23, Mathematics, Physics and Chemistry 18.

   (*i*) How many students are enrolled in at least one of the subjects ?

   (*ii*) How many students are not studying any of the subjects ?
4. In a survey it was found that 21 people liked product *A*, 26 liked product *B* and 29 liked product *C*. If 14 people liked products *A* and *B*, 12 people liked products *C* and *A*, 14 people liked products *B* and *C* and 8 liked all the three products. Find how many liked product *C* only.

5. In a survey of 100 persons it was found that 28 read magazine A, 30 read magazine B, 42 read magazine C, 8 read magazines A and B, 10 read magazines A and C, 5 read magazines B and C and 3 read all the three magazines.

   (*i*) How many read none of three magazines ?

   (*ii*) How many read magazine C only ?

6. In a class, 36 students offered physics, 46 offered chemistry and 48 offered mathematics. Of these, 26 are in both chemistry and mathematics, 24 are in both physics and chemistry, 22 are in both mathematics and physics. 12 have offered all the three subjects. Find :

   (*i*) the number of students in the class.

   (*ii*) the number of students who have offered mathematics but not chemistry.

   (*iii*) the number of students who have offered exactly one subject.

7. In a survey of 60 people, it was found that 25 people read Newspaper H, 26 read Newspaper T, 26 read Newspaper I, 9 read both H and I, 11 read both H and T, 8 read both T and I, 3 read all three newspapers. Find :

   (*i*) the number of people who read at least one of the newspapers

   (*ii*) the number of people who read exactly one newspaper.

8. In a survey of 100 students, the number of students studying the various languages were found to be : English only 18, English but not Hindi 23, English and Sanskrit 8, English 26, Sanskrit 48, Sanskrit and Hindi 8, no language 24.

   (*i*) How many students were studying Hindi ?

   (*ii*) How many students were studying English and Hindi ?

Answers

1. 8	**2.** 246	**3.** (*i*) 153	(*ii*) 22
4. 11	**5.** (*i*) 20	(*ii*) 30	
6. (*i*) 70	(*ii*) 22 (*iii*) 22	**7.** (*i*) 52	(*ii*) 30
8. (*i*) 18	(*ii*) 3.		

2

Countability of Sets

2.1. INTRODUCTION

In this chapter, we shall study finite sets and infinite sets (*i.e.*, not finite sets) both. We know that a function f from a set A to a set B is called a *bijective function* if f is both one-one and onto. We will see that a finite set is always a countable set but an infinite set may or may not be a countable set.

2.2. EQUIPOTENT SETS

Two sets A and B are called **equipotent sets** or to have the **same cardinality** or to have the **same number of elements** if there exists a bijective function from the set A onto the set B.

If A and B are equipotent sets then we write $A \approx B$ and say that there is *one-to-one correspondence* between the sets A and B.

Illustrations (*i*) Let $A = \{2, 3, 7\}$ and $B = \{a, b, c\}$. A and B are equipotent sets for $f : A \to B$ defined by $f(2) = a$, $f(3) = b$, $f(7) = c$ is a bijective function.

(*ii*) Let $A = \{1, 3, 5\}$ and $B = \{1, 2, 3, 7\}$. The sets A and B cannot be equipotent sets because we cannot define an onto function from A to B.

(*iii*) Let $A = \{1, 2, 3, 4, 5\}$ and $B = \{a, b, l, m\}$. The sets A and B cannot be equipotent sets because we cannot define a one-one function from A to B.

Thus, we see that if A and B are finite sets then these sets can be equipotent sets only when the number of elements in A and B are equal.

Example 1. *Show that the set of all positive even integers and the set of all natural numbers have the same cardinality.*

Sol. Let $E = \{2, 4, 6, 8, ...\}$. Define $f : \mathbf{N} \to E$ by $f(n) = 2n$, $n \in \mathbf{N}$

f is 1 – 1. Let $n_1, n_2 \in \mathbf{N}$ and $f(n_1) = f(n_2)$

$$\Rightarrow \quad 2n_1 = 2n_2 \Rightarrow n_1 = n_2 \Rightarrow f \text{ is } 1-1$$

f is onto. Let $2k \in E$

$$\therefore \quad k \in \mathbf{N} \text{ and } f(k) = 2k$$

$\therefore$ f is onto. $\therefore$ f is a bijection.

$\therefore$ E and $\mathbf{N}$ have the same cardinality.

Remark. The one-to-one correspondence between the elements of E and $\mathbf{N}$ can be shown as follows :

N :	1	2	3	4	5	......
	↓	↓	↓	↓	↓	
E :	2	4	6	8	10	......

Example 2. *Show that the interval (a, b), where a < b, has the same cardinality as that of the interval (0, 1).*

Sol. Define $f: (0, 1) \to (a, b)$ by $f(x) = (b - a)x + a, \quad x \in (0, 1)$

$$0 < x < 1 \Rightarrow 0 < (b - a)x < b - a \qquad (\because \; b - a > 0)$$
$$\Rightarrow a < (b - a)x + a < (b - a) + a$$
$$\Rightarrow a < f(x) < b$$

$\therefore$ 'f' is well defined.

f is 1 – 1. Let $x_1, x_2 \in (0, 1)$ and $f(x_1) = f(x_2)$.

$$\Rightarrow \qquad (b - a)x_1 + a = (b - a)x_2 + a$$
$$\Rightarrow \qquad x_1 = x_2 \Rightarrow f \text{ is } 1 - 1.$$

f is onto. Let $y \in (a, b)$

$$\Rightarrow a < y < b \Rightarrow a - a < y - a < b - a \Rightarrow 0 < \frac{y-a}{b-a} < 1$$

Also, $$f\left(\frac{y-a}{b-a}\right) = (b-a).\left(\frac{y-a}{b-a}\right) + a = y$$

$\therefore$ f is onto.

$\therefore$ f is a bijection.

$\therefore$ The intervals (a, b) and $(0, 1)$ have the same cardinality.

Theorem. Prove that the relation of being equipotent is an equivalence relation on any collection of sets.

Proof. Let A, B, C be any sets in a given collection of sets.

Relation ≈ is reflexive. The identity map $i : A \to A$ is a bijection. $\therefore$ $A \approx A$

$\therefore$ ≈ is reflexive.

Relation ≈ is symmetric. Let $A \approx B$.

$\therefore$ There exists a bijection map $f : A \to B$.

$\therefore$ f has an inverse function $f^{-1} : B \to A$ and this function is also bijective. $\therefore$ $B \approx A$

$\therefore$ ≈ is symmetric.

Relation ≈ is transitive. Let $A \approx B$ and $B \approx C$.

$\therefore$ There exist bijection maps $f : A \to B$ and $g : B \to C$.

The composition gof of mappings from A to C is also 1 – 1 and onto.

$\therefore$ $gof : A \to C$ is a bijection. $\therefore$ $A \approx C$

$\therefore$ ≈ is transitive.

$\therefore$ The relation ≈ is an equivalence relation.

Illustration. The intervals (4, 7) and (2, 8) are equipotent to the interval (0, 1).

$$\therefore \qquad (4, 7) \approx (0, 1) \quad \text{and} \quad (0, 1) \approx (2, 8)$$

$\therefore$ By transitivity, $(4, 7) \approx (2, 8)$

$\therefore$ The intervals (4, 7) and (2, 8) have the same cardinality.

Example 3. *If sets A and C have the same cardinality and the sets B and D have the same cardinality then show that the sets A × B and C × D have the same cardinality.*

Sol. Let $f : A \to C$ and $g : B \to D$ be bijections.

Define $h : A \times B \to C \times D$ by

$$h(a, b) = (f(a), g(b)), \text{ where } (a, b) \in A \times B$$

h is 1 – 1. Let $(a, b), (a', b') \in A \times B$ and $h(a, b) = h(a', b')$

$\Rightarrow \quad (f(a), g(b)) = (f(a')\, g(b'))$

$\Rightarrow \quad f(a) = f(a'), g(b) = g(b')$

$\Rightarrow \quad a = a', b = b' \quad \Rightarrow \quad (a, b) = (a', b') \quad \Rightarrow \quad h$ is $1 - 1$.

h is onto. Let $(c, d) \in C \times D$.

$$c \in C \Rightarrow \exists\, \alpha \in A : f(\alpha) = c$$
$$d \in D \Rightarrow \exists\, \beta \in B : g(\beta) = d$$

$\therefore \quad (\alpha, \beta) \in A \times B$ and $h(\alpha, \beta) = (f(\alpha), g(\beta)) = (c, d)$

$\therefore$ h is onto.

$\therefore$ h is a bijection.

$\therefore$ The sets $A \times B$ and $C \times D$ have the same cardinality.

Example 4. *Show that the set* **R** *of real numbers and the open interval* $(-1, 1)$ *have the same cardinality.*

Sol. Define $f : \mathbf{R} \to (-1, 1)$ by $f(x) = \dfrac{x}{1+|x|}$, $x \in \mathbf{R}$

$$x < 0 \Rightarrow f(x) = \frac{x}{1-x} = -\frac{(-x)}{1+(-x)} \in (-1, 0)$$

$$x = 0 \Rightarrow f(x) = \frac{0}{1+0} = 0$$

$$x > 0 \Rightarrow f(x) = \frac{x}{1+x} \in (0, 1)$$

$\therefore \quad f(x) \in (-1, 1)\ \forall\, x \in \mathbf{R}$

f is 1 – 1. Let $x_1, x_2 \in \mathbf{R}$ and $f(x_1) = f(x_2)$.

$$\Rightarrow \quad \frac{x_1}{1+|x_1|} = \frac{x_2}{1+|x_2|} \qquad \text{...(1)}$$

Case I. $\mathbf{x_1 < 0, x_2 < 0}$

$$(1) \Rightarrow \quad \frac{x_1}{1-x_1} = \frac{x_2}{1-x_2} \Rightarrow x_1 - x_1x_2 = x_2 - x_1x_2 \Rightarrow x_1 = x_2$$

Case II. $\mathbf{x_1 \geq 0, x_2 \geq 0}$

$$(1) \Rightarrow \quad \frac{x_1}{1+x_1} = \frac{x_2}{1+x_2} \Rightarrow x_1 + x_1x_2 = x_2 + x_1x_2 \Rightarrow x_1 = x_2$$

Case III. $\mathbf{x_1 < 0, x_2 \geq 0}$

Here $\quad f(x_1) = \dfrac{x_1}{1-x_1} < 0$ and $f(x_2) = \dfrac{x_2}{1+x_2} \geq 0$

$\therefore$ This case is impossible.

Similarly $x_1 \geq 0, x_2 < 0$ is impossible.

$\therefore$ f is 1 – 1.

f is onto. Let $y \in (-1, 1)$

Case I. y ∈ (– 1, 0)

Here $$\frac{y}{1+y} < 0 \text{ and } f\left(\frac{y}{1+y}\right) = \frac{\frac{y}{1+y}}{1+\left|\frac{y}{1+y}\right|} = \frac{\frac{y}{1+y}}{1-\frac{y}{1+y}} = y$$

Case II. y = 0

Here $$f(0) = \frac{0}{1+|0|} = 0$$

Case III. y ∈ (0, 1)

Here $$\frac{y}{1-y} > 0 \text{ and } f\left(\frac{y}{1-y}\right) = \frac{\frac{y}{1-y}}{1+\left|\frac{y}{1-y}\right|} = \frac{\frac{y}{1-y}}{1+\frac{y}{1-y}} = y$$

$\therefore$ f is onto.

$\therefore$ f is a bijection

$\therefore$ **R** and (– 1, 1) have the same cardinality.

Example 5. *Let X be any set. A function from X to the set {0, 1} is called a* ***characteristic function*** *of X. If C(X) denotes the family of characteristic functions of X then show that the sets C(X) and P(X) have the same cardinality, where P(X) is the power set of X.*

Sol. Let $A \in P(X)$. Define

$f : P(X) \to C(X)$ by $f(A) = \chi_A$, where χ_A is a characteristic function of X defined by

$$\chi_A(x) = \begin{cases} 0 & \text{if } x \in X - A \\ 1 & \text{if } x \in A \end{cases}$$

f is 1 – 1. Let $A, B \in P(X)$ and $f(A) = f(B)$.

$$\Rightarrow \quad \chi_A = \chi_B \Rightarrow \chi_A(x) = \chi_B(x) \ \forall \ x \in X$$

$x \in A \Rightarrow \chi_A(x) = 1 \Rightarrow \chi_B(x) = 1 \Rightarrow x \in B \Rightarrow A \subseteq B$

$x \in B \Rightarrow \chi_B(x) = 1 \Rightarrow \chi_A(x) = 1 \Rightarrow x \in A \Rightarrow B \subseteq A$

$\therefore \quad A = B \quad \therefore \quad f$ is 1 – 1.

f is onto. Let ϕ be a characteristic function of X.

Let $A = \{x \in X : \phi(x) = 1\}$

$\therefore$ $A \in P(X)$ and $\chi_A = \phi$

Also $f(A) = \chi_A = \phi$

$\therefore$ f is onto.

$\therefore$ f is a bijection.

$\therefore$ $P(X)$ and $C(X)$ have the same cardinality.

EXERCISE 2.1

1. Show that the sets $\{x \in \mathbf{R} : x^2 - 5x + 6 = 0\}$ and $\{3, 4\}$ are equipotent.
2. Show that the intervals (0, 1) and (5, 9) are equipotent.
3. If $a < b$ and $c < d$, show that the intervals (a, b) and (c, d) have the same cardinality.
4. For any sets A and B, show that the sets $A \times B$ and $B \times A$ have the same cardinality.
5. For any sets A, B and C, show that the sets $(A \times B) \times C$ and $A \times (B \times C)$ have the same cardinality.

2.3. DENUMERABLE SET

A is said to be a **denumerable set** if it has the same cardinality as **N.**

In otherwords, a set A is denumerable if there exists a bijection from **N** onto A.

A denumerable set is also called a **countably infinite set**.

For example, if $a_1, a_2, a_3, \ldots$ is an infinite sequence of *distinct elements* then the set $\{a_1, a_2, a_3, \ldots\ldots\}$ is denumerable. This is so, because the function $f : \mathbf{N} \to \{a_1, a_2, a_3, \ldots\ldots\}$ defined by $f(n) = a_n$ is a bijection.

Example 1. *Show that the set containing the elements of the infinite sequence* $\{(-1)^n n\}$ *is denumerable.*

Sol. The given sequence is $\{(-1)^n n\}$. Let $a_n = (-1)^n n$

The terms $a_1, a_2, a_3, \ldots\ldots$ are all distinct.

Define $f : \mathbf{N} \to \{a_1, a_2, a_3, \ldots\ldots\}$ by $f(n) = a_n = (-1)^n n, n \in \mathbf{N}$

Since the elements of the set $\{a_1, a_2, a_3, \ldots\ldots\}$ are distinct and the set itself is infinite, the function 'f' is a bijection.

$\therefore$ The set containing the elements of the given sequence is denumerable.

Example 2. *Show that the set* $\mathbf{N} \times \mathbf{N}$ *is denumerable.*

Sol. The elements of $\mathbf{N} \times \mathbf{N}$ can be written as follows :

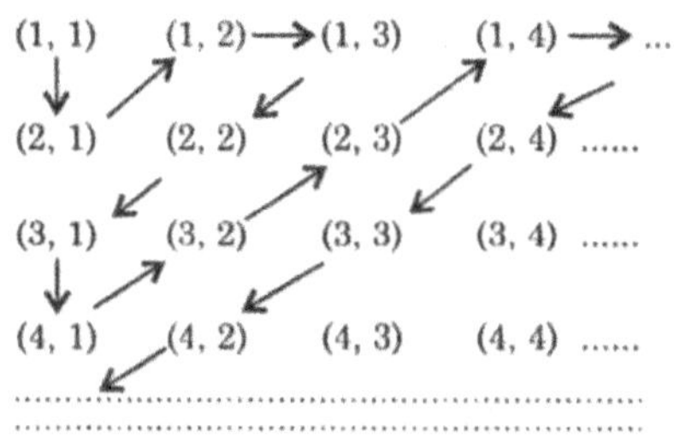

By following the arrows, the set $\mathbf{N} \times \mathbf{N}$ can be expressed as $\{(1, 1), (2, 1), (1, 2), (1, 3), (2, 2), \ldots\ldots\}$. The terms (1, 1), (2, 1), (1, 2), (1, 3), (2, 2), ... of the corresponding infinite sequence are distinct.

$\therefore$ The sets $\mathbf{N} \times \mathbf{N}$ and **N** have the same cardinality.

$\therefore$ The set $\mathbf{N} \times \mathbf{N}$ is denumerable.

Alternative method

Define $f : \mathbf{N} \to \mathbf{N} \times \mathbf{N}$ as follows. Let $a \in \mathbf{N}$.

$\therefore$ There exist unique $l, m \in \mathbf{N}$ such that $a = 2^{l-1}(2m - 1)$.

(For example, $30 = 2^{2-1}(2(8) - 1)$)

Define $$f(a) = f(2^{l-1}(2m - 1)) = (l, m) \in \mathbf{N} \times \mathbf{N}$$

Since l, m corresponding to the natural number $a = 2^{l-1}(2m-1)$ are unique, the function 'f' is well defined.

f is 1 – 1. Let $2^{l-1}(2m-1), 2^{r-1}(2s-1) \in \mathbf{N}$ and $f(2^{l-1}(2m-1)) = f(2^{r-1}(2s-1))$.

$\Rightarrow \quad (l, m) = (r, s) \quad \Rightarrow \quad l = r, m = s$

$\Rightarrow \quad 2^{l-1}(2m-1) = 2^{r-1}(2s-1)$

$\therefore \quad f$ is 1 – 1.

f is onto. Let $(u, v) \in \mathbf{N} \times \mathbf{N}$

$\Rightarrow \quad u, v \in \mathbf{N}$ and $2^{u-1}(2v-1) \in \mathbf{N}$ and $f(2^{u-1}(2v-1)) = (u, v)$

$\therefore \quad f$ is onto. $\therefore \quad f$ is a bijection.

$\therefore$ The sets $\mathbf{N} \times \mathbf{N}$ and $\mathbf{N}$ have the same cardinality.

$\therefore$ The set $\mathbf{N} \times \mathbf{N}$ is denumerable.

Example 3. *The set* **Z** *of all integers is denumerable.*

Sol. Define $f : \mathbf{N} \to \mathbf{Z}$ by

$$f(n) = \begin{cases} \dfrac{n}{2} & \text{if } n \text{ is even} \\ -\dfrac{n-1}{2} & \text{if } n \text{ is odd} \end{cases}$$

f is 1 – 1. Let $n_1, n_2 \in \mathbf{N}$ and $f(n_1) = f(n_2)$.

Case I. n_1, n_2 are even

$$f(n_1) = f(n_2) \Rightarrow \frac{n_1}{2} = \frac{n_2}{2} \Rightarrow n_1 = n_2$$

Case II. n_1, n_2 are odd

$$f(n_1) = f(n_2) \Rightarrow -\frac{n_1-1}{2} = -\frac{n_2-1}{2} \Rightarrow n_1 = n_2$$

Case III. n_1 is even and n_2 is odd

$$f(n_1) = f(n_2) \Rightarrow \frac{n_1}{2} = -\frac{n_2-1}{2} \Rightarrow n_1 + n_2 = 1$$

This is impossible.

Case IV. n_1 is odd and n_2 is even

$$f(n_1) = f(n_2) \Rightarrow -\frac{n_1-1}{2} = \frac{n_2}{2} \Rightarrow n_1 + n_2 = 1$$

This is impossible.

$\therefore \quad f$ is 1 – 1.

f is onto. Let $n \in \mathbf{Z}$.

Case I. n > 0

$\therefore \quad 2n \in \mathbf{N}$ and is even.

$$\therefore \quad f(2n) = \frac{2n}{2} = n$$

Case II. n ≤ 0

$\Rightarrow \quad 2n \le 0 \Rightarrow -2n \ge 0 \Rightarrow 1 - 2n > 0$

∴ $1 - 2n \in$ **N** and is odd.

$$\therefore \quad f(1 - 2n) = -\frac{(1-2n)-1}{2} = n$$

∴ **f is onto.** ∴ f is a bijection.

∴ The set **Z** of integers is denumerable.

Remark 1. The function $f : \mathbf{N} \to \mathbf{Z}$ defined above gives the following one-to-one correspondence between the elements of **N** and **Z**.

N :	1	2	3	4	5	6 ......
	↓	↓	↓	↓	↓	↓
Z :	0	1	– 1	2	– 2	3 ......

Remark 2. The set **N** is a proper subset of **Z** and still these sets are having the same cardinality.

Theorem. Prove that every infinite set contains a denumerable subset.

Proof. Let A be any infinite set.

∴ The set A is not finite.

Let $P(A)$ be the power set of A. Let $f : P(A) \to A$ be a choice function* defined as follows :

Let $f(A) = a_1$

$f(A - \{a_1\}) = a_2$

$f(A - \{a_1, a_2\}) = a_3$

....

$f\{A - \{a_1, a_2, ..., a_{n-1}\}) = a_n$

....

Since A is infinite, the set $A - \{a_1, a_2, ..., a_{n-1}\}$ is not empty for every $n \in$ **N**.

∴ The sequence $a_1, a_2, a_3, ...$ is infinite.

Also, $a_2 \in A - \{a_1\}$ implies $a_2 \ne a_1$.

$a_3 \in A - \{a_1, a_2\}$ implies $a_3 \ne a_1, a_2$.

Similarly, $a_n \ne a_i \ \forall \ i < n$.

∴ The elements of the sequence $a_1, a_2, a_3, ...$ are distinct.

∴ The function $\phi : \mathbf{N} \to \{a_1, a_2, a_3, ...\}$ defined by $\phi(n) = a_n$, $n \in$ **N** is a bijection.

∴ The set $\{a_1, a_2, a_3, ...\}$ is denumerable.

Also, $a_i \in A \ \forall \ i$

∴ $\{a_1, a_2, a_3, ...\}$ is a denumerable subset of the infinite set A.

EXERCISE 2.2

1. Show that the set containing the elements of the infinite sequence $\{(-1)^{n-1} n\}$ is denumerable.
2. Show that the set containing the elements of the infinite sequence $\{(-1)^{n+3}\}$ is not denumerable.
3. Prove that a subset of a denumerable set is either finite or denumerable.
4. If F is a finite subset of an infinite set A, then show that the sets $A - F$ and A have the same cardinality.

***Choice function.** Let $\{A_i : i \in I\}$ be a set of subsets of a set A. A function $f : \{A_i : i \in I\} \to A$ is called a **choice function** if $f(A_i) \in A_i \ \forall \ i \in I$.

Hints

3. Let $A = \{a_1, a_2, a_3, ...\}$ be a denumerable set and let B be a subset of A. If B is not finite, let b_1 be the first element in A such that $b_1 \in B$. Let b_2 be the first element which follows b_1 in the sequence $a_1, a_2, a_3, ...$ and $b_2 \in B$. Similarly, we find b_3 and so on.

   $\therefore$ $B = \{b_1, b_2, b_3, ...\}$. and it is an infinite sequence of distinct elements and hence denumerable.

4. Let $F = \{a_1, a_2, ..., a_m\}$. Let $f : P(A) \to A$ be a choice function defined as follows :

   $f(A) = a_1, f(A - \{a_1\}) = a_2, f(A - \{a_1, a_2\}) = a_3, ..., f(A - \{a_1, a_2, ..., a_{m-1}\}) = a_m, f(A - F) = a_{m+1}$, where a_{m+1} is any element in $A - F$, $f(A - (F \cup \{a_{m+1}\})) = a_{m+2}, ...$

   Since A is infinite, the set $A - (F \cup \{a_{m+1}, a_{m+2}, ...\})$ is non-empty.

   $\therefore$ The sequence $a_1, a_2, ..., a_m, a_{m+1}, ...$ is infinite and its elements are distinct.

   Let $D = \{a_1, a_2, ...\}$. $\therefore$ D is a denumerable subset of A. In D, the first m elements are from F.

   $\therefore$ $$F \subseteq D \quad \text{and} \quad A - D \subseteq A - F$$

   Define $\phi : A \to A - F$ as follows :

   For $a \in A - D$, $\phi(a) = a \in A - F$

   For $a_n \in D$, $\phi(a_n) = a_{m+n} \in A - F$

   Show that ϕ is a bijection.

2.4. COUNTABLE SET

A set is said to be a **countable set** if it is either finite or denumerable.

Illustrations (*i*) The set $\{2, 3, 7\}$ is countable because every finite set is countable.

(*ii*) The set $\left\{1, \frac{1}{2}, \frac{1}{3}, \frac{1}{4}, ...\right\}$ is countable because every denumerable set is countable.

Theorem 1. Prove that the countable union of finite sets is countable.

Proof. Let $S_1, S_2, S_3, ...$ be a sequence of finite sets. If the sequence $S_1, S_2, S_3, ...$ is finite then their union $\cup S_i$ will contain finitely many elements.

$\therefore$ $\cup S_i$ is finite and hence countable.

Let the sequence $S_1, S_2, S_3, ...$ be infinite. Let $C = \bigcup_{i \in \mathbf{N}} S_i$. If $C = \phi$, then C is finite and hence countable. Let $C \neq \phi$.

Define $$A_1 = S_1, \quad A_2 = S_2 - S_1, \quad A_3 = S_3 - (S_1 \cup S_2), ...$$

$\therefore$ The sets $A_1, A_2, A_3, ...$ are finite sets and are also pairwise disjoint.

Let $$A_1 = \{a_{11}, a_{12}, ..., a_{1n_1}\}, \quad A_2 = \{a_{21}, a_{22}, ..., a_{2n_2}\}, ...$$

Let $B = \bigcup_{i \in \mathbf{N}} A_i$

$\therefore$ $$B = \{a_{11}, a_{12}, ..., a_{1n_1}, a_{21}, a_{22}, ..., a_{2n_2}, a_{31}, ...\}$$

Define $f : B \to \mathbf{N}$ by

$$f(a_{ij}) = (n_1 + n_2 + ... + n_{i-1}) + j, \quad a_{ij} \in B$$

Since $A_1, A_2, ...$ are pairwise disjoint, the definition of 'f' is well defined.

For example, $$f(a_{11}) = 0 + 1 = 1, \quad f(a_{1n_1}) = 0 + n_1 = n_1,$$
$$f(a_{21}) = n_1 + 1, \qquad f(a_{22}) = n_1 + 2, ...$$

f is $1 - 1$, because distinct elements of B have distinct images.

f is onto. Let $n \in \mathbf{N}$. We can find unique numbers i and j such that
$n = (n_1 + n_2 + \ldots\ldots + n_{i-1}) + j$

$\therefore \quad f(a_{ij}) = (n_1 + n_2 + \ldots\ldots + n_{i-1}) + j = n$

$\therefore$ f is onto. $\therefore$ f is a bijection.

$\therefore$ B is denumerable and hence countable.

Since $B = C = \bigcup_{i \in \mathbf{N}} S_i$, the countable union $\bigcup_{i \in \mathbf{N}} S_i$ is also countable.

$\therefore$ The countable union of finite sets is countable.

Theorem 2. Prove that the denumerable union of denumerable sets is denumerable.

Proof. Let $S_1, S_2, S_3, \ldots\ldots$ be an infinite sequence of denumerable sets.

Let $\quad S_1 = \{a_{11}, a_{12}, \ldots\ldots\},\ S_2 = \{a_{21}, a_{22}, \ldots\ldots\},\ \ldots\ldots$

For $n = 2, 3, 4, \ldots\ldots$ define

$$A_n = \{a_{ij} : i + j = n\}$$

$\therefore \quad A_2 = \{a_{11}\},\quad A_3 = \{a_{12}, a_{21}\},\quad A_4 = \{a_{13}, a_{22}, a_{31}\}, \ldots\ldots$

$\therefore$ Each A_n is a finite set.

$\therefore$ $\cup \{A_n : n > 1\}$ being denumerable union of finite sets is denumerable.

Since, $\quad \cup S_i = \cup \{A_n : n > 1\}$, the set $\cup S_i$ is also denumerable.

$\therefore$ The denumerable union of denumerable sets is denumerable.

Theorem 3. Prove that the countable union of countable sets is countable.

Proof. Let $S_1, S_2, S_3, \ldots\ldots$ be a countable sequence of countable sets. We know that a countable set is either finite or denumerable.

Case I. Each S_i is finite.

$\therefore$ $\cup S_i$ is the countable union of finite sets.

We have already proved the result that the countable union of finite sets is countable.

Case II. There is at least one S_i which is denumerable.

Let a_{ij} represent the general element of the set S_i, where $i, j \in \mathbf{N}$.

For $n = 2, 3, 4, \ldots\ldots$ define

$$A_n = \{a_{ij} : i + j = n\}.$$

The definition of $A_2, A_3, A_4, \ldots\ldots$ is valid because at least one S_i is denumerable.

The set $A_2, A_3, A_4, \ldots\ldots$ are all finite sets.

$\therefore$ $\cup \{A_n : n > 1\}$ being denumerable union of finite sets is denumerable and hence countable.

Since, $\cup S_i = \cup\{A_n : n > 1\}$, the set $\cup S_i$ is also countable.

$\therefore$ The countable union of countable sets is countable.

Corollary. *Prove that the union of two countable sets is countable.* (KU 2006 C)

Proof. Take countable sets to be S_1 and S_2 and follow the above proof.

EXERCISE 2.3

1. Prove that a subset of a countable set is countable.
2. Prove that the denumerable union of pairwise disjoint denumerable sets is denumerable.
3. Prove that the finite union of denumerable sets is denumerable.
4. Show that the set **Q** of rational numbers is countable. *(KU 2006)*
5. Show that the set of rational numbers in the interval [2, 7] is countable.

Hints

1. Let B be a subset of a countable set A. We discuss here only that case in which B is not finite and A is denumerable. Let $A = \{a_1, a_2, ...\}$.
   Let b_1 be the first element in the sequence $a_1, a_2, ...$ such that $b_1 \in B$. Let b_2 be the first element which follows b_1 in the sequence such that $b_2 \in B$. Similarly, we find b_3 and so on.
   $\therefore \quad B = \{b_1, b_2, b_3, ...\}$. This set is denumerable.
2. Let S_1, S_2 ...... be an infinite sequence of pairwise disjoint denumerable sets.
   Let $\qquad S_i = \{a_{i1}, a_{i2}, ...\}, i \in \mathbf{N}$ and $C = \cup S_i, i \in \mathbf{N}$
   $\therefore \qquad C = \{a_{ij} : (i, j) \in \mathbf{N} \times \mathbf{N}\}$
   Define $f : C \to \mathbf{N} \times \mathbf{N}$ by $f(a_{ij}) = (i, j)$. Show that f is a bijection.
   $\therefore \qquad C = \cup S_i$ is denumerable, because $\mathbf{N} \times \mathbf{N}$ is so.
3. Let $S_1, S_2, ..., S_k$ be k denumerable sets.
   Let $\quad S_i = \{a_{i1}, a_{i2}, ...\}, \quad 1 \le i \le k$.
   For $n > 1$, define $A_n = \{a_{ij} : i + j = n\}$
   $\therefore$ Each A_n is a finite set.
   $\underset{n>1}{\cup} A_n$ being denumerable union of finite sets is denumerable.
   Also $C = \underset{n>1}{\cup} A_n$, so $\overset{k}{\underset{i=1}{\cup}} S_i$ (= C) is also denumerable.
4. We have $\mathbf{Q} = \mathbf{Q}^+ \cup \{0\} \cup \mathbf{Q}^-$
   Define $f : \mathbf{Q}^+ \to \mathbf{N} \times \mathbf{N}$ by $f\left(\frac{p}{q}\right) = (p, q)$, where $p/q \in \mathbf{Q}^+$ and $(p, q) = 1$.
   f is a bijection. $\therefore$ $\mathbf{Q}^+$ is denumerable
   $\therefore$ $\mathbf{Q}^-$ is also denumerable.
   $\therefore$ **Q** being union of countable sets is countable.

2.5. UNCOUNTABLE SET

A set is said to be an **uncountable set** if it is not countable.

Thus, an uncountable set is neither finite nor denumerable.

$\therefore$ A finite set is always countable and not uncountable.

An infinite set may or may not be denumerable. An uncountable set is always infinite and it does not have the same cardinality as **N.**

An uncountable set is also called a **nondenumerable set.**

Example 1. *Show that every super set of an uncountable set is uncountable.*

Sol. Let A be an uncountable set.

Let B be any super set of A *i.e.,* $B \supseteq A$.

If possible, let the set B be not uncountable.

$\therefore$ The set B is countable.

Since $A \subseteq B$ and every subset of a countable set is countable, the set A is also countable.

This is impossible.

$\therefore$ Our supposition is wrong,

$\therefore$ The set B is uncountable.

$\therefore$ Every super set of an uncountable set is uncountable.

Example 2. *Show that the set of all sequences whose terms are the digits 0 and 1 is uncountable.*

Sol. Let S be the set of all sequences whose terms are the digits 0 and 1.

If possible, let the set S be countable. The set S is not finite.

$\therefore$ The set S is denumerable.

Let $S = \{S_1, S_2, S_3, \ldots\}$, where the terms of the sequences $S_1, S_2, S_3, \ldots$ are digits 0 and 1.

Let $T = \{t_n\}$ be the sequence defined as follows :

$$t_n = \begin{cases} 0 & \text{if } n\text{th term of } S_n \text{ is } 1 \\ 1 & \text{if } n\text{th term of } S_n \text{ is } 0 \end{cases}$$

$\therefore$ $t_n \neq n$th term of S_n $\forall$ $n \in \mathbf{N}$

$\therefore$ The sequence T is different from the sequences $S_1, S_2, S_3, \ldots$ $\therefore$ $T \notin S$

Since each term of T is either 0 or 1, the sequence T lies in the set S.

This is impossible.

$\therefore$ Our supposition is wrong.

$\therefore$ The set S is uncountable.

Theorem 1. Prove that the interval [0, 1] is uncountable.y *(KU 2007)*

Proof. If possible let the interval [0, 1] be countable. Clearly [0, 1] is not finite.

$\therefore$ [0, 1] is denumerable.

Let $$[0, 1] = \{x_1, x_2, x_3, \ldots\}$$

We write 1 as 0.9999 ...... in the form of a decimal. The other numbers in [0, 1] are also written in the form of decimals. Those numbers in [0, 1] which can be written in the form of a decimal in two ways, we choose to write the number as a decimal in which infinite number of nines appears.

For example,

$\frac{1}{2}$ is written as 0.4999 ..... and not as 0.5000 ......

$\frac{1}{4}$ is written as 0.2499 .... and not as 0.2500 ......

This type of representation ensures unique representation of all numbers in [0, 1]. We have $$[0, 1] = \{x_1, x_2, x_3, \ldots\}$$

Let
$$\begin{aligned}
x_1 &= 0.a_{11}\, a_{12}\, a_{13} \ldots\ldots a_{1n} \ldots\ldots \\
x_2 &= 0.a_{21}\, a_{22}\, a_{23} \ldots\ldots a_{2n} \ldots\ldots \\
x_3 &= 0.a_{31}\, a_{32}\, a_{33} \ldots\ldots a_{3n} \ldots\ldots \\
&\ldots\ldots\ldots\ldots\ldots\ldots\ldots\ldots \\
x_n &= 0.a_{n1}\, a_{n2}\, a_{n3} \ldots\ldots a_{nn} \ldots\ldots \\
&\ldots\ldots\ldots\ldots\ldots\ldots\ldots\ldots
\end{aligned}$$

where $a_{ij} \in \{0, 1, 2, \ldots\ldots, 9\}$ and each decimal expression contains infinite number of non-zero digits.

Let $$y = 0.b_1 b_2 b_3 \ldots\ldots b_n \ldots\ldots$$

where $b_i \in \{1, 2, 3, \ldots\ldots, 9\}\ \forall\ i$ and

$$b_1 \neq a_{11},\ b_2 \neq a_{22},\ b_3 \neq a_{33},\ \ldots\ldots,\ b_n \neq a_{nn},\ \ldots\ldots$$

Since b contains infinite number of non-zero digits, we have $y \in [0, 1]$.

Also
$$y \neq x_1 \text{ because } b_1 \neq a_{11}$$
$$y \neq x_2 \text{ because } b_2 \neq a_{22}$$
$$y \neq x_3 \text{ because } b_3 \neq a_{33}$$
$$\ldots\ldots\ldots\ldots\ldots\ldots\ldots\ldots\ldots\ldots\ldots$$

$\therefore$ $$y \neq x_n\ \forall\ n \in \mathbf{N}$$

This is impossible because $y \in [0, 1]$.

$\therefore$ Our supposition is wrong.

$\therefore$ The interval $[0, 1]$ is uncountable.

Theorem 2. Prove that the removal of a denumerable set from an uncountable set does not change its cardinality.

Proof. Let A be an uncountable set and let $B = \{b_1, b_2, b_3, \ldots\ldots\}$ be a denumerable subset of A. The set $A - B$ cannot be countable for otherwise the set $(A - B) \cup B = A$ would become countable.

$\therefore$ The set $A - B$ is uncountable.

$\therefore$ $A - B$ is an infinite set.

Since every infinite set has a denumerable subset, let $D = \{d_1, d_2, d_3, \ldots\ldots\}$ be a denumerable subset of $A - B$. Let $A^* = A - (B \cup D)$.

$\therefore$ $$A = A^* \cup D \cup B = A^* \cup \{d_1, d_2, d_3, \ldots\ldots\} \cup \{b_1, b_2, b_3, \ldots\ldots\}$$

and $$A - B = A^* \cup D \qquad = A^* \cup \{d_1, d_2, d_3, \ldots\ldots\}$$

Define $f : A \to A - B$ by

$$f(a) = a \qquad \text{if } a \in A^*$$
$$f(d_n) = d_{2n-1} \qquad \text{if } n \in \mathbf{N}$$
$$f(b_n) = d_{2n} \qquad \text{if } n \in \mathbf{N}$$

The function 'f' is a bijection.

$\therefore$ The sets A and $A - B$ have the same cardinality.

Remark. The one-to-one correspondence between the elements of A and $A - B$ is as follows :

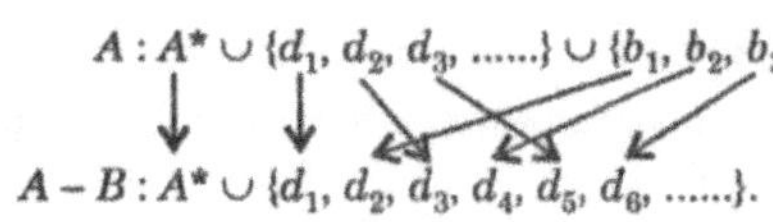

2.6. CARDINALITY OF SETS

We shall assign a symbol to each set A, called **cardinality** of A or **cardinal number** of A in such a way that two sets A and B are assigned the same cardinality if and only if A and B are equipotent sets.

The cardinality of set A is denoted by $n(A)$, $|A|$ or card (A). We shall prefer using the notation $n(A)$.

According to the definition, $n(A) = n(B)$ if and only if $A \approx B$.

If A is a finite set then its cardinality $n(A)$ is defined as the number of element in A.

For example, the cardinality of the set $\{4, 5, 9\}$ is 3.

The cardinality of the infinite set **N** is denoted by $\aleph_0$. The symbol $\aleph_0$ is read aleph-nought and the symbol $\aleph$ is the first letter *aleph* of Hebrew alphabet.

$\therefore$ Every denumerable set has cardinality $\aleph_0$.

For example, $n(\mathbf{Z}) = \aleph_0$, $n(\mathbf{Q}) = \aleph_0$.

The cardinality of the uncountable set $[0, 1]$ is denoted by **c**.

2.7. POWER OF THE CONTINUUM

A set A is said to have the **power of the continuum** if A has the same cardinality as the unit interval $[0, 1]$.

In other words, set A has the power of the continuum if $A \approx [0, 1]$.

We shall be able to prove that every interval has the power of the continuum.

Theorem 1. Prove that the cardinality of the interval (0, 1) is c.

Proof. Let $$A = [0, 1] - \left\{0, 1, \frac{1}{2}, \frac{1}{3}, \ldots\ldots\right\}$$

$\therefore$ A can also be written as $(0, 1) - \left\{\frac{1}{2}, \frac{1}{3}, \ldots\ldots\right\}$.

$\therefore$ $$[0, 1] = A \cup \left\{0, 1, \frac{1}{2}, \frac{1}{3}, \ldots\ldots\right\} \quad \text{and} \quad (0, 1) = A \cup \left\{\frac{1}{2}, \frac{1}{3}, \ldots\ldots\right\}$$

Define $f : [0, 1] \to (0, 1)$ by

$$f(x) = \begin{cases} x & \text{if } x \in A \\ \dfrac{1}{2} & \text{if } x = 0 \\ \dfrac{1}{n+2} & \text{if } x = \dfrac{1}{n}, n \in \mathbf{N} \end{cases}$$

The function 'f' is a bijection.

$\therefore$ The intervals $[0, 1]$ and $(0, 1)$ have the same cardinality.

Since $n([0, 1]) = \mathbf{c}$, we have $n((0, 1)) = \mathbf{c}$.

Remark 1. Under 'f', the one-to-one correspondence between $[0, 1]$ and $(0, 1)$ is as follows :

$$[0, 1] : A \cup \left\{0,\ 1,\ \frac{1}{2},\ \frac{1}{3}, \ldots\ldots\right\}$$

$$\downarrow \quad \downarrow \quad \downarrow \quad \downarrow \quad \downarrow$$

$$(0, 1) : A \cup \left\{\frac{1}{2}, \frac{1}{3}, \frac{1}{4}, \frac{1}{5}, \ldots\ldots\right\}.$$

Remark 2. It can be proved that each of the intervals $[0, 1)$ and $(0, 1]$ has the power of the continuum.

Theorem 2. If a < b, then the interval [a, b] has the power of the continuum.

Proof. We know that the interval $[0, 1]$ has the power of the continuum.

Define $f : [0, 1] \to [a, b]$ by $f(x) = (b - a)\,x + a, \quad x \in [0, 1]$

We can show that f is a bijection.

$\therefore$ $[0, 1]$ and $[a, b]$ have the same cardinality.

$\therefore$ The cardinality of $[a, b]$ is **c**.

$\therefore$ The interval $[a, b]$ has the power of the continuum.

Remark. For $a < b$, it can be proved that the each of the intervals (a, b), $[a, b)$, $(a, b]$ has the power of the continuum.

Theorem 3. Prove that the set R, of real numbers, has the power of the continuum.

Proof. Define $f : \mathbf{R} \to (-1, 1)$ by

$$f(x) = \frac{x}{1+|x|}, x \in \mathbf{R}.$$

We can show that 'f' is a bijection.

$\therefore$ **R** and $(-1, 1)$ have the same cardinality.

Define $g : (0, 1) \to (-1, 1)$ by

$$g(x) = (1 - (-1))x + (-1) = 2x - 1, \quad x \in (0, 1).$$

We can also show that 'g' is a bijection.

$\therefore$ $(0, 1)$ and $(-1, 1)$ have the same cardinality.

$\therefore$ $(0, 1)$ and **R** have the same cardinality.

Since cardinality of $(0, 1)$ is **c**, the set **R** has the power of the continuum.

EXERCISE 2.4

1. Show that the interval $[0, 1)$ has the power of the continuum.
2. Show that the interval $(0, 1]$ has the power of the continuum.
3. If $a < b$, show that the interval (a, b) is uncountable.
4. If $a < b$, show that the interval $(a, b]$ is uncountable.
5. Show that the interval $[3, 7]$ is uncountable.
6. Show that the set $\mathbf{R}^+$ of positive real numbers has the power of the continuum.
7. Show that the set of all irrational numbers is uncountable.

Hints

1. Let $A = [0, 1] - \left\{0, 1, \frac{1}{2}, \frac{1}{3}, \ldots\ldots\right\}$

$\therefore$ A can also be written as $[0, 1) - \left\{0, \frac{1}{2}, \frac{1}{3}, \ldots\ldots\right\}$

$\therefore$ $$[0, 1] = A \cup \left\{0, 1, \frac{1}{2}, \frac{1}{3}, \ldots\ldots\right\} \text{ and } [0, 1) = A \cup \left\{0, \frac{1}{2}, \frac{1}{3}, \ldots\ldots\right\}$$

Define $f : [0, 1] \to [0, 1)$ by

$$f(x) = \begin{cases} x & \text{if } x \in A \\ 0 & \text{if } x = 0 \\ \dfrac{1}{n+1} & \text{if } x = \dfrac{1}{n}, n \in \mathbf{N} \end{cases}$$

The function f is a bijection.

2. Let $A = [0, 1] - \left\{0, 1, \frac{1}{2}, \frac{1}{3}, \ldots\ldots\right\}$

$\therefore$ A can also be written as $(0, 1] - \left\{1, \frac{1}{2}, \frac{1}{3}, \ldots\ldots\right\}$

$\therefore$ $[0, 1] = A \cup \left\{0, 1, \frac{1}{2}, \frac{1}{3}, \ldots\ldots\right\}$ and $(0, 1] = A \cup \left\{1, \frac{1}{2}, \frac{1}{3}, \ldots\ldots\right\}$

Define $f : [0, 1] \to (0, 1]$ by

$$f(x) = \begin{cases} x & \text{if } x \in A \\ 1 & \text{if } x = 0 \\ \dfrac{1}{n+1} & \text{if } x = \dfrac{1}{n}, n \in \mathbf{N} \end{cases}$$

The function f is a bijection.

6. Define $f : \mathbf{R} \to \mathbf{R}^+$ by

$$f(x) = \begin{cases} \dfrac{1}{1-x} & \text{if } x < 0 \\ x+1 & \text{if } x \geq 0 \end{cases}.$$

Show that f is 1– 1. Let $y \in \mathbf{R}^+$.

$$y \in (0, 1) \Rightarrow \frac{y-1}{y} \in \mathbf{R} \text{ and } \frac{y-1}{y} < 0$$

$$\Rightarrow \qquad f\left(\frac{y-1}{y}\right) = \frac{1}{1 - \dfrac{y-1}{y}} = y$$

$$y \in [1, \infty) \Rightarrow y - 1 \geq 0 \Rightarrow f(y-1) = (y-1) + 1 = y.$$

$\therefore$ f is also onto.

7. If possible, let the set S of irrational numbers be countable.

$\therefore$ $\mathbf{R} = \mathbf{Q} \cup S$ is countable, which is impossible.

3

Partially Ordered Sets

3.1. INTRODUCTION

We know that a *relation* from a set A to a set B is a subset of $A \times B$. Thus if R is a relation from A to B, then R is a set of ordered pairs where each first element comes from A and each second element comes from B. If $(a, b) \in R$ then we say "a is R-related to b" and write aRb. In particular if $B = A$, then we say that R is a relation on A. In the present chapter, our main emphasis would be on a special type of relation called *'partial ordering relation'* defined on a set. A set with a *partial ordering relation* forms a *lattice* under certain conditions.

3.2. DIRECTED GRAPH OF A RELATION ON A SET

A directed graph is a pictorial representation of a relation defined on a *finite* set. Let R be a relation on a finite set A. We draw a small circle for each element of the set A and label these circles with the corresponding elements of the set A. These circles are called the **vertices** of the directed graph of R. We draw an arrow from each vertex a_i to each vertex a_j whenever a_iRa_j *i.e.*, $(a_i, a_j) \in R$. These arrows are called the **edges**. An edge from a vertex to itself is called a **loop.** This diagram is called the **directed graph** of the relation R. A directed graph is also called a **digraph.**

Illustration. Let $A = \{1, 2, 3, 5\}$.

Let $$R = \{(1, 1), (1, 2), (3, 5), (2, 3), (3, 1), (2, 2)\}$$
be a relation on A.

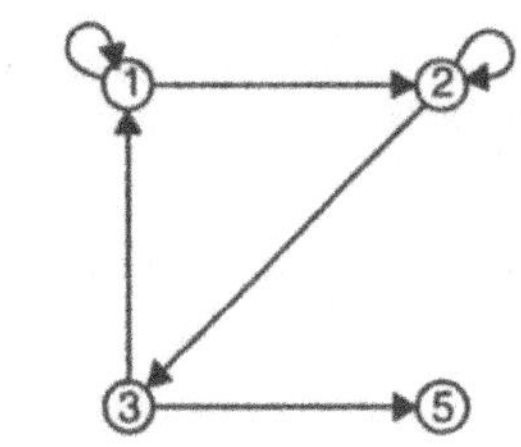

The elements in A are 1, 2, 3 and 5.

$\therefore$ We draw four vertices and mark these as 1, 2, 3, 5.

$(1, 1) \in R \Rightarrow$ There is a loop from 1 to itself.

$(1, 2) \in R \Rightarrow$ There is an edge from 1 to 2.

$(3, 5) \in R \Rightarrow$ There is an edge from 3 to 5.

$(2, 3) \in R \Rightarrow$ There is an edge from 2 to 3.

$(3, 1) \in R \Rightarrow$ There is an edge from 3 to 1.

$(2, 2) \in R \Rightarrow$ There is a loop from 2 to itself.

Example 1. *Find the relation corresponding to the given directed graph.*

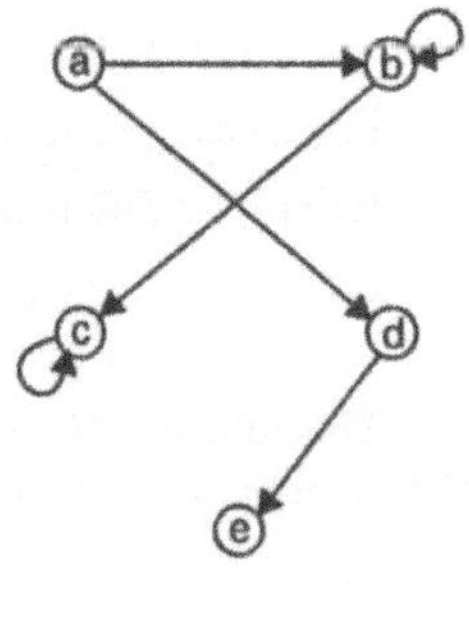

Sol. In the directed graph, there are five vertices marked a, b, c, d and e. Let $A = \{a, b, c, d, e\}$. Let R be the relation on A whose directed graph is given.

Arrow from a to $b \Rightarrow (a, b) \in R$

Loop from b to itself $\Rightarrow (b, b) \in R$

Arrow from b to c $\Rightarrow$ $(b, c) \in R$
Arrow from a to d $\Rightarrow$ $(a, d) \in R$
Loop from c to itself $\Rightarrow$ $(c, c) \in R$
Arrow from d to e $\Rightarrow$ $(d, e) \in R$.

$\therefore$ R = **{(a, b), (b, b), (b, c), (a, d), (c, c), (d, e)}.**

Also, this relation is defined on the set **{a, b, c, d, e}.**

3.3. INVERSE OF A RELATION

Let R be a relation from a set A to set B. The **inverse relation** of R, denoted by R^{-1}, is the relation from B to A given by

$$R^{-1} = \{(b, a) : (a, b) \in R\}.$$

For example if $R = \{(1, 4), (2, 3), (3, 3), (2, 4)\}$ is a relation on the set $A = \{1, 2, 3, 4, 5\}$ then

$$R^{-1} = \{(4, 1), (3, 2), (3, 3), (4, 2)\}.$$

3.4. TYPE OF RELATIONS

(*i*) **Reflexive Relation.** A relation R on a set A is called a **reflexive relation** if aRa for every $a \in A$ *i.e.*, $(a, a) \in R$ for every $a \in A$.

Thus a relation R on a set A is not reflexive if there exist at least one $a \in A$ for which $(a, a) \notin R$.

For example if $A = \{2, 4, 7\}$, then the relation $\{(2, 2), (2, 4), (4, 4), (7, 7)\}$ on A is reflexive.

(*ii*) **Symmetric Relation.** A relation R on a set A is called a **symmetric relation** if whenever aRb then bRa *i.e.*, if whenever $(a, b) \in R$ then $(b, a) \in R$.

Thus a relation R on a set A is not symmetric if there exist $a, b \in A$ such that $(a, b) \in R$ but $(b, a) \notin R$.

For example if $A = \{2, 4, 7\}$ then the relation $\{(2, 4), (4, 2), (7, 7)\}$ on A is symmetric.

(*iii*) **Antisymmetric Relation.** A relation R on a set A is called an **antisymmetric relation** if whenever aRb and bRa then $a = b$ *i.e.*, if whenever $(a, b) \in R$ and $(b, a) \in R$ then $a = b$.

Thus a relation R on a set A is not antisymmetric if there exist $a, b \in A$ such that $(a, b) \in R$ and $(b, a) \in R$ but $a \neq b$.

For example if $A = \{2, 4, 7\}$, then the relation $\{(2, 4), (4, 4)\}$ on A is antisymmetric whereas the relation $\{(2, 4), (4, 2)\}$ on A is not antisymmetric.

(*iv*) **Transitive Relation.** A relation R on a set A is called a **transitive relation** if whenever aRb and bRc then aRc *i.e.*, if whenever $(a, b) \in R$ and $(b, c) \in R$ then $(a, c) \in R$.

Thus a relation R on a set A is not transitive if there exist $a, b, c \in A$ such that $(a, b) \in R$ and $(b, c) \in R$ but $(a, c) \notin R$.

For example if $A = \{2, 4, 7\}$ then the relation $\{(2, 4), (4, 7), (2, 7), (4, 4)\}$ on A is transitive.

Remark. The properties of being symmetric and antisymmetric are not negatives of each other.

Let $A = \{1, 2, 3\}$

and $R = \{(1, 1), (2, 2)\}$ and $R' = \{(1, 2), (2, 1), (1, 3)\}$.

The relation R on A is both symmetric as well as antisymmetric whereas the relation R' on A is neither symmetric nor antisymmetric.

Example 2. *Consider the relation*

$$R = \{(1, 1), (2, 2), (2, 3), (3, 2), (4, 2), (4, 4)\} \text{ on } A = \{1, 2, 3, 4\}.$$

Show that R is neither reflexive nor symmetric nor antisymmetric nor transitive.

Sol. We have $A = \{1, 2, 3, 4\}$

and $R = \{(1, 1), (2, 2), (2, 3), (3, 2), (4, 2), (4, 4)\}$.

$3 \in A$ and $(3, 3) \notin R$

$\therefore$ R is not reflexive.

$(4, 2) \in R$ but $(2, 4) \notin R$

$\therefore$ R is not symmetric.

$(2, 3) \in R$ and $(3, 2) \in R$ but $2 \neq 3$

$\therefore$ R is not antisymmetric.

$(4, 2) \in R$ and $(2, 3) \in R$ but $(4, 3) \notin R$

$\therefore$ R is not transitive.

Example 3. *Each of the following defines a relation on* **N** :

(*i*) $x > y$ (*ii*) xy *is a square*

(*iii*) $x + y = 10$ (*iv*) $x + 4y = 10$.

Determine which relations are reflexive, symmetric, antisymmetric, transitive.

Sol. (*i*) Let R_1 be the relation on **N** given by '$x > y$'.

$\therefore$ $R_1 = \{(x, y) : x > y \text{ and } x, y \in \mathbf{N}\}$

$1 \in \mathbf{N}$ and $1 \not> 1$. $\therefore$ $(1, 1) \notin R_1$ $\therefore$ R_1 is not reflexive.

$(3, 1) \in R_1$ because $3 > 1$ but $(1, 3) \notin R_1$ because $1 \not> 3$ $\therefore$ R_1 is not symmetric.

For $x \neq y$, either $x > y$ or $y > x$

$\Rightarrow$ For $x \neq y$, either $(x, y) \in R_1$ or $(y, x) \in R_1$ $\therefore$ R_1 is antisymmetric.

Let $(x, y), (y, z) \in R_1$.

$\Rightarrow$ $x > y, \; y > z \Rightarrow x > z \Rightarrow (x, z) \in R_1$ $\therefore$ R_1 is transitive.

(*ii*) Let R_2 be the relation on **N** given by 'xy is a square'.

$\therefore$ $R_2 = \{(x, y) : xy \text{ is a square and } x, y \in \mathbf{N}\}$

$x \in \mathbf{N} \Rightarrow x^2$ is a square $\Rightarrow (x, x) \in R_2$ $\therefore$ R_2 is reflexive.

$(x, y) \in R_2 \Rightarrow xy$ is a square $\Rightarrow yx$ is a square $\Rightarrow (y, x) \in R_2$

$\therefore$ R_2 is symmetric.

$(2, 8), (8, 2) \in R_2$ because $2 \times 8 = (4)^2, 8 \times 2 = (4)^2$

Since $2 \neq 8$, R_2 is not antisymmetric.

Let $(x, y), (y, z) \in R_2$

$\Rightarrow$ xy and yz are squares.

Let $xy = \alpha^2, \; yz = \beta^2$ for some $\alpha, \beta \in \mathbf{N}$.

$$\Rightarrow \quad (xy)(yz) = \alpha^2\beta^2 \Rightarrow xy^2z = \alpha^2\beta^2 \Rightarrow xz = \frac{\alpha^2\beta^2}{y^2} = \left(\frac{\alpha\beta}{y}\right)^2$$

$$xz \in \mathbf{N} \Rightarrow \left(\frac{\alpha\beta}{y}\right)^2 \in \mathbf{N} \Rightarrow \frac{\alpha\beta}{y} \in \mathbf{N}$$

$\Rightarrow$ xz is a square $\Rightarrow$ $(x, z) \in R_2$ $\therefore$ R_2 is transitive.

(*iii*) Let R_3 be the relation on **N** given by '$x + y = 10$'.

$\therefore$ $$R_3 = \{(x, y) : x + y = 10 \text{ and } x, y \in \mathbf{N}\}$$

$1 \in \mathbf{N}$ and $1 + 1 \neq 10$ $\therefore$ $(1, 1) \notin R_3$ $\therefore$ R_3 is not reflexive.

$$(x, y) \in R_3 \Rightarrow x + y = 10 \Rightarrow y + x = 10 \Rightarrow (y, x) \in R_3$$

$\therefore$ R_3 is symmetric.

$$(1, 9), (9, 1) \in R_3 \text{ because } 1 + 9 = 10, 9 + 1 = 10$$

Since $1 \neq 9$, the relation R_3 is not antisymmetric.

$$(2, 8), (8, 2) \in R_3 \text{ because } 2 + 8 = 10, 8 + 2 = 10.$$

Since $2 + 2 \neq 10$, we have $(2, 2) \notin R_3$. $\therefore$ R_3 is not transitive.

(*iv*) Let R_4 be the relation on **N** given by '$x + 4y = 10$'.

$\therefore$ $$R_4 = \{(x, y) : x + 4y = 10 \text{ and } x, y \in \mathbf{N}\}$$

$$1 \in \mathbf{N} \text{ and } 1 + 4(1) \neq 10$$

$\therefore$ $(1, 1) \notin R_4$ $\therefore$ R_4 is not reflexive.

$$(6, 1) \in R_4 \text{ because } 6 + 4(1) = 10$$

Also, $(1, 6) \notin R_4$ because $1 + 4(6) \neq 10$ $\therefore$ R_4 is not symmetric.

R_4 contains only two elements (6, 1) and (2, 2).

$\therefore$ R_4 is trivially antisymmetric and transitive.

3.5. EQUIVALENCE RELATION

A relation R on a set A is called an **equivalence relation** if R is reflexive, symmetric and transitive.

In other words, a relation R on a set A is called an **equivalence relation** if :

(*i*) **(Reflexive)** : aRa for all $a \in A$.

(*ii*) **(Symmetric)** : If aRb then bRa.

(*iii*) **(Transitive)** : If aRb and bRc then aRc.

Example 4. *Prove that in the set* **Z** *of integers* $R = \{(a, b) : a, b \in \mathbf{Z}, 5/(a - b)\}$ *is an equivalence relation. Find all the equivalence classes.* (KU 2006)

Sol. $$a \in \mathbf{Z} \Rightarrow a - a = 0 = 5(0) \Rightarrow 5/(a - a)$$

$\therefore$ $$aRa \ \forall a \in \mathbf{Z} \quad \therefore \quad R \text{ is reflexive.}$$

Let aRb.

$\therefore$ $$a - b = 5m \text{ for some } m \in \mathbf{Z}$$

$\Rightarrow$ $$b - a = 5(-m) \Rightarrow bRa \quad \therefore \quad R \text{ is symmetric.}$$

Let aRb and bRc.

$\Rightarrow$ $$a - b = 5r \text{ and } b - c = 5s \text{ for some integers } r, s$$

$\Rightarrow$ $$(a - b) + (b - c) = 5r + 5s \Rightarrow a - c = 5(r + s) \Rightarrow aRc$$

$\therefore$ R is transitive. $\therefore$ R is an equivalence relation.

Equivalence classes in **Z** are [0], [1], [2], [3] and [4].

$$[0] = \{x \in \mathbf{Z} : xR0\} = \{\ldots\ldots, -10, -5, 0, 5, 10, \ldots\ldots\}$$
$$[1] = \{x \in \mathbf{Z} : xR1\} = \{\ldots\ldots, -9, -4, 1, 6, 11, \ldots\ldots\}$$
$$[2] = \{x \in \mathbf{Z} : xR2\} = \{\ldots\ldots, -8, -3, 2, 7, 12, \ldots\ldots\}$$
$$[3] = \{x \in \mathbf{Z} : xR3\} = \{\ldots\ldots, -7, -2, 3, 8, 13, \ldots\ldots\}$$
$$[4] = \{x \in \mathbf{Z} : xR4\} = \{\ldots\ldots, -6, -1, 4, 9, 14, \ldots\ldots\}$$

Example 5. *Show that the relation $\subseteq$ of set inclusion on the power set of a set is not an equivalence relation.*

Sol. Let P represent the power set of the given set.

The relation '$\subseteq$' is reflexive because $A \subseteq A \ \forall \ A \in P$.

The relation '$\subseteq$' is transitive because $A \subseteq B$, $B \subseteq C$ implies $A \subseteq C$.

The relation '$\subseteq$' is not symmetric because $A \subseteq B$ need not imply $B \subseteq A$.

$\therefore$ The relation '$\subseteq$' is not an equivalence relation.

3.6. PARTIAL ORDERING RELATION

A relation R on a set A is called a **partial ordering relation** if R is reflexive, antisymmetric and transitive.

In other words, a relation R on a set A is a **partial ordering relation** if :

(*i*) **(Reflexive)** : aRa for all $a \in A$.

(*ii*) **(Antisymmetric)** : If aRb and bRa then $a = b$.

(*iii*) **(Transitive)** : If aRb and bRc then aRc.

In practice a partial ordering relation is frequently denoted by '$\leq$'. The statement $a \leq b$ is read "a precedes b" or simply as "a is less than or equal to b". With the use of the symbol "$\leq$" the properties of a partial ordering relation look quite natural. Under this notation, the above three properties of a partial ordering relation appear in the following form :

(*i*) **(Reflexive)** : $a \leq a$ for all $a \in A$.

(*ii*) **(Antisymmetric)** : If $a \leq b$ and $b \leq a$ than $a = b$.

(*iii*) **(Transitive)** : If $a \leq b$ and $b \leq c$ then $a \leq c$.

Remark. A partial ordering relation on a set A is also referred as *a partial order relation* or simply as a *partial order.*

3.7. PARTIALLY ORDERED SETS

If R is a partial ordering relation on a set A then the set A with the partial ordering relation R is called a **partially ordered set** or simply a **poset**.

Thus a partially ordered set consists of two things, a set A and a partial order relation defined on A. A partially ordered set A with partial order relation '$\leq$' is denoted by $(A, \leq)$. A partially ordered set $(A, \leq)$ is also referred as A if the partial ordering is fixed in the context of the discussion.

Illustrations I. The relation '$\leq$' (usual less than or equal to) is a partial ordering on the set **N** of natural numbers.

$\therefore$ The pair $(\mathbf{N}, \leq)$ is a partially ordered set.

Similarly $(\mathbf{N}, \geq)$ is a partially ordered set.

The pairs $(\mathbf{Z}, \leq)$, $(\mathbf{Z}, \geq)$, $(\mathbf{Q}, \leq)$, $(\mathbf{Q}, \geq)$, $(\mathbf{R}, \leq)$, $(\mathbf{R}, \geq)$ are also partially ordered sets.

II. Let S be any class of sets. The relation '$\subseteq$' of set inclusion is a partially ordering on S, because

(*i*) $A \subseteq A \ \forall \ A \in S$

(*ii*) $A \subseteq B, B \subseteq A \ \Rightarrow \ A = B$

(*iii*) $A \subseteq B, B \subseteq C \ \Rightarrow \ A \subseteq C$

$\therefore$ $(S, \subseteq)$ is a partially ordered set.

III. Let R be the *relation of divisibility* on the set **N** of natural numbers.

$\therefore$ For $a, b \in \mathbf{N}$, we have aRb if a/b *i.e.*, if there exists a natural number c such that $b = ac$.

We have

(*i*) $a/a \ \forall \ a \in \mathbf{N} \ \therefore \ aRa \ \forall \ a \in \mathbf{N}$

$\therefore$ The relation R is reflexive.

(*ii*) Let a/b and b/a, where $a, b \in \mathbf{N}$.

$\therefore$ There exist natural numbers c and d such that $b = ac$ and $a = bd$.

$$\Rightarrow \quad b = (bd)c \ \Rightarrow \ cd = 1 \ \Rightarrow \ c = 1, d = 1$$

$$\therefore \quad a = b(1) = b$$

$\therefore$ The relation R is antisymmetric.

(*iii*) Let a/b and b/c, where $a, b, c \in \mathbf{N}$.

$\therefore$ There exist natural numbers d and e such that $b = ad$ and $c = be$.

$$\Rightarrow \quad c = (ad)e = a(de) \ \Rightarrow \ a/c \text{ i.e., } aRc$$

$\therefore$ The relation R is transitive.

$\therefore$ R is a partial ordering on **N**.

$\therefore$ (**N**, R) is a partially ordered set.

IV. The relation of divisibility is not a partial order relation on the set **Z** of integers. We have $4/(-4)$ and $(-4)/4$ but $4 \neq -4$.

$\therefore$ The relation of divisibility is not antisymmetric.

$\therefore$ This relation is not a partial order relation on **Z**.

Example 6. *Let R be the relation on* **Z** *defined as follows :*

$$aRb \text{ if } b = a^r \text{ for some } r \in \mathbf{N}.$$

Show that (**Z**, *R*) *is a partially ordered set.*

Sol. We have aRb if $b = a^r$ for some $r \in \mathbf{N}$.

R is reflexive. $a \in \mathbf{Z} \ \Rightarrow \ a = a^1 \ \Rightarrow \ aRa$

$\therefore$ aRa for all $a \in \mathbf{Z}$

$\therefore$ R is reflexive.

R is antisymmetric. Let aRb and bRa.

$\therefore$ There exists $r, s \in \mathbf{N}$ such that $b = a^r$ and $a = b^s$.

$$\Rightarrow \quad a = (a^r)^s = a^{rs} \text{ i.e., } a^{rs} = a$$

$$\therefore \quad \text{Either } a = 1 \ \text{ or } \ a = -1 \ \text{ or } \ rs = 1.$$

$$a = 1 \ \Rightarrow \ b = (1)^r = 1 \ \Rightarrow \ a = b$$

$$a = -1 \Rightarrow \ b = (-1)^r \ \Rightarrow \ b = \pm 1 \ \Rightarrow \ b = -1,$$

because $b = 1$ would imply $a = (1)^s = 1$.

$$\therefore \quad a = -1 \Rightarrow \ b = -1 \ \Rightarrow \ \text{i.e., } a = b$$

$$rs = 1 \ \Rightarrow \ r = 1, s = 1 \qquad (\because \ r, s \in \mathbf{N})$$

$$\therefore \quad b = (a)^1 = a \ \text{ i.e., } a = b$$

$\therefore$ R is antisymmetric.

R is transitive. Let aRb and bRc.

$\therefore$ There exist $r, s \in \mathbf{N}$ such that $b = a^r$ and $c = b^s$.

$\Rightarrow \quad c = (a^r)^s = a^{rs} \quad \Rightarrow \quad aRc \qquad (\because \quad rs \in \mathbf{N})$

$\therefore$ R is transitive.

$\therefore$ R is a partial ordering on **Z**.

$\therefore$ (**Z**, R) is a partially ordered set.

Example 7. *Let (A, R) and (B, R') be partially ordered sets. Define a relation R'' on $A \times B$ by $(a, b)\, R''\, (a', b')$ if aRa' and $bR'b'$. Show that $(A \times B, R'')$ is also a partially ordered set.*

Sol. R'' on $A \times B$ is defined by $(a, b)\, R''\, (a', b')$ if aRa' and $bR'b'$.

R″ is reflexive. Let $(a, b) \in A \times B$.

$\Rightarrow \quad a \in A, b \in B \quad \Rightarrow \quad aRa, bR'b \quad \Rightarrow \quad (a, b)\, R''\, (a, b)$

$\therefore$ R'' is reflexive.

R″ is antisymmetric. Let $(a, b)\, R''\, (c, d)$ and $(c, d)\, R''\, (a, b)$.

$(a, b)\, R''\, (c, d) \quad \Rightarrow \quad a\,Rc$ and $bR'd$. Also $(c, d)\, R''\, (a, b) \quad \Rightarrow \quad cRa$ and $dR'b$

$\therefore \quad a = c$ and $b = d$ because R and R' are both antisymmetric.

$\therefore \quad (a, b) = (c, d)$

$\therefore$ R'' is antisymmetric.

R″ is transitive. Let $(a, b)\, R''\, (c, d)$ and $(c, d)\, R''\, (e, f)$.

$(a, b)\, R''\, (c, d) \quad \Rightarrow \quad aRc$ and $bR'd$

$(c, d)\, R''\, (e, f) \quad \Rightarrow \quad cRe$ and $dR'f$

$\therefore$ aRe and $bR'f$ because R and R' are both transitive.

$\therefore$ $(a, b)\, R''\, (e, f)$

$\therefore$ R'' is transitive.

$\therefore$ R'' is a partial ordering on $A \times B$.

$\therefore$ $(A \times B, R'')$ is a partially ordered set.

3.8. DUAL PARTIALLY ORDERED SET

Let (A, R) be a partially ordered set. The inverse relation R^{-1} of R is given by

$$R^{-1} = \{(b, a) : (a, b) \in R\}.$$

We shall show that R^{-1} is also a partial ordering on A.

R⁻¹ is reflexive. Let $a \in A$.

$\Rightarrow \quad (a, a) \in R \quad \Rightarrow \quad (a, a) \in R^{-1}$

$\therefore$ R^{-1} is reflexive.

R⁻¹ is antisymmetric. Let $(a, b), (b, a) \in R^{-1}$.

$\Rightarrow \quad (b, a), (a, b) \in R \quad \Rightarrow \quad a = b$

$\therefore$ R^{-1} is antisymmetric.

R⁻¹ is transitive. Let $(a, b), (b, c) \in R^{-1}$.

$\Rightarrow \quad (b, a), (c, b) \in R \quad \Rightarrow \quad (c, b), (b, a) \in R \quad \Rightarrow \quad (c, a) \in R$

$\Rightarrow \quad (a, c) \in R^{-1}$

$\therefore$ R^{-1} is transitive.

$\therefore$ R^{-1} is a partial ordering on A.

$\therefore$ (A, R^{-1}) is a partially ordered set.

The partially ordered set (A, R^{-1}) is called the **dual partially ordered set** of the partially ordered set (A, R).

3.9. PARTIALLY ORDERED SUBSETS OF A PARTIALLY ORDERED SET

Let (A, R) be a partially ordered set and let B be a subset of A.

Define a relation R_B on B as $bR_B\, b'$ whenever bRb' as elements of A. It is easy to check that R_B is a partial order relation on B. This partial order relation on B is called the **restriction** of R to B or the **induced partial order relation** on B. The subset B of A with partial order relation R_B is called a **partially ordered subset** of the partially ordered set A.

3.10. QUASI-ORDER RELATION

A relation R on a set A is called a **quasi-order relation** on A if :

(*i*) **(Irreflexive) :** $a \not R a$ for all $a \in A$.

(*ii*) **(Transitive) :** If aRb and bRc then aRc.

3.11. STRICT PARTIAL ORDERING RELATION

A relation R on a set A is called a **strict partial ordering relation** on A if :

(*i*) **(Asymmetric) :** For $a, b \in A$, aRb and bRa cannot hold simultaneously.

(*ii*) **(Transitive) :** If aRb and bRc then aRc.

3.12. NOTATION

Let $(A, \leq)$ be a partially ordered set. If $a \leq b$, then we say that '*a precedes b*' or '*a is less than or equal to b*'.

In the context of the symbol '$\leq$', we define the symbols : $<, \geq, >$.

If $a \leq b$ and $a \neq b$ then we write $a < b$ and say '*a* **strictly precedes** *b*' or '*a* **is strictly less than** *b*'.

If $a \leq b$ then we write $b \geq a$ and say '*b* **succeeds** *a*' or '*b* **is greater than or equal to** *a*'.

If $a \leq b$ and $a \neq b$ then we write $b > a$ and say '*b* **strictly succeeds** *a*' or '*b* **is strictly greater than** *a*'.

From now onward, we shall make frequent use of above defined symbols.

Example 8. *Let A be any set. Show that '$\leq$' is a partial order on A if and only if '$<$' is a quasi-order relation on A.*

Sol. Let '$\leq$' be a partial order on the set A.

$<$ is irreflexive. For $a \in A$, we have $a \leq a$.

Since $a = a$, we have $a \nless a$.

$\therefore$ '$<$' is irreflexive.

$<$ is transitive. Let $a < b$ and $b < c$.

$\Rightarrow \quad a \leq b, a \neq b$ and $b \leq c, b \neq c \Rightarrow a \leq c$

If possible, let $a = c$.

$\therefore \quad b \leq c$ implies $b \leq a$

$\therefore \quad a \leq b$ and $b \leq a$ implies $a = b$, which is impossible.

$\therefore \quad a \leq c$ and $a \neq c \quad \therefore \quad a < c$

$\therefore$ '$<$' is transitive.

$\therefore$ '$<$' is a quasi-order relation on A.

Conversely, let $<$ be a quasi-order relation on A.

We know that $a \leq b$ if either $a < b$ or $a = b$.

$\leq$ is reflexive. For $a \in A$, we have $a = a$. $\therefore$ $a \leq a$

$\therefore$ '$\leq$' is reflexive.

$\leq$ is antisymmetric. Let $a \leq b$ and $b \leq a$.

$\Rightarrow$ $a < b$ or $a = b$ and $b < a$ or $b = a$

If possible, let $a \neq b$.

By transitivity of '<', $a < b$ and $b < a$ implies $a < a$.

This is impossible because '<' is irreflexive.

$\therefore$ $a = b$ $\therefore$ '$\leq$' is antisymmetric.

$\leq$ is transitive. Let $a \leq b$ and $b \leq c$.

$\Rightarrow$ $a < b$ or $a = b$ and $b < c$ or $b = c$.

Case I. a < b and b < c

Since '<' is transitive, we have $a < c$.

$\Rightarrow$ $a \leq c$

Case II. a = b and b = c

$$a = b,\ b = c \Rightarrow a = c \Rightarrow a \leq c$$

Case III. a < b and b = c

$$a < b,\ b = c \Rightarrow a < c \Rightarrow a \leq c$$

Case IV. a = b and b < c

$$a = b,\ b < c \Rightarrow a < c \Rightarrow a \leq c$$

$\therefore$ In all cases, $a \leq c$

$\therefore$ '$\leq$' is transitive.

$\therefore$ '$\leq$' is a partial order on A.

Example 9. *Let A be any set. Show that if '$\leq$' is a partial order on A then '<' is a strict partial order on A.*

Sol. Let '$\leq$' be a partial order on the set A.

< is asymmetric. Let $a < b$ and $b < a$ for some $a, b \in A$.

$\Rightarrow$ $a \leq b$ and $b \leq a \Rightarrow a = b$ ($\because$ $\leq$ is antisymmetric)

This is impossible because $a < b$ implies $a \neq b$.

$\therefore$ For $a, b \in A$, $a < b$ and $b < a$ cannot hold simultaneously.

$\therefore$ < is asymmetric.

< is transitive. Let $a < b$ and $b < c$.

$\Rightarrow$ $a \leq b, a \neq b$ and $b \leq c, b \neq c \Rightarrow a \leq c$

If possible, let $a = c$

$\therefore$ $b \leq c \Rightarrow b \leq a$

$\therefore$ $a \leq b$ and $b \leq a$ implies $a = b$, which is impossible.

$\therefore$ $a \leq c$ and $a \neq c$ $\therefore$ $a < c$

$\therefore$ < is transitive.

$\therefore$ '<' is a strict partial ordering on A.

3.13. COMPARABILITY OF ELEMENTS

Let $(A, \leq)$ be a partially ordered set. Two distinct elements a and b of A are said to be **comparable elements** if either

$$a < b \quad \text{or} \quad b < a.$$

If a and b are distinct elements of A and $a \nless b$ and $b \nless a$ then a and b are called **noncomparable elements** and we write $a \parallel b$.

In a partially ordered set, two distinct elements may or may not be comparable. This justifies the use of the word 'partial' in defining a partially ordered set.

For example, let $A = \{1, 2, 3, 4\}$.

and $R = \{(1, 1), (2, 2), (3, 3), (4, 4), (1, 2) (2, 3), (1, 3)\}$ be a partial order relation on A.

Here 1 and 2 are comparable because $(1, 2) \in R$.

1 and 4 are noncomparable because $(1, 4) \notin R$, $(4, 1) \notin R$.

2 and 4 are noncomparable because $(2, 4) \notin R$, $(4, 2) \notin R$.

Example 10. *Let $A = \{1, 2, 3, 4\}$ and $P(A)$ be the power set of A. $P(A)$ is partially ordered by 'set inclusion'. Find which of the following elements of $P(A)$ are comparable :*

(*i*) $\{2\}, \{1, 3, 4\}$ (*ii*) $\{2, 3\}, \{1, 3\}$

(*iii*) $\{3, 4\}, \{4\}$ (*iv*) $\{2, 4\}, \{2, 3, 4\}$.

Sol. (*i*) $\{2\}$ and $\{1, 3, 4\}$ are **noncomparable** because neither

$$\{2\} \subset \{1, 3, 4\} \text{ nor } \{1, 3, 4\} \subset \{2\}.$$

(*ii*) $\{2, 3\}$ and $\{1, 3\}$ are **noncomparable** because neither $\{2, 3\} \subset \{1, 3\}$ nor $\{1, 3\} \subset \{2, 3\}$.

(*iii*) $\{3, 4\}$ and $\{4\}$ are **comparable** because $\{4\} \subset \{3, 4\}$.

(*iv*) $\{2, 4\}$ and $\{2, 3, 4\}$ are **comparable** because $\{2, 4\} \subset \{2, 3, 4\}$.

3.14. TOTALLY ORDERED SETS

We know that a partially ordered set is a set equipped with a partial ordering relation defined on the set under consideration. In a partially ordered set A, two distinct elements of A may or may not be comparable.

Let $A = \{1, 2, 3\}$

and $R = \{(1, 1) (2, 2) (3, 3), (1, 2) (1, 3) (2, 3)\}$

(A, R) is a partially ordered set. Here every pair of distinct elements of A are comparable elements. Such a partially ordered set deserves a special name.

A partially ordered set A is called a **totally ordered set** if A has no noncomparable elements.

Thus if A is a totally ordered set then for every distinct pair a, b of elements of A, we have either $a < b$ or $b < a$.

A totally ordered set is also called a **linearly ordered set.**

For example, the set **N** of natural numbers is a totally ordered set with the relation 'usual $\leq$ in numbers'.

Example 11. *Let A be a set containing at least two elements. Show that the power set $P(A)$ of A is not a totally ordered set under the relation $\leq$ of set inclusion.*

Sol. The set $P(A)$ is a partially ordered set w.r.t. the relation '$\leq$'.

Let $a, b \in A$.

$\therefore \quad \{a\}, \{b\} \in P(A)$

Here $\quad \{a\} \not\subset \{b\}$ and $\{b\} \not\subset \{a\}$ $\therefore$ Neither $\{a\} < \{b\}$ nor $\{b\} < \{a\}$.

$\therefore$ $\{a\}$ and $\{b\}$ are noncomparable elements of $P(A)$.

$\therefore$ $P(A)$ is not a totally ordered set.

3.15. CHAINS

We know that a partially ordered set may or may not be a totally ordered set. Let A be a partially ordered set. If B is a subset of A such that B is a totally ordered subset of A with the induced partial order relation on B, then the totally ordered subset B of A is called a **chain** in A. The number of elements in a chain is called the **length** of the chain.

It may be noted that every subset of a totally ordered set is a **chain.**

Illustrations I. The set **N** of natural numbers is a partially ordered set with the relation of divisibility.

N is not a totally ordered set because 2, 5 $\in$ **N** and neither 2/5 nor 5/2.

Let $\quad A = \{5, 10, 20, 80\}$.

A is a chain in **N** because 5/10, 5/20, 5/80, 10/20, 10/80, 20/80.

II. Let A be a set containing at least two elements. We have seen that the power set $P(A)$ of A is not a totally ordered set under the relation $\subseteq$ of set inclusion.

For $a \in A$, the set $\{\phi, \{a\}, A\}$ forms a chain in $P(A)$.

III. The set **N** of natural numbers is a totally ordered set with the relation "usual $\leq$ in numbers".

$\therefore$ Every subset of **N** is a chain in **N**.

Example 12. *Let the set of natural numbers* **N** *be partially ordered by divisibility. State whether each of the following subsets of* **N** *are chains in* **N** :

(*i*) {*10, 5, 30*} (*ii*) {*2, 12, 48, 384, 6, 24, 96*}

(*iii*) {*2, 3*} (*iv*) {*4*}.

Sol. (*i*) Let $\quad A = \{10, 5, 30\}$

$\therefore \quad A = \{5, 10, 30\}$

We have 5/10 and 10/30.

$\therefore$ Any two distinct elements of A are comparable.

$\therefore$ A is a chain in **N**.

Note. We need not check 5 divides 30 or not. This will follow by using transitivity.

(*ii*) Let $\quad B = \{2, 12, 48, 384, 6, 24, 96\}$

$\therefore \quad B = \{2, 6, 12, 24, 48, 96, 384\}$

We have 2/6, 6/12, 12/24, 24/48, 48/96, 96/384.

$\therefore$ Any two distinct elements of B are comparable.

$\therefore$ B is a chain in **N**.

(*iii*) Let $\quad C = \{2, 3\}$.

Here $2 \nmid 3$ and $3 \nmid 2$.

$\therefore$ 2 and 3 are not comparable.

$\therefore$ C is not a chain in **N**.

(*iv*) Let $D = \{4\}$.

Since D has only one element, it is trivially a chain in **N**.

3.16. ANTICHAINS

Let A be a partially ordered set. If B is a subset of A such that no two elements of B are comparable, then the subset B of A is called an **antichain** in A.

For example, let A be a set containing at least two elements. The power set $P(A)$ of A is a partially ordered set with the relation $\subseteq$ of set inclusion. Let $a, b \in A$.

$\therefore$ $\{a\}, \{b\} \in P(A)$ and neither $\{a\} \subseteq \{b\}$ nor $\{b\} \subseteq \{a\}$.

$\therefore$ The subset $\{\{a\}, \{b\}\}$ of $P(A)$ is an antichain in $P(A)$.

EXERCISE 3.1

1. Let $A = \{2, 3, 5, 9\}$. Draw the directed graph of the relation $\{(2, 3), (3, 5), (5, 5), (2, 2)\}$ defined on A.
2. Find the relation corresponding to the given directed graph.

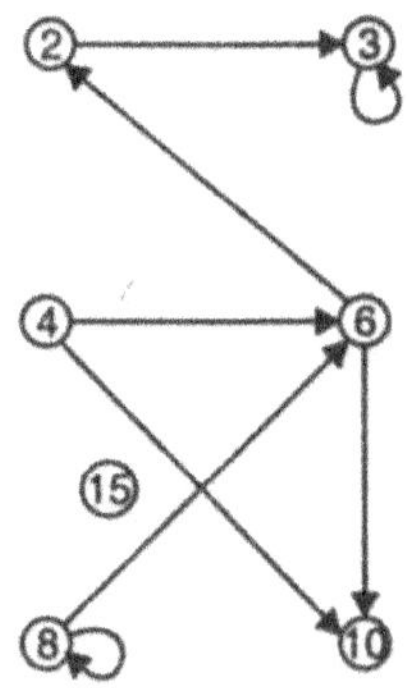

Q. 2

3. Let $A = \{2, 4, 5, 6\}$. Show that :
   (*i*) The relation $\{(2, 4), (5, 5)\}$ on A is not reflexive.
   (*ii*) The relation $\{(2, 5), (2, 2), (4, 4), (5, 5), (6, 6)\}$ on A is not symmetric.
   (*iii*) The relation $\{(2, 4), (4, 2)\}$ on A is symmetric.
   (*iv*) The relation $\{(2, 4), (4, 6), (6, 2)\}$ on A is not transitive.
   (*v*) The relation $\{(2, 4), (4, 6), (2, 6)\}$ on A is transitive.
   (*vi*) The relation $\{(2, 4), (4, 2)\}$ on A is not an equivalence relation.
   (*vii*) The inverse relation of the relation $\{(2, 5), (5, 6)\}$ on A is not $\{(5, 2), (5, 6)\}$.
4. Let $A = \{1, 2, 3, 4\}$. Show that the relation :
   (*i*) $R = \{(1, 1), (1, 2), (2, 2), (2, 3), (3, 3), (4, 4)\}$ on A is reflexive but neither symmetric nor transitive.
   (*ii*) $R = \{(1, 2), (2, 3)\}$ on A is neither reflexive nor symmetric nor transitive.
   (*iii*) $R = \{(1, 1), (2, 2), (3, 3), (4, 4), (1, 3), (3, 1), (1, 4), (4, 1)$ on A is not an equivalence relation.
5. Prove that in the set **Z** of integers, $R = \{(a, b) : a, b \in \mathbf{Z}, 6/(a - b)\}$ is an equivalence relation. Find all the equivalence classes. (*KU 2006 C*)
6. Let $A = \{1, 2, 3, 4\}$. Determine which of the following relations on A are antisymmetric :
   (*i*) $R = \{(1, 1), (1, 2), (3, 3), (4, 1)\}$ (*ii*) $R = \{(1, 2), (1, 3), (2, 1), (1, 4)\}$
   (*iii*) $R = A \times A$ (*iv*) $R = \phi$.
7. Let $A = \{1, 2, 3\}$. Determine which of the following relations on A are partial ordering of A :
   (*i*) $R_1 = \{(1, 1), (2, 1), (2, 2), (3, 3)\}$ (*ii*) $R_2 = \{(1, 1), (2, 2)\}$
   (*iii*) $R_3 = \{(1, 1), (2, 3), (3, 1), (2, 1), (2, 2), (3, 3)\}$.
   (*iv*) $R_4 = \{(1, 1), (1, 2), (2, 2), (3, 3), (2, 3), (3, 1)\}$.
8. Let A be the set of all lines in a plane. Let R be a relation on A defined as $l_1 R l_2$ if the lines l_1 and l_2 are parallel. Show that (A, R) is not a partially ordered set.
9. Let R be a relation on **N** defined by xRy if x and y share a common factor other than 1. Show that (N, R) is not a poset.
10. The set of natural numbers **N** is partially ordered by divisibility. Find which of the following numbers are comparable. (*i*) 2 and 7 (*ii*) 5 and 15 (*iii*) 17 and 33 (*iv*) 18 and 90.

11. Let $A = \{1, 2, 3, 5, 6, 10, 15, 30\}$. The set A is a partially ordered set with the relation 'divisibility'. Determine all comparable and noncomparable elements of A.
12. Show that the partially ordered set N of natural numbers partially ordered by divisibility is not a totally ordered set.
13. Show that the partially ordered set $A = \{2, 6, 12, 36, 72\}$ partially ordered by divisibility is a totally ordered set.
14. Let the set of natural numbers N be partially ordered by divisibility. State whether each of the following is a chain in N or not :
(*i*) {3, 6, 12} (*ii*) {1, 2, 3} (*iii*) {7}
(*iv*) {2, 4, 8, 17} (*v*) {3, 6, 18, 72} (*vi*) N.

Answers

1. 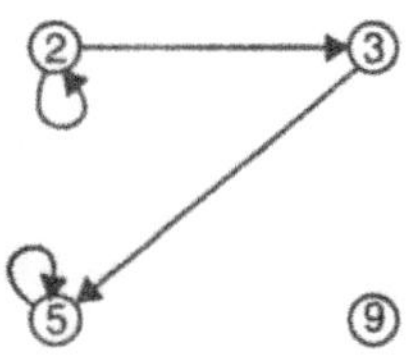

2. {(2, 3), (3, 3), (6, 2), (4, 6), (4, 10), (6, 10), (8, 6), (8, 8)} on {2, 3, 4, 6, 8, 10, 15}
5. [0] = {......, – 12, – 6, 0, 6, 12, ...}
[1] = {......, – 11, – 5, 1, 7, 13, ...}
[2] = {......, – 10, – 4, 2, 8, 14, ...}
[3] = {......, – 9, – 3, 3, 9, 15, ...}
[4] = {......, – 8, – 2, 4, 10, 16, ...}
[5] = {......, – 7, – 1, 5, 11, 17, ...}
6. (*i*), (*iv*) **7.** (*i*), (*iii*) **10.** (*ii*), (*iv*)
11. Pairs of comparable elements are (1, 2), (1, 3), (1, 5), (1, 6), (1, 10), (1, 15), (1, 30), (2, 6), (2, 10), (2, 30), (3, 6), (3, 15), (3, 30), (5, 10), (5, 15), (5, 30), (6, 30), (10, 30), (15, 30)
Pairs of noncomparable elements are (2, 3), (2, 5), (2, 15), (3, 5), (3, 10), (5, 6), (6, 10), (6, 15), (10, 15)
14. (*i*), (*iii*), (*v*).

Hints

8. R is not antisymmetric. **9.** 10R15, 15R10 but $10 \neq 15$.

3.17. HASSE DIAGRAM

Let A be a partially ordered set and let '$\leq$' denote the partial order relation on A. Let $a, b \in A$. a is said to be an **immediate predecessor** of b if $a < b$ and there exists no element c in A such that $a < c < b$.

In other words, a is an immediate predecessor of b if $a < b$ and no element in A lies between a and b.

If a is an immediate predecessor of b then b is said to be an **immediate successor** of a or a **cover** of a.

If a is an immediate predecessor of b then we write $a << b$.

For example, the set N of natural numbers is a partially ordered set with the relation of divisibility. We have $4 < 8$ because $4 \neq 8$ and 4/8. Here $4 << 8$ because there is no element c in N such that $4 < c < 8$ *i.e.*, $4/c$ and $c/8$.

$\therefore$ 4 is an immediate predecessor of 8.

It is wrong to say that $5 << 20$ because $10 \in$ N and we have $5 < 10 < 20$ *i.e.*, 5/10/20.

A Hasse diagram is a pictorial representation of a finite partially ordered set. Let A be a finite partially ordered set. The elements of A are represented by small dots and these are marked with the correspoding elements. These dots are called the **vertices** of the Hasse diagram of A. Whenever $a << b$ in A *i.e.*, $a < b$ and there is no element c in A such that $a < c < b$, we place b higher than a and draw a line between them. Such lines are called the **edges** of the Hasse diagram. Thus in a Hasse diagram, movement upward indicates succession. This diagram is called the **Hasse diagram** of the partially ordered set A. It is very important to note that we have $x < y$ in A if and only if there is a directed path from the vertex x to the vertex y in the Hasse diagram.

Remark. If a partially ordered set A is totally ordered *i.e.*, for distinct $x, y \in A$, we have either $x < y$ or $y < x$ then the Hasse diagram of A will consist of only one path.

This also implies that the Hasse diagram of a chain in a partially ordered set will also consist of only one path.

Illustration. Let $A = \{1, 2, 3, 4, 5, 6\}$ be a partially ordered set with the relation 'divisibility'. We have $1 \leq 1, 2 \leq 2, 3 \leq 3, 4 \leq 4, 5 \leq 5, 6 \leq 6, 1 \leq 2, 1 \leq 3, 1 \leq 4, 1 \leq 5, 1 \leq 6, 2 \leq 4, 2 \leq 6, 3 \leq 6$.

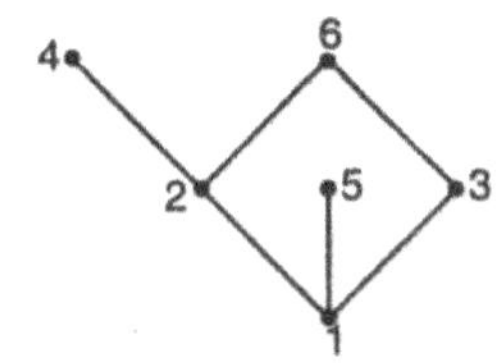

For the Hasse diagram of A, we observe that

$$1 << 2, 1 << 3, 1 << 5, 2 << 4, 2 << 6, 3 << 6.$$

$1 << 4$ is not true because $1 < 2 < 4$.

$1 << 6$ is not true because $1 < 3 < 6$.

The Hasse diagram of A is shown in the figure.

We observe that $1 < 4$ and there is a directed path from the vertex 1 to the vertex 4. Similarly there is a directed path from the vertex 1 to the vertex 6, so we must have $1 < 6$.

Example 1. *Let $A = \{4, 5, 6, 7\}$ be the partially ordered set with the relation 'usual $\leq$ in numbers'. Draw the directed graph and the Hasse diagram.*

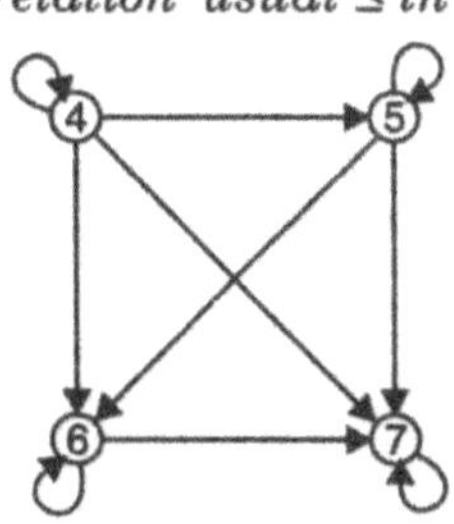

Sol. We have

$$A = \{4, 5, 6, 7\}$$

and $\quad 4 \leq 4, \quad 5 \leq 5, \quad 6 \leq 6, \quad 7 \leq 7, \quad 4 \leq 5,$

$4 \leq 6, \quad 4 \leq 7, \quad 5 \leq 6, \quad 5 \leq 7, \quad 6 \leq 7.$

The directed graph is shown in the figure.

To draw the Hasse diagram, we are to locate pairs $a, b \in A$ such that $a << b$ *i.e.*, $a < b$ and there is no element c in A such that $a < c < b$. $a << b$ can hold only when $a \neq b$. $4 << 5$ is true. $4 << 6$ is not true because $4 < 5 < 6$. $4 << 7$ is not true because $4 < 5 < 7$. $5 << 6$ is true. $5 << 7$ is not true because $5 < 6 < 7$. $6 << 7$ is true.

$\therefore$ In the Hasse diagram, we are to display only $4 << 5$, $5 << 6$ and $6 << 7$.

The Hasse diagram is shown in the figure,

Remark. The partially ordered set in the above problem is also totally ordered.

Example 2. *Let $A = \{1, 2, 3, 4, 12\}$ be the partially ordered set with the relation $a \leq b$ if and only if a/b. Draw the directed graph and the Hasse diagram.*

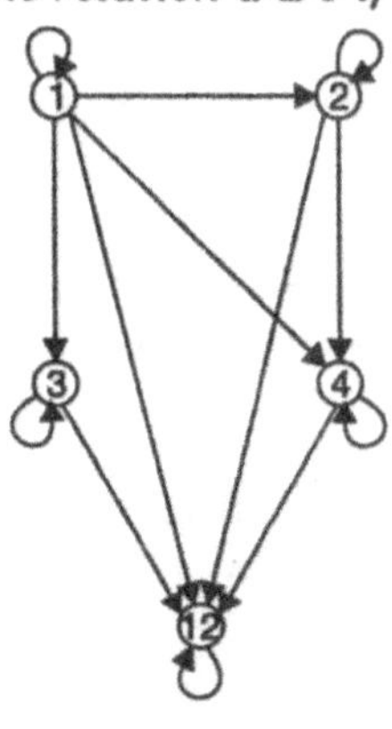

Sol. We have

$$A = \{1, 2, 3, 4, 12\}$$

and $\quad 1 \leq 1, \quad 1 \leq 2, \quad 1 \leq 3, \quad 1 \leq 4, \quad 1 \leq 12,$

$2 \leq 2, \quad 2 \leq 4, \quad 2 \leq 12, \quad 3 \leq 3, \quad 3 \leq 12,$

$4 \leq 4, \quad 4 \leq 12, \quad 12 \leq 12.$

The directed graph is shown in the figure.

To draw the Hasse diagram, we are to locate pairs $a, b \in A$ such that $a << b$. For $a << b$, we must have $a \neq b$. $1 << 2$ is true. $1 << 3$

is true. 1 << 4 is not true because 1 < 2 < 4. 1 << 12 is not true because 1 < 2 < 12. 2 << 4 is true. 2 << 12 is not true because 2 < 4 < 12. 3 << 12 is true. 4 << 12 is true.

∴ In the Hasse diagram, we are to display only 1 << 2, 1 << 3, 2 << 4, 3 << 12, 4 << 12.

The Hasse diagram is shown in the figure.

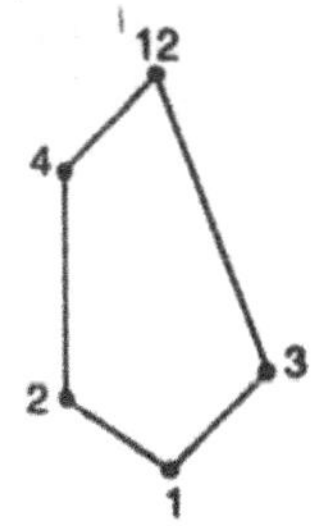

Example 3. *Let A = {1, 2, 3}. Draw the Hasses diagram of the partially ordered set P(A) with the relation of 'set inclusion'.*

Sol. We have $A = \{1, 2, 3\}$.

∴ $P(A) = \{\phi, \{1\}, \{2\}, \{3\}, \{1, 2\}, \{2, 3\}, \{1, 3\}, \{1, 2, 3\}\}$

For a << b we must have $a \neq b$.

a << b with a = ϕ

ϕ << {1}, ϕ << {2}, ϕ << {3} are true. ϕ << {1, 2}, ϕ << {2, 3}, ϕ << {1, 3}, ϕ << {1, 2, 3} are not true because we have ϕ < {1} < {1, 2} in case of ϕ << {1, 2}.

a << b with a = {1}

{1} << {1, 2}, {1} << {1, 3} are true. {1} << {1, 2, 3} is false because {1} < {1, 2} < {1, 2, 3}.

a << b with a = {2}

{2} << {1, 2}, {2} << {2, 3} are true. {2} << {1, 2, 3} is false because {2} < {2, 3} < {1, 2, 3}.

a << b with a = {3}

{3} << {2, 3}, {3} << {1, 3} are true.

{3} << {1, 2, 3} is not true because {3} < {2, 3} < {1, 2, 3}.

a << b with a = {1, 2}

Only {1, 2} << {1, 2, 3} is true.

a << b with a = {2, 3}

Only {2, 3} << {1, 2, 3} is true.

a << b with a = {1, 3}

Only {1, 3} << {1, 2, 3} is true.

The Hasse diagram is shown in the figure.

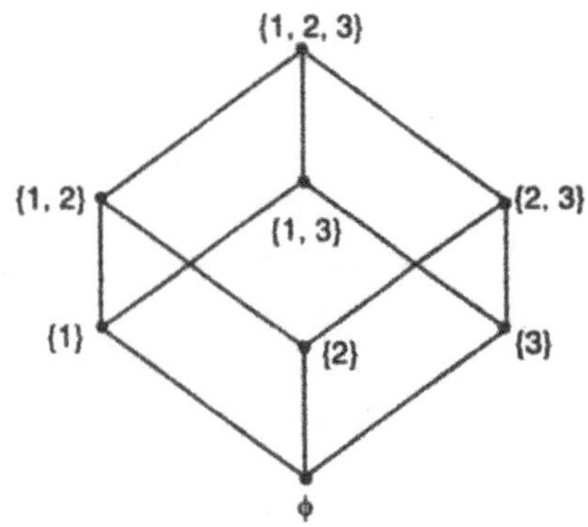

Example 4. *Let D_{12} denote the set of positive divisors of 12 and partially ordered by divisibility. Draw the Hasse diagram of this partially ordered set.*

Sol. We have $D_{12} = \{1, 2, 3, 4, 6, 12\}$

and $1 \leq 1, \quad 1 \leq 2, \quad 1 \leq 3, \quad 1 \leq 4, \quad 1 \leq 6, \quad 1 \leq 12,$

$2 \leq 2, \quad 2 \leq 4, \quad 2 \leq 6, \quad 2 \leq 12,$

$3 \leq 3, \quad 3 \leq 6, \quad 3 \leq 12, \quad 4 \leq 4, \quad 4 \leq 12,$

$6 \leq 6, \quad 6 \leq 12, \quad 12 \leq 12.$

We find pairs of distinct elements a, b such that a << b.

All possible cases of a << b are :

1 << 2, 1 << 3, 2 << 4, 2 << 6, 3 << 6, 4 << 12, 6 << 12.

The Hasse diagram is shown in the figure.

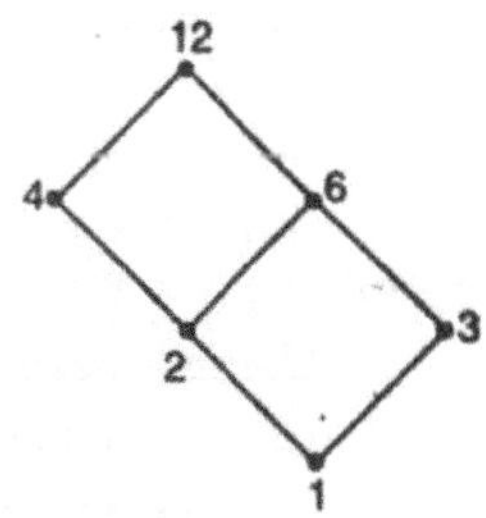

Example 5. *Let D_{24} denote the set of positive divisors of 24 and partially ordered by divisibility. Draw the Hasse diagram of this partially ordered set.*

Sol. We have $D_{24} = \{1, 2, 3, 4, 6, 8, 12, 24\}$

and $1 \le 1$, $1 \le 2$, $1 \le 3$, $1 \le 4$, $1 \le 6$, $1 \le 8$, $1 \le 12$, $1 \le 24$,
$2 \le 2$, $2 \le 4$, $2 \le 6$, $2 \le 8$, $2 \le 12$, $2 \le 24$,
$3 \le 3$, $3 \le 6$, $3 \le 12$, $3 \le 24$, $4 \le 4$, $4 \le 8$, $4 \le 12$, $4 \le 24$,
$6 \le 6$, $6 \le 12$, $6 \le 24$, $8 \le 8$, $8 \le 24$,
$12 \le 12$, $12 \le 24$, $24 \le 24$.

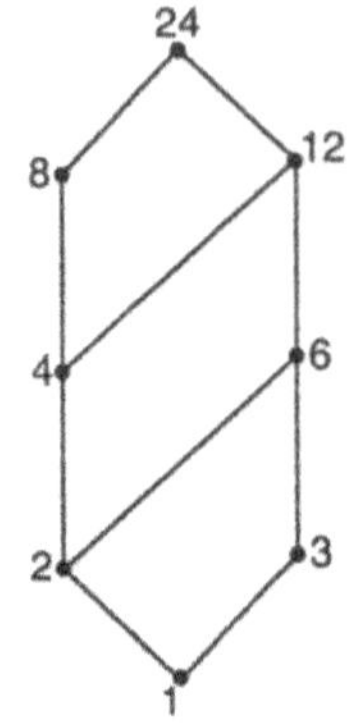

We find pairs of distinct elements a, b such that $a << b$.

All possible cases of $a << b$ are :

$1 << 2$, $1 << 3$, $2 << 4$, $2 << 6$, $3 << 6$, $4 << 8$, $4 << 12$, $6 << 12$, $8 << 24$, $12 << 24$.

The Hasse diagram is shown in the figure.

Example 6. *Let A = {1, 2, 3, 4, 6, 8, 9, 12, 18, 24} be the partially ordered set with the relation 'x divides y'. Draw the Hasse diagram of this partially ordered set.*

Sol. We have $A = \{1, 2, 3, 4, 6, 8, 9, 12, 18, 24\}$.

For $a << b$, we must have $a \ne b$.

a << b with a = 1

$1 << 2$, $1 << 3$, are true.

$1 << 4$, $1 << 6$, $1 << 8$, $1 << 9$, $1 << 12$, $1 << 18$, $1 << 24$ are not true because we have $1 < 2 < 8$ in case of $1 << 8$.

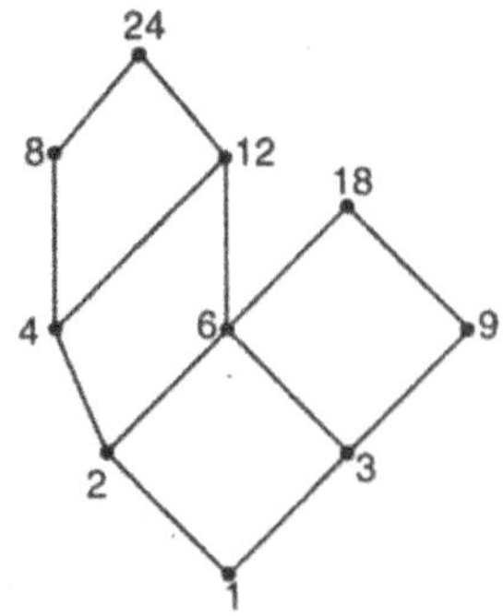

a << b with a = 2

$2 << 4$, $2 << 6$ are true.

$2 << 8$, $2 << 12$, $2 << 18$, $2 << 24$ are not true.

a << b with a = 3 We have $3 << 6$, $3 << 9$.

a << b with a = 4 We have $4 << 8$, $4 << 12$.

a << b with a = 6 We have $6 << 12$, $6 << 18$.

a << b with a = 8 We have $8 << 24$.

a << b with a = 9 We have $9 << 18$.

a << b with a = 12 We have $12 << 24$.

There are no other values of a and b for which $a << b$. The Hasse diagram is shown in the figure.

Example 7. *Let A = {a, b, c, d, e} be a partially ordered set with the given Hasse diagram. Find the set {(x, y) : x ≤ y and x, y ∈ A}.*

Sol. We have $A = \{a, b, c, d, e\}$.

Since the relation '≤' is reflexive, we have $a \le a$, $b \le b$, $c \le c$, $d \le d$, $e \le e$.

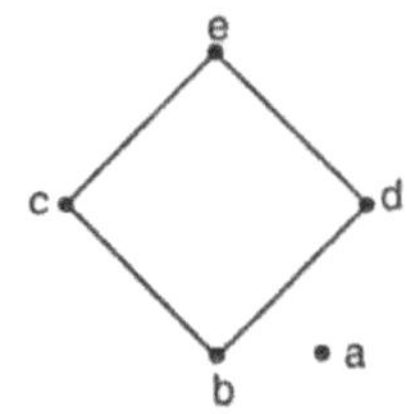

Using Hasse diagram, we have $b < c$, $b < d$, $c < e$, $d < e$.

Also $b < e$ because there is a directed path from b to e.

$\therefore$ $\{(x, y) : x \le y \text{ and } x, y \in A\}$

= **{(a, a), (b, b), (c, c), (d, d), (e, e), (b, c), (b, d), (c, e), (d, e) (b, e)}.**

Example 8. *Let $A = \{1, 2, 6, 8\}$ be a partially ordered set with the given Hasse diagram. Show that the relation '≤' on A is defined as $x \leq y$ iff x/y.*

Sol. We have $A = \{1, 2, 6, 8\}$.

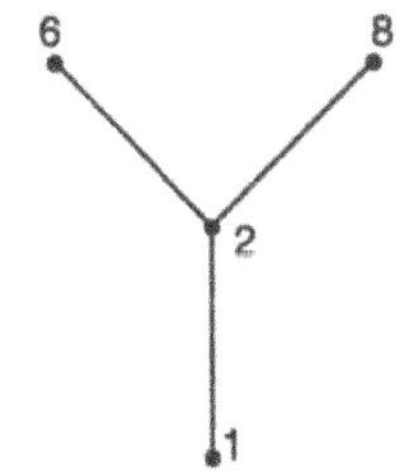

Using Hasse diagram, we have $1 < 2$, $2 < 6$, $2 < 8$.

Also $1 < 6$, $1 < 8$ because there are directed paths from 1 to 6 and from 1 to 8 respectively. Since A is partially ordered, we have $1 \leq 1$, $2 \leq 2$, $6 \leq 6$, $8 \leq 8$.

$$\therefore \quad \{(x, y) : x \leq y \text{ and } x, y \in A\} = \{(1, 1), (2, 2), (6, 6), (8, 8), (1, 2), (2, 6), (2, 8), (1, 6), (1, 8)\}$$

This set is also equal to $\{(x, y) : x/y \text{ and } x, y \in A\}$.

$$\therefore \quad x \leq y \text{ if and only if } x/y.$$

Example 9. *Draw the Hasse diagram of the partially ordered set with the given directed graph.*

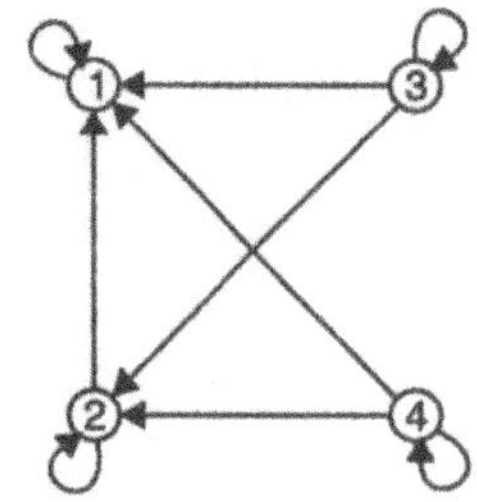

Sol. Let A be the required partially ordered set with the relation '≤'. The elements shown in the digraph are 1, 2, 3, 4.

$$\therefore \quad A = \{1, 2, 3, 4\}$$

Using the graph, we have

$$1 \leq 1,\ 2 \leq 2,\ 3 \leq 3,\ 4 \leq 4,\ 2 \leq 1,\ 3 \leq 1,\ 3 \leq 2,\ 4 \leq 2,\ 4 \leq 1.$$

$$\therefore \quad 2 << 1,\ 3 << 2,\ 4 << 2$$

$3 << 1$ is not true because $3 < 2 < 1$.

$4 << 1$ is not true because $4 < 2 < 1$.

The Hasse diagram of A is shown in the figure.

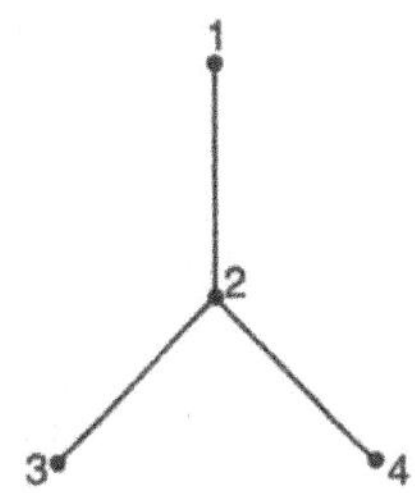

Remark. The symbol '≤' used above has nothing to do with the usual meaning of the symbol '≤' in numbers.

It may also be recalled that $a < b$ means $a \leq b$ and $a \neq b$.

Example 10. *Let m be a positive integer. A set of positive integers with sum m is called a* **partition** *of m. For example, 3-2-1-1 is a partition of 7 because $3 + 2 + 1 + 1 = 7$. Let A be the set of all partitions of the positive integer m. Define a relation ≤ on A as follows : $P_1 \leq P_2$ if the integers in P_1 can be added to obtain the integers in P_2. The relation ≤ is a partial order on A. For $m = 5$, draw the Hasse diagram of the partially ordered set A.*

Sol. We have $\quad m = 5$.

$\therefore \quad A = \{5, 4\text{-}1, 3\text{-}2, 2\text{-}2\text{-}1, 3\text{-}1\text{-}1, 2\text{-}1\text{-}1\text{-}1, 1\text{-}1\text{-}1\text{-}1\text{-}1\}$

Equivalently $\quad A = \{1\text{-}1\text{-}1\text{-}1\text{-}1, 2\text{-}1\text{-}1\text{-}1, 3\text{-}1\text{-}1, 2\text{-}2\text{-}1, 3\text{-}2, 4\text{-}1, 5\}$

For $a << b$, we must have $a \neq b$.

a << b with a = 1-1-1-1-1

1-1-1-1-1 << 2-1-1-1 is true.

1-1-1-1-1 << 3-1-1, 1-1-1-1-1 << 2-2-1, 1-1-1-1-1 << 3-2, 1-1-1-1-1 << 4-1, 1-1-1-1-1 << 5 are not true because we have 1-1-1-1-1 < 2-1-1-1 < 3-1-1 in case of 1-1-1-1-1 << 3-1-1.

a << b with a = 2-1-1-1

2-1-1-1 << 3-1-1, 2-1-1-1 << 2-2-1 are true.

2-1-1-1 << 3-2, 2-1-1-1 << 4-1, 2-1-1-1 << 5 are not true.

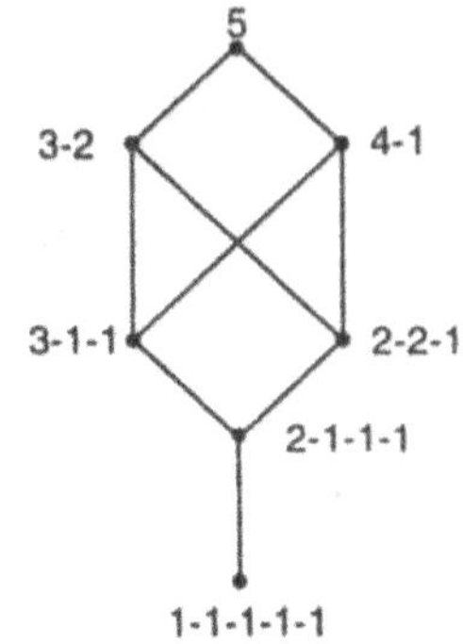

a << b with a = 3-1-1

3-1-1 << 3-2, 3-1-1 << 4-1 are true.

3-1-1 << 5 is not true.

a << b with a = 2-2-1

2-2-1 << 3-2, 2-2-1 << 4-1 are true.

2-2-1 << 5 is not true.

a << b with a = 3-2

Only 3-2 << 5 is true.

a << b with a = 4-1

Only 4-1 << 5 is true.

The Hasse diagram is shown in the figure.

Example 11. *Let $A = \{a, b, c, d, e\}$ be a partially ordered set whose Hasse diagram is given. For each $x \in A$, define $p(x) = \{y \in A : y \leq x\}$ i.e. $p(x)$ is the set of all predecessors of x.*

Let $p(A) = \{p(x) : x \in A\}$. Draw the Hasse diagram of the partially ordered set $p(A)$ partially ordered by 'set inclusion'.

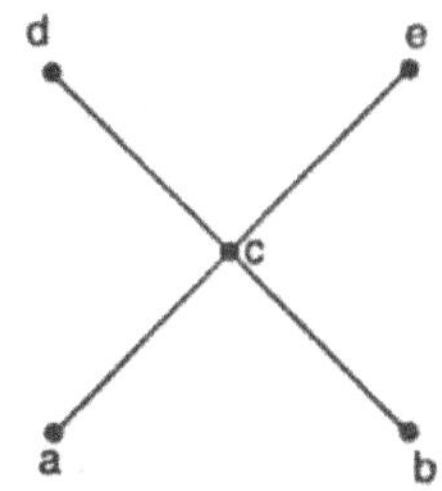

Sol. We have $A = \{a, b, c, d, e\}$

Since the relation '$\leq$' is reflexive, we have

$$a \leq a,\ b \leq b,\ c \leq c,\ d \leq d,\ e \leq e.$$

Using Hasse diagram of A, we have

$$a < c,\ c < e,\ b < c,\ c < d.$$

Also, $a < e$, $b < d$, $a < d$, $b < e$ because there are directed paths between the elements concerned.

$\therefore$ $\{(x, y) : x \leq y \text{ and } x, y \in A\} = \{(a, a), (b, b), (c, c), (d, d), (e, e), (a, c), (c, e), (b, c), (c, d), (a, e), (b, d), (a, d), (b, e)\}$

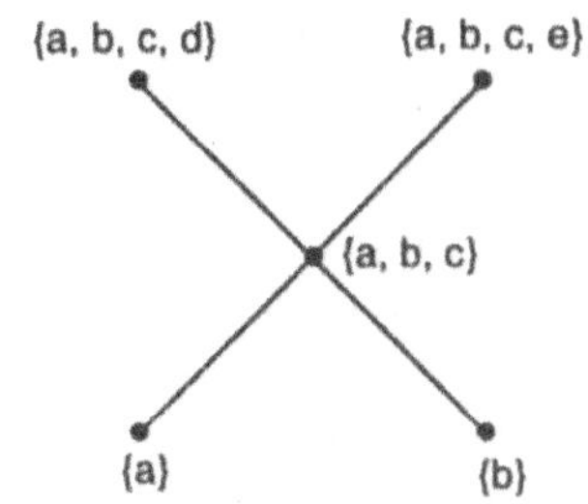

Now, $p(a) = \{x : x \leq a\} = \{a\}$

$p(b) = \{x : x \leq b\} = \{b\}$

$p(c) = \{x : x \leq c\} = \{c, a, b\}$

$p(d) = \{x : x \leq d\} = \{d, c, b, a\}$

$p(e) = \{x : x \leq e\} = \{e, c, a, b\}$

$\therefore$ $p(A) = \{p(x) : x \in A\}$

$= \{\{a\}, \{b\}, \{a, b, c\}, \{a, b, c, d\}, \{a, b, c, e\}\}$,

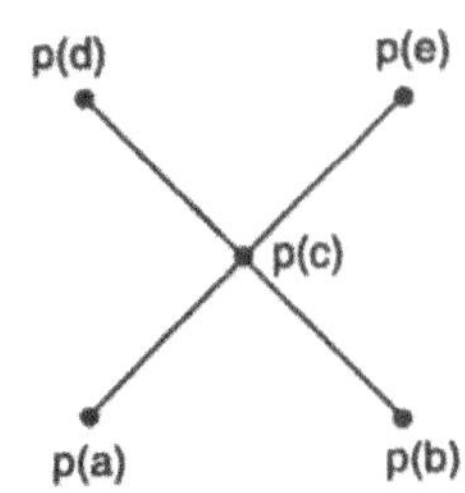

Under the partial order 'set inclusion', we have

$\{a\} << \{a, b, c\}$, $\{b\} << \{a, b, c\}$, $\{a, b, c\} << \{a, b, c, d\}$,

$\{a, b, c\} << \{a, b, c, e\}$

i.e., $p(a) << p(c)$, $p(b) << p(c)$, $p(c) << p(d)$, $p(c) << p(e)$.

The Hasse diagram of $p(A)$ is shown in the figure.

Remark. The Hasses diagram of $p(A)$ can also be drawn as shown in the adjoining figure. It is interesting to note that the Hasse diagrams of A and $p(A)$ are identical except for the labelling of the vertices.

EXERCISE 3.2

1. Let $P(A)$ be the power set of the set $A = \{2, 3, 4, 5, 6, 7, 8, 9, 10\}$. The set $P(A)$ is a partially ordered set with the relation of set inclusion. Which of the following are true ?

   (*i*) $\phi << \{2\}$ (*ii*) $\{2\} << \{2, 3, 4\}$

   (*iii*) $\{2, 4, 7, 8\} << \{2, 4, 7\}$ (*iv*) $\{2, 3, 7, 8\} << \{2, 3, 4, 7, 8\}$

   (*v*) $\phi << A$ (*vi*) $A - \{3\} << A$.

2. The set **N** of natural numbers is a partially ordered set with the relation of divisibility. In this partially ordered set, show that :

   (*i*) $1 << 2 << 4 << 8 << 16 << 32$ is true.

   (*ii*) $1 << 3 << 12$ is false.

   (*iii*) $17 << 51 << 153 << 306 << 918$ is true.

   (*iv*) $5 << 6$ is false.

3. Draw the Hasse diagram of the partially ordered set $\{2, 3, 4, 6\}$ under the relation of 'divisibility'.

4. Draw the Hasse diagram of the partially ordered set $\{2, 3, 4, 6\}$ under the relation 'usual $\leq$ of numbers'.

5. Let $A = \{x, y\}$. Draw the Hasse diagram of the partially ordered set $P(A)$ with the relation of 'set inclusion'.

6. Let $A = \{a, b, c, d, e, f, g\}$ be a partially ordered set with the given Hasse diagram. Find the set $\{(x, y) : x \leq y \text{ and } x, y \in A\}$.

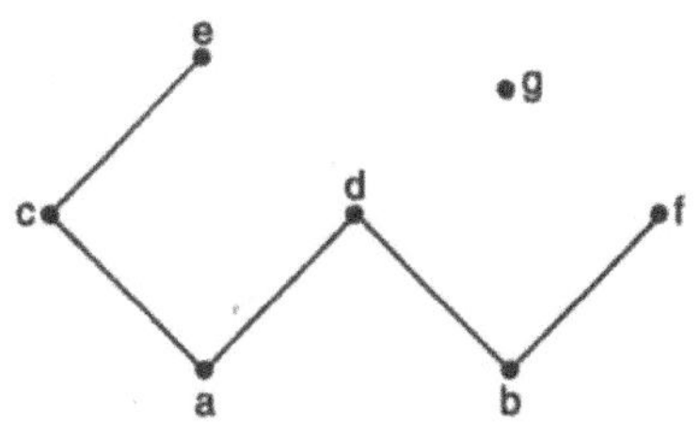

Q. 6

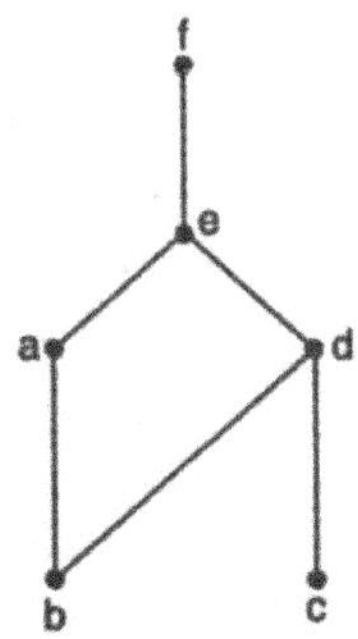

Q. 7

7. Let $A = \{a, b, c, d, e, f\}$ be a partially ordered set with the given Hasse diagram. Find the set $\{(x, y) : x \leq y \text{ and } x, y \in A\}$.

8. Draw the Hasse diagram of the partially ordered set with the given directed graph.

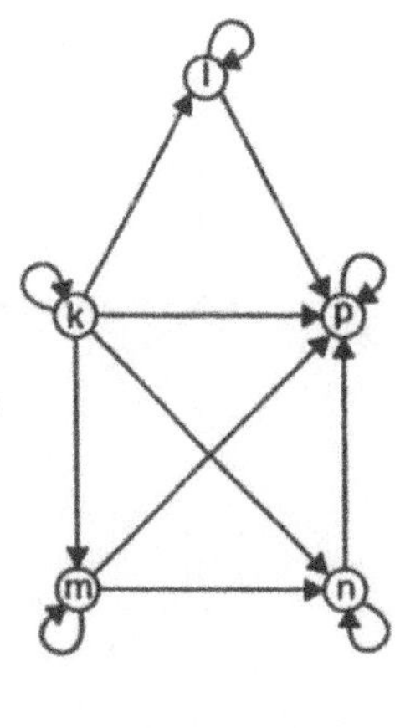

9. Let D_n denote the set of positive divisors of n and partially ordered by divisibility. Draw the Hasse diagram of :

(i) D_5 (ii) D_8
(iii) D_{15} (iv) D_{30}
(v) D_{36}.

10. Let A be the set of all partitions of 4. Let a partition P_1 precedes a partition P_2 if the integers in P_1 can be added to obtain the integers in P_2. With this relation on A, show that A is a partially ordered set. Also draw the Hasse diagram of A.

11. Let $A = \{1, 2, 3, 4, 5\}$ be a partially ordered set whose Hasse diagram is given. For each $a \in A$ define $p(a) = \{x \in A : x \leq a\}$.
Let $p(A) = \{p(a) : a \in A\}$. Draw the Hasse diagram of the partially ordered set $p(A)$ partially ordered by set inclusion.

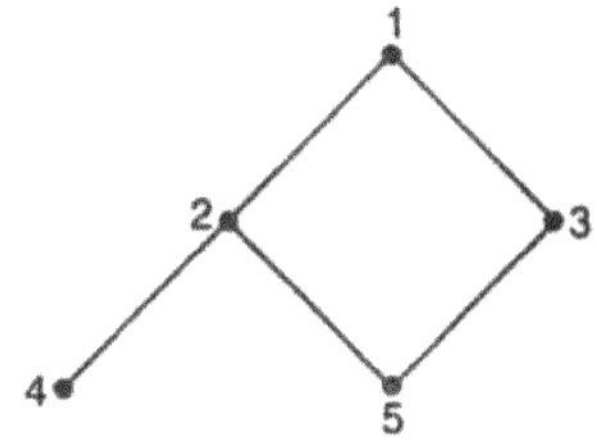

Answers

1. (i), (iv), (vi)

3. 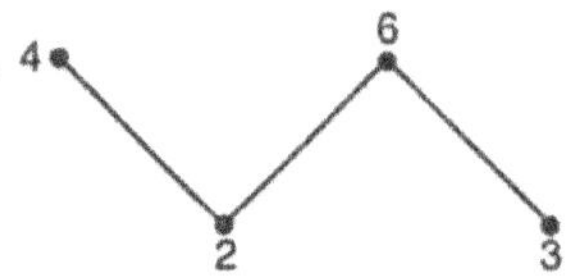

4. 

5. 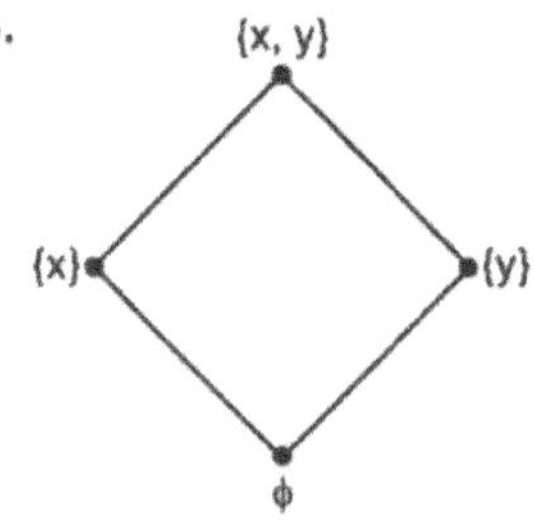

6. $\{(a, a), (b, b), (c, c), (d, d), (e, e), (f, f), (g, g), (a, c), (c, e), (a, d), (b, d)\ (b, f), (a, e)\}$

7. $\{(a, a), (b, b), (c, c), (d, d), (e, e), (f, f), (b, a), (a, e), (e, f), (d, e), (c, d), (b, e), (c, e), (a, f), (d, f), (b, f), (b, d), (c, f)\}$

8. 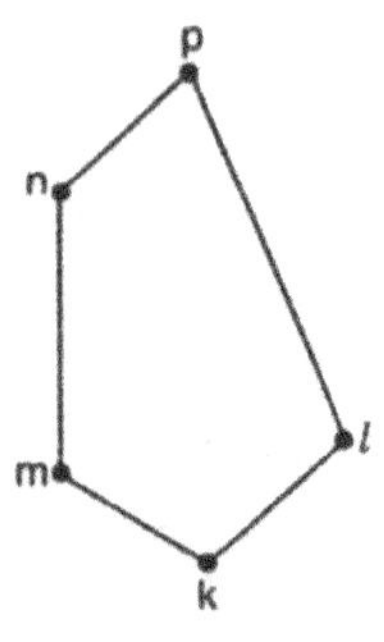

9. (i) 

(*ii*)

(*iii*)

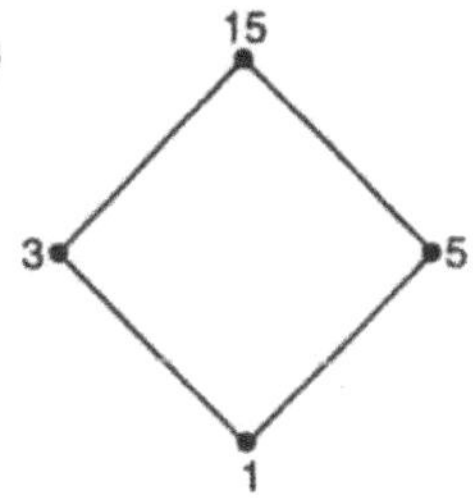

(*iv*)

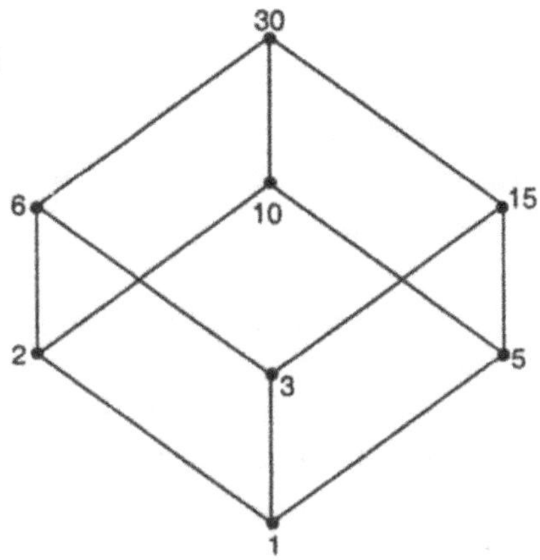

(*v*)

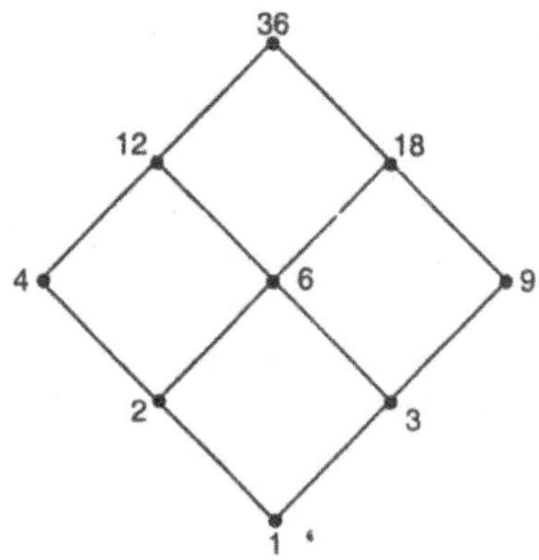

10.

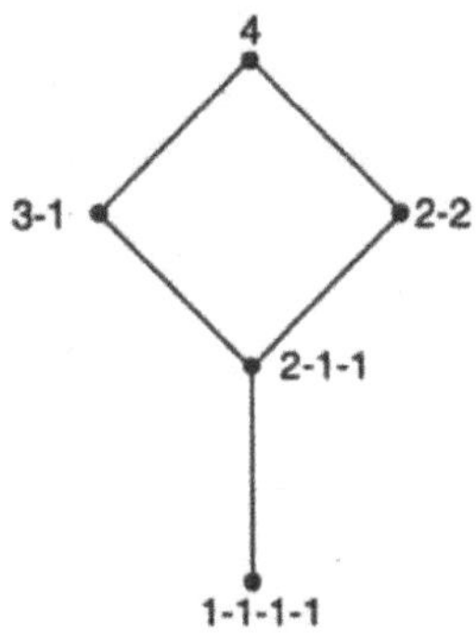

11.

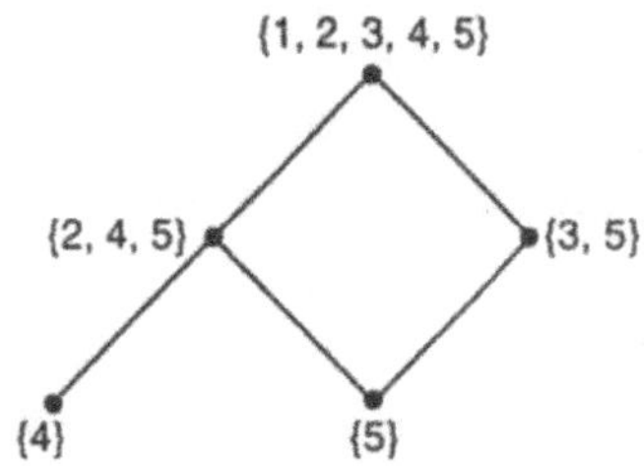

3.18. MINIMAL ELEMENT AND MAXIMAL ELEMENT

Let A be a partially ordered set.*

An element $a \in A$ is called a **minimal element** of A if $x \leq a$ implies $x = a$ that is, if no element of A strictly precedes a.

---

*When we say that A is a partially ordered set then it is always understood that there is a relation defined on A which is reflexive, antisymmetric and transitive and this relation is generally denoted by '$\leq$'.

Also, if $a \leq b$ and $a \neq b$ then we write $a < b$.

If $a \leq b$ then we say that a precedes b or b succeeds a.

If $a < b$ then we say that a strictly precedes b or b strictly succeeds a.

Similarly, an element $a \in A$ is called a **maximal element** of A if $x \geq a$ implies $x = a$ that is, no element of A strictly succeeds a.

If A is a finite partially ordered set then the Hasse diagram of A can be used very conveniently to identify the minimal elements and maximal elements of A. The process is as follows : a is a minimal element of A if no edge enters a from below.

b is a maximal element of A if no edge leaves b in the upward direction.

A finite partially ordered set always admits of at least one minimal element and one maximal element.

Illustration. Let A be a finite partially ordered set with the Hasse diagram shown in the figure.

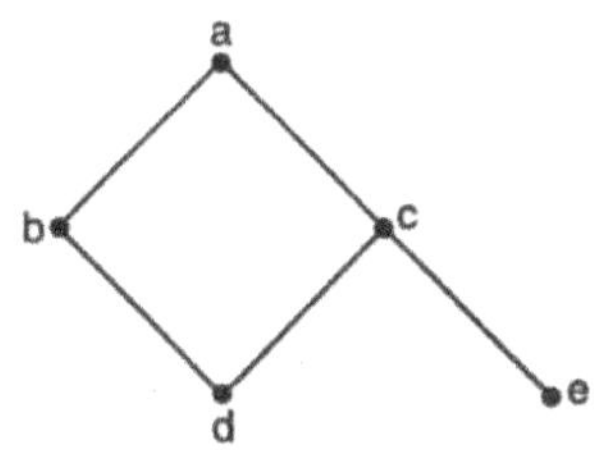

$\therefore \qquad A = \{a, b, c, d, e\}$

A has two minimal elements, d and e, because no edge enters d or e from below. A has one maximal element a because no edge leaves a in the upward direction.

Thus we see that a Hasse diagram is very useful in locating minimal elements and maximal elements.

If a partially ordered set A is infinite then A may or may not have minimal elements and maximal elements.

Illustration. The set **Z** of integers is a partially ordered set with the relation 'usual $\leq$ of numbers.' This partially ordered set has neither minimal elements nor maximal elements.

For if $a \in \mathbf{Z}$ is a minimal element, then we have $a - 1 \leq a$ and $a - 1 \neq a$.

Similarly if $a \in \mathbf{Z}$ is a maximal element then we have $a + 1 \geq a$ and $a + 1 \neq a$.

Theorem. Let A be a finite partially ordered set. Prove that A has at least one minimal element and one maximal element.

Proof. Let a be any element of A. If a is not a minimal element of A then there exists $a_1 \in A$ such that $a_1 < a$. If a_1 is a minimal element, we are done. If not so, there exists $a_2 \in A$ such that $a_2 < a_1$.

$\therefore \qquad a_2 < a_1 < a$

This argument cannot be continued indefinitely since A is finite. Ultimately, we get a finite chain

$a_m < a_{m-1} < \ldots\ldots a_2 < a_1 < a$, which cannot be extended further.

Thus for any $b \in A$, we cannot have $b < a_m$.

$\therefore$ No element of A strictly precedes a_m.

$\therefore$ a_m is a minimal element of A.

The proof of the other part of the theorem is left as an exercise for the reader.

3.19. FIRST ELEMENT AND LAST ELEMENT

Let A be a partially ordered set.

An element $a \in A$ is called a **first element** of A if $a \leq x$ for every $x \in A$ that is, if a precedes every element in A.

An element $b \in A$ is called a **last element** of A if $y \leq b$ for every $y \in A$ that is, if b succeeds every element in A.

Illustration. Let $A = \{1, 2, 3, 4, 6, 8, 9, 12, 18, 24\}$ be the partially ordered set with the relation $x \le y$ if x/y. The Hasse diagram of A is given in the figure.

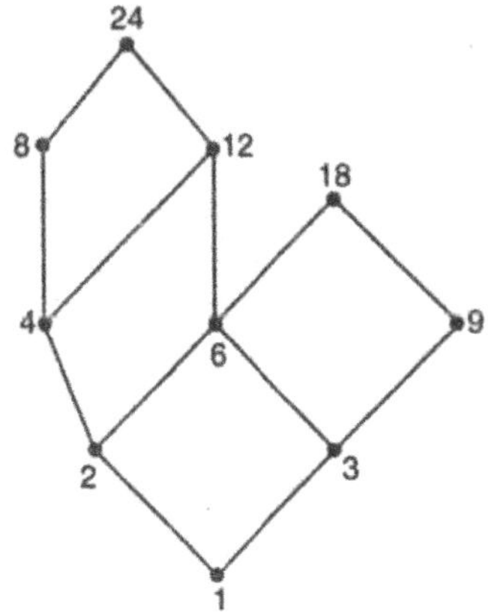

A has only one minimal element 1 because no edge enters 1 from below.

1 is also the first element of A because we have $1 \le x$ for every $x \in A$.

A has two maximal elements, 18 and 24, because no edge leaves 18 or 24 in the upward direction.

18 is not the last element because $12 \le 18$ is not true. 24 is not the last element because $18 \le 24$ is not true.

$\therefore$ There is no $b \in A$ such that $x \le b$ for every $x \in A$.

$\therefore$ A has no last element.

Theorem. Prove that a partially ordered set can have at most one first element and one last element.

Proof. Let A be a partially ordered set. Let a and a' be two first elements of A.

$\therefore$ $a \le x$, $a' \le x$ for every $x \in A$.

$\therefore$ In particular, $a \le a'$, $a' \le a$

$\therefore$ By antisymmetry, $a = a'$.

$\therefore$ A cannot have more than one first element.

The proof of the other part of the theorem is left as an exercise for the reader.

Note 1. It is possible that a partially ordered set may have neither a first element nor a last element.

Note 2. The first element and the last element of a partially ordered set are also called the **least element** and the **greatest element** respectively.

Example 1. *In a partially ordered set A admitting of first element and last element, show that :*

(*i*) *the first element is also a minimal element*

(*ii*) *the last element is also a maximal element.*

Sol. (*i*) Let a be the first element of A.

$\therefore$ $a \le x$ for every $x \in A$

$\therefore$ No element of A strictly precedes a, *i.e.*, $y \le a$ implies $y = a$.

$\therefore$ a is a minimal element of A.

(*ii*) Let b be the last element of A.

$\therefore$ $x \le b$ for every $x \in A$

$\therefore$ No element of A strictly succeeds b, *i.e.*, $y \ge b$ implies $y = b$.

$\therefore$ b is a maximal element of A.

Example 2. *In a totally ordered set A admitting of at least one minimal element and one maximal element, show that :*

(*i*) *a minimal element is also the first element.*

(*ii*) *a maximal element is also the last element.*

Sol. (*i*) Let a be a minimal element of A.

$\therefore$ $x \le a$ implies $x = a$. If possible, let a be not the first element of A.

$\therefore$ There exists at least one element $y \in A$ such that $a \le y$ is not true.

$\Rightarrow$ Neither $a < y$ nor $a = y$. Since A is totally ordered and $a \neq y$, we have $a < y$ or $y < a$.

$\therefore$ We must have $y < a$.

$\Rightarrow$ $y \leq a$ and $y \neq a$.

This is impossible because a is a minimal element of A.

$\therefore$ a must be the first element of A.

(*ii*) Let a be a maximal element of A.

$\therefore$ $x \geq a$ implies $x = a$. If possible, let a be not the last element of A.

$\therefore$ There exists at least one element $y \in A$ such that $y \leq a$ in not true.

$\Rightarrow$ Neither $y < a$ nor $y = a$. Since A is totally ordered and $y \neq a$, we have $y < a$ or $a < y$.

$\therefore$ We must have $a < y$.

$\Rightarrow$ $a \leq y$ and $y \neq a$

This is impossible because a is a maximal element of A.

$\therefore$ a must be the last element of A.

Remark. A totally ordered set has at most one maximal element and at most one minimal element.

Example 3. *Let $A = \{1, 2, 3, 4, 5, 6\}$ be a partially ordered set with the given Hasse diagram.*

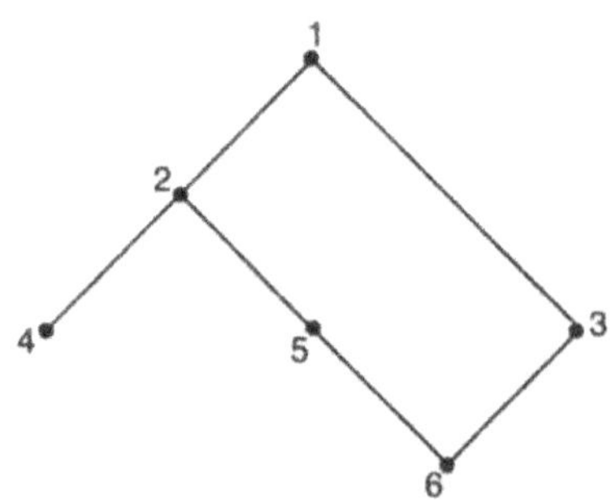

(*i*) *Find all minimal and maximal elements of A.*

(*ii*) *Find the first element and last element of A, if either exists.*

(*iii*) *Find all totally ordered subset of A containing at least three elements.*

Sol. (*i*) No edge enters 4 or 6 from below.

$\therefore$ **4** and **6** are minimal elements.

No edge leaves 1 in the upward direction.

$\therefore$ **1** is a maximal element.

(*ii*) There is no $a \in A$ for which $a \leq x$ for every $x \in A$.

$\therefore$ There is no first element.

We have $1 \leq 1, 2 \leq 1, 3 \leq 1, 4 \leq 1, 5 \leq 1, 6 \leq 1$.

$\therefore$ **1** is the last element of A.

Note. 6 could have been the first element of A provided we have $6 \leq 4$ *i.e.*, an edge from 6 to 4.

(*iii*) We have $4 \leq 2 \leq 1, 6 \leq 5 \leq 2, 6 \leq 5 \leq 2 \leq 1, 5 \leq 2 \leq 1, 6 \leq 3 \leq 1$.

$\therefore$ The totally ordered subsets of A with at least three elements are :

{1, 2, 4}, {2, 5, 6}, {1, 2, 5}, {1, 3, 6}, {1, 2, 5, 6}.

EXERCISE 3.3

1. Let A be a finite partially ordered set. Prove that A has at least one maximal element.
2. Show that the set **N** of natural numbers partially ordered by the relation 'usual $\leq$' has one minimal element and no maximal element.
3. Show that the set **N** of natural numbers partially ordered by the relation 'usual $\geq$' has one maximal element and no minimal element.

4. Show that the set $P(A)$ of the subset of a finite set A, partially ordered by the relation 'usual $\subseteq$', has minimal element ϕ and maximal element A.
5. Prove that a partially ordered set can have at most one last element.
6. Let $A = \{a, b, c, d, e\}$ be a partially ordered set such that $a << b$, $b << c$, $c << d$, $d << e$. Show that :
   (*i*) A has only one minimal element and this is also the first element of A.
   (*ii*) A has only one maximal element and this is also the last element of A.
7. Show that the set **N** of natural numbers partially ordered by the relation 'usual $\leq$' has first element '1' and no last element.
8. Let $A = \{1, 2, 3, 4, 5, 6\}$ be a partially ordered set with the given Hasse diagram :
   (*i*) Find all minimal and maximal elements of A.
   (*ii*) Find the first element and last element of A, if either exists.
   (*iii*) Find all totally ordered subsets of A containing at least three elements.

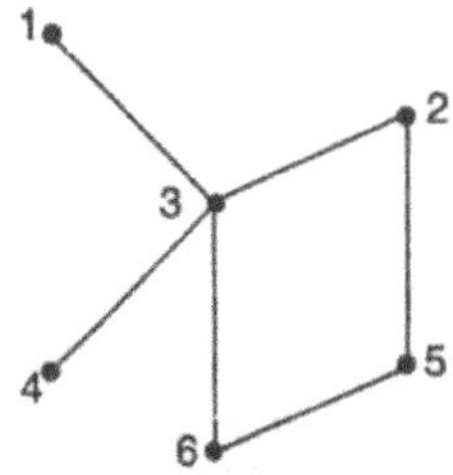

Answer

8. (*i*) Minimal elements : 4 and 6 ; maximal elements : 1 and 2
   (*ii*) No first element ; No last element
   (*iii*) $\{1, 3, 4\}$, $\{2, 3, 6\}$, $\{2, 5, 6\}$, $\{2, 3, 4\}$, $\{1, 3, 6\}$.

3.20. UPPER BOUND AND LOWER BOUND

Let A be a partially ordered set and let B be a subset of A.

An element $a \in A$ is called an **upper bound** of B if $b \leq a$ for every $b \in B$ that is, if a succeeds every element of B.

Similarly, an element $a \in A$ is called a **lower bound** of B if $a \leq b$ for every $b \in B$ that is, if a precedes every element of B.

A subset of a partially ordered set may or may not have upper bounds and lower bounds.

If a subset B of a partially ordered set A has an upper bound, we say B is **bounded above** and if B has a lower bound, we say B is **bounded below.** If B is both bounded above and bounded below, we say B is **bounded**.

Remark. If a is an upper bound of a subset B of a partially ordered set A then whenever $a \leq a'$, a' is also an upper bound of B.

Similarly, if a is a lower bound of B then whenever $a' \leq a$, a' is also a lower bound of B.

3.21. SUPREMUM AND INFIMUM

Let A be a partially ordered set and let B be a subset of A.

An element $a \in A$ is called the **supremum** of B if :

(*i*) a is an upper bound of B *i.e.*, $b \leq a$ for every $b \in B$

(*ii*) a precedes every upper bound of B *i.e.*, $a \leq a'$ whenever a' is an upper bound of B.

Similarly, an element $a \in A$ is called the **infimum** of B if :

(*i*) a is a lower bound of B *i.e.*, $a \leq b$ for every $b \in B$

(*ii*) a succeeds every lower bound of B *i.e.*, $a' \leq a$ whenever a' is a lower bound of B.

The supremum of B is denoted by 'sup (B)' and its infimum is denoted by 'inf (B)'.

The term **least upper bound** is also used for the term supremum and we write sup (B) as lub (B).

Similarly, the term **greatest lower bound** is also used for the term infimum and we write inf (B) as glb (B).

Remarks 1. The supremum and infimum of a subset of a partially ordered set may or may not exist.

2. The supremum and infimum of a subset of a partially ordered set may or may not belong to that subset.

Theorem. Let B be a subset of a partially ordered set A. Prove that sup (B) and inf (B) are unique, provided they exist.

Proof. Let $a, a' \in A$ be sup (B).

$\therefore$ (*i*) a, a' are upper bounds of B.

(*ii*) a, a' precede every upper bound of B.

(*ii*) $\Rightarrow$ $a \le a'$ ($\because$ a' is an upper bound of B)

Also $a' \le a$ ($\because$ a is an upper bound of B)

$\therefore$ By antisymmetry, $a = a'$

$\therefore$ Sup (B) is unique.

The other part of the theorem is left for the reader as an exercise.

Example 1. *Let $A = \{a, b, c, d, e, f, g, h\}$ be a partially ordered set with given Hasse diagram. Find the upper bounds and lower bounds of the sets (i) $\{a, b\}$ (ii) $\{c, d, e\}$.*

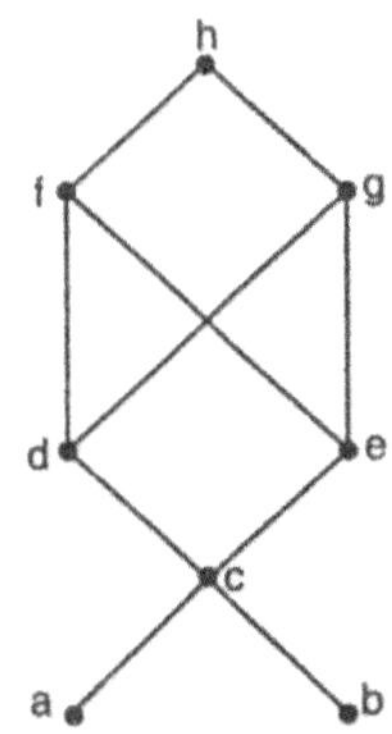

Sol. We have $A = \{a, b, c, d, e, f, g, h\}$.

(*i*) The given set is $\{a, b\}$, say B_1.

We have $a \le a$ and $b \le a$ is not true.

$\therefore$ a is not an upper bound of B_1.

Similarly b is not an upper bound of B_1.

$a \le c, b \le c \Rightarrow c$ is an upper bound of B_1

$a \le d, b \le d \Rightarrow d$ is an upper bound of B_1

($\because$ There are directed paths from a to d and from b to d).

Similarly e, f, g, h are upper bounds of B_1.

$\therefore$ The upper bounds of the set $\{a, b\}$ are **c, d, e, f, g, h.**

There is no $x \in A$ such that $x \le a$ and $x \le b$.

$\therefore$ The set $\{a, b\}$ has no lower bound.

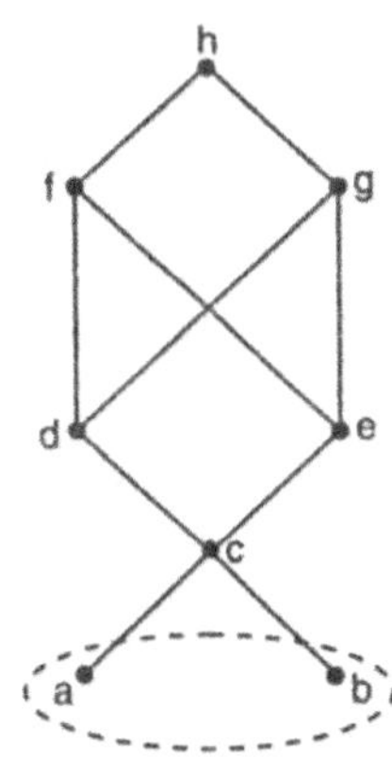

(*ii*) The given set is $\{c, d, e\}$, say B_2. Neither a nor b can succeed every element of B_2. For example, $c \le a$, $c \le b$ are false.

$\therefore$ a and b are not upper bounds of B_2.

$c \le c$ is true but $d \le c$, $e \le c$ are false.

$\therefore$ c is not an upper bound of B_2.

$c \le d$, $d \le d$ are true but $e \le d$ is not true.

$\therefore$ d is not an upper bound of B_2. Similarly e is not an upper bound of B_2.

$c \le f,\quad d \le f,\quad e \le f \quad \Rightarrow f$ is an upper bound of B_2

$c \le g,\quad d \le g,\quad e \le g \quad \Rightarrow g$ is an upper bound of B_2

$c \le h,\quad d \le h,\quad e \le h \quad \Rightarrow h$ is an upper bound of B_2

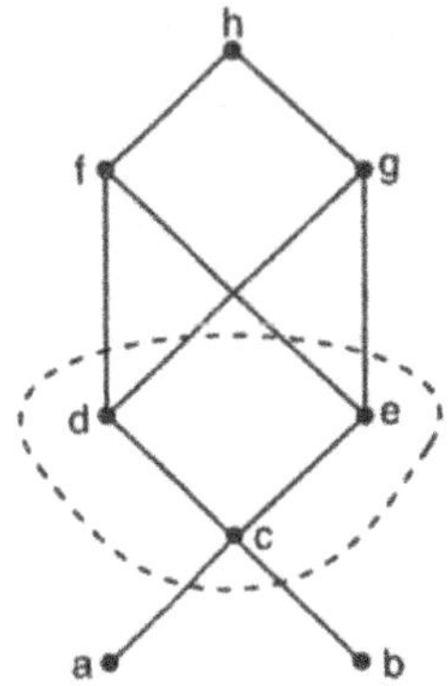

∴ The upper bounds of the set {c, d, e} are **f, g, h.**

$a \le c,\ a \le d,\ a \le e \quad \Rightarrow \quad a$ is a lower bound of B_2

$b \le c,\ b \le d,\ b \le e \quad \Rightarrow \quad b$ is a lower bound of B_2

$c \le c,\ c \le d,\ c \le e \quad \Rightarrow \quad c$ is a lower bound of B_2

$d \le c$ and $d \le e$ are false.

∴ d is not a lower bound of B_2.

$e \le c$ and $e \le d$ are false.

∴ e is not a lower bound of B_2.

$f \le c, f \le d, f \le e$ are all false.

∴ f is not a lower bound of B_2.

Similarly, g and h are not lower bounds of B_2.

∴ The lower bounds of the set {c, d, e} are **a, b, c.**

Example 2. *Let D_{50} be the set of all positive divisors of 50 and partially ordered by the relation of divisibility. Find the supremum and infimum of the subset {5, 10} of D_{50}, if either exists.*

Sol. We have $\quad D_{50} = \{1, 2, 5, 10, 25, 50\}$

and $\quad 1 << 2, 1 << 5, 2 << 10, 5 << 10, 5 << 25, 10 << 50, 25 << 50.$

∴ The Hasse diagram of the partially ordered set D_{50} is shown in the figure.

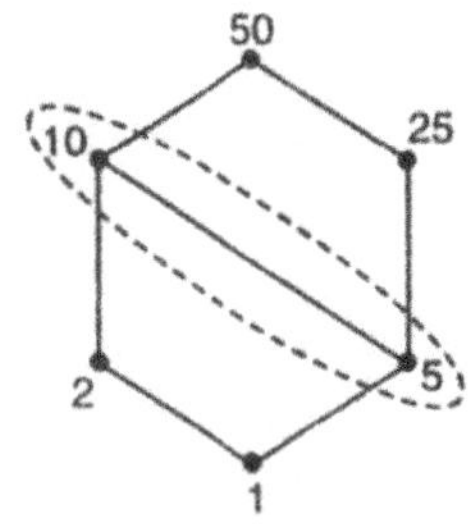

Let $B = \{5, 10\}$. The upper bounds of B are 10, 50 because only 10 and 50 succeed every element in B.

Also $\quad 10 \le 10$ and $10 \le 50$

∴ 10 precedes every upper bound of B.

∴ $\quad$ **sup (B) = 10.**

The lower bounds of B are 1, 5 because only 1 and 5 precede every element in B.

Also $\quad 1 \le 5$ and $5 \le 5$

∴ 5 succeeds every lower bound of B.

∴ $\quad$ **inf (B) = 5.**

Example 3. *Let $A = \{1, 2, 3, \ldots\ldots, 8\}$ be a partially ordered set with the Hasse diagram shown in the figure. Find the supremum and infimum of the subset {4, 5, 7}, if either exists.*

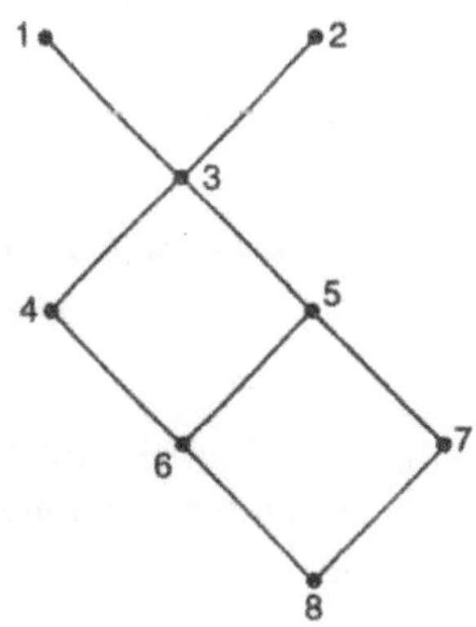

Sol. We have $\quad A = \{1, 2, 3, \ldots\ldots 8\}$

Let $\quad B = \{4, 5, 7\}$

The upper bounds of B are 1, 2, 3 because each of these succeed every element in B.

Also $\quad 3 \le 1, 3 \le 2, 3 \le 3$

∴ 3 precedes every upper bound of B.

∴ $\quad$ **sup (B) = 3.**

The only lower bound of B is 8 because it precedes every element of B.

Since $8 \leq 8$, we have

$$\inf (B) = 8.$$

Note. 6 is not a lower bound of B because $6 \leq 4$, $6 \leq 5$ are true but $6 \leq 7$ is not true.

Similarly 7 is not a lower bound of B because $7 \leq 7$, $7 \leq 5$ are true but $7 \leq 4$ is not true.

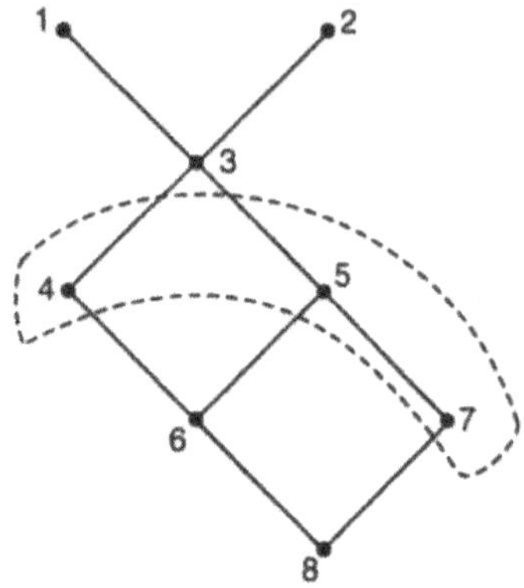

3.22. SUPREMUM AND INFIMUM OF SETS CONTAINING TWO ELEMENTS

The case of finding supremum and infimum of a set containing two elements deserves special attention. Let A be a partially ordered set and let a, b be in A. The supremum and infimum of the set $\{a, b\}$ are denoted by sup (a, b) and inf (a, b) respectively, provided either exist.

Theorem. Let A be a partially ordered set and a, b ∈ A. If a ≤ b, then prove that sup (a, b) = b and inf (a, b) = a.

Proof. We have $a \leq b$.

$a \leq b, b \leq b \Rightarrow b$ is an upper bound of $\{a, b\}$

Let c be any upper bound of $\{a, b\}$.

$\Rightarrow \quad a \leq c, b \leq c \Rightarrow$ The upper bound b precedes c.

$\therefore$ b is the supremum of $\{a, b\}$.

$\therefore \quad \sup (a, b) = b.$

The other part of the theorem is left for the reader as an exercise.

Corollary. If A is a totally ordered set and a, b ∈ A then sup (a, b) and inf (a, b) always exist.

Proof. Since A is totally ordered and $a, b \in A$, we have either $a \leq b$ or $b \leq a$.

$$a \leq b \Rightarrow \sup (a, b) = b \quad \text{and} \quad \inf (a, b) = a$$
$$b \leq a \Rightarrow \sup (a, b) = a \quad \text{and} \quad \inf (a, b) = b.$$

$\therefore$ The result holds.

Example 4. *Let the set* **N** *of natural numbers be partially ordered by divisibility. For* $a, b \in$ **N**, *show that :*

$$\sup (a, b) = \text{lcm } (a, b) \quad \text{and} \quad \inf (a, b) = \gcd (a, b)$$

Sol. Let $lcm(a, b) = m$.

$\therefore$ m is the least common multiple of a, b.

$\therefore$ (*i*) a/m, b/m and

(*ii*) if a/n, b/n then m/n.

(*i*) $\Rightarrow \quad a \leq m, b \leq m$

$\Rightarrow$ m is an upper bound of $\{a, b\}$

Let u be any upper bound of $\{a, b\}$.

$\Rightarrow \quad a \leq u, b \leq u \Rightarrow a/u, b/u$

$\therefore$ (ii) $\Rightarrow$ m/u $\Rightarrow$ $m \le u$

$\therefore$ m precedes every upper bound of $\{a, b\}$.

$\therefore$ sup $(a, b) = m$ *i.e.*, **sup (a, b) = lcm(a, b).**

Let $gcd(a, b) = d$.

$\therefore$ d is the greatest common divisor of a, b.

$\therefore$ (i) d/a, d/b and

(ii) if n/a, n/b then n/d.

(i) $\Rightarrow$ $d \le a$, $d \le b$

$\Rightarrow$ d is a lower bound of $\{a, b\}$

Let l be any lower bound of $\{a, b\}$.

$\Rightarrow$ $l \le a$, $l \le b$ $\Rightarrow$ l/a, l/b

$\therefore$ (ii) $\Rightarrow$ l/d $\Rightarrow$ $l \le d$

$\therefore$ d succeeds every lower bound of $\{a, b\}$.

$\therefore$ inf $(a, b) = d$ *i.e.*, **inf (a, b) = gcd (a, b).**

Remark. In the above example, sup(a, b) and inf (a, b) always exist for $a, b \in$ **N**. It is worth mentioning here that the partially ordered set under consideration is not totally ordered.

EXERCISE 3.4

1. Let $A = \{a, b, c, d\}$ be a partially ordered set with the given Hasse diagram. Find the upper bounds and lower bounds of the sets (i) $\{b\}$ (ii) $\{c\}$.

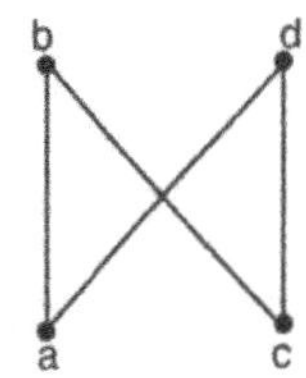

2. Let B be a subset of a partially ordered set A. Prove that inf (B) is unique, provided it exists.
3. Let $A = \{a, b, c, d, e, f\}$ be a partially ordered set with the Hasse diagram shown in the figure. Find the upper bounds, lower bounds, supremum and infimum of the subset $\{b, c, d\}$ of A, if either exists.

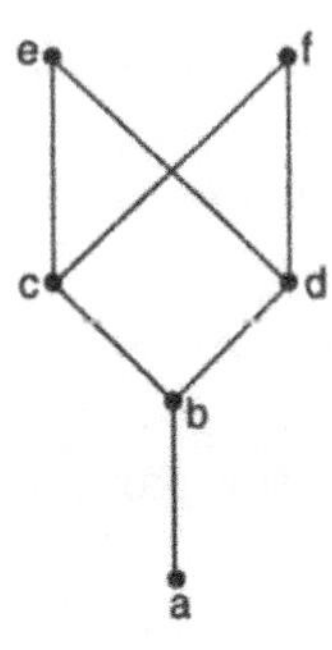

Q. 3

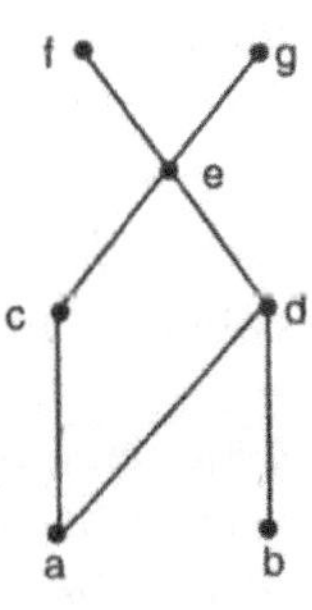

Q. 4

4. Let $A = \{a, b, c, d, e, f, g\}$ be a partially ordered set with the Hasse diagram shown in the figure. Find the upper bounds, lower bounds, supremum and infimum of the subset $\{c, d, e\}$ of A, if either exists.

5. Let $A = \{1, 2, 3, \ldots\ldots, 8\}$ be a partially ordered set with the given Hasse diagram. Find the upper bounds, lower bounds, supremum and infimum of the subset $\{1, 2, 5\}$ of A, if either exists.

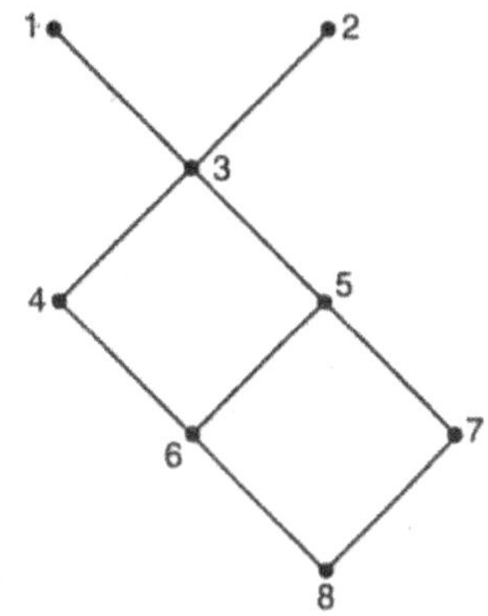

6. Let A be a partially ordered set and $a, b \in A$. If $a \leq b$, then prove that inf $(a, b) = a$.
7. If the set **N** of natural numbers is partially ordered by divisibility then show that sup (a, b) and inf (a, b) exist for any pair of elements a, b in **N**.
8. Let A be a non-empty set with at least two elements. Let $P(A)$ be the power set of A partially ordered by set inclusion.

   If $B, C \in P(A)$, then show that sup $(B, C) = B \cup C$ and inf $(B, C) = B \cap C$.

Answers

1. (*i*) Upper bound is b, lower bounds are a, b, c.

   (*ii*) Upper bounds are b, c, d, lower bound is c.
3. Upper bounds are e and f, lower bounds are a and b, supremum does not exist, infimum is b.
4. Upper bounds are e, f, g, lower bound is a, supremum is e, infimum is a.
5. No upper bound, lower bounds are 5, 6, 7, 8, no supremum, infimum is 5.

3.23. ISOMORPHISM OF PARTIALLY ORDERED SETS

Let A and B be partially ordered sets. A function $f : A \to B$ is called a **similarity mapping** from A into B if :

(*i*) f is one-one

(*ii*) $a_1 \leq a_2$ in A iff $f(a_1) \leq f(a_2)$ in B.

Condition (*ii*) is equivalent to say that f preserves the partial order relation.

Two partially ordered sets A and B are called **order isomorphic** (or *similar*) if there exists a function $f : A \to B$ such that :

(*i*) f is one-one (*ii*) f is onto

(*iii*) $a_1 \leq a_2$ in A iff $f(a_1) \leq f(a_2)$ in B.

The function f is called an **order isomorphism** between A and B.

Equivalently, partially ordered sets A and B are **order isomorphic** if there is a function $f : A \to B$ which is bijective and preserving the partial order relation. If the partially ordered sets A and B are order isomorphic then we write $A \approx B$.

Remark. The symbol '$\leq$' in '$a_1 \leq a_2$ in A' represents the partial order relation on A and in $f(a_1) \leq f(a_2)$ in B, it represents the partial order relation on B.

Illustration. Let E be the set of all positive even integers.

$\therefore$ **N** and E are both partially ordered sets under usual '$\leq$' of numbers.

Define $f : \mathbf{N} \to E$ by $f(n) = 2n$, $n \in \mathbf{N}$.

f is one-one and onto both.

Also, $n_1 \le n_2$ in $\mathbf{N} \Leftrightarrow 2n_1 \le 2n_2 \Leftrightarrow f(n_1) \le f(n_2)$ in E.

$\therefore$ f is an order isomorphism between **N** and E.

$\therefore$ Partially ordered sets **N** and E are order isomorphic.

Theorem 1. Let A and B be order isomorphic partially ordered sets. If A is totally ordered then prove that B is also totally ordered.

Proof. Let $f : A \to B$ be the order isomorphism between A and B.

$\therefore$ f is bijective and preserves the partial order relation.

Let b_1 and b_2 be any elements of B.

$\therefore$ There exists a_1 and a_2 in A such that $f(a_1) = b_1$ and $f(a_2) = b_2$.

Since A is totally ordered and $a_1, a_2 \in A$, we have either $a_1 \le a_2$ or $a_2 \le a_1$.

$\Rightarrow \quad f(a_1) \le f(a_2) \quad$ or $\quad f(a_2) \le f(a_1) \quad$ ($\because$ f is an order isomorphism)

$\Rightarrow \quad b_1 \le b_2 \quad$ or $\quad b_2 \le b_1$

$\therefore$ Every two elements of B are comparable.

$\therefore$ B is also totally ordered.

Theorem 2. Let A and B be order isomorphic partially ordered sets. Prove that

(*i*) a ∈ A is the first element of A iff f(a) is the first element of B.

(*ii*) a ∈ A is the last element of A iff f(a) is the last element of B.

(*iii*) a ∈ A is a minimal element of A iff f(a) is a minimal element of B.

(*iv*) a ∈ A is a maximal element of A iff f(a) is a maximal element of B.

Proof. Let $f : A \to B$ be the order isomorphism between partially ordered sets A and B.

$\therefore$ f is bijective and preserves partial order relation.

(*i*) Let $a \in A$ be the first element of A. If possible let $f(a)$ be not the first element of B.

$\Rightarrow$ $f(a) \le b$ for every $b \in B$ is false

$\therefore$ There exists $y \in B$ such that $f(a) \le y$ is not true.

$\Rightarrow$ $f(a) \le f(x)$ is not true. (letting $f(x) = y$)

$\Rightarrow$ $a \le x$ is not true for otherwise we would have $f(a) \le f(x)$.

$\Rightarrow$ a is not the first element of A. This is impossible.

$\therefore$ $f(a)$ is the first element of B.

Similarly, the converse hold.

(*iii*) Let $a \in A$ be a minimal element of A. If possible let $f(a)$ be not a minimal element of B.

$\therefore$ There exists at least one $x \in A$ such that $f(x) \le f(a)$ and $f(x) \ne f(a)$.

$\Rightarrow \quad x \le a$ and $x \ne a \Rightarrow x$ cannot be a minimal element of A. This is impossible.

$\therefore$ $f(a)$ is a minimal element of B.

Similarly, the converse hold.

The proofs of parts (*ii*) and (*iv*) are left for the reader as exercises.

Example 1. *Let A and B be order isomorphic partially ordered sets. Let $f : A \to B$ be an order isomorphism between A and B.*

Show that $a \ll a'$ in A if and only if $f(a) \ll f(a')$ in B.

Sol. $\quad a \ll a' \Rightarrow a < a' \Rightarrow a \le a', a \ne a'$

$$\Rightarrow f(a) \le f(a'), f(a) \ne f(a') \Rightarrow f(a) < f(a')$$

Let $x \in A$ be such that $f(a) < f(x) < f(a')$.

$$\Rightarrow \quad f(a) \le f(x) \le f(a'), f(a) \ne f(x), f(x) \ne f(a')$$

$$\Rightarrow \quad a \le x \le a', a \ne x, x \ne a' \Rightarrow a < x < a'$$

This is impossible because $a << a'$.

$\therefore$ There is no element between $f(a)$ and $f(a')$. $\therefore$ $f(a) << f(a')$.

Conversely, let $f(a) << f(a')$.

$\Rightarrow$ $f(a) < f(a')$ $\Rightarrow$ $f(a) \leq f(a'), f(a) \neq f(a')$

$\Rightarrow$ $a \leq a', a \neq a'$ $\Rightarrow$ $a < a'$

Let $x \in A$ be such that $a < x < a'$.

$\Rightarrow$ $a \leq x \leq a', a \neq x, x \neq a'$

$\Rightarrow$ $f(a) \leq f(x) \leq f(a'), f(a) \neq f(x), f(x) \neq f(a')$

$\Rightarrow$ $f(a) < f(x) < f(a')$

This is impossible because $f(a) << f(a')$.

$\therefore$ There is no element between a and a'. $\therefore$ $a << a'$.

Remark. The result of above example is used in drawing the Hasse diagram of a partially ordered set provided the Hasse diagram of an order isomorphic partially ordered set is given.

Example 2. *Let $A = \{a, b, c, d, e\}$ be a partially ordered set with the given Hasse diagram. Let the partially ordered set $B = \{1, 2, 3, 4, 5\}$ be order isomorphic to A. Let $f : A \to B$ defined by $f(a) = 1, f(b) = 3, f(c) = 5, f(d) = 2, f(e) = 4$ be an order isomorphism. Draw the Hasse diagram of the partially ordered set B.*

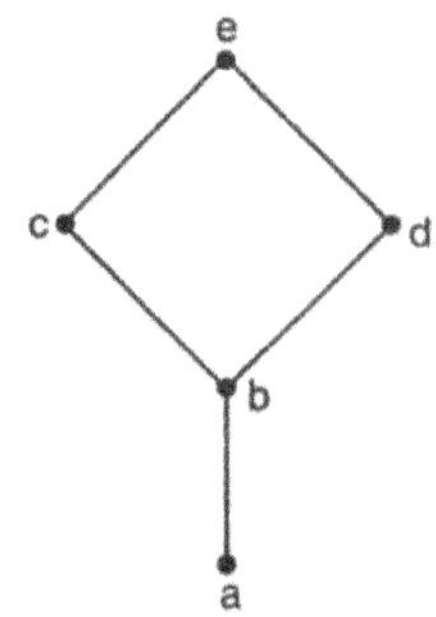

Sol. Using the Hasse diagram of A, we have

$$a << b,\ b << c,\ c << e,\ b << d,\ d << e.$$

Since f is an order isomorphism, the only immediate predecences in B are

$f(a) << f(b), f(b) << f(c), f(c) << f(e), f(b) << f(d), f(d) << f(e)$

i.e., $1 << 3$, $3 << 5$, $5 << 4$, $3 << 2$, $2 << 4$.

Using these immediate precedences, the Hasse diagram of partially ordered set B can be drawn.

The same is shown in the figure.

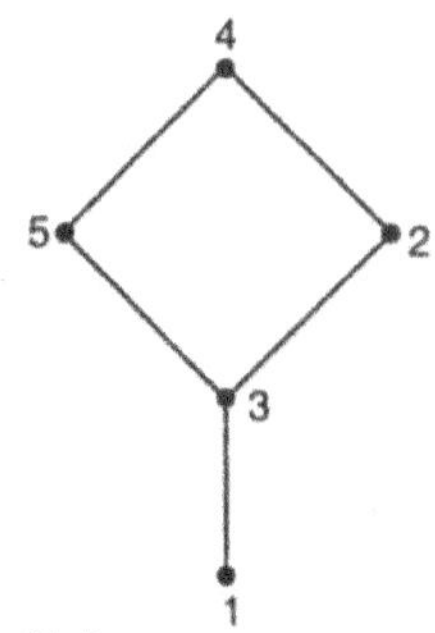

Theorem 3. Let A and B be partially ordered sets. Let $f : A \to B$ be a one-one function from A into B. Prove that the following are equivalent :

(*i*) $a_1 \leq a_2$ in A $\Leftrightarrow$ $f(a_1) \leq f(a_2)$ in B

(*ii*) $a_1 \leq a_2$ in A $\Rightarrow$ $f(a_1) \leq f(a_2)$ in B and $a_1 \parallel a_2$ in A $\Rightarrow$ $f(a_1) \parallel f(a_2)$ in B.

Proof. Let (*i*) hold.

$\therefore$ $a_1 \leq a_2$ in A $\Rightarrow$ $f(a_1) \leq f(a_2)$ in B

Let $a_1 \parallel a_2$ in A.

$\Rightarrow$ $a_1 \neq a_2, a_1 \nleq a_2, a_2 \nleq a_1$

$a_1 \neq a_2$ $\Rightarrow$ $f(a_1) \neq f(a_2)$ ($\because$ f is one-one)

If possible, let $f(a_1) < f(a_2)$.

$\Rightarrow$ $f(a_1) \leq f(a_2)$ $\Rightarrow$ $a_1 \leq a_2$ [Using (*i*)]

$\Rightarrow$ $a_1 < a_2$, which is impossible.

$\therefore$ $f(a_1) \nleq f(a_2)$. Similarly $f(a_2) \nleq f(a_1)$. $\therefore$ $f(a_1) \parallel f(a_2)$ in B.

$\therefore$ (*i*) $\Rightarrow$ (*ii*)

Let (*ii*) hold.

$\therefore$ $a_1 \le a_2$ in A $\Rightarrow$ $f(a_1) \le f(a_2)$ in B ...(1)

Let $f(a_1) \le f(a_2)$ in B. ...(2)

If possible, let $a_1 \nleq a_2$.

$\Rightarrow$ $a_1 \ne a_2$ and $a_1 \nless a_2$

$a_1 \ne a_2 \Rightarrow f(a_1) \ne f(a_2)$ ($\because$ f is one-one)

$\therefore$ (2) $\Rightarrow$ $f(a_1) < f(a_2)$

$\Rightarrow$ $f(a_1)$ and $f(a_2)$ are comparable.

$\Rightarrow$ a_1 and a_2 must be comparable, because $a_1 \parallel a_2$ would imply $f(a_1) \parallel f(a_2)$.

$\Rightarrow$ $a_1 < a_2$ or $a_2 < a_1$

$\Rightarrow$ $a_2 < a_1$ ($\because$ $a_1 \nless a_2$)

$\Rightarrow$ $a_2 \le a_1 \Rightarrow f(a_2) \le f(a_1)$ [Using (1)]

$\Rightarrow$ $f(a_2) < f(a_1)$

This is impossible, because we have $f(a_1) < f(a_2)$.

$\therefore$ Our supposition in wrong.

$\therefore$ $a_1 \le a_2$

$\therefore$ $f(a_1) \le f(a_2)$ in B $\Rightarrow$ $a_1 \le a_2$ in A

$\therefore$ $a_1 \le a_2$ in A $\Leftrightarrow$ $f(a_1) \le f(a_2)$ in B.

$\therefore$ (*ii*) $\Rightarrow$ (*i*)

$\therefore$ The result holds.

Example 3. *Let A be a partially ordered set. For* $a \in A$, *define* $p(a) = \{x \in A : x \le a\}$. *Let* $p(A) = \{p(a) : a \in A\}$. $p(A)$ *is a partially ordered set with respect to the relation 'set inclusion. Show that the partially ordered sets A and p(A) are order isomorphic.*

Sol. Define $f : A \to p(A)$ by

$$f(a) = p(a) = \{x : x \le a\},\ a \in A.$$

f is one-one. Let $a, b \in A$ and $a \ne b$. We shall show that $f(a) \ne f(b)$.

$a \ne b \Rightarrow a < b$ or $b < a$ or $a \parallel b$

$a < b \Rightarrow b < a$ is false $\Rightarrow b \notin p(a)$

$\Rightarrow p(a) \ne p(b) \Rightarrow f(a) \ne f(b)$ ($\because$ $b \le b \Rightarrow b \in p(b)$)

$b < a \Rightarrow a < b$ is false $\Rightarrow a \notin p(b)$

$\Rightarrow p(a) \ne p(b) \Rightarrow f(a) \ne f(b)$

$a \parallel b \Rightarrow a < b$ is false $\Rightarrow a \notin p(b)$

$\Rightarrow p(a) \ne p(b) \Rightarrow f(a) \ne f(b)$

$\therefore$ $a \ne b \Rightarrow f(a) \ne f(b)$

$\therefore$ f is one-one.

f is onto. Let $y \in p(A)$.

$\therefore$ $y = p(a)$ for some $a \in A$

$\therefore$ $f(a) = p(a) = y$ $\therefore$ f is onto.

f preserves order. Let $a \le b$ in A.

$x \in p(a) \Rightarrow x \le a \Rightarrow x \le b$ ($\because$ $a \le b$)

$\Rightarrow x \in p(b) \Rightarrow p(a) \subseteq p(b)$

$\Rightarrow$ $p(a) \le p(b) \Rightarrow f(a) \le f(b)$

$\therefore$ $a \le b$ in A $\Rightarrow f(a) \le f(b)$ in $p(A)$.

Let $a, b \in A$ and $a \parallel b$ *i.e.*, a and b are not comparable.

$\Rightarrow$ $a \le b$ is false $\Rightarrow a \notin p(b) \Rightarrow p(a) \not\subseteq p(b) \Rightarrow p(a) \le p(b)$ is false

$\Rightarrow$ $f(a) \le f(b)$ is false.

Also, '$b \le a$' is false.

$\Rightarrow$ $f(b) \le f(a)$ is false

$\therefore$ $f(a) \parallel f(b)$. $\therefore$ f preserves order.

$\therefore$ f is an order isomorphism. $\therefore$ A and $p(A)$ are order isomorphic.

EXERCISE 3.5

1. Let A and B be totally ordered sets and $f : A \to B$ be a one-one function. Show that f is a similarity mapping from A into B if $a_1 \le a_2$ in $A \Rightarrow f(a_1) \le f(a_2)$ in B.
2. Let A and B be partially ordered sets and $f : A \to B$ be a one-one function. Show that f is a similarity mapping from A into B if and only if
   (*i*) $a_1 \le a_2$ in $A \Rightarrow f(a_1) \le f(a_2)$ in B
   (*ii*) $a_1 \parallel a_2$ in $A \Rightarrow f(a_1) \parallel f(a_2)$ in B.
3. Let $A = \{a, b, c, d, e\}$ be a partially ordered set with the given Hasse diagram.
   Let $B = \{l, m, n, o, p\}$ be order isomorphic to A. Let $f : A \to B$ defined by $f(a) = o$, $f(b) = m$, $f(c) = l$, $f(d) = n$, $f(e) = p$ be an order isomorphism. Draw the Hasse diagram of the partially ordered set B.

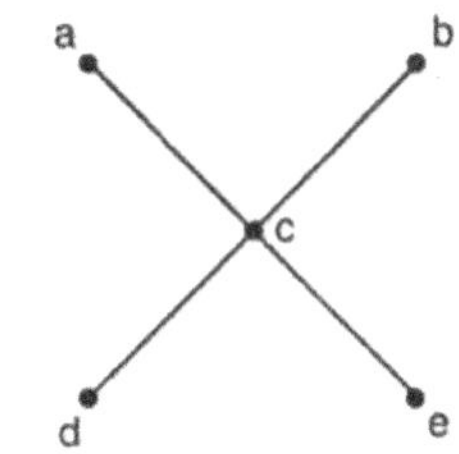

4. Let $A = \{a, b, c, d, e\}$ be a partially order set with the given Hasse diagram.
   Let $B = \{1, 2, 3, 4, 5\}$ be order isomorphic to A. Let $f : A \to B$ defined by $f(a) = 2$, $f(b) = 1$, $f(c) = 3$, $f(d) = 4$, $f(e) = 5$ be an order isomorphism. Draw the Hasse diagram of the partially ordered set B.

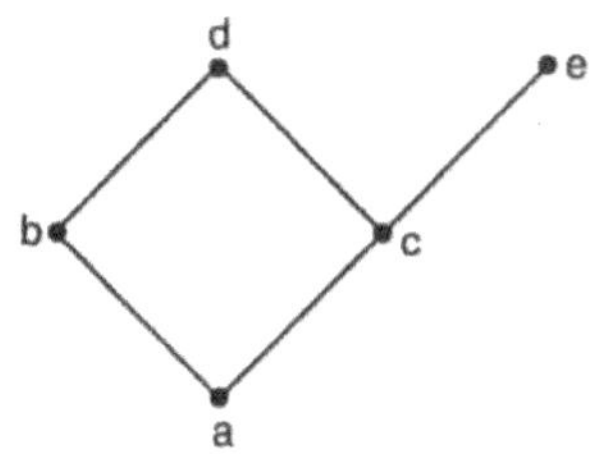

Answers

3.

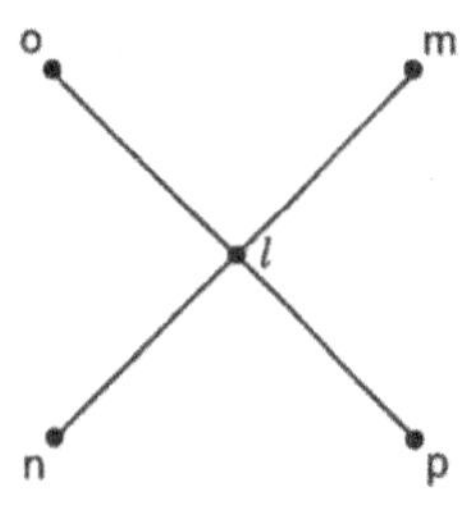

4.

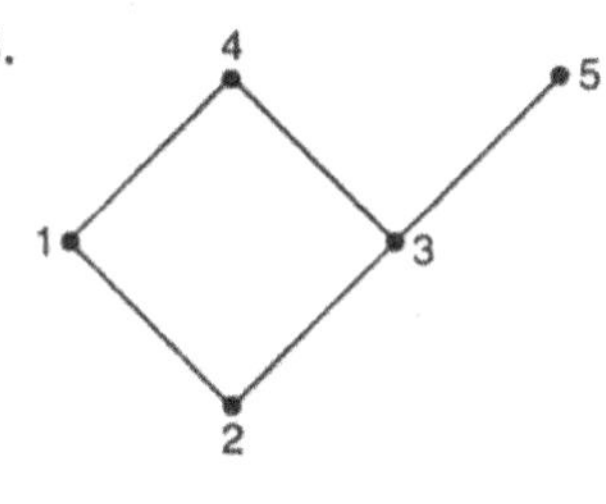

Hint

1. Let $f(a_1) \le f(a_2)$ in B.

$\therefore$ $f(a_1) = f(a_2)$ or $f(a_1) < f(a_2)$

$f(a_1) = f(a_2) \Rightarrow a_1 = a_2 \Rightarrow a_1 \le a_2$. Let $f(a_1) < f(a_2)$

$a_1, a_2 \in A \Rightarrow a_1 \le a_2$ or $a_2 \le a_1$ ($\because$ A is totally ordered)

If $a_1 \le a_2$, we are done. Let $a_2 \le a_1$

$\Rightarrow f(a_2) \le f(a_1) \Rightarrow f(a_1) = f(a_2)$ ($\because$ $\le$ is antisymmetric)

This is impossible.

$\therefore$ $f(a_1) \le f(a_2) \Rightarrow a_1 \le a_2$.

4

Pigeonhole Principle

4.1. INTRODUCTION

Pigeonhole principle is a well known proof technique in mathematics. This principle is also known as the *shoe box argument* or *Dirichlet drawer principle.*

Suppose these are 80 rooms in a hostel and these rooms are occupied by 85 hostelers. Here one can easily judge that at least one room must have been occupied by more than one hosteler. Pigeonhole principle is just a generalisation of situations of this type.

4.2. PIGEONHOLE PRINCIPLE

Statement. If m pigeonholes are occupied by m + 1 or more pigeons then prove that at least one pigeonhole is occupied by more than one pigeon.

Proof. Let the number of pigeons be n, where $n > m$. We label the m pigeonholes with numbers 1 through m and n pigeons with numbers 1 through n. Now starting with pigeon with label 1, we assign each pigeon, a pigeonhole labelled with the same number.

Thus m pigeons are assigned all the pigeonholes. Since $n > m$, some pigeons (*i.e.*, $n - m$ in number) are left and they have not been assigned any pigeonhole. Thus at least one pigeonhole will be occupied by more than one pigeon.

Example 1. *There are 13 students in a class. Show that at least two students were born in the same month.*

Sol. Number of students (pigeons) = 13

Number of months (pigeonholes) = 12

Since 13 > 12, by **pigeonhole principle**, at least one pigeonhole is occupied by more than one pigeon.

$\therefore$ At least two students were born in the same month.

Example 2. *Show that a function from one finite set to a smaller finite set cannot be one-one.*

Sol. Let A and B be finite sets, where $n(B) < n(A)$. Let $f : A \to B$ be a function.

$\therefore$ To each element (pigeon) in A, the function f associate a unique element (pigeonhole) in B. Since $n(A) > n(B)$, the number of pigeons is greater than the number of pigeonholes.

$\therefore$ By **pigeonhole principle,** at least one pigeonhole is occupied by more than one pigeon.

$\therefore$ There exist at least one element in B which is the image of more than one element in A.

$\therefore$ By definition, the function f cannot be one-one.

Example 3. *Suppose there are five points lying on or inside an equilateral triangle of side 4 cm. Show that at least two of these points must not be more than 2 cm apart.*

Sol. Let ABC be an equilateral triangle with each side equal to 4 cm. Let D, E, F be the mid-points of the sides BC, CA, AB respectively. The triangle ABC is divided in four equilateral triangles t_1, t_2, t_3 and t_4 and the sides of these triangles are 2 cm each.

Number of small triangles (pigeonholes) = 4

Number of points (pigeons) in $\Delta ABC = 5$

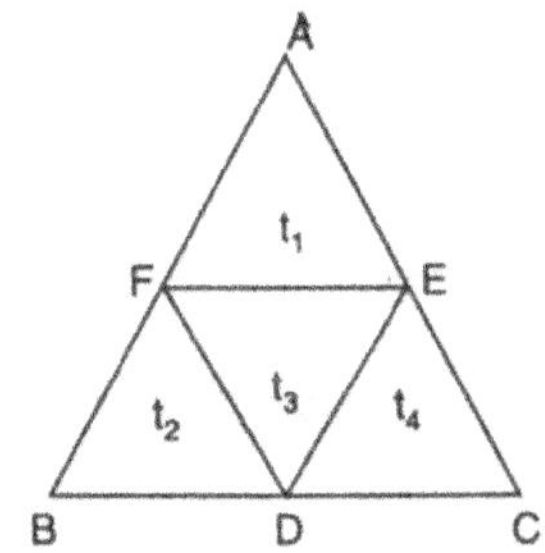

Since 5 > 4, by **pigeonhole principle**, at least one pigeonhole is occupied by more than one pigeon.

∴ At least one of the triangles t_1, t_2, t_3, t_4 contain more than one point.

∴ There are at least two points in a same small triangle. Since side of each small triangle is 2 cm, so there are at least two points which are not more than 2 cm apart.

Example 4. *Find the minimum number of elements that one needs to take from the set {1, 2, 3, ..., 9} so that the sum of two of the numbers is even.*

Sol. Let $A = \{1, 2, 3, ..., 9\}$.

We know that the sum of two even integers or two odd integers is even.

We consider the partition : {2, 4, 6, 8}, {1, 3, 5, 7, 9} of A.

There are 2 sets (pigeonholes) in this partition of A.

Let B be a subset of A containing 3 elements (pigeons).

[∵ 3 is the minimum number which is greater than 2.]

∴ By **pigeonhole principle**, at least one pigeonhole is occupied by more than one pigeon.

∴ One of the sets in the partition will have more than one element of B.

∴ B has either two even numbers or two odd numbers ranging from 1 to 9.

∴ There exists two numbers in B whose sum is even.

∴ Any subset of A of **3 elements** will contain two elements whose sum is even.

Example 5. *Find the minimum number of elements that one needs to take from the set {1, 2, 3, ..., 9} to be sure that two of the numbers add up to 10.*

Sol. Let $A = \{1, 2, 3, ..., 9\}$.

We observe that $1 + 9 = 10$, $2 + 8 = 10$, $3 + 7 = 10$, $4 + 6 = 10$.

We consider the partition : {1, 9}, {2, 8} , {3, 7}, {4, 6}, {5} of A.

There are 5 sets (pigeonholes) in this partition of A.

Let B be a subset of A containing 6 elements (pigeons).

[∵ 6 is the minimum number which is greater than 5]

∴ By **pigeonhole principle**, at least one pigeonhole is occupied by more than one pigeon.

∴ At least one of the sets in the partition will have more than one element of B. Since first four sets in the partition have two elements each and the last set has only one element, at least one of the first four sets will contain two elements of B.

∴ There exists two elements in B whose sum is 10.

∴ Any subset of A of **6 elements** will contain two elements whose sum is 10.

Example 6. *The publication list of a professor consists of 45 research papers each marked either 'Foreign journal' or 'Indian journal'. The number of papers in Foreign journals is 26. Show that there are at least 2 research papers marked 'Foreign journal' in the list having exactly 5 research papers between them.*

Sol. Let a_i denote the position of the ith paper marked Foreign journal, $1 \le i \le 26$.

$\therefore$ $a_1, a_2, \ldots\ldots, a_{26}$ are distinct number and $a_1, a_2, \ldots\ldots, a_{26} \le 45$

$\therefore$ The numbers $a_1 + 6, a_2 + 6, \ldots\ldots, a_{26} + 6$ are also all distinct and

$$a_1 + 6, a_2 + 6, \ldots\ldots, a_{26} + 6 \le 45 + 6 = 51$$

$\therefore$ $a_1, a_2, \ldots\ldots, a_{26}, a_1 + 6, a_2 + 6, \ldots\ldots a_{26} + 6$ are 52 (= 26 × 2) positive numbers.

Also the value of these 52 numbers (pigeons) ranges from 1 to 51.

$\therefore$ There can be 51 distinct values (pigeonholes) of these numbers.

Since 52 > 51, by **pigeonhole principle**, at least one pigeonhole is occupied by more than one pigeon.

$\therefore$ There exists at least two numbers having same value.

Since $a_1, a_2, \ldots\ldots, a_{26}$ are distinct and $a_1 + 6, a_2 + 6, \ldots\ldots, a_{26} + 6$ are also distinct, we have

$$a_i = a_j + 6 \quad \text{for some } i, j$$

$$\Rightarrow \quad a_i - a_j = 6$$

$\therefore$ a_jth and a_ith 'Foreign journal' papers have 5(= 6 – 1) research papers between them.

Remark. Between a_jth and a_ith 'Foreign journal' papers, the five papers are $(a_j + 1)$th, $(a_j + 2)$th, $(a_j + 3)$th, $(a_j + 4)$th and $(a_j + 5)$th.

Example 7. *A chess player wants to prepare for a championship match by playing some practice games in 77 days. He wants to play at least one game a day but not more than 132 games altogether. Show that there is a period of consecutive days within which he plays exactly 21 games.*

Sol. Let a_i denote the total number of games he plays upto the ith day. On each day the player has to play at least one game.

$$\therefore \quad a_1 \ge 1, a_2 \ge a_1 + 1, a_3 \ge a_2 + 1, \ldots\ldots, a_{77} \ge a_{76} + 1$$

$$\therefore \quad a_1 \ge 1, a_2 > a_1, a_3 > a_2, \ldots\ldots, a_{77} > a_{76}$$

Total number of practice games is not more than 132.

$$\therefore \quad a_{77} \le 132$$

$$\therefore \text{ We have } \quad 1 \le a_1 < a_2 < a_3 < \ldots\ldots < a_{76} < a_{77} \le 132$$

$$\text{or} \quad 1 \le a_1 < a_2 < \ldots\ldots < a_{77} \le 132 \qquad \ldots(1)$$

$$(1) \Rightarrow \quad a_1 + 21 < a_2 + 21 < \ldots\ldots < a_{77} + 21 \le 153$$

$\therefore$ $a_1, a_2, \ldots\ldots, a_{77}, a_1 + 21, a_2 + 21, \ldots\ldots, a_{77} + 21$ are 154 (= 77 × 2) positive numbers.

Also the values of these 154 numbers (pigeons) ranges from 1 to 153.

$\therefore$ There can be 153 distinct values (pigeonholes) of these numbers.

Since 154 > 153, by **pigeonhole principle** at least one pigeonhole is occupied by more than one pigeon.

$\therefore$ There exists at least two numbers having the same value.

Since $a_1 < a_2 < \ldots\ldots < a_{77}$, no two of $a_1, a_2, \ldots\ldots, a_{77}$ can be equal.

Similarly no two of $a_1 + 21, a_2 + 21, \ldots\ldots, a_{77} + 21$ can be equal.

$\therefore \quad a_i = a_j + 21$ for some i, j

$\Rightarrow \quad a_i - a_j = 21$

$\therefore$ From $(j + 1)$th day to the ith day, he played 21 matches.

EXERCISE 4.1

1. 15 balls are put in 12 boxes. Show that at least one box contain more than one ball.
2. There are 30 students in a class. Show that at least two students will have same first letter of the alphabet in their names.
3. If 15 shoes are selected from 14 pairs of shoes. Show that there must be a pair of matched shoes among the selection.
4. Let X and Y be finite sets. A function $f : X \to Y$ is an injection. Show that X cannot have more elements than Y.
5. Suppose there are more than 16 points lying on or inside an equilateral triangle of side 3 cm. Show that at least two of these points must not be more than 0.75 cm apart.
6. Show that if any of four numbers from 1 to 6 are chosen, then two of them will add up to 7.
7. Find the minimum number of elements that we need to take from the set {1, 2, 3, ..., 9} so that the difference of two numbers is 5.
8. A doctor gives prescription of 20 tablets to a patient with the instructions to take at least one tablet per day for 15 days. Show that there is a period of consecutive days during which the patient takes a total of 9 tablets.

Answer

7. 6.

4.3. GENERALISED PIGEONHOLE PRINCIPLE

Statement. If m pigeonholes are occupied by n or more pigeons, where n > m, then prove that at least one pigeonhole is occupied by more than $\left[\frac{n-1}{m}\right]$ pigeons.

Proof. If possible, let no pigeonhole is occupied by more than $\left[\frac{n-1}{m}\right]$ pigeons.

$\therefore$ Every pigeonhole contains at most $\left[\frac{n-1}{m}\right]$ pigeons.

$\therefore$ No. of pigeons in any pigeonhole $\le \left[\frac{n-1}{m}\right]$

$\therefore$ No. of pigeons in m pigeonholes $\le m \times \left[\frac{n-1}{m}\right] \le m \times \left(\frac{n-1}{m}\right) = n - 1 < n$

$\therefore$ No. of pigeons $< n$

This is impossible because the number of pigeons is n or more.

$\therefore$ Our supposition is wrong.

$\therefore$ There is at least one pigeonhole which is occupied by more than $\left[\frac{n-1}{m}\right]$ pigeons.

Remark. [ ] is representing the greatest integer function.

Corollary. If m pigeonholes are occupied by km + 1 or more pigeons, where $k \in N$, then at least one pigeonhole is occupied by more than $\left[\frac{(mk+1)-1}{m}\right] = [k] = k$ pigeons.

Example 1. *How many persons at least in a group of 85 persons have the same last initial in their names ?* (KU 2006 C)

Sol. Number of persons (pigeons) = 85

Number of letters in the alphabet (pigeonholes) = 26

Since 85 > 26, by **generalised pigeonhole principle**, at least one pigeonhole is occupied by more than $\left[\frac{85-1}{26}\right] = \left[\frac{42}{13}\right] = \left[3\frac{3}{13}\right] = 3$ pigeons.

$\therefore$ More than 3 persons will have the same last initial in their names.

$\therefore$ At least 4(= 3 + 1) persons will have the same last initial in their names.

Example 2. *Six friends on a tour found that they have a total of Rs. 7206 with them. Show that at least one of them has minimum Rs. 1201.*

Sol. Number of friends (pigeonholes) = 6

Number of rupees (pigeons) = 7206

Since 7206 > 6, by **generalised pigeonhole principle**, at least one pigeonhole is occupied by more than $\left[\frac{7206-1}{6}\right] = \left[\frac{7205}{6}\right] = \left[1200\frac{5}{6}\right] = 1200$ pigeons.

$\therefore$ At least one of the friends has more than Rs. 1200.

$\therefore$ At least one of the friends has minimum **Rs. 1201.**

Example 3. *How many persons among 3,00,000 persons are born at the same time (hour, minute, second) ?*

Sol. Number of persons (pigeons) = 3,00,000

Number of hours (pigeonholes) in a day = 24

$\therefore$ By **generalised pigeonhole principle,** at least one pigeonhole is occupied by more than $\left[\frac{3,00,000-1}{24}\right] = 12499$ pigeons.

$\therefore$ At least 12500 (= 12499 + 1) persons are born during the same hour.

Now 1 hour = 60 minutes

Let $m = 60$ and $n = 12500$

$$\therefore \quad \left[\frac{n-1}{m}\right] = \left[\frac{12500-1}{60}\right] = 208$$

$\therefore$ By **generalised pigeonhole principle,** more than 208 persons are born during the same minute.

$\therefore$ At least 209 (= 208 + 1) persons are born during the same minute.

Also, 1 minute = 60 seconds

Let $m = 60$ and $n = 209$

$$\therefore \quad \left[\frac{n-1}{m}\right] = \left[\frac{209-1}{60}\right] = 3$$

∴ By **generalised pigeonhole principle**, more than 3 persons are born during the same second.

∴ At least 4(= 3 + 1) persons are born during the same second.

∴ At least **4 persons** are born at the same time.

Example 4. *Find the minimum number of students in a class to be sure that three of them are born in the same month.*

Sol. Number of months (pigeonholes) = 12

Let the number of students (pigeons) be n.

Let three students are born in the same month.

∴ More than two students are born in the same month.

∴ At least one pigeonhole is occupied by more than 2 pigeons.

Also, by **generalised pigeonhole principle,** at least one pigeonhole is occupied by more than $\left[\frac{n-1}{12}\right]$ pigeons.

$$\therefore \quad \left[\frac{n-1}{12}\right] = 2$$

$$\Rightarrow \quad \frac{n-1}{12} \geq 2 \Rightarrow n \geq 25$$

∴ Minimum value of $n = 25$

∴ Minimum number of students = **25.**

Example 5. *Find the minimum number of persons selected so that at least eight of them will have birthdays on the same day of the week.*

Sol. Number of days (pigeonholes) in a week = 7

Let the number of persons (pigeons) be n.

Let at least eight persons have birthdays on the same day of the week.

∴ More than seven persons have birthdays on the same day of the week.

∴ At least one pigeonhole is occupied by more than seven pigeons.

Also, by **generalised pigeonhole principle,** at least one pigeonhole is occupied by more than $\left[\frac{n-1}{7}\right]$ pigeons.

$$\therefore \quad \left[\frac{n-1}{7}\right] = 7$$

$$\Rightarrow \quad \left[\frac{n-1}{7}\right] \geq 7 \Rightarrow n \geq 50.$$

∴ Minimum value of $n = 50$

Minimum number of persons = **50.**

EXERCISE 4.2

1. Show that at least 3 students in a group of 65 students have the same first initial in their names ?
2. Show that if 25 books in a library contain a total of 52,042 pages, then one of the books must contain at least 2,082 pages. *(KU 2006)*
3. Show that if 30 dictionaries in a library contain a total of 61,327 pages, then one of the dictionaries must have at least 2,045 pages.
4. Ten friends on a tour found that they have a total of Rs. 85,607 with them. Show that at least one of them has minimum Rs. 8,561.
5. Seven members of a family have a total of Rs. 2886 in their pockets. Show that at least one of them must have at least Rs. 413 in his pocket.
6. Show that among one lakh people, there are at least two persons who were born at the same time (hour, minute, second).
7. How many people among 2,00,000 people are born at the same time (hour, minute, second) ?
8. How many people among 4,00,000 people are born at the same time ? *(KU 2007)*
9. Find the minimum number of persons selected so that at least nine of them will have birthdays on the same days of the week.

Answers

7. 3 8. 5 9. 57.

5
Probability

I. PROBABILITY OF AN EVENT

5.1. INTRODUCTION

The words 'Probability' and 'Chance' are quite familiar to everyone. Many a times, we come across statements like, "Probably it may rain today", "Chances of his visit to the university are very few", "It is possible that he may pass the examination with good marks". In the above statements, the words probably, chance, possible etc. convey the sense of uncertainty about the occurrence of some event. Ordinarily, it may appear that there cannot be any exact measurement for these uncertainties, but in Statistics, we do have methods for calculating the degree of certainty of events in numerical value, provided certain conditions are satisfied.

5.2. RANDOM EXPERIMENT

When we perform experiments in science and engineering, repeatedly under very nearly identical conditions, we get almost the same result. Such experiments are called **deterministic experiments**.

There also exist experiments in which the results may not be essentially the same even if the experiment is performed under very nearly identical conditions. Such experiments are called **random experiments**. If we toss a coin, we may get 'head' or 'tail'. This is a random experiment. Throwing of a die is also a random experiment as any of the six faces of the die may come up. In this experiment, there are six possibilities (1 or 2 or 3 or 4 or 5 or 6).

Remark 1. A *die* is a small cube used in gambling. On its six faces, dots are marked as :

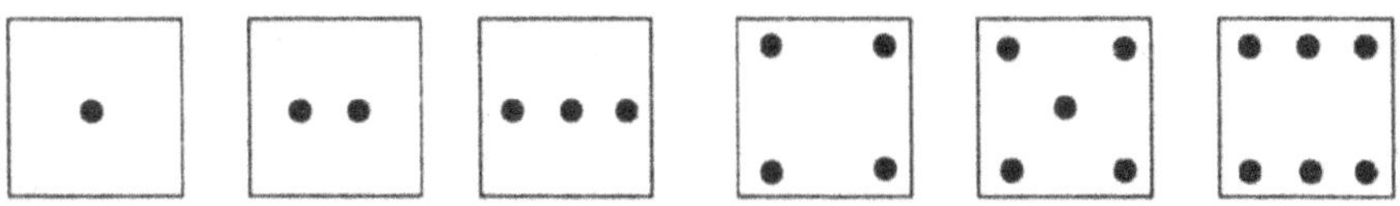

Numbers on a die

Plural of the word die is *dice*. The outcome of throwing a die is the number of dots on its upper most face.

Remark 2. A *pack of cards* consists of four suits called *Spades, Hearts, Diamonds* and *Clubs*. Each suit consists of 13 cards, of which nine cards are numbered from 2 to 10, an ace, a king, a queen and a jack (or knave). Spades and clubs are black faced cards, while hearts and diamonds are red faced cards. The kings, queens and jacks are called *face* cards.

5.3. SAMPLE SPACE

The **sample space** of a random experiment is defined as the set of all possible outcomes of the experiment. The possible outcomes are called **sample points.** The sample space is generally denoted by the letter S. We list the sample space of some random experiments :

Random experiment	*Sample space*
1. Throwing of a fair die	$S = \{1, 2, 3, 4, 5, 6\}$
2. Tossing of an unbiased coin	$S = \{H, T\}$
3. Tossing of two unbiased coins	$S = \{HH, HT, TH, TT\}$
4. A family of two children	$S = \{BB, BG, GB, GG\}$

5.4. TREE DIAGRAM

A **Tree diagram** is a device used to enumerate all the logical possibilities of a sequence of steps where each step can occur in a finite number of ways. A tree diagram is constructed from left to right and the number of branches at any point corresponds to the number of ways the next step can occur.

Illustration 1. The tree diagram of the sample space of the toss of two coins is shown in the figure.

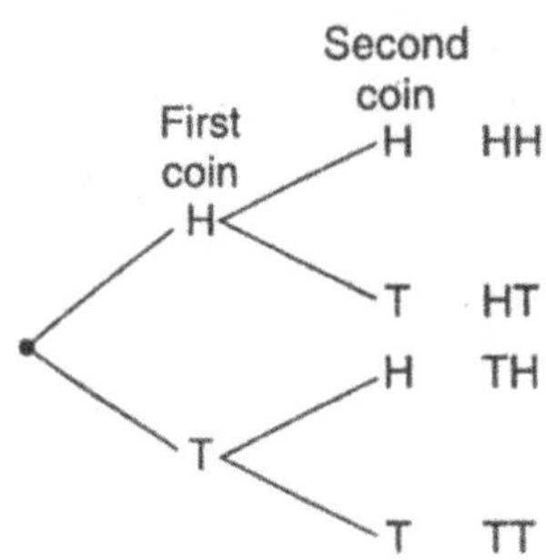

Illustration 2. The tree diagram of the sample space of the two tosses of a coin is shown in the figure.

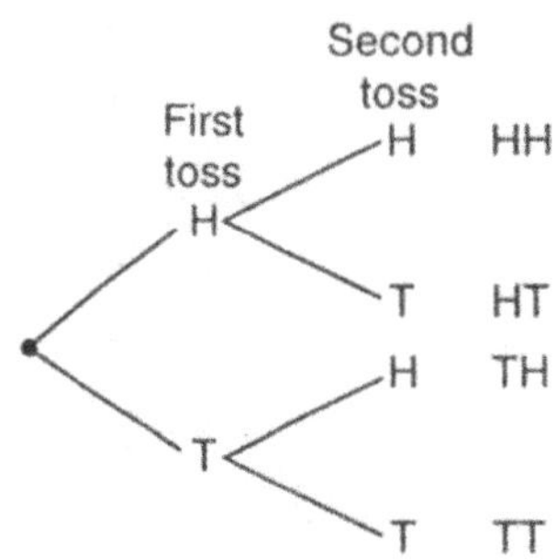

5.5. EVENT

An **event** is defined as a subset of the sample space. An event is called an **elementary (or simple) event** if it contains only one sample point. In the experiment of rolling a die, the event A of getting '3' is a simple event. We write $A = \{3\}$. An event is called an **impossible event** if it can never occur. In the above example, the event $B = \{7\}$ of getting '7' is an impossible event. On the other hand, an event which is sure to occur is called a **certain event.** In the above example of rolling a die, the event C of getting a number less than 7 is a certain event.

In the throwing of two dice, the cases favourable to getting sum 7 are 6 *viz.* (1, 6), (2, 5), (3, 4), (4, 3), (5, 2) and (6, 1).

5.6. ALGEBRA OF EVENTS

We know that the events of a random experiment are sets, being subsets of the sample space. Thus, we can use set operations to form new events.

Let A and B be any two events associated with a random experiment.

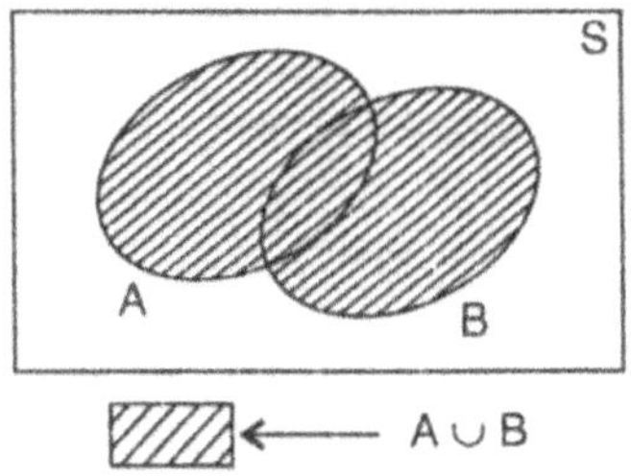

The event of occurrence of either A or B or both is written as 'A or B' and is denoted by the subset $A \cup B$ of the sample space. The event of occurrence of both A and B is written as 'A and B' and is denoted by the subset $A \cap B$ of the sample space. For simplicity, the event $A \cap B$ is also denoted by 'AB'.

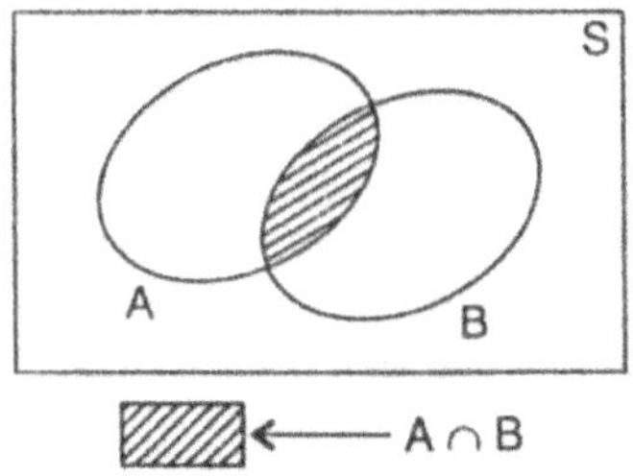

The event of non-occurrence of event A is written as 'not A' and is denoted by the subset A', which is the complement of set A. The event A' is called the **complementary event** of the event A.

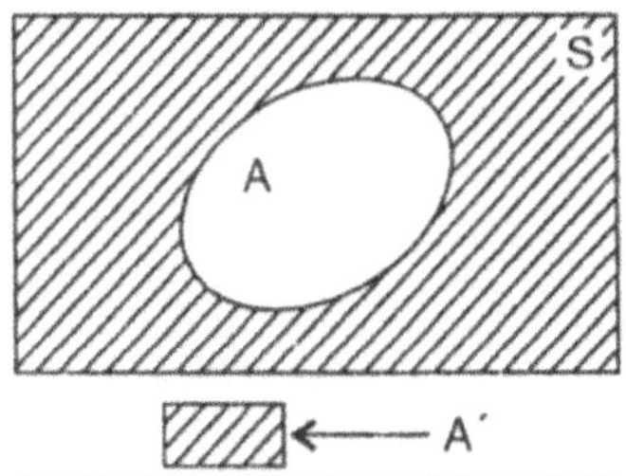

Illustration. Let a die be tossed.

$\therefore$ $S = \{1, 2, 3, 4, 5, 6\}$

Let A = event of getting an even number

and B = event of getting a number less than 5

$\therefore$ $A = \{2, 4, 6\}$ and $B = \{1, 2, 3, 4\}$

Here

$A \cup B$ = event of occurrence of either A or B or both

$= \{1, 2, 3, 4, 6\}$ *(See figure)*

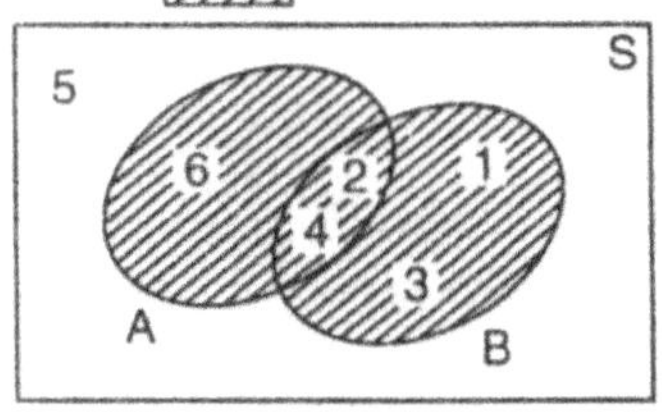

$A \cap B$ = event of occurrence of both A and B

$= \{2, 4\}$ *(See figure)*

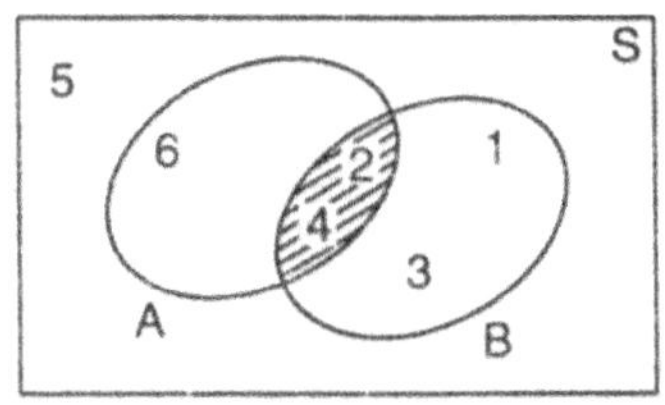

Also, A' = event of non-occurrence of $A = \{1, 3, 5\}$

and B' = event of non-occurrence of $B = \{5, 6\}$.

5.7. EQUALITY LIKELY OUTCOMES

The outcomes of a random experiment are called **equally likely,** if all of these have equal preferences. In the experiment of tossing a unbiased coin, the outcomes 'Head' and 'Tail' are equally likely.

In our discussion, we shall always assume the outcomes of a random experiment to be equally likely.

5.8. EXHAUSTIVE OUTCOMES

The outcomes of a random experiment are called **exhaustive,** if these cover all the possible outcomes of the experiment. In the experiment of rolling a die, the outcomes 1, 2, 3, 4, 5, 6 are exhaustive.

5.9. DEFINITION OF PROBABILITY

Suppose in a random experiment, there are n exhaustive, equally likely outcomes. Let A be an event and there are m outcomes (cases) favourable to the happening of it. Then the **probability** $P(A)$ of the happening of the event A is defined as

$$P(A) = \frac{\text{Total no. of cases favourable to the happening of } A}{\text{Total no. of exhaustive, equally likely cases}} = \frac{m}{n}.$$

It may be observed from this definition, that $0 \le m \le n$.

$$\therefore \quad 0 \le \frac{m}{n} \le 1 \text{ or } \mathbf{0 \le P(A) \le 1.}$$

The number of cases favourable to the non-happening of the event A is $n - m$.

$$\therefore \quad P(\text{not } A) = \frac{n-m}{n} = \frac{n}{n} - \frac{m}{n} = 1 - \frac{m}{n} = 1 - P(A).$$

$\therefore \quad P(A) + P(\text{not } A) = 1.$ *i.e.*, $P(A) + P(\bar{A}) = 1.$

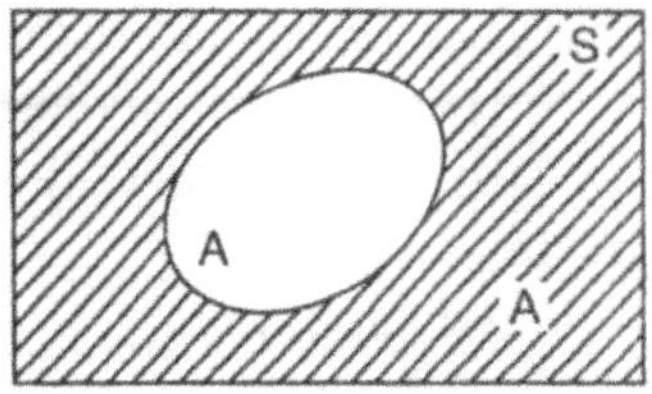

If A is a *sure event*, the $P(A) = \frac{n}{n} = 1$ and if A happen to be an *impossible event*, then

$$P(A) = \frac{0}{n} = 0.$$

5.10. 'ODDS IN FAVOUR' AND 'ODDS AGAINST' AN EVENT

The ratio of cases in favour of A and cases against A is called the **odds in favour** of A. Similarly, the ratio of cases against A and cases in favour of A is called the **odds against** A.

If odds in favour of A are $m : n$, then (*i*) odds against A are $n : m$ and (*ii*) probability of $A = \frac{m}{m+n}$.

If odds against A are $m : n$, then (*i*) odds in favour of A are $n : m$ and (*ii*) probability of $A = \frac{n}{m+n}$.

Illustration. Let $P(A) = \frac{3}{5}$.

Let total number of cases be 5λ.

$\therefore \quad P(A) = \frac{3\lambda}{5\lambda}$ and this implies that cases in favour of A are 3λ and cases against A are $5\lambda - 3\lambda = 2\lambda$.

$\therefore$ odds in favour of A are $3\lambda : 2\lambda$ or **3 : 2** and odds against A are $2\lambda : 3\lambda$ or **2 : 3.**

Example 1. *Find the probability of getting the sum 10 in a single throw of two dice.*

Sol. Here $\quad S = \{(1, 1), (1, 2), (1, 3), \ldots\ldots, (6, 5), (6, 6)\}$.

$\therefore$ No. of possible outcomes $= 6 \times 6 = 36$.

Let A be the event of getting sum 10.

$$\therefore \quad A = \{(4, 6), (5, 5), (6, 4)\}$$

$$P(A) = \frac{3}{36} = \mathbf{\frac{1}{12}}.$$

Remark. In the above experiment, the sample point (a, b) means that 'a' is on the first die and 'b' is on the second die.

Example 2. *Find the probability of getting sum 10 in two throws of a die.*

Sol. Here $S = \{(1, 1), (1, 2), (1, 3), ..., (6, 5), (6, 6)\}$

$\therefore$ No. of possible outcomes $= 6 \times 6 = 36$

Let A be the event of getting sum 10

$\therefore$ $A = \{(4, 6), (5, 5), (6, 4)\}$

$$P(A) = \frac{3}{36} = \frac{1}{12}.$$

Remark. In the above experiment, the sample point (a, b) means that 'a' occurred in the first toss and 'b' occurred in the second toss.

Example 3. *Find the probability of getting 53 sundays in a :*

(*a*) *non-leap year* (*b*) *leap year.*

Sol. (*a*) There are 365 days in a non-leap year.

Now $365 = 7 \times 52 + 1$

$\therefore$ A non-leap year will contain at least 52 Sundays. The possibilities for the remaining one day are :

(*i*) Sunday (*ii*) Monday

(*iii*) Tuesday (*iv*) Wednesday

(*v*) Thursday (*vi*) Friday

(*vii*) Saturday.

Let A be the events of getting 53 Sundays in a non-leap year. Therefore only (*i*) possibility is favourable to the event A.

$$\therefore \quad P(A) = \frac{1}{7}.$$

(*b*) There are 366 days in a leap year.

Now $366 = 7 \times 52 + 2$

$\therefore$ A leap year will contain at least 52 Sundays. The possible combinations for the remaining two days are :

(*i*) Sunday and Monday (*ii*) Monday and Tuesday

(*iii*) Tuesday and Wednesday (*iv*) Wednesday and Thursday

(*v*) Thursday and Friday (*vi*) Friday and Saturday

(*vii*) Saturday and Sunday.

Let A be the event of getting 53 Sundays in a leap year. Therefore, only those combinations will be favourable to the event A which contain 'Sunday'.

$\therefore$ The combinations (*i*) and (*vii*) are favourable to the happening of A.

$$\therefore \quad P(A) = \frac{2}{7}.$$

Example 4. *Three coins are tossed. Find the probability of getting at least two heads.*

Sol. Let S be the sample space.

$\therefore$ $S = (HHH, HHT, HTH, THH, HTT, THT, TTH, TTT)$

Let A be the event of getting at least two heads.

$\therefore$ $A = \{HHH, HHT, HTH, THH\}$

$\therefore$ P(at least two heads)

$$= P(A) = \frac{\text{No. of cases favourable to } A}{\text{Total no. of exhaustive, equally likely cases}} = \frac{4}{8} = \frac{1}{2}.$$

Example 5. *Find the probability of getting a 'King' or a 'Queen' in a single draw from a well-shuffled pack of playing cards.*

Sol. Let A be the event of getting a king or a queen in the draw.

$\therefore$ No. of favourable cases for the happening of the event $A = 4 + 4 = 8$

Total no. of cases $= 52$

$$\therefore \quad P(\text{king or queen}) = P(A) = \frac{8}{52} = \mathbf{\frac{2}{13}}.$$

Example 6. *Two unbiased dice are thrown. Find the probability that the total of the numbers on the dice is greater than 8.*

Sol. Let S be the sample space.

$$\therefore \quad S = \{(1, 1), (1, 2), (1, 3), \ldots\ldots, (6, 5), (6, 6)\}$$

There are $6 \times 6 = 36$ exhaustive, equally likely outcomes.

Let A be the event of getting total greater than 8.

$$\therefore \quad A = \{(3, 6), (4, 5), (5, 4), (6, 3), (4, 6), (5, 5), (6, 4), (5, 6), (6, 5), (6, 6)\}$$

$$\therefore \quad P(\text{sum} > 8) = P(A) = \frac{\text{No. of cases favourable to A}}{\text{Total no. of cases}} = \frac{10}{36} = \mathbf{\frac{5}{18}}.$$

Example 7. *The following table gives a distribution of wages :*

Weekly wage (in Rs.)	*No. of workers*
30—35	*9*
35—40	*108*
40—45	*488*
45—50	*230*
50—55	*112*
55—60	*30*
60—65	*16*
65—70	*7*

An individual is taken at random from the above group. Find the probability (i) his wage is under Rs. 40 (ii) his wage is Rs. 55 or above (iii) his wage is either between Rs. 45—50 or Rs. 35—40.

Sol. Total number of workers $= 9 + 108 + 488 + 230 + 112 + 30 + 16 + 7 = 1000$

(*i*) Number of workers having wage under $40 = 9 + 108 = 117$

$$\therefore \quad P(\text{wage under Rs. 40}) = \frac{\text{No. of workers having wage under 40}}{\text{Total number of workers}} = \frac{117}{1000} = \mathbf{0.117.}$$

(*ii*) Number of workers having wage 55 or above $= 30 + 16 + 7 = 53$

$$\therefore \quad P(\text{wage Rs. 55 or above}) = \frac{53}{1000} = \mathbf{0.053.}$$

(*iii*) Number of workers having wage either between Rs. 45 – 50 or Rs. 35 – 40

$$= 230 + 108 = 338$$

$\therefore$ P(wage either between Rs. 45—50 or Rs. 35—40) $= \dfrac{338}{1000} = \mathbf{0.338.}$

Example 8. *Three unbiased dice are thrown simultaneously. Find the probability of getting (i) sum not greater than 5 (ii) sum at least 15 (iii) sum equal to 8.*

Sol. Let S be the sample space.

$\therefore$ $S = \{(1, 1, 1), (1, 1, 2), ..., (6, 6, 5), (6, 6, 6)\}$

There are $6 \times 6 \times 6 = 216$ exhaustive equally likely outcomes.

(*i*) Let A = event that sum is not greater than 5.

$\therefore$ $A = \{(1, 1, 1), (1, 1, 2)\ (1, 2, 1), (2, 1, 1), (1, 1, 3),$
$(1, 2, 2), (1, 3, 1), (2, 1, 2), (2, 2, 1), (3, 1, 1)\}$

$$\therefore \quad P(A) = \frac{n(A)}{n(S)} = \frac{10}{216} = \mathbf{\frac{5}{108}}$$

(*ii*) Let B = event that sum is at least 15

$\therefore$ $B = \{(3, 6, 6), (4, 5, 6), (4, 6, 5), (5, 4, 6), (5, 5, 5),$
$(5, 6, 4), (6, 3, 6), (6, 4, 5), (6, 5, 4), (6, 6, 3),$
$(4, 6, 6), (5, 5, 6), (5, 6, 5), (6, 4, 6), (6, 5, 5),$
$(6, 6, 4), (5, 6, 6), (6, 5, 6), (6, 6, 5), (6, 6, 6)\}$

$$\therefore \quad P(B) = \frac{n(B)}{n(S)} = \frac{20}{216} = \mathbf{\frac{5}{54}}$$

(*iii*) Let C = event of getting sum 8

$\therefore$ $C = \{(1, 1, 6), (1, 2, 5), (1, 3, 4), (1, 4, 3), (1, 5, 2),$
$(1, 6, 1), (2, 1, 5), (2, 2, 4), (2, 3, 3), (2, 4, 2),$
$(2, 5, 1), (3, 1, 4), (3, 2, 3), (3, 3, 2), (3, 4, 1),$
$(4, 1, 3), (4, 2, 2), (4, 3, 1), (5, 1, 2), (5, 2, 1), (6, 1, 1)\}$

$$\therefore \quad P(C) = \frac{n(C)}{n(S)} = \frac{21}{216} = \mathbf{\frac{7}{72}}.$$

Example 9. *A player tosses two fair coins. He wins Rs. 5 if 2 heads appear, Rs. 2 if one head appear and Re. 1 if no head occur. Find his expected value of the amount.*

Sol. The expected value of the amount won by the player is the sum of products of various amounts with corresponding probabilities.

In the given example, let x denote the 'number of heads'.

$\therefore$ The possible values of x are 0, 1 and 2.

Here $S = \{TT, TH, HT, HH\}$

$$\therefore \quad p_0 = P(x = 0) = P(\text{no head}) = \frac{1}{4}$$

$$p_1 = P(x = 1) = P(\text{one head}) = \frac{2}{4} = \frac{1}{2}$$

$$p_2 = P(x = 2) = P(\text{two heads}) = \frac{1}{4}$$

$\therefore$ Expected value

$$= (p_0 \times 1) + (p \times 2) + (p_2 \times 5) = \left(\frac{1}{4} \times 1\right) + \left(\frac{1}{2} \times 2\right) + \left(\frac{1}{4} \times 5\right)$$

$$= 0.25 + 1 + 1.25 = \textbf{Rs. 2.5.}$$

Example 10. *Find the probability that in a random arrangement of the letters of the word* **DAUGHTER**, *the letter* **D** *occupies the first place.*

Sol. The word **DAUGHTER** contains 8 letters and all are different.

$\therefore$ Total no. of possible arrangements

$$= 8\,! = \mathbf{40320}$$

Let A be the event of getting a word with **D** at the first place.

$\therefore$ Favourable cases to the event A

$$= 1 \times \text{no. of ways of arranging 7 letters (except D).}$$
$$= 1 \times 7\,! = 5040$$

$\therefore$ P(**D** occupies first place)

$$P(A) = \frac{5040}{40320} = \frac{1}{8}.$$

Example 11. *Find the probability that in a random arrangement of letters of the word* **INSTITUTION**, *the three* **T**'*s are together.*

Sol. The word **INSTITUTION** contains 3**I**'s 2 **N**'s, 1 **S**, 3 **T**'s, 1 **U**'s and 1 **O**.

$$\therefore \quad \text{Total no. of arrangements} = \frac{11!}{3!2!1!3!1!1!} = 554400$$

For finding the no. of favourable cases to the event under consideration, we shall consider 3 **T**'s as one block. So in arranging 3 **I**'s, 2 **N**'s, 1 **S**, 1 (**TTT**)'s, 1 **U** and 1 **O**, three **T** will remain together.

$$\therefore \quad \text{No. of favourable cases} = \frac{9!}{3!2!1!1!1!1!} = 30240$$

$$\therefore \quad P(3\text{T's are together}) = \frac{30240}{554400} = \mathbf{0.0545}.$$

Example 12. *Find the probability that in a random arrangement of letters of the word* **MATHEMATICS,** *the consonants occur together.*

Sol. The word **MATHEMATICS**, contains 2 **M**'s, 2 **A**'s, 2 **T**'s, 1 **H**, 1 **E**, 1 **I**, 1 **C** and 1 **S**.

$\therefore$ Total no. of exhaustive cases

$$= \frac{11!}{2!2!2!} = 4989600$$

For finding the no. of favourable cases to the event under consideration, we shall consider all consonents **M, T, H, M, T, C, S** as one block. So in arranging **2 A**'s, **1 E**'s, **1 I**'s, **1 (MTHMTCS)**, all the consonents will occur together. The consonents **MTHMTCS** can arrange themselves in

$$\frac{7!}{2!2!} = 1260 \text{ ways}$$

$$\therefore \quad \text{No. of favourable cases} = \frac{5!}{2!1!1!1!} \times 1260 = 75600$$

$$\therefore \quad P(\text{consonants are together}) = \frac{75600}{4989600} = \mathbf{0.01515}.$$

Example 13. *A bag contains 10 red and 8 black balls. Two balls are drawn at random. Find the probability that :*

(*i*) *both balls are red.*

(*ii*) *one ball is red and the other is black.*

Sol. Total number of balls = 10 + 8 = 18

(*i*) $P(\text{both red balls}) = \dfrac{\text{No. of selections of 2 out of 10 red balls}}{\text{No. of selections of 2 out of 18 balls}}$

$$= \frac{^{10}C_2}{^{18}C_2} = \frac{10 \times 9}{1 \times 2} \div \frac{18 \times 17}{1 \times 2} = \frac{10 \times 9}{18 \times 17} = \mathbf{\frac{5}{17}}.$$

(*ii*) $P(\text{one red ball and one black ball})$

$$= \frac{\begin{pmatrix}\text{No. of selections of} \\ \text{1 out of 10 red balls}\end{pmatrix}\begin{pmatrix}\text{No. of selections of} \\ \text{1 out of 8 black balls}\end{pmatrix}}{\text{No. of selections of 2 out of 18 balls.}} = \frac{^{10}C_1 \times {^8C_1}}{^{18}C_2}$$

$$= (10 \times 8) \div \frac{18 \times 17}{2} = \frac{10 \times 8}{9 \times 17} = \mathbf{\frac{80}{153}}.$$

Example 14. *Five cards are drawn from a pack of playing cards. Find the probability that :*

(*i*) *4 are aces and 1 is a queen* (*ii*) *3 are eights and 2 are kings*

(*iii*) *at least one queen is drawn.*

Sol. (*i*) Let A be the event of drawing 4 aces and 1 queen.

$$\therefore \quad P(A) = \frac{\begin{pmatrix}\text{No. of selections of} \\ \text{4 out of 4 aces}\end{pmatrix}\begin{pmatrix}\text{No. of selections of} \\ \text{1 out of 4 queens}\end{pmatrix}}{\text{No. of selections of 5 out 52 cards}}$$

$$= \frac{^4C_4 \times {^4C_1}}{^{52}C_5} = \frac{1 \times 4}{2598960} = \frac{1}{649740} = \mathbf{0.0000015.}$$

(*ii*) Let B be the event of getting 3 eights and 2 kings.

$$\therefore \quad P(B) = \frac{\begin{pmatrix}\text{No. of selections of} \\ \text{3 out of 4 eights}\end{pmatrix}\begin{pmatrix}\text{No. of selections of} \\ \text{2 out of 4 kings}\end{pmatrix}}{\text{No. of selections of 5 out 52 cards}} = \frac{^4C_3 \times {^4C_2}}{^{52}C_5}$$

$$= \frac{4 \times 6}{2598960} = \frac{1}{108290} = \mathbf{0.0000092.}$$

(*iii*) Let C be the event of getting at least one queen

$$\therefore \quad P(C) = 1 - \text{Prob. of getting no queen}$$

$$= 1 - \frac{\text{No. of selections of 5 out of 48 non-queen cards}}{\text{No. of selections of 5 out 52 cards}}$$

$$= 1 - \frac{^{48}C_5}{^{52}C_5} = 1 - \frac{1712304}{2598960} = 1 - 0.6588 = \mathbf{0.3412.}$$

Example 15. *From a group of 8 children (5 boys and 3 girls) three children are selected at random. Calculate the probabilities that the selected group contains :*

(*i*) *no girl* (*ii*) *only one girl*

(*iii*) *only one particular girl.*

Sol. No. of girls $= 3$

No. of boys $= 5$

Let A_0, A_1, A_2 and A_3 be the events that there are no girl, one girl, two girls and three girls in the random selection of three children.

(*i*) Required probability $= P(A_0) = \dfrac{{}^5C_3}{{}^8C_3} = \dfrac{\mathbf{5}}{\mathbf{28}}$.

(*ii*) Required probability $= P(A_1) = \dfrac{{}^3C_1 \times {}^5C_2}{{}^8C_3} = \dfrac{\mathbf{15}}{\mathbf{28}}$.

(*iii*) The particular girl can be selected in only one way. The remaining two children in the selection must be boys.

$\therefore$ Required probability $= \dfrac{1 \times {}^5C_2}{{}^8C_3} = \dfrac{\mathbf{5}}{\mathbf{28}}$.

Example 16. *A coin is tossed n times. What is the probability that head will appear an odd number of times ?*

Sol. In each toss, we get either head or tail.

$\therefore$ No. of possible outcomes in one toss $= 2$

$\therefore$ No. of possible outcomes in n tosses $= 2^n$

Let A = event of getting head odd number of times.

$\therefore$ A will happen if head appears 1 time, 3 times, 5 times, ..., in n tosses.

$$\therefore \quad n(A) = {}^nC_1 + {}^nC_3 + {}^nC_5 + \ldots\ldots = 2^{n-1}$$

$\therefore$ P(having head odd number of times)

$$= \frac{n(A)}{2^n} = \frac{2^{n-1}}{2^n} = \frac{\mathbf{1}}{\mathbf{2}}.$$

Example 17. *Given a group of four persons, find the probability that :*

(*i*) *no two of them have birthday on the same day of the year.*

(*ii*) *all of them have same birthday.*

Sol. Here $S = \{(1, 1, 1, 1), (1, 1, 1, 2), \ldots\ldots, (365, 365, 365, 364)\ (365, 365, 365, 365)\}$.

$\therefore$ No. of possible outcomes

$$= 365 \times 365 \times 365 \times 365 = (365)^4$$

(*i*) No. of choices for first person, second person, third person and fourth person are 365, 364, 363 and 362 respectively.

$\therefore$ No. of favourable cases $= 365 \times 364 \times 363 \times 362$

$$\therefore \quad P(\text{of having distinct birthdays}) = \frac{365 \times 364 \times 363 \times 362}{(365)^4} = \frac{\mathbf{364 \times 363 \times 362}}{\mathbf{(365)^3}}.$$

(*ii*) Let A be the event that their birthdays are on the same day.

$$\therefore \quad A = \{(1, 1, 1, 1), (2, 2, 2, 2), \ldots\ldots, (365, 365, 365, 365)\}$$

$\therefore$ P(of having same birthday)

$$= \frac{n(A)}{n(S)} = \frac{365}{(365)^4} = \frac{\mathbf{1}}{\mathbf{(365)^3}}.$$

EXERCISE 5.1

1. A single letter is selected at random from the word **"PROBABILITY"**. What is the probability that it is a vowel ?
2. Find the probability of getting at most two heads in a single toss of three coins.
3. From a pack of 52 playing cards, one card is drawn at random. What is the probability that it will be a queen of club or king of diamond ?
4. Two unbiased dice are thrown. Find the probability that both dice show the same number.
5. Two unbiased dice are thrown. Find the probability that the total of the numbers on the dice is 8.
6. The following table gives a distribution of marks :

Marks	*No. of students*
0—10	7
10—20	10
20—30	15
30—40	28
40—50	23
50—60	45
60—70	12
70—80	7
80—90	2
90—100	1

An individual is taken at random from the above group. Find the probability that :

(*i*) his marks are below 50 (*ii*) his marks are below 70.

(*iii*) his marks are between 30 and 60 (*iv*) his marks are above 70.

7. Find the probability that in a random arrangement of the letters of the word **MALENKOV**, the letter **A** occupies the third place.
8. Find the probability that in a random arrangement of letters the word **BHARAT**, the two **A** occupies the first two places.
9. Find the probability that in a random arrangement of letters of the word **INDIA**, the two **I** are together.
10. Find the probability that in a random arrangement of the letters of the word **STATISTICS**, the three **T** are in the beginning.
11. Find the probability that in a random arrangement of the letters of the word **STATISTICS**, the three **T** are together.
12. Two cards are drawn from a pack of 52 cards, at random. What is the probability that one is a king and the other is a queen ?
13. Four cards are drawn from a pack of playing cards. Find the probability that none is a king.
14. From a bag containing 4 red and 5 black balls, four balls are drawn at random. Find the probability of getting two red and two black balls.
15. A bag contains 7 white, 5 red and 8 black balls. Two balls are drawn at random. Calculate the probability that none is white.
16. A bag contains 7 white, 5 red and 8 black balls. Three balls are drawn at random. Calculate the probability that there is one ball of each colour.

Answers

1. $\frac{4}{11}$	**2.** $\frac{7}{8}$	**3.** $\frac{1}{26}$	**4.** $\frac{1}{6}$
5. $\frac{5}{36}$	**6.** (*i*) 0.5533	(*ii*) 0.9333	(*iii*) 0.64
(*iv*) 0.0667	**7.** 0.125	**8.** 0.0667	**9.** 0.4
10. 0.0083	**11.** 0.0667	**12.** $\frac{8}{663}$	**13.** = 0.7187
14. $\frac{10}{21}$	**15.** $\frac{39}{95}$	**16.** $\frac{14}{57}$.	

5.11. MUTUALLY EXCLUSIVE EVENTS

Two events associated with a random experiment are said to be **mutually exclusive** if both cannot occur together in the same trial. In the experiment of throwing a die, the events $A = \{1, 4\}$ and $B = \{2, 5, 6\}$ are mutually exclusive events. In the same experiment, the events $A = \{1, 4\}$ and $C = \{2, 4, 5, 6\}$ are not mutually exclusive because if 4 appear on the die, then it is favourable to both events A and C. The definition of mutually exclusive events can also be extended to more than two events. We say that more than two events are mutually exclusive if the happening of one of these rules out the happening of all other events. The events $A = \{1, 2\}$, $B = \{3\}$ and $C = \{6\}$, are mutually exclusive in connection with the experiment of throwing a single die.

5.12. ADDITION THEOREM (FOR MUTUALLY EXCLUSIVE EVENTS)

If A and B are two mutually exclusive events associated with a random experiment, then prove that

$$\mathbf{P(A \text{ or } B) = P(A) + P(B).}$$

Proof. Let n be the total number of exhaustive, equally like cases of the experiment.

Let m_1 and m_2 be the number of cases favourable to the happening of the events A and B respectively.

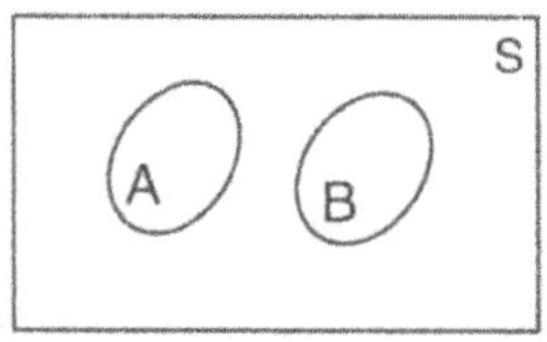

$$\therefore \qquad P(A) = \frac{m_1}{n}$$

and

$$P(B) = \frac{m_2}{n}$$

Since the events are given to be mutually exclusive, therefore there cannot be any sample point common to both events A and B.

$\therefore$ The event A or B can happen in exactly $m_1 + m_2$ ways.

$$\therefore \qquad P(A \text{ or } B) = \frac{m_1 + m_2}{n} = \frac{m_1}{n} + \frac{m_2}{n} = P(A) + P(B)$$

Hence, $\qquad$ **P(A or B) = P(A) + P(B).**

This theorem can also be extended to even more than two events.

If $A_1, A_2, ..., A_k$ are pairwise m.e. events, then

$$\mathbf{P(A_1 \text{ or } A_2 \text{ or } ...... \text{ or } A_k) = P(A_1) + P(A_2) + ... + P(A_k).}$$

Example 1. *The probability that a company executive will travel by plane is 2/3 and that he will travel by train is 1/5. Find the probability of his travelling by plane or train.*

Sol. Let A = event that the company executive travel by plane

$\therefore$ $P(A) = 2/3$.

Let B = event that the company executive travel by train

$\therefore$ $P(B) = 1/5$

The events A and B are mutually exclusive, because he cannot travel by plane and train at the same time.

$\therefore$ Prob. of travelling of the executive by plane or train

$$= P\,(A \text{ or } B) = P(A) + P(B) = \frac{2}{3} + \frac{1}{5} = \mathbf{\frac{13}{15}}.$$

Example 2. *In a single throw of 2 dice, determine the probability of getting total 7 or 11.*

Sol. Here $S = \{(1, 1), (1, 2), ..., (6, 5), (6, 6)\}$.

Let A and B be the events of getting total 7 and 11 respectively.

$\therefore$ $A = \{(1, 6), (2, 5), (3, 4), (4, 3), (5, 2), (6, 1)\}$

and $B = \{(5, 6), (6, 5)\}$

The events A and B are mutually exclusive and

$$P(A) = \frac{6}{36} = \frac{1}{6} \quad \text{and} \quad P(B) = \frac{2}{36} = \frac{1}{18}$$

$\therefore$ *By addition theorem,*

$P(\text{total is 7 or 9}) = P(A \cup B) = P(A) + P(B)$

$$= \frac{1}{6} + \frac{1}{18} = \frac{3+1}{18} = \mathbf{\frac{2}{9}}.$$

Example 3. *One of the two events must happen. Given that the chance of one is two-third of the other, find the odds in favour of the other.*

Sol. Let the given events be A, B and $P(A) = \frac{2}{3}\,P(B)$.

Let $P(B) = x$. $\therefore$ $P(A) = \frac{2}{3}\,x$.

The events A and B are exhaustive.

$\therefore$ $P(A \text{ or } B) = 1 \Rightarrow P(A) + P(B) = 1$ ($\because$ A and B are *m.e.*)

$$\Rightarrow \quad \frac{2}{3}\,x + x = 1 \Rightarrow \frac{5x}{3} = 1 \Rightarrow x = \frac{3}{5}$$

$\therefore$ $P(B) = \frac{3}{5}$.

$P(B) = \frac{3}{5}$ implies that the odds in favour of B are $3 : 5 - 2$ *i.e.*, **3 : 2.**

Example 4. *In a single throw of two dice, find the probability that neither a doublet nor a total of 9 will appear.*

Sol. Here $S = \{(1, 1), (1, 2), ... , (6, 5), (6, 6)\}$

$\therefore$ Number of possible outcomes $= 6 \times 6 = 36$

Let E = event that doublet is occurred

and F = event that sum is 9

$\therefore$ $E = \{(1, 1), (2, 2), (3, 3), (4, 4), (5, 5), (6, 6)\}$

and $F = \{(3, 6), (4, 5), (5, 4), (6, 3)\}$

$\therefore$ $P(E) = \frac{6}{36} = \frac{1}{6}$ and $P(F) = \frac{4}{36} = \frac{1}{9}$

P(neither a doublet nor a total of 9)

$$= P(E^c \cap F^c) = P((E \cup F)^c) = 1 - P(E \cup F) \qquad ...(1)$$

The events E and F are *m.e.*

$\therefore$ By *addition theorem*,

$$P(E \cup F) = P(E) + P(F) = \frac{1}{6} + \frac{1}{9} = \frac{5}{18}$$

$\therefore$ (1) $\Rightarrow$ $P(E^c \cap F^c) = 1 - \frac{5}{18} = \frac{13}{18}$.

EXERCISE 5.2

1. (*i*) A box contains 4 red balls, 4 green balls and 7 white balls. What is the probability that a ball drawn is either green or white ?

   (*ii*) A bag contains 5 white, 7 black and 6 red balls. A ball is drawn at random. Find the probability that it is a red or a white ball. (*KU 2007*)
2. Find the probability that a card drawn from a pack of playing cards is either a 'queen' or a 'king'.
3. Let A and B be two mutually exclusive events of an experiment.

   If P('not A') = 0.65, $P(A \cup B) = 0.65$, find the value of $P(B)$. (*KU 2006*)
4. In a class of 25 students with roll numbers 1 to 25, a student is picked up at random to answer a question. Find the probability that the roll number of the selected student is either a multiple of 5 or of 7.
5. From a set of 17 cards, numbered 1, 2, 3, ..., 16, 17, one is drawn at random. Show that the chance that its number is divisible by 3 or 7 is 7/17.
6. A, B, C are three pairwise mutually exclusive and exhaustive events associated with a random experiment. Find $P(A)$, given that $P(B) = \frac{3}{2} P(A)$ and $P(C) = \frac{1}{2} P(B)$.
7. Find the probability of getting the sum 9 or 11 in a single throw of two dice.
8. Find the probability of getting the sum 8 or 11 or 12 in a single throw of two dice.
9. In a single throw of three dice, find the probability of getting a total of 17 or 18.

Answers

1. (*i*) $\frac{11}{15}$ (*ii*) $\frac{11}{18}$ 2. $\frac{2}{13}$ 3. 0.3 4. $\frac{8}{25}$

6. $\frac{4}{13}$ 7. $\frac{1}{6}$ 8. $\frac{2}{9}$ 9. $\frac{1}{54}$.

5.13. ADDITION THEOREM (GENERAL)

If A and B are two events, not necessarily mutually exclusive, associated with a random experiments, then prove that

$$\mathbf{P(A \text{ or } B) = P(A) + P(B) - P(AB).}$$

Proof. Let n be the total number of exhaustive equally likely cases of the experiment.

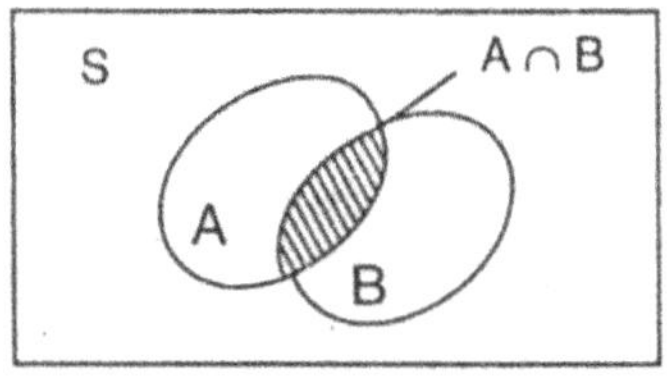

Let m_1 and m_2 be the number of cases favourable to the happening of the events A and B respectively.

$$\therefore \qquad P(A) = \frac{m_1}{n} \quad \text{and} \quad P(B) = \frac{m_2}{n}$$

Since the events are given to be not necessarily nonmutually exclusive, there may be some sample points common to both events A and B.

Let m_3 be number of these common sample points. m_3 will be zero in case A and B are mutually exclusive.

$$\therefore \qquad P(AB) = \frac{m_3}{n}$$

The m_3 sample points which are common to both events A and B, are included in the events A and B separately.

$\therefore$ Number of sample points in the event A or B

$$= m_1 + m_2 - m_3$$

m_3 is subtracted from $m_1 + m_2$ to avoid counting of common sample points twice.

$$\therefore \qquad P(A \text{ or } B) = \frac{m_1 + m_2 - m_3}{n} = \frac{m_1}{n} + \frac{m_2}{n} - \frac{m_3}{n} = P(A) + P(B) - P(AB)$$

Hence, $\quad \mathbf{P(A \text{ or } B) = P(A) + P(B) - P(AB).}$

Corollary 1. If events A and B happen to be mutually exclusive events, then $P(AB) = 0$ and in this case *addition theorem* implies

$$P(A \text{ or } B) = P(A) + P(B) - P(AB) = P(A) + P(B) - 0 = P(A) + P(B)$$

$$\therefore \qquad P(A \text{ or } B) = P(A) + P(B).$$

This is the same as the **addition theorem** for mutually exclusive events.

Corollary 2. If A, B, C are three events associated with a random experiment, then

$$\mathbf{P(A \text{ or } B \text{ or } C) = P(A) + P(B) + P(C) - P(BC) - P(CA) - P(AB) + P(ABC).}$$

Example 1. *Find the probability that a card drawn from a pack of playing cards is either a 'queen' or a 'spade'.*

Sol. Total number of cases (cards) = 52

Let $\qquad A$ = event of drawing a 'queen'

$\qquad B$ = event of drawing a 'spade'

$$\therefore \qquad P(A) = \frac{4}{52} \quad \text{and} \quad P(B) = \frac{13}{52}$$

Here the events are not mutually exclusive as drawing of the card 'queen of spade' is common to both events.

$$\therefore \qquad P(AB) = \frac{1}{52}$$

By *addition theorem*, the probability of getting either a queen or a spade is

$$P(A \text{ or } B) = P(A) + P(B) - P(AB) = \frac{4}{52} + \frac{13}{52} - \frac{1}{52} = \frac{16}{52} = \frac{4}{13}.$$

Example 2. *A bag contains 30 balls numbered from 1 to 30. One ball is drawn at random. Find the probability that the number of the ball is a multiple of 5 or 6.*

Sol. Let A = event that the number of ball is a multiple of 5.

$$A = \{5, 10, 15, 20, 25, 30\}$$

$$\therefore \quad P(A) = \frac{6}{30}$$

Let B = event that the number of ball is a multiple of 6

$$\therefore \quad B = \{6, 12, 18, 24, 30\}$$

$$\therefore \quad P(B) = \frac{5}{30}$$

The events A and B are not *m.e.* because the sample point 30 is common to both.

$$\therefore \quad AB = \{30\}$$

$$\therefore \quad P(AB) = \frac{1}{30}$$

By *addition theorem*, the probability of getting a multiple of 5 or 6 is

$$P(A \text{ or } B) = P(A) + P(B) - P(AB) = \frac{6}{30} + \frac{5}{30} - \frac{1}{30} = \frac{10}{30} = \frac{1}{3}.$$

Example 3. *One number is drawn from numbers 1 to 150. Find the probability that it is divisible by either 3 or 5.*

Sol. Here $S = \{1, 2, 3, \ldots\ldots, 149, 150\}$.

Let A = event that the number is divisible by 3

$$\therefore \quad A = \{3, 6, 9, \ldots\ldots, 147, 150\}$$

$$\therefore \quad P(A) = \frac{50}{150}$$

Let B = event that the number is divisible by 5

$$\therefore \quad B = \{5, 10, 15 \ldots\ldots, 145, 150\}$$

$$\therefore \quad P(B) = \frac{30}{150}$$

The events A and B are not *m.e.* because the sample points 15, 30, 45, ..., 150 are common to both.

$$\therefore \quad AB = \{15, 30, 45, \ldots\ldots, 135, 150\}$$

$$\therefore \quad P(AB) = \frac{10}{150}$$

By *addition theorem*, the required probability of getting a multiple of either 3 or 5 is

$$P(A \text{ or } B) = P(A) + P(B) - P(AB) = \frac{50}{150} + \frac{30}{150} - \frac{10}{150} = \frac{70}{150} = \frac{7}{15}.$$

Example 4. *Find the probability of 4 turning for at least once in two tosses of a fair die.*

Sol. Here $S = \{(1, 1), (1, 2), ..., (6, 5), (6, 6)\}$.

Let A = event of getting 4 on the first die

and B = event of getting 4 on the second die.

$\therefore$ $A = \{(4, 1), (4, 2), (4, 3), (4, 4), (4, 5), (4, 6)\}$

and $B = \{(1, 4), (2, 4), (3, 4), (4, 4), (5, 4), (6, 4)\}$

$$\therefore \quad P(A) = \frac{n(A)}{n(S)} = \frac{6}{36} = \frac{1}{6} \quad \text{and} \quad P(B) = \frac{n(B)}{n(S)} = \frac{6}{36} = \frac{1}{6}$$

The events A and B are not *m.e.* because the sample (4, 4) is common to both.

$\therefore$ $AB = \{(4, 4)\}$

$$\therefore \quad P(AB) = \frac{1}{36}$$

By *addition theorem*, the required probability of getting four at least once is

$$P(A \text{ or } B) = P(A) + P(B) - P(AB) = \frac{1}{6} + \frac{1}{6} - \frac{1}{36} = \mathbf{\frac{11}{36}}.$$

Example 5. *A student applies for a job in two firms X and Y. The probability of his being selected in firm X is 0.7 and being rejected in the firm Y is 0.5. The probability of at least one of his application being rejected is 0.6. What is the probability that he will be selected in one of the firms ?*

Sol. Let A = event of getting selected in X

B = event of getting selected in Y

$\therefore$ $P(A) = 0.7, \quad P(B) = 1 - 0.5 = 0.5$

$P(AB) = 1 - P$ (rejecting in at least one firm) $= 1 - 0.6 = 0.4$

$\therefore$ P(selected in one of the firm)

$= P(A \cup B) = P(A) + P(B) - P(AB)$

$= 0.7 + 0.5 - 0.4 = \mathbf{0.8}$.

EXERCISE 5.3

1. Find the probability that a card drawn from a pack of playing cards is either a 'king' or a 'club'.
2. One card is drawn from a pack of playing cards, each of 52 cards being equally likely to be drawn. Find the probability that :
   (*i*) the card drawn is either red or a king.
   (*ii*) the card drawn is either diamond or red.
3. A drawer contains 50 bolts and 150 nuts. Half of the bolts and half of the nuts are rusted. If one item is chosen at random, what is the probability that it is rusted or is a bolt ?
4. Two dice are thrown simultaneously. Find the probability that at least one of the following two events occurs :
   E : an even total is thrown
   F : a total larger than 8 is thrown.

5. From 30 tickets marked with first 30 numerals, one is drawn at random. Find the probability that it is :
(*i*) a multiple of 5 or 7 (*ii*) a multiple of 3 or 7.

6. Find the probability of getting an even number on the first die or a total of 8, in a single throw of two dice.

7. A construction company is bidding for two contracts A and B. The probability that the company will get contract A is 3/5, the probability that the company will get contract B is 1/3 and the probability that the company will get both the contracts is 1/8. What is the probability that the company will get contract A or B ?

8. A card is drawn at random from a well-suffled pack of playing cards. What is the probability that it is either a heart or a jack ?

9. The probability that a contractor will get a plumbing contract is 2/3 and the probability that he will not get an electric contract is 5/9. The probability of getting at least one contract is 4/5. What is the probability that he will get both contracts ?

10. A chartered accountant applied for a job in two firms x and y. He estimates that the probability of his being selected in firm x is $\frac{7}{10}$ and being rejected in y is $\frac{1}{2}$ and the probability of his being selected in both the firms is $\frac{2}{5}$. What is the probability that he will be selected in at least one of the firms ?

11. Find the probability of getting at least one five, in a single throw of two dice.

12. There are 1 to 100 pages in a book. A page is opened at random. Find the probability of getting page of number which is a multiple of 6 or of 10.

Answers

1. $\frac{4}{13}$	2. (*i*) $\frac{7}{13}$	(*ii*) $\frac{1}{2}$	3. $\frac{5}{8}$
4. $\frac{2}{3}$	5. (*i*) $\frac{1}{3}$	(*ii*) $\frac{13}{30}$	6. $\frac{5}{9}$
7. $\frac{97}{120}$	8. $\frac{4}{13}$	9. $\frac{14}{45}$	10. $\frac{4}{5}$
11. $\frac{11}{36}$	12. 0.23.		

5.14. CONDITIONAL PROBABILITY

Let us consider the random experiment of throwing a die. Let A be the event of getting an odd number on the die.

$$\therefore \quad S = \{1, 2, 3, 4, 5, 6\} \quad \text{and} \quad A = \{1, 3, 5\}$$

$$\therefore \quad P(A) = \frac{3}{6} = \frac{1}{2}$$

Let $B = \{2, 3, 4, 5, 6\}$

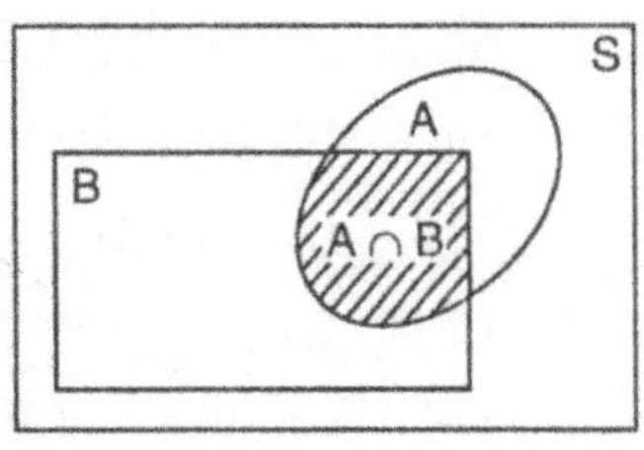

If, after the die is thrown, we are given the information that the event B has occurred, then the probability of event A will no more be 1/2, because in this case, the favourable cases are three and the total number of possible outcomes will be five and not six. The probability of event A, with the condition that event B has happened will be 3/5. This conditional probability is denoted as $P(A/B)$. Let us define the concept of conditional probability in a formal manner.

Let A and B be any two events associated with a random experiment. The probability of occurrence of event A when the event B has already occurred is called the **conditional probability** of A when B is given and is denoted as $P(A/B)$. The conditional probability $P(A/B)$ is meaningful only when $P(B) \neq 0$, *i.e.*, when B is not an impossible event.

By definition,

$$P(A/B) = \text{probability of occurrence of event } A \text{ when the event } B \text{ has already occurred}$$

$$= \frac{\text{no. of cases favourable to } B \text{ which are also favourable to } A}{\text{no. of cases favourable to } B}$$

$$\therefore \quad \mathbf{P(A/B)} = \frac{\textbf{no. of cases favourable to } \mathbf{A \cap B}}{\textbf{no. of cases favourable to B}}.$$

$$\text{Also,} \quad P(A/B) = \frac{\dfrac{\text{no. of cases favourable to } A \cap B}{\text{no. of cases in the sample space}}}{\dfrac{\text{no. of cases favourable to } B}{\text{no. of cases in the sample space}}}$$

$$\therefore \quad \mathbf{P(A/B)} = \frac{\mathbf{P(A \cap B)}}{\mathbf{P(B)}}.$$

Remark 1. If $P(A) \neq 0$, the $P(B/A) = \dfrac{P(B \cap A)}{P(A)} = \dfrac{P(A \cap B)}{P(A)}$.

Remark 2. If A and B are *m.e.* events, then

$$P(A/B) = \frac{P(A \cap B)}{P(B)} = \frac{0}{P(B)} = 0 \quad \text{and} \quad P(B/A) = \frac{P(B \cap A)}{P(A)} = \frac{0}{P(A)} = 0.$$

$\therefore$ **If A and B are m.e. events, then A/B and B/A are impossible events.**

For an illustration, let us consider the random experiment of throwing two coins.

$$\therefore \quad S = \{HH, HT, TH, TT\}$$

$$\text{Let} \quad A = \{HH, HT\},\ B = \{HH, TH\},\ C = \{HH, HT, TH\} \text{ and } D = \{TT\}.$$

$$\therefore \quad P(A) = \frac{2}{4} = \frac{1}{2},\quad P(B) = \frac{2}{4} = \frac{1}{2},\quad P(C) = \frac{3}{4},\quad P(D) = \frac{1}{4}.$$

A/B is the event of getting A with the condition that B has occurred.

$$\therefore \quad P(A/B) = \frac{n(A \cap B)}{n(B)} = \frac{n\{HH\}}{n\{HH, TH\}} = \frac{1}{2}.$$

$$\text{Similarly,} \quad P(A/C) = \frac{n\{HH, HT\}}{n\{HH, HT, TH\}} = \frac{2}{3} \quad \text{and} \quad P(B/C) = \frac{n\{HH, TH\}}{n\{HH, HT, TH\}} = \frac{2}{3}.$$

We observe that $P(A/C) \neq P(A)$, $P(B/C) \neq P(B)$.

The events A and D are *m.e.* and we have

$$P(A/D) = \frac{n(A \cap D)}{n(D)} = \frac{0}{1} = 0 \quad \text{and} \quad P(D/A) = \frac{n(D \cap A)}{n(A)} = \frac{0}{2} = 0.$$

Example 1. *If $P(E) = 0.40$, $P(F) = 0.35$ and $P(E \cup F) = 0.55$, find $P(E/F)$.*

Sol. We have $\quad P(E) = 0.40,\ P(F) = 0.35,\ P(E \cup F) = 0.55.$

By *addition theorem*,

$$P(E \cup F) = P(E) + P(F) - P(E \cap F)$$

$$\therefore \quad 0.55 = 0.40 + 0.35 - P(E \cap F)$$

$$\therefore \quad P(E \cap F) = 0.75 - 0.55 = 0.20.$$

Required probability, $P(E/F) = \dfrac{P(E \cap F)}{P(F)} = \dfrac{0.20}{0.35} = \mathbf{\dfrac{4}{7}}$.

Example 2. *If A and B are events associated with a random experiment such that* $P(A) = \dfrac{1}{3}$, $P(B) = \dfrac{1}{4}$, $P(A \cap B) = \dfrac{1}{5}$, *find :*

(*i*) $P(A/B)$ (*ii*) $P(B/A)$

(*iii*) $P(A \cup B)$ (*iv*) $P(\overline{A}/\overline{B})$

(*v*) $P(\overline{B}/\overline{A})$.

Sol. We have $P(A) = \dfrac{1}{3}$, $P(B) = \dfrac{1}{4}$ and $P(A \cap B) = \dfrac{1}{5}$.

(*i*) We have $P(A/B) = \dfrac{P(A \cap B)}{P(B)} = \dfrac{\frac{1}{5}}{\frac{1}{4}} = \mathbf{\dfrac{4}{5}}$

(*ii*) $$P(B/A) = \frac{P(B \cap A)}{P(A)} = \frac{P(A \cap B)}{P(A)} = \frac{\frac{1}{5}}{\frac{1}{3}} = \mathbf{\frac{3}{5}}$$

(*iii*) $$P(A \cup B) = P(A) + P(B) - P(A \cap B)$$

$$= \frac{1}{3} + \frac{1}{4} - \frac{1}{5} = \frac{20 + 15 - 12}{60} = \mathbf{\frac{23}{60}}$$

(*iv*) $$P(\overline{A} / \overline{B}) = \frac{P(\overline{A} \cap \overline{B})}{P(\overline{B})} = \frac{P(\overline{A \cup B})}{P(\overline{B})} = \frac{1 - P(A \cup B)}{1 - P(B)}$$

$$= \frac{1 - (23/60)}{1 - (1/4)} = \frac{37/60}{3/4} = \mathbf{\frac{37}{45}}$$

(*v*) $$P(\overline{B} / \overline{A}) = \frac{P(\overline{B} \cap \overline{A})}{P(\overline{A})} = \frac{P(\overline{B \cup A})}{P(\overline{A})} = \frac{1 - P(B \cup A)}{1 - P(A)}$$

$$= \frac{1 - (23/60)}{1 - (1/3)} = \frac{37/60}{2/3} = \mathbf{\frac{37}{40}}.$$

Example 3. *A coin is tossed twice and the four possible outcomes are assumed to be equally likely. If E is the event "both head and tail have occurred", and F the event : "at most one tail is observed", find P(E), P(F), P(E/F) and P(F/E).*

Sol. We have $S = \{HH, HT, TH, TT\}$

$E = \{HT, TH\}$ and $F = \{HH, HT, TH\}$.

$$\therefore \quad E \cap F = \{HT, TH\}$$

$$\therefore \quad P(E) = \frac{n(E)}{n(S)} = \frac{2}{4} = \mathbf{\frac{1}{2}}, \qquad P(F) = \frac{n(F)}{n(S)} = \mathbf{\frac{3}{4}},$$

$$P(E/F) = \frac{n(E \cap F)}{n(F)} = \mathbf{\frac{2}{3}} \quad \text{and} \quad P(F/E) = \frac{n(E \cap F)}{n(E)} = \frac{2}{2} = \mathbf{1}.$$

Example 4. *A die is rolled. If the outcome is an even number, what is the probability that it is a prime number ?*

Sol. Let S be the sample space of the experiment.

$\therefore \quad S = \{1, 2, 3, 4, 5, 6\}$

Let $\quad A$ = event of getting an even number

and $\quad B$ = event of getting a prime number.

$\therefore \quad A = \{2, 4, 6\}$ and $B = \{2, 3, 5\}$

Also, $\quad A \cap B = \{2\}$

Required probability = Probability of getting a prime number given that the outcome is an even number

$$= P(B/A) = \frac{P(B \cap A)}{P(A)} = \frac{P(A \cap B)}{P(A)} = \frac{1/6}{3/6} = \frac{1}{3}.$$

Alternatively,

$$P(B/A) = \frac{\text{no. of cases favourable to '} A \cap B\text{'}}{\text{no. of cases favourable to } A} = \frac{1}{3}.$$

Example 5. *A die is thrown twice and the sum of the number appearing is observed to be 6. What is the conditional probability that the number 4 has appeared at least once ?*

Sol. Let S be the sample space of the experiment.

$\therefore \quad S = \{(1, 1), (1, 2), \ldots\ldots, (6, 5), (6, 6)\}.$

Let $\quad A$ = event of getting sum 6

and $\quad B$ = event of getting 4 at least once.

$\therefore \quad A = \{(1, 5), (2, 4), (3, 3), (4, 2), (5, 1)\}$

and $\quad B = \{(4, 1), (4, 2), (4, 3), (4, 4), (4, 5), (4, 6), (1, 4), (2, 4), (3, 4), (5, 4), (6, 4)\}.$

$$\therefore \quad P(A) = \frac{5}{36} \text{ and } P(B) = \frac{11}{36}$$

Also $\quad A \cap B = \{(4, 2), (2, 4)\} \quad \therefore \quad P(A \cap B) = \frac{2}{36}$

Now, required probability

= Probability of getting 4 on at least one die given that sum in 6

$$= P(B/A) = \frac{P(B \cap A)}{P(A)} = \frac{P(A \cap B)}{P(A)} = \frac{2/36}{5/36} = \frac{2}{5}.$$

Alternatively,

$$P(B/A) = \frac{\text{no. of cases favourable to '} A \cap B\text{'}}{\text{no. of cases favourable to } A} = \frac{2}{5}.$$

Remark 1. In practical problems, it would be easier to use the formulae :

(*i*) $P(A/B) = \dfrac{\text{no. of favourable to '} A \cap B\text{'}}{\text{no. of cases favourable to } B}$

(*ii*) $P(B/A) = \dfrac{\text{no. of cases favourable to '} B \cap A\text{' } i.e., \text{'} A \cap B\text{'}}{\text{no. of cases favourable to } A}.$

Remark 2. The event $A \cap B$ is same as $B \cap A$ and each consists of sample points which are common to both A and B.

Example 6. *Find the probability that a single throw of a die results in a number less than 6 if (i) no other information is given (ii) it is given that the throw resulted in an odd number.*

Sol. Here $S = \{1, 2, 3, 4, 5, 6\}$.

Let A = event of getting number less than 6

and B = event of getting an odd number.

$\therefore$ $A = \{1, 2, 3, 4, 5\}$ and $B = \{1, 3, 5\}$.

(*i*) $P(\text{no. is less than 6}) = P(A) = \dfrac{\mathbf{5}}{\mathbf{6}}$

(*ii*) P(no. is less than 6 given that the number obtained is odd)

$$= P(A/B) = \frac{P(A \cap B)}{P(B)} = \frac{3/6}{3/6} = \mathbf{1}. \qquad (\because\ A \cap B = \{1, 3, 5\})$$

Alternatively,

$$P(A/B) = \frac{\text{no. of cases favourable to } A \cap B}{\text{no. of cases favourable to } B} = \frac{n\{1,3,5,\}}{n\{1,3,5\}} = \frac{3}{3} = 1.$$

Example 7. *Weather records show that the probability of high barometric pressure is 0.82 and the probability of rain and high barometric pressure is 0.20. Find the probability of rain given high barometric pressure.*

Sol. Let A and B denote the events 'rain' and 'high barometric pressure' respectively

$\therefore$ $P(B) = 0.82,\ P(A \cap B) = 0.20$

$\therefore$ Required probability

$$= P(A/B) = \frac{P(A \cap B)}{P(B)} = \frac{0.20}{0.82} = \frac{\mathbf{10}}{\mathbf{41}}.$$

Example 8. *One card in drawn from a well shuffled pack of 52 cards. If E is the event "the card drawn is either a king or an ace" and F is the event "the card drawn is either an ace or a jack", then find the probability of the conditional event E/F.*

Sol. There are 4 kings and 4 aces in the pack.

$$\therefore \quad P(E) = \frac{4+4}{52} = \frac{2}{13}$$

There are 4 aces and 4 jacks in the pack.

$$\therefore \quad P(F) = \frac{4+4}{52} = \frac{2}{13}$$

The event $E \cap F$ contains 4 aces.

$$\therefore \quad P(E \cap F) = \frac{4}{52} = \frac{1}{13}$$

$$\therefore \quad \text{Required probability} = P(E/F) = \frac{P(E \cap F)}{P(F)} = \frac{1/13}{2/13} = \frac{\mathbf{1}}{\mathbf{2}}.$$

Example 9. *In a certain school, 20% of the students failed in English, 15% of the students failed in Mathematics, and 10% of the students failed both in English and Mathematics. A student is selected at random. If he failed in English, what is the probability that he also failed in Mathematics ?*

Sol. Let E and M be the events that a students is fail in English and Mathematics respectively.

$\therefore$ $$P(E) = \frac{20}{100} = \frac{1}{5}, \quad P(M) = \frac{15}{100} = \frac{3}{20} \quad \text{and} \quad P(E \cap M) = \frac{10}{100} = \frac{1}{10}.$$

Required probability $= P(M/E)$

$$= \frac{P(M \cap E)}{P(E)} = \frac{P(E \cap M)}{P(E)} = \frac{1/10}{1/5} = \frac{1}{2}.$$

Example 10. *Three fair coins are tossed. Find the probability that they are all tails, if :*

(i) at least one of the coins show tail *(ii) two coins show tail*

(iii) at least two coins show head *(iv) at most one coin show head.*

Sol. Here $S = \{HHH, HHT, HTH, THH, HTT, THT, TTH, TTT\}$.

Let A = event of getting all tails $\therefore$ $A = \{TTT\}$

(*i*) Let B = event that at least one of the coins show tail

$\therefore$ $B = \{HHT, HTH, THH, HTT, THT, TTH, TTT\}$

$\therefore$ $A \cap B = \{TTT\}$

$\therefore$ Required probability $= P(A/B) = \dfrac{n(A \cap B)}{n(B)} = \dfrac{1}{7}$.

(*ii*) Let B = event that two coins show tail

$\therefore$ $B = \{HTT, THT, TTH\}$

$\therefore$ $A \cap B = \phi$

$\therefore$ Required probability $= \dfrac{n(A \cap B)}{n(B)} = \dfrac{0}{3} = \mathbf{0}$.

(*iii*) Let B = event that at least two coins show head

$\therefore$ $B = \{HHH, HHT, HTH, THH\}$

$\therefore$ $A \cap B = \phi$

$\therefore$ Required probability $= P(A/B) = \dfrac{n(A \cap B)}{n(B)} = \dfrac{0}{4} = \mathbf{0}$.

(*iv*) Let B = event that at most one coin show head.

$\therefore$ $B = \{HTT, THT, TTH, TTT\}$

$\therefore$ $A \cap B = \{TTT\}$

$\therefore$ Required probability $= P(A/B) = \dfrac{n(A \cap B)}{n(B)} = \dfrac{1}{4}$.

Remark. The value of P(A/B) is equal to $\dfrac{n(A \cap B)}{n(B)}$ which is also equal to $\dfrac{P(A \cap B)}{P(B)}$.

EXERCISE 5.4

1. If P (not A) = 0.7, $P(B) = 0.7$ and $P(B/A) = 0.5$, then find $P(A/B)$ and $P(A \cup B)$.
2. If $P(A) = 0.4, P(B) = 0.8, P(B/A) = 0.6$, find $P(A/B)$ and $P(A \cup B)$.
3. For two events A and B, $P(A) = 0.5, P(B) = 0.6$ and $P(A \cap B) = 0.3$. Find the conditional probabilities $P(A/B)$ and $P(B/A)$.

4. (i) If $A \subseteq B$, show that $P(B/A) = 1$ (ii) If $A \cap B = \phi$, show that $P(B/A) = 0$.
(iii) If $A \subseteq B$, show that $P(A/C) \leq P(B/C)$.

5. A die is thrown. Find the probability that the number obtained is greater than 2 if :
(i) no other information is given (ii) it is given that the number obtained is less than 5.

6. Find the probability of drawing a king, when a card is drawn from a well suffled pack of cards. It is also given that the card drawn is a face card.

7. A pair of dice is thrown. If the sum is seven, find the probability that one of the dice shows 3.

8. A pair of fair dice is thrown. If the two numbers appearing are different, find the probability that the sum is 4 or less.

9. A die is thrown twice and the sum of the numbers appearing is observed to be 7. What is the conditional probability that the number 2 has appeared at least once ?

10. The probability that a person stopping at a petrol pump will ask to have his tyres checked is 0.12, the probability that he will ask to have his oil checked is 0.29 and the probability that he will ask to have both of them checked is 0.07.
(i) What is the probability that a person who has oil checked will also have tyre checked ?
(ii) What is the probability that a person stopping at the petrol pump will have either tyres or oil checked ?

11. A fair coin is tossed four times. Find the probability that there are all heads if the first two tosses result in head. *(KU 2006 C)*

Answers

1. 0.2143, 0.85	2. 0.3, 0.96	3. 0.5, 0.6	
5. (i) $\frac{2}{3}$	(ii) $\frac{1}{2}$	6. $\frac{1}{3}$	7. $\frac{1}{3}$
8. $\frac{2}{15}$	9. $\frac{1}{3}$	10. (i) $\frac{7}{29}$ (ii) $\frac{17}{50}$	11. $\frac{1}{4}$.

5.15. INDEPENDENT EVENTS

Let A and B be two events associated with a random experiment. We know that

$$P(B/A) = \frac{P(B \cap A)}{P(A)} = \frac{P(A \cap B)}{P(A)}$$

$$\therefore \quad P(A \cap B) = P(A)\, P(B/A).$$

In general $P(B/A)$ may or may not be equal to $P(B)$. When $P(B/A)$ and $P(B)$ are equal, then the events A and B are of special importance.

Two events associated with a random experiment are said to be **independent events** if the occurrence or non-occurrence of one event does not affect the probability of the occurrence of the other event. For example, the events A and B are independent events when $P(A/B) = P(A)$ and $P(B/A) = P(B)$.

Theorem. Let A and B be events associated with a random experiment. Prove that the events A and B are independent if and only if

$$\mathbf{P(A \cap B) = P(A)\ P(B).}$$

Proof. Let A and B be independent events.

$$\therefore \quad P(A \cap B) = \frac{P(A \cap B)}{P(B)} \times P(B) = P(A/B)\, P(B) = P(A)\, P(B) \quad [\because \ P(A/B) = P(A)]$$

$$\therefore \quad P(A \cap B) = P(A)\, P(B).$$

Conversely, let $P(A \cap B) = P(A)\,P(B)$.

$$\therefore \quad P(A/B) = \frac{P(A \cap B)}{P(B)} = \frac{P(A)\,P(B)}{P(B)} = P(A)$$

and
$$P(B/A) = \frac{P(B \cap A)}{P(A)} = \frac{P(A \cap B)}{P(A)} = \frac{P(A)\,P(B)}{P(A)} = \text{P(B)}.$$

$$\therefore \quad P(A/B) = P(A) \text{ and } P(B/A) = P(B).$$

$\therefore$ A and B are independent events.

Remark 1. $P(A \cap B) = P(A)\,P(B)$ is the necessary and sufficient condition for the events A and B to be independent.

Remark 2. Let A and B be events associated with a random experiment.

(*i*) Let A and B be *m.e.* $\therefore$ $P(A \cap B) = 0$

$\therefore$ $P(A \cap B) \neq P(A)\,P(B)$ *i.e.*, A and B are not independent events.

$\therefore$ **Mutually exclusive events cannot be independent.**

(*ii*) Let A and B be independent.

$\therefore$ $P(A \cap B) = P(A)\,P(B)$ *i.e.*, $P(A \cap B) \neq 0$.

$\therefore$ A and B are not *m.e.* events.

$\therefore$ **Independent events cannot be mutually exclusive.**

Important observation. If A and B be any two events associated with a random experiment, then their physical description is not sufficient to decide if A and B are independent events or not. A and B are declared to be independent events only when we have

$$P(A \cap B) = P(A)\,P(B).$$

5.16. DEPENDENT EVENTS

Let A and B be two events associated with a random experiment. If A and B are not independent events, then these are called **dependent events.**

$\therefore$ In case of dependent events, we have $\mathbf{P(A \cap B) = P(A)\,P(B/A)}$.

Theorem. Let A and B be events associated with a random experiment. If A and B are independent, then show that the events (*i*) $\overline{\mathbf{A}}$, B (*ii*) A, $\overline{\mathbf{B}}$ (*iii*) $\overline{\mathbf{A}}, \overline{\mathbf{B}}$ are also independent.

Proof. The events A and B are independent.

$$\therefore \quad P(A \cap B) = P(A)\,P(B) \quad \text{...(1)}$$

(*i*) $(A \cap B) \cap (\overline{A} \cap B) = (A \cap \overline{A}) \cap (B \cap B) = \phi \cap B = \phi$

and $(A \cap B) \cup (\overline{A} \cap B) = (A \cup \overline{A}) \cap B = S \cap B = B$

$\therefore$ The events $A \cap B$ and $\overline{A} \cap B$ are *m.e.* and their union in B.

$\therefore$ By *addition theorem*, we have

$$P(B) = P(A \cap B) + P(\overline{A} \cap B).$$

$$\Rightarrow \quad P(\overline{A} \cap B) = P(B) - P(A \cap B) = P(B) - P(A)\,P(B) \quad \text{[Using (1)]}$$

$$= (1 - P(A))\,P(B) = P(\overline{A})\,P(B).$$

$\therefore$ $\mathbf{P(\overline{A} \cap B) = P(\overline{A})\,P(B)}$ *i.e.*, $\overline{A}$ and B are independent.

(*ii*) $(A \cap B) \cap (A \cap \overline{B}) = (A \cap A) \cap (B \cap \overline{B}) = A \cap \phi = \phi$

and $(A \cap B) \cup (A \cap \overline{B}) = A \cap (B \cup \overline{B}) = A \cap S = A$

$\therefore$ The events $A \cap B$ and $A \cap \overline{B}$ are *m.e.* and their union is A.

$\therefore$ By *addition theorem*, we have

$$P(A) = P(A \cap B) + P(A \cap \bar{B}).$$

$$\Rightarrow \quad P(A \cap \bar{B}) = P(A) - P(A \cap B) = P(A) - P(A)\,P(B) \qquad \text{[Using (1)]}$$

$$= P(A)(1 - P(B)) = P(A)\,P(\bar{B}).$$

$\therefore$ $\mathbf{P(A \cap \bar{B}) = P(A)\,P(\bar{B})}$ *i.e.*, A and $\bar{B}$ are independent.

(*iii*) $(\bar{A} \cap B) \cap (\bar{A} \cap \bar{B}) = (\bar{A} \cap \bar{A}) \cap (B \cap \bar{B}) = \bar{A} \cap \phi = \phi$

and $\quad (\bar{A} \cap B) \cup (\bar{A} \cap \bar{B}) = \bar{A} \cap (B \cup \bar{B}) = \bar{A} \cap S = \bar{A}$

$\therefore$ The events $\bar{A} \cap B$ and $\bar{A} \cap \bar{B}$ are *m.e.* and their union is $\bar{A}$.

$\therefore$ By *addition theorem*, we have

$$P(\bar{A}) = P(\bar{A} \cap B) + P(\bar{A} \cap \bar{B}) \qquad ...(1)$$

$$\Rightarrow \quad P(\bar{A} \cap \bar{B}) = P(\bar{A}) - P(\bar{A} \cap B) = P(\bar{A}) - P(\bar{A})\,P(B) \qquad \text{[Using part } (i)]$$

$$= P(\bar{A})\,(1 - P(B)) = P(\bar{A})\,P(\bar{B}).$$

$\therefore$ $\quad \mathbf{P(\bar{A} \cap \bar{B}) = P(\bar{A})\,P(\bar{B})}$ *i.e.*, $\bar{A}$ and $\bar{B}$ are independent.

Example 1. *If A and B are independent events such that $P(A \cup B) = 0.6$ and $P(A) = 0.2$, find $P(B)$.*

Sol. We have $P(A \cup B) = 0.6$ and $P(A) = 0.2$.

By *addition theorem*, we have

$$P(A \cup B) = P(A) + P(B) - P(A \cap B)$$

$$\therefore \quad P(A \cup B) = P(A) + P(B) - P(A)\,P(B) \qquad (\because \; A \text{ and } B \text{ are independent})$$

$$\Rightarrow \quad 0.6 = 0.2 + P(B) - (0.2)\,P(B)$$

$$\Rightarrow \quad 0.4 = P(B)\,(1 - 0.2)$$

$$\Rightarrow \quad (0.8)\,P(B) = 0.4 \;\Rightarrow\; P(B) = \frac{0.4}{0.8} = \frac{1}{2} = \mathbf{0.5}.$$

Example 2. *A pair of dice is thrown. A is the event : "the sum is 8" and B is the event :"at least one odd number is obtained". Show that the events A and B are dependent.*

Sol. Let S be the sample space.

$$\therefore \quad S = \{(1, 1), (1, 2), ..., (6, 5), (6, 6)\}.$$

$$A = \{(2, 6), (3, 5), (4, 4), (5, 3), (6, 2)\}$$

$B = \{(1, 1), (1, 2), (1, 3), (1, 4), (1, 5), (1, 6), (3, 1), (3, 2), (3, 3), (3, 4), (3, 5), (3, 6), (5, 1), (5, 2), (5, 3), (5, 4), (5, 5), (5, 6), (2, 1), (2, 3), (2, 5), (4, 1), (4, 3), (4, 5), (6, 1), (6, 3), (6, 5)\}$.

$$\therefore \quad P(A) = \frac{5}{36} \quad \text{and} \quad P(B) = \frac{27}{36} = \frac{3}{4}$$

Also, $\quad A \cap B = \{(3, 5), (5, 3)\}$. $\quad \therefore \quad P(A \cap B) = \dfrac{2}{36} = \dfrac{1}{18}$

$$P(A)\,P(B) = \frac{5}{36} \times \frac{3}{4} = \frac{5}{48} \neq P(A \cap B).$$

The events A and B are **dependent.**

Example 3. *A coin is tossed thrice and all the eight outcomes are assumed equally likely. In which of the following cases are the events E and F independent ?*

(i) E : the first throw results in head.

F : the last throw results in tail.

(ii) E : the number of heads is two.

F : the last throw results in head.

Sol. Let S be the sample space.

$\therefore \quad S = \{HHH, HHT, HTH, THH, HTT, THT, TTH, TTT\}$.

(*i*) Here $\quad E = \{HHH, HHT, HTH, HTT\}$ and $F = \{HHT, HTT, THT, TTT\}$.

$$\therefore \quad P(E) = \frac{4}{8} = \frac{1}{2} \quad \text{and} \quad P(F) = \frac{4}{8} = \frac{1}{2}$$

There are 2 cases favourable to the event $E \cap F$, namely HHT and HTT.

$$\therefore \quad P(E \cap F) = \frac{2}{8} = \frac{1}{4}$$

Also, $P(E)P(F) = \frac{1}{2} \times \frac{1}{2} = \frac{1}{4} = P(E \cap F)$.

$\therefore$ The events E and F are **independent.**

(*ii*) Here $\quad E = \{HHT, HTH, THH\}$ and $F = \{HHH, HTH, THH, TTH\}$.

$$\therefore \quad P(E) = \frac{3}{8} \quad \text{and} \quad P(F) = \frac{4}{8} = \frac{1}{2}$$

There are 2 cases favourable to the event $E \cap F$, namely HTH and THH.

$$\therefore \quad P(E \cap F) = \frac{2}{8} = \frac{1}{4}$$

Also, $P(E)\,P(F) = \frac{3}{8} \times \frac{1}{2} = \frac{3}{16} \neq P(E \cap F)$.

$\therefore$ The events E and F are **not independent.**

EXERCISE 5.5

1. The events A and B are given to be independent. Find $P(B)$ if it is given that $P(A) = 0.35$ and $P(A \cup B) = 0.60$.
2. For two events A and B, let $P(A) = 0.4$ and $P(B) = p$ and $P(A \cup B) = 0.6$.
   (*i*) Find p so that A and B are mutually exclusive.
   (*ii*) Find p so that A and B are independent.
3. (*i*) If $P(\text{not } B) = 0.65$, $P(A \cup B) = 0.85$ and A and B are independent events, find $P(A)$.
   (*ii*) If $P(\text{not } A) = 0.4$, $P(A \cup B) = 0.75$ and A, B are given to be independent events, find the value of $P(B)$.
4. A coin is tossed twice and all possible outcomes are assumed to be equally likely. E is the event : "at most one head has occurred" and F is the event : "at most one tail has occurred". Show that the events E and F are not independent.

5. A coin is tossed three times and all possible outcomes are assumed to be equally likely. E is the event : "both heads and tails have occurred", and F is the event : "at most one tail has occurred". Show that the events E and F are independent.
6. One card is drawn from a pack of 52 cards so that each card is equally likely to be selected. A is the event : "the card is a heart" and B is the event : "the card is a king." Show that A and B are independent.
7. One card is drawn from a pack of 52 cards so that each card is equally likely to be selected. A is the event : "the card is a spade" and B is the event : "the card is red". Show that A and B are not independent.
8. A die is thrown and the 6 possible outcomes are assumed to be equally likely. If E is the event : "the number appearing is a multiple of 3", and F is the event : "the number appearing is even". Show that the events E and F are independent.
9. One card is drawn from a pack of 52 cards so that each card is equally likely to be selected. In which of the following cases are the events E and F are independent ?

   (*i*) E : the card drawn is a spade.
   F : the card drawn is an ace.

   (*ii*) E : the card drawn is a king or queen.
   F : the card drawn is a queen or jack.

Answers

1. $\frac{5}{13}$ 2. (*i*) $\frac{1}{5}$ (*ii*) $\frac{1}{3}$

3. (*i*) $\frac{10}{13}$ (*ii*) $\frac{3}{8}$ 9. (*i*).

5.17. INDEPENDENT EXPERIMENTS

Two random experiments are said to be **independent experiments** if the occurrence or non-occurrence of an event in one experiment does not in any way affect the probability of occurrence of any event in the other experiment. For example, two tosses of a coin are independent experiments.

5.18. MULTIPLICATION THEOREM

If A and B be events associated with independent random experiments E_1 and E_2 respectively, then prove that

$$P(AB) = P(A)P(B).$$

Proof. Since the random experiments E_1 and E_2 are independent, the sample spaces of the experiments are not affected by the events.

Let n_1 and n_2 be the numbers of exhaustive, equally likely cases in the first and second experiment respectively.

Let m_1 be the number of cases favourable to the happening of the event A out of n_1 cases of the first experiment.

$$\therefore \quad P(A) = \frac{m_1}{n_1}$$

Let m_2 be the number of cases favourable to the happening of the event B out of n_2 cases of the second experiment.

$$\therefore \quad P(B) = \frac{m_2}{n_2}$$

By the **fundamental principle of events**, the number of cases favourable to the happening of the event AB in this specified order is m_1m_2. Also the number of exhaustive, equally likely cases in the combined experiment is n_1n_2.

$$\therefore \qquad P(AB) = \frac{m_1m_2}{n_1n_2} = \frac{m_1}{n_1} \cdot \frac{m_2}{n_2} = P(A)P(B)$$

$$\therefore \qquad \mathbf{P(AB) = P(A)\ P(B).}$$

The theorem can also be extended to even more than two events.

Let $A_1, A_2, ..., A_k$ be k events associated with k independent random experiments $E_1, E_2, ..., E_k$ with probabilities $p_1, p_2, ..., p_k$ respectively, then

$$P(A_1A_2, ..., A_k) = P(A_1)\ P(A_1) ......P(A_k) = p_1\, p_2 ...... p_k.$$

Also $\quad P(\text{not } A_1) = P(\bar{A}_1) = 1 - p_1$

$P(\text{not } A_2) = P(\bar{A}_2) = 1 - p_2$

....

....

$P(\text{not } A_k) = P(\bar{A}_k) = 1 - p_k.$

Since $A_1, A_2, ..., A_k$ are associated with independents experiments, therefore, the events $\bar{A}_1, \bar{A}_2, ..., \bar{A}_k$ are also associated with independent experiments.

Now, P(event of happening of at least one of $A_1, A_2, ..., A_k$)

$= 1 - P$(event of happening of none of $A_1, A_2, ..., A_k$)

$= 1 - P(\bar{A}_1 \bar{A}_2 ...... \bar{A}_k) = 1 - P(\bar{A}_1)\, P(\bar{A}_2) ......P(\bar{A}_k)$

(*By Multiplication theorem*)

$= 1 - (1 - p_1)(1 - p_2) ...... (1 - p_k).$

Remark. If A and B are events associated with experiments which are not independent, then the probability of the event 'AB' is found by using the result :

$$\mathbf{P(AB) = P(A)\ P(B/A).}$$

This result can also be extended to more than two experiments.

Example 1. *A husband and a wife appear in an interview for two vacancies for the same post. The probability of husband's selection is 3/5 and that of wife's selection is 1/5. Find the probability that :*

(*i*) *both are selected* (*ii*) *exactly one is selected*

(*iii*) *none is selected.*

Sol. The random experiments 'interview of husband' and 'interview of wife' are independent.

Let $\quad H$ = event that 'Husband' is selected

and $\quad W$ = event that 'Wife' is selected.

$$\therefore \qquad P(H) = \frac{3}{5} \quad \text{and} \quad P(W) = \frac{1}{5}$$

Also, $\quad P(\bar{H}) = 1 - P(H) = 1 - \frac{3}{5} = \frac{2}{5} \quad$ and $\quad P(\bar{W}) = 1 - P(W) = 1 - \frac{1}{5} = \frac{4}{5}$

(*i*) P(both are selected)

$= P(HW) = P(H)P(W)$ (Using multiplication theorem)

$$= \frac{3}{5} \times \frac{1}{5} = \mathbf{\frac{3}{25}}.$$

(*ii*) P(exactly one is selected)

$$= P(H\overline{W} \cup \overline{H}W) = P(H\overline{W}) + P(\overline{H}W) \quad \text{(Using addition theorem)}$$

$$= P(H)P(\overline{W}) + P(\overline{H})P(W) \quad \text{(Using multiplication theorem)}$$

$$= \frac{3}{5} \times \frac{4}{5} + \frac{2}{5} \times \frac{1}{5} = \mathbf{\frac{14}{25}}.$$

(*iii*) P(none is selected) $= P(\overline{H}\,\overline{W}) = P(\overline{H})\,P(\overline{W})$ (Using multiplication theorem)

$$= \frac{2}{5} \times \frac{4}{5} = \mathbf{\frac{8}{25}}.$$

Note. In this example, it is worth while to note that $\frac{3}{25} + \frac{14}{25} + \frac{8}{25} = \frac{25}{25} = 1.$

This has happened because the events : both are selected, exactly one is selected and none is selected are *mutually exclusive* and *exhaustive*.

Example 2. *The odds in favour of one student passing a test are 3 : 7. The odds against another student passing it are 3 : 5. What is the probability that both pass the test ?*

Sol. Let A = event that first pass the test

$$\therefore \quad P(A) = \frac{3}{3+7} = \frac{3}{10}$$

Let B = event the second pass the test

$$\therefore \quad P(B) = \frac{5}{3+5} = \frac{5}{8}$$

The random experiments of results of students are independent.

$\therefore$ P(both pass the test) $= P(AB) = P(A)P(B) = \frac{3}{10} \times \frac{5}{8} = \mathbf{\frac{3}{16}}.$

Example 3. *A class consists of 100 students. Out of these 25 are girls and 75 are boys. 10 of them are rich, remaining poor. Out of 100, only 20 students are fair complexioned. What is the probability of selecting a fair complexioned rich girl ?*

Sol. Let A = event that a girl student is selected

B = event that a rich student is selected

C = event that a fair complexioned student is selected.

$$\therefore \quad P(A) = \frac{25}{100} = \frac{1}{4}, \quad P(B) = \frac{10}{100} = \frac{1}{10} \quad \text{and} \quad P(C) = \frac{20}{100} = \frac{1}{5}.$$

Now, ABC is the event of selecting a rich, fair, complexioned girls student. The random experiments of selections are independent.

$$\therefore \quad P(ABC) = P(A)\,P(B)\,P(C) = \frac{1}{4} \times \frac{1}{10} \times \frac{1}{5} = \frac{1}{200} = \mathbf{0.005}.$$

Example 4. *A speaks truth in 60% of the cases and B in 90% of the cases. In what percentage of cases, are they likely to contradict each other in stating the same fact ?*

Sol. The random experiments of speeches of A and B are independent.

Let E = event of A speaking truth

and F = event of B speaking truth.

$\therefore \qquad P(E) = \frac{60}{100} = \frac{6}{10} \quad \text{and} \quad P(F) = \frac{90}{100} = \frac{9}{10}$

Probability of A and B contradicting each other $= P(E\bar{F} \text{ or } \bar{E}F)$

$$= P(E\bar{F}) + P(\bar{E}F) = P(E)\,P(\bar{F}) + P(\bar{E})\,P(F)$$

$$= P(E)(1 - P(F)) + (1 - P(E))P(F)$$

$$= \frac{6}{10}\left(1 - \frac{9}{10}\right) + \left(1 - \frac{6}{10}\right)\frac{9}{10} = \frac{6}{10} \times \frac{1}{10} + \frac{4}{10} \times \frac{9}{10} = \frac{42}{100} = 42\%$$

$\therefore$ A and B are likely to contradicts each other in **42%** cases.

Example 5. *A problem in statistics is given to five students A, B, C, D and E. Their chances of solving the problem are* $\frac{1}{2}, \frac{1}{3}, \frac{1}{4}, \frac{1}{5}$ *and* $\frac{1}{6}$ *respectively. What is the probability that the problem will be solved ?*

Sol. The random experiments of trying the problem by the given students are independent.

Let A_1 = event that A fails to solve the problem
A_2 = event that B fails to solve the problem
A_3 = event that C fails to solve the problem
A_4 = event that D fails to solve the problem
A_5 = event that E fails to solve the problem.

$\therefore \qquad P(A_1) = 1 - \frac{1}{2} = \frac{1}{2} \qquad P(A_2) = 1 - \frac{1}{3} = \frac{2}{3}$

$P(A_3) = 1 - \frac{1}{4} = \frac{3}{4} \qquad P(A_4) = 1 - \frac{1}{5} = \frac{4}{5}$

$P(A_5) = 1 - \frac{1}{6} = \frac{5}{6}$

Now, P(event that the problem is solved by at least one student)

$= 1 - P$(event that the problem is not solved by any of the five students)

$$= 1 - P(A_1 A_2 A_3 A_4 A_5)$$

$$= 1 - P(A_1)\,P(A_2)\,P(A_3)\,P(A_4)\,P(A_5) \qquad \text{(By Multiplication Theorem)}$$

$$= 1 - \left(\frac{1}{2} \times \frac{2}{3} \times \frac{3}{4} \times \frac{4}{5} \times \frac{5}{6}\right) = 1 - \frac{1}{6} = \frac{\mathbf{5}}{\mathbf{6}}.$$

Example 6. *Three bags contain 7 white 8 red, 9 white 6 red and 5 white 7 red balls respectively. One ball, at random, is drawn from each bag. Find the probability that all of them are of the same colour.*

Sol. The three random experiments of drawing balls from given bags are independent.

Bag I	Bag II	Bag III
7 White	9 White	5 White
8 Red	6 Red	7 Red

Let W_i and R_i be the events of drawing white ball and red ball respectively from the ith bag, $i = 1, 2, 3$.

Required probability $= P$(all balls of same colour)

$$= P(W_1 W_2 W_3 \text{ or } R_1 R_2 R_3)$$

$$= P(W_1 W_2 W_3) + P(R_1R_2R_3)$$
$$= P(W_1)P(W_2)P(W_3) + P(R_1)P(R_2)P(R_3)$$
$$= \frac{7}{7+8} \times \frac{9}{9+6} \times \frac{5}{5+7} + \frac{8}{7+8} \times \frac{6}{9+6} \times \frac{7}{5+7}$$
$$= \frac{7}{15} \times \frac{9}{15} \times \frac{5}{12} + \frac{8}{15} \times \frac{6}{15} \times \frac{7}{12} = \frac{651}{2700} = \mathbf{\frac{217}{900}}.$$

Example 7. *A can hit a target 3 times in 5 shots, B 2 times in 5 shots, C 3 times in 4 shots. Each fire a volley. What is the probability that 2 shots hit the target ?*

Sol. The three random experiments of hitting target by A, B, C are independent.

Let A, B, C be the events that A, B, C respectively hit the target.

$$\therefore \quad P(A) = \frac{3}{5}, \quad P(B) = \frac{2}{5} \quad \text{and} \quad P(C) = \frac{3}{4}$$

$$\therefore \quad P(\overline{A}) = 1 - \frac{3}{5} = \frac{2}{5}, \quad P(\overline{B}) = 1 - \frac{2}{5} = \frac{3}{5} \quad \text{and} \quad P(\overline{C}) = 1 - \frac{3}{4} = \frac{1}{4}$$

Required probability $= P(\text{2 shots hit the target})$

$$= P(AB\overline{C} \text{ or } A\overline{B}C \text{ or } \overline{A}BC)$$
$$= P(AB\overline{C}) + P(A\overline{B}C) + P(\overline{A}BC)$$
$$= P(A)P(B)P(\overline{C}) + P(A)P(\overline{B})P(C) + P(\overline{A})P(B)P(C)$$
$$= \frac{3}{5} \times \frac{2}{5} \times \frac{1}{4} + \frac{3}{5} \times \frac{3}{5} \times \frac{3}{4} + \frac{2}{5} \times \frac{2}{5} \times \frac{3}{4}$$
$$= \frac{6}{100} + \frac{27}{100} + \frac{12}{100} = \mathbf{\frac{9}{20}}.$$

Example 8. *A bag has 4 red and 5 black balls, a second bag has 3 red and 7 black balls. One ball is drawn from the first and two from the second. Find the probability that out of three balls, two are black and one is red.*

Sol. The random experiments of 'drawing one ball from first bag' and 'drawing two balls from second bag' are independent.

We are to find the probability two black balls and one red ball.

Bag I	Bag II
4 Red 5 Black	3 Red 7 Black

$\therefore$ Required probability

$$= P(F_r\, S_{bb} \text{ or } F_b\, S_{rb}),$$

where F_r is the event of drawing red ball from the first bag etc.

$$= P(F_rS_{bb}) + P(F_bS_{rb}) = P(F_r)\,P(S_{bb}) + P(F_b)\,P(S_{rb})$$
$$= \frac{4}{4+5} \times \frac{{}^7C_2}{{}^{3+7}C_2} + \frac{5}{4+5} \times \frac{{}^3C_1\,{}^7C_1}{{}^{3+7}C_2}$$
$$= \frac{4}{9} \times \frac{\frac{7\times 6}{1\times 2}}{\frac{10\times 9}{1\times 2}} + \frac{5}{9} \times \frac{3\times 7}{\frac{10\times 9}{1\times 2}} = \frac{4}{9} \times \frac{7}{15} + \frac{5}{9} \times \frac{7}{15} = \frac{7}{15}.$$

Example 9. *In a hockey match, the probability of winning of Indian team against Pakistani team in 1/4. Three matches are played. Find the probability that :*

(*i*) *India looses all the matches.*

(*ii*) *India wins at least one match.*

(*iii*) *India wins two matches.*

Sol. The random experiments of matches between Indian team and Pakistani team are independent.

Let A_i to the event of winning of Indian team in the ith match, $i = 1, 2, 3$.

$$\therefore \qquad P(A_i) = \frac{1}{4} \quad \text{and} \quad P(\overline{A}_i) = 1 - \frac{1}{4} = \frac{3}{4}$$

(*i*) P(India loosing all the matches)

$$= P(\overline{A}_1\ \overline{A}_2\ \overline{A}_3) = P(\overline{A}_1)\ P(\overline{A}_2)\ P(\overline{A}_3) = \frac{3}{4} \times \frac{3}{4} \times \frac{3}{4} = \mathbf{\frac{27}{64}}$$

(*ii*) P(India winning at least one match)

$$= 1 - \text{P(India loosing all the matches)}$$

$$= 1 - \frac{27}{64} = \mathbf{\frac{37}{64}} \qquad \text{[Using part (}i\text{)]}$$

(*iii*) P(India winning two matches)

$$= P(A_1\ A_2\ \overline{A}_3 \text{ or } A_1 \overline{A}_2\ A_3 \text{ or } \overline{A}_1\ A_2\ A_3)$$

$$= P(A_1\ A_2\ \overline{A}_3) + P(A_1\ \overline{A}_2\ A_3) + P(\overline{A}_1\ A_2\ A_3)$$

$$= P(A_1)\ P(A_2)\ P(\overline{A}_3) + P(A_1)\ P(\overline{A}_2)\ P(A_3) + P(\overline{A}_1)\ P(A_2)\ P(A_3)$$

$$= \frac{1}{4} \times \frac{1}{4} \times \frac{3}{4} + \frac{1}{4} \times \frac{3}{4} \times \frac{1}{4} + \frac{3}{4} \times \frac{1}{4} \times \frac{1}{4} = \mathbf{\frac{9}{64}}.$$

Example 10. *Three groups of children consists of respectively 3 girls and 1 boy, 2 girls and 2 boys, 1 girl and 3 boys. One child is selected at random from each group. Find the chance that three selected children comprise 1 girl and 2 boys.*

Sol. The random experiments of drawing one child from each group are independent.

3 Girls, 1 Boy	2 Girls, 2 Boys	1 Girl, 3 Boys
Group I	Group II	Group III

Let G_i and B_i be the events of selecting a girl and a boy respectively from the ith group, $i = 1, 2, 3$.

$\therefore \quad P$(1 girl and 2 boys)

$$= P(G_1B_2B_3 \text{ or } B_1G_2B_3 \text{ or } B_1B_2G_3)$$

$$= P(G_1B_2B_3) + P(B_1G_2B_3) + P(B_1B_2G_3)$$

$$= P(G_1)P(B_2)P(B_3) + P(B_1)P(G_2)P(B_3) + P(B_1)P(B_2)P(G_3)$$

$$= \left(\frac{3}{4} \times \frac{2}{4} \times \frac{3}{4}\right) + \left(\frac{1}{4} \times \frac{2}{4} \times \frac{3}{4}\right) + \left(\frac{1}{4} \times \frac{2}{4} \times \frac{1}{4}\right) = \frac{18 + 6 + 2}{64} = \mathbf{\frac{13}{32}}.$$

Example 11. *A six faced die is so biased that it is twice as likely to show an even number than an odd number, when thrown. It is thrown twice. What is the probability that the sum of the two numbers thrown is even ?*

Sol. Let E and O be the events of getting an even number (2 or 4 or 6) and an odd number (1 or 3 or 5) respectively.

$$\therefore \qquad P(E) + P(O) = 1 \qquad \text{and} \quad P(E) = 2 \times P(O)$$

$$\therefore \qquad P(E) = 2/3 \qquad \text{and} \quad P(O) = 1/3$$

Now, the sum of two numbers is even if either both are even or both are odd.

The random experiments of throwing die twice are independent experiments.

Let E_i, O_i be the events of throwing an even number, an odd number respectively in the ith throw, $i = 1, 2$.

$$P(\text{sum is even}) = P(E_1E_2 \text{ or } O_1O_2)$$
$$= P(E_1E_2) + P(O_1O_2) = P(E_1)P(E_2) + P(O_1)P(O_2)$$
$$= \left(\frac{2}{3} \times \frac{2}{3}\right) + \left(\frac{1}{3} \times \frac{1}{3}\right) = \frac{4}{9} + \frac{1}{9} = \mathbf{\frac{5}{9}}.$$

Example 12. *A bag contains 6 white ball and 4 black balls. Two balls are drawn at random one by one without replacement. Find the probability that both balls are white.*

Sol. Since the first ball is not replaced before the second draw, the random experiments of drawing balls are not independent.

Let W_1 = event that first ball is white

W_2 = event that second ball is white

P(both balls are white)

$$= P(W_1W_2) = P(W_1)P(W_2/W_1)$$
$$= \frac{6}{6+4} \times \frac{6-1}{(6-1)+4} = \frac{6}{10} \times \frac{5}{9} = \mathbf{\frac{1}{3}}.$$

Example 13. *Three balls are drawn successively from an urn containing 6 red balls, 4 white ball and 5 blue balls. Find the probability that these are drawn in the order red, white and blue, if each ball is not replaced.*

Sol. Since the balls are not replaced before successive draws, the three random experiments of drawing balls are not independent.

Let us define events :

A = first ball is red, B = second ball is white and C = third ball is blue.

Now ABC is the event of drawing three balls in the order red, white and blue.

By Multiplication rule,

$$P(ABC) = P(A)\, P(B/A)\, P(C/AB).$$

Now $$P(A) = \frac{\text{no. of red balls}}{\text{total no. of balls}} = \frac{6}{6+4+5} = \frac{6}{15} = \frac{2}{5}$$

$$P(B/A) = \frac{\text{no. of white balls}}{\text{total no. of balls left}} = \frac{4}{5+4+5} = \frac{4}{14} = \frac{2}{7}$$

$$P(C/AB) = \frac{\text{no. of blue balls}}{\text{total no. of balls left}} = \frac{5}{5+3+5} = \frac{5}{13}$$

$$\therefore \quad P(ABC) = \frac{2}{5} \times \frac{2}{7} \times \frac{5}{13} = \mathbf{\frac{4}{91}}.$$

Example 14. *From a pack of playing cards, two cards are drawn one by one without replacement. Find the probability that :*

(i) first is king and second is queen

(ii) one is king and other in queen.

Sol. Since the first card is not replaced before the second draw, the random experiments of drawing cards are not independent.

Let K_i and Q_i be the events of drawing a king and a queen respectively in the ith draw, $i = 1, 2$.

(*i*) P(first is king and second is queen)

$$= P(K_1Q_2) = P(K_1)\,P(Q_2/K_1)$$

$$= \frac{4}{52} \times \frac{4}{51} = \frac{4}{663}.$$

(*ii*) P(one king and one queen)

$$= P(K_1Q_2 \text{ or } Q_1K_2) = P(K_1\,Q_2) + P(Q_1\,K_2)$$

$$= P(K_1)\,P(Q_2/K_1) + P(Q_1)\,P(K_2/Q_1)$$

$$= \frac{4}{52} \times \frac{4}{51} + \frac{4}{52} \times \frac{4}{51} = \frac{8}{663}.$$

Example 15. *A box contains 9 tickets numbered 1 to 9 inclusive. If three tickets are drawn from the box one at a time without replacements, find the probability that the numbers on the tickets are alternatively either odd, even, odd or even, odd, even.*

Sol. Since the tickets are not replaced before successive draws, the random experiments of drawing tickets are not independent.

Let O_i and E_i be the events of drawing ticket of odd number and even number respectively in the ith draw, $i = 1, 2, 3$.

Required probability

$$= P(O_1E_2O_3 \text{ or } E_1O_2E_3) = P(O_1E_2O_3) + P(E_1O_2E_3)$$

$$= P(O_1)P(E_2/O_1)P(O_3/O_1E_2) + P(E_1)P(O_2/E_1)P(E_3/E_1O_2)$$

$$= \frac{5}{9} \times \frac{4}{8} \times \frac{4}{7} + \frac{4}{9} \times \frac{5}{8} \times \frac{3}{7} = \frac{10}{63} + \frac{5}{42} = \frac{5}{18}.$$

Example 16. *An urn contains 8 red, 3 white and 9 blue balls. Three balls are drawn at random one by one without replacement. Determine the probability that :*

(*i*) *all three are red* (*ii*) *all three are blue*

(*iii*) *balls are drawn in the order red, white, blue*

(*iv*) *one of each colour is drawn.*

Sol. The random experiments of drawing balls without replacements are not independent.

8 Red
3 White
9 Blue

Let R_i, W_i, B_i be the events of drawing red ball, white ball, blue ball respectively in the ith draw, $i = 1, 2, 3$.

(*i*) P(all three are red) $= P(R_1R_2R_3) = P(R_1)\,P(R_2/R_1)\,P(R_3/R_1R_2)$

$$= \frac{8}{20} \times \frac{8-1}{20-1} \times \frac{8-2}{20-2} = \frac{8}{20} \times \frac{7}{19} \times \frac{6}{18} = \frac{14}{285}$$

(*ii*) P(all three are blue) $= P(B_1B_2B_3) = P(B_1)P(B_2/B_1)P(B_3/B_1B_2)$

$$= \frac{9}{20} \times \frac{9-1}{20-1} \times \frac{9-2}{20-2} = \frac{9}{20} \times \frac{8}{19} \times \frac{7}{18} = \frac{7}{95}$$

(*iii*) P(balls are drawn in the order red, white, blue)

$$= P(R_1W_2B_3) = P(R_1)P(W_2/R_1)P(B_3/R_1W_2)$$

$$= \frac{8}{20} \times \frac{3}{19} \times \frac{9}{18} = \frac{3}{95}$$

(*iv*) P(one of each colour)

$$= P(R_1W_2B_3 \text{ or } R_1B_2W_3 \text{ or } B_1R_2W_3 \text{ or } B_1W_2R_3 \text{ or } W_1R_2B_3 \text{ or } W_1B_2R_3).$$

$$= P(R_1W_2B_3) + P(R_1B_2W_3) + P(B_1R_2W_3) + P(B_1W_2R_3) + P(W_1R_2B_3) + P(W_1B_2R_3)$$

$$= \left(\frac{8}{20}\times\frac{3}{19}\times\frac{9}{18}\right) + \left(\frac{8}{20}\times\frac{9}{19}\times\frac{3}{18}\right) + \left(\frac{9}{20}\times\frac{8}{19}\times\frac{3}{18}\right) + \left(\frac{9}{20}\times\frac{3}{19}\times\frac{8}{18}\right) + \left(\frac{3}{20}\times\frac{8}{19}\times\frac{9}{18}\right) + \left(\frac{3}{20}\times\frac{9}{19}\times\frac{8}{18}\right)$$

$$= 6\left(\frac{8\times3\times9}{20\times19\times18}\right) = \frac{\mathbf{18}}{\mathbf{95}}.$$

Example 17. *A bag contains 8 red and 5 white balls. Two successive drawings of three balls are made such that (i) balls are replaced before second trial, (ii) balls are not replaced before second trial. Find the probability that 1st drawing will give 3 white and the 2nd 3 red balls.*

Sol. Let W_1 and R_2 be the events of getting 3 white balls in the first draw and 3 red balls in the second draw respectively.

8 Red 5 White

(*i*) Since the balls of first draw are replaced, the random experiments are independent.

$$\therefore \quad P(W_1R_2) = P(W_1)\,P(R_2) = \frac{{}^5C_3}{{}^{13}C_3}\times\frac{{}^8C_3}{{}^{13}C_3} = \frac{10}{286}\times\frac{56}{286} = \mathbf{0.0068}$$

(*ii*) Since the balls of first draw are not replaced, the random experiments are not independent.

$$\therefore \quad P(W_1R_2) = P(W_1)\,P(R_2/W_1) = \frac{{}^5C_3}{{}^{13}C_3}\times\frac{{}^8C_3}{{}^{10}C_3} = \frac{10}{286}\times\frac{56}{120} = \mathbf{0.0163}.$$

Example 18. *Let p be the probability that a man aged x years will die in a year. Find the probability that out of n men $A_1, A_2, \ldots\ldots, A_n$ each aged x, A_1 will die in the year and be the first to die.*

Sol. Let E_i be the event that A_i will die in a year, $i = 1, 2, \ldots\ldots, n$.

$$\therefore \quad P(E_i) = p \quad i = 1, 2, \ldots\ldots, n.$$

P(at least one will die)

$$= P(E_1 \cup E_2 \cup \ldots\ldots \cup E_n) = 1 - P(\overline{E_1 \cup E_2 \cup \ldots\ldots \cup E_n})$$

$$= 1 - P(\bar{E}_1 \cap \bar{E}_2 \cap \ldots\ldots \cap \bar{E}_n) = 1 - P(\bar{E}_1)\,P(\bar{E}_2)\ldots\ldots P(\bar{E}_n)$$

$$= 1 - (1-p)(1-p)\ldots\ldots(1-p) = 1-(1-p)^n$$

Since each of $A_1, A_2, \ldots\ldots, A_n$ is equally likely to be the first to die, the probability that A_1 will be first to die

$$= \frac{1}{n}\times(1-(1-p)^n) = \frac{\mathbf{1-(1-p)^n}}{\mathbf{n}}.$$

Example 19. *In each of a set of games, it is 2 to 1 in favour of the winner of the previous game. What is the chance that the player who wins the first game shall win at least three of the next four games ?*

Sol. Let W_i be the event that the winner of the first game wins the ith game, $i = 2, 3, 4, 5$.

$\therefore$ P(Winner of the first game wins at least 3 out of the next 4 games)

$$= P(W_2\, W_3\, W_4\, \overline{W_5} \text{ or } W_2\, W_3\, \overline{W_4}\, W_5 \text{ or } W_2\, \overline{W_3}\, W_4\, W_5 \text{ or } \overline{W_2}\, W_3\, W_4\, W_5 \text{ or } W_2\, W_3\, W_4\, W_5)$$

$$= P(W_2\, W_3\, W_4\, \overline{W_5}) + P(W_2\, W_3\, \overline{W_4}\, W_5) + P(W_2\, \overline{W_3}\, W_4\, W_5) + P(\overline{W_2}\, W_3\, W_4\, W_5) + P(W_2\, W_2\, W_3\, W_4)$$

$$= P(W_2)\, P(W_3/W_2)\, P(W_4/W_2\, W_3)\, P(\overline{W_5}/W_2\, W_3\, W_4)$$
$$+ P(W_2)\, P(W_3/W_2)\, P(\overline{W_4}/W_2\, W_3)\, P(W_5/W_2\, W_3\, \overline{W_4})$$
$$+ P(W_2)\, P(\overline{W_3}/W_2)\, P(W_4/W_2\, \overline{W_3})\, P(W_5/W_2\, \overline{W_3}\, W_4)$$
$$+ P(\overline{W_2})\, P(W_3/\overline{W_2})\, P(W_4/\overline{W_2}\, W_3)\, P(W_5/\overline{W_2}\, W_3\, W_4)$$
$$+ P(W_2)\, P(W_3/W_2)\, P(W_4/W_2\, W_3)\, P(W_5/W_2\, W_3\, W_4)$$

$$= \frac{2}{3}\times\frac{2}{3}\times\frac{2}{3}\times\frac{1}{3} + \frac{2}{3}\times\frac{2}{3}\times\frac{1}{3}\times\frac{1}{3} + \frac{2}{3}\times\frac{1}{3}\times\frac{1}{3}\times\frac{2}{3} + \frac{1}{3}\times\frac{1}{3}\times\frac{2}{3}\times\frac{2}{3} + \frac{2}{3}\times\frac{2}{3}\times\frac{2}{3}\times\frac{2}{3}$$

$$= \frac{8+4+4+4+16}{81} = \frac{36}{81}.$$

Example 20. *A bag contains n white and n red balls. Pair of balls are drawn without replacement until the bag is empty. Show that the probability that each pair consists of one white and one red ball is* $\frac{2^n}{{}^{2n}C_n}$.

Sol. The random experiments of drawing pairs of balls without replacement are not independent.

n White n Red

Let A_i be the event of drawing a pair of a white ball and a red ball in the ith draw, $1 \le i \le n$.

$\therefore$ P(each pair contains white and red balls).

$$P(A_1\, A_2\, A_3 \,\ldots\ldots\, A_n)$$
$$= P(A_1)\, P(A_2/A_1)\, P(A_3/A_1A_2) \,\ldots\ldots\, P(A_n/A_1A_2A_3 \,\ldots\ldots\, A_{n-1})$$
$$= \frac{n \times n}{{}^{2n}C_2} \times \frac{(n-1)(n-1)}{{}^{2n-2}C_2} \times \frac{(n-2)(n-2)}{{}^{2n-4}C_2} \times \ldots\ldots \times \frac{1 \times 1}{{}^{2}C_2}$$
$$= \frac{n^2 (n-1)^2 (n-2)^2 \,\ldots\ldots\, 1^2}{{}^{2n}C_2 \,.\, {}^{2n-2}C_2 \,.\, {}^{2n-4}C_2 \,\ldots\ldots\, {}^{2}C_2}$$
$$= \frac{(n\,!)^2}{\dfrac{2n\,!}{2\,!(2n-2)\,!} \,.\, \dfrac{(2n-2)\,!}{2\,!(2n-4)\,!} \,.\, \dfrac{(2n-4)\,!}{2\,!(2n-6)\,!} \ldots\ldots 1}$$
$$= \frac{(n\,!)^2}{\dfrac{(2n)\,!}{2^n}} = \frac{2^n}{\dfrac{(2n)\,!}{n\,!\,n\,!}} = \frac{2^n}{{}^{2n}C_n}.$$

EXERCISE 5.6

1. Two cards are drawn from a pack of cards in succession (with replacement). Find the probability that the first card is spade and the second is a black king.
2. A husband and a wife appear in an interview for two vacancies for the same post. The probability of husband's selection is $\frac{2}{5}$ and that of wife is $\frac{4}{5}$. What is the probability that both of them will be selected ?
3. A person is known to hit the target in 3 out of 4 shots, whereas another person is known to hit the target in 2 out of 3 shots. Find the probability of the target being hit at all, when both try.
4. A man wants to marry a girl having qualities : white complexion—the probability of getting such a girl is one in twenty, handsome dowry — the probability of getting this is one in thirty. Find the probability of his getting married to a white complexioned girl who may also bring handsome dowry.
5. (*i*) The probabilities of hitting a target by three persons are $\frac{1}{3}, \frac{1}{4}$ and $\frac{1}{5}$. Find the probability that the target is hit, when they all fire simultaneously.

   (*ii*) A problem in mathematics is given to three students whose chances of solving the problem are $\frac{1}{2}, \frac{1}{3}$ and $\frac{1}{4}$. What is the probability that the problem is solved ? *(K U 2006)*
6. A candidate is called for interview by three companies. For the first company there are 12 candidates, for the second there are 15 candidates and for the third there are 10 candidates. What is the probability of his getting selected in at least one of the companies ?
7. The probability of A winning a race is $\frac{1}{5}$ and the probability of B winning the race is $\frac{1}{6}$. Find the probability that none will win the race.
8. (*i*) The odds in favour of '*A*' solving a problem are 7 : 6, and the odds against '*B*' solving the same problem are 11 : 8. What is the probability that the problem will be solved, if both try the problem ?

   (*ii*) A can solve 90% of problems given in a book and B can solve 70%. What is the probability that at least one of them will solve the problem, selected at random ?
9. The odds in favour of '*X*' solving a problem are 1 : 5 and the odds in favour of '*Y*' solving the same problem are 2 : 3. What is the probability that exactly one of them would be able to solve the problem ?
10. The odds in favour of first speaking the truth are 3 : 2. The odds in favour of second speaking the truth are 5 : 3. In what percentage of cases are they likely to contradict each other on an identical point ?
11. There are two papers in Mathematics in a certain examination. The probability that a candidate passes in paper I is 60% and that he passes in paper II is 50%. What is the probability that the candidate passes only in one paper ?
12. What is the probability of throwing 6 with a die at least once in 3 attempts ?
13. If 1 out of 10 bulbs is defective, find the probability that out of sample of 3 bulbs, with replacement :

    (*i*) 2 bulbs are defective. (*ii*) no bulb is defective.
14. Find the probability of drawing a heart on each of the two consecutive draws of one card from a well-shuffled pack of playing cards, if the card is not replaced after the first draw.
15. An urn contains two black and three white balls. A ball is drawn at random and then a second ball is drawn at random from those left. What is the probability that both balls are white ?

16. A bag contains 10 red and 6 black balls. 4 balls are drawn successively one by one and are not replaced. What is the probability that these are alternatively of different colours, beginning with red ?

17. A bag contains 10 red and 6 black balls, 4 balls are drawn successively one by one and are not replaced. What is the probability that these are alternatively of different colours ?

18. Two cards are drawn one by one from a well-shuffled pack of 52 cards. Find the probability that both are aces, if the first card is not replaced.

19. A bag contains 13 balls numbered from 1 to 13. Suppose an even number is considered as a success. Two balls are drawn one by one without replacement. Find the probability of getting one success.

20. Find the probability of drawing a queen, a king and a knave in that order from a pack of cards in three consecutive draws, the cards are not being replaced.

Answers

1. $\frac{1}{104}$	**2.** $\frac{8}{25}$	**3.** $\frac{11}{12}$	**4.** $\frac{1}{600}$
5. (*i*) $\frac{3}{5}$	(*ii*) $\frac{3}{4}$	**6.** 0.23	**7.** $\frac{2}{3}$
8. (*i*) $\frac{181}{247}$	(*ii*) $\frac{97}{100}$	**9.** $\frac{13}{30}$	**10.** 47.5%
11. $\frac{1}{2}$	**12.** $\frac{91}{216}$	**13.** (*i*) $\frac{27}{1000}$	(*ii*) $\frac{729}{1000}$
14. $\frac{1}{17}$	**15.** $\frac{3}{10}$	**16.** $\frac{45}{728}$	**17.** $\frac{45}{364}$
18. $\frac{1}{221}$	**19.** $\frac{7}{13}$	**20.** $\frac{8}{16575}$.	

5.19. TOTAL PROBABILITY RULE

Let $E_1, E_2, \ldots\ldots, E_n$ be n mutually exclusive and exhaustive events, with non-zero probabilities, of a random experiment. If A be any arbitrary event of the sample space of the above random experiment with P(A) > 0, then prove that

$$P(A) = P(E_1)P(A/E_1) + P(E_2)P(A/E_2) + \ldots\ldots + P(E_n)P(A/E_n).$$

Proof. Let S be the sample space of the random experiment.

Since $E_1, E_2, \ldots\ldots, E_n$ are exhaustive, we have

$$S = E_1 \cup E_2 \cup \ldots\ldots \cup E_n.$$

Now $\quad A = S \cap A = (E_1 \cup E_2 \cup \ldots\ldots \cup E_n) \cap A$

$\Rightarrow \quad A = (E_1 \cap A) \cup (E_2 \cap A) \cup \ldots\ldots \cup (E_n \cap A) \qquad \ldots(1)$

Since $E_1, E_2, \ldots\ldots, E_n$ are mutually exclusive, we have $E_i \cap E_j = \phi$ for $i \neq j$.

Now $(E_i \cap A) \cap (E_j \cap A) = (E_i \cap E_j) \cap A = \phi \cap A = \phi$

$\therefore \quad E_1 \cap A, E_2 \cap A, \ldots\ldots, E_n \cap A$ are also mutually exclusive.

By **addition theorem,** (1) implies

$$P(A) = P(E_1 \cap A) + P(E_2 \cap A) + \ldots\ldots + P(E_n \cap A)$$

$$\Rightarrow \quad \mathbf{P(A) = P(E_1)P(A/E_1) + P(E_2)P(A/E_2) + \ldots.. P(E_n)P(A/E_n).}$$

Remark. In practical problems, it is found convenient to write as follows :

$$P(A) = P(E_1A \text{ or } E_2A \text{ or } \ldots\ldots \text{ or } E_nA) = P(E_1A) + P(E_2A) + \ldots\ldots + P(E_nA)$$

$$\therefore \quad \mathbf{P(A) = P(E_1)P(A/E_1) + P(E_2)P(A/E_2) + \ldots\ldots + P(E_n)P(A/E_n).}$$

Corollary. In particular if $n = 2$, we have

$$\mathbf{P(A) = P(E_1)P(A/E_1) + P(E_2)P(A/E_2).}$$

Example 1. *A university purchased computers from three firms. The percentage of computer purchased and percentage of defective computers is shown in the table below :*

Firm

	HCL	*WIPRO*	*IBM*
Percent Purchase	45	25	30
Percent Defective	2	3	1

Let A be the event "Computer purchased from HCL"
B be the event "Computer purchased from WIPRO"
C be the event "Computer purchased from IBM"
D be the event "Computer was defective".
Find P(A), P(B), P(C), P(D/A), P(D/B), P(D/C) and P(D).

Sol. Total number of computers

$$= 45 + 25 + 30 = 100$$

$$\therefore \quad P(A) = \frac{45}{100} = 0.45, \quad P(B) = \frac{25}{100} = 0.25, \quad P(C) = \frac{30}{100} = 0.30,$$

$$P(D/A) = \frac{2}{100} = 0.02, \quad P(D/B) = \frac{3}{100} = 0.03, \quad P(D/C) = \frac{1}{100} = 0.01$$

$$P(D) = P(AD \text{ or } BD \text{ or } CD) = P(AD) + P(BD) + P(CD)$$
$$= P(A)\,P(D/A) + P(B)\,P(D/B) + P(C)\,P(D/C)$$
$$= (0.45)\,(0.02) + (0.25)\,(0.03) + (0.30)\,(0.01) = \mathbf{0.0195.}$$

Example 2. *A bag contains 5 red and 3 black balls and another bag contains 4 red and 5 black balls. One of these is selected at random and two balls are drawn. What is the probability that one of them is red and the other black ?*

Sol. Let E_1 and E_2 be the events of selecting the first bag and the second bag respectively.

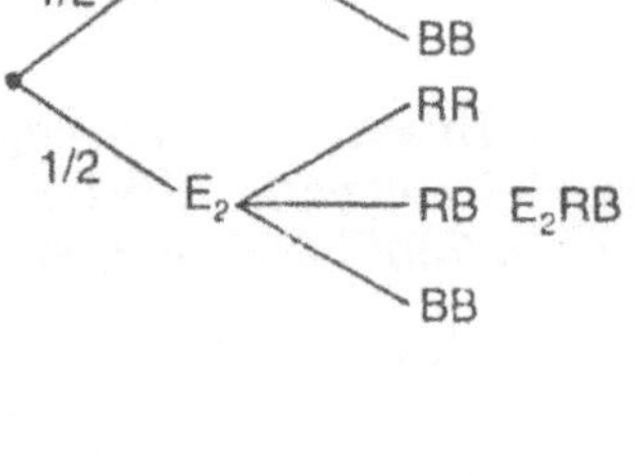

$$\therefore \quad P(E_1) = \frac{1}{2}, \; P(E_2) = \frac{1}{2}$$

Let RB be the event of drawing one red and one black ball.

$$\therefore \quad P(RB) = P(E_1RB \text{ or } E_2RB)$$
$$= P(E_1RB) + P(E_2RB)$$
$$= P(E_1)P(RB/E_1) + P(E_2)P(RB/E_2)$$
$$= \frac{1}{2} \times \frac{{}^5C_1 \times {}^3C_1}{{}^8C_2} + \frac{1}{2} \times \frac{{}^4C_1 \times {}^5C_1}{{}^9C_2}$$
$$= \frac{1}{2} \times \frac{5 \times 3}{28} + \frac{1}{2} \times \frac{4 \times 5}{36} = \mathbf{\frac{275}{504}}.$$

Example 3. *An urn contains 4 tickets numbered 1, 2, 3, 4 and another contains 6 tickets numbered 2, 4, 6, 7, 8, 9. If one of the two urns is chosen at random and a ticket is drawn at random from the urn chosen, find the probability that the ticket drawn bears the number 2 or 4.*

Sol. Let E_1 and E_2 be the events of choosing the urn I and the urn II respectively.

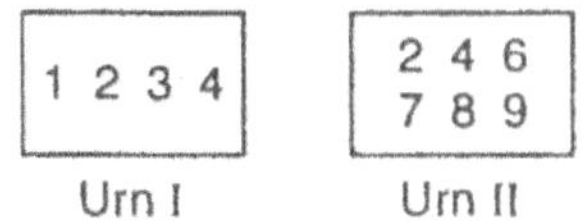

$$\therefore \quad P(E_1) = \frac{1}{2}, \; P(E_2) = \frac{1}{2}$$

Let A be the event of drawing a ticket bearing the number 2 or 4.

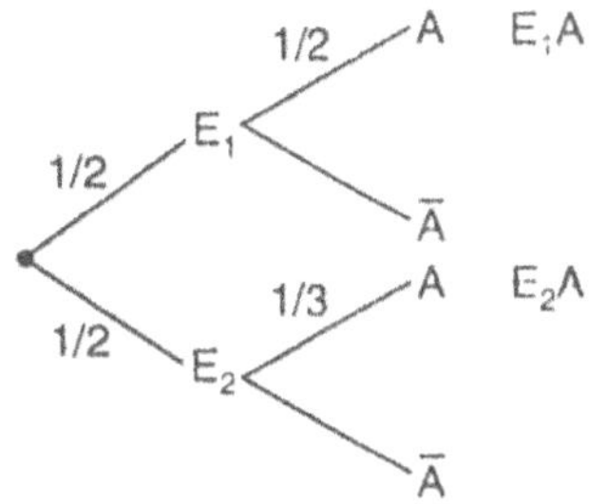

$$\therefore \quad P(A/E_1) = \frac{2}{4} = \frac{1}{2}$$

and $$P(A/E_2) = \frac{2}{6} = \frac{1}{3}$$

Required probability $= P(A)$

$= P(E_1A \text{ or } E_2A)$

$= P(E_1A) + P(E_2A)$

$= P(E_1)P(A/E_1) + P(E_2)P(A/E_2)$

$$= \frac{1}{2} \times \frac{1}{2} + \frac{1}{2} \times \frac{1}{3} = \frac{1}{4} + \frac{1}{6} = \frac{5}{12}.$$

Example 4. *Suppose that 5 men out of 100 men and 25 women out of 1000 women are good oretor. Assuming that there are equal number of men and women, find the probability, of choosing an oretor.*

Sol. Let E_1 and E_2 be the events of choosing a man and a woman respectively

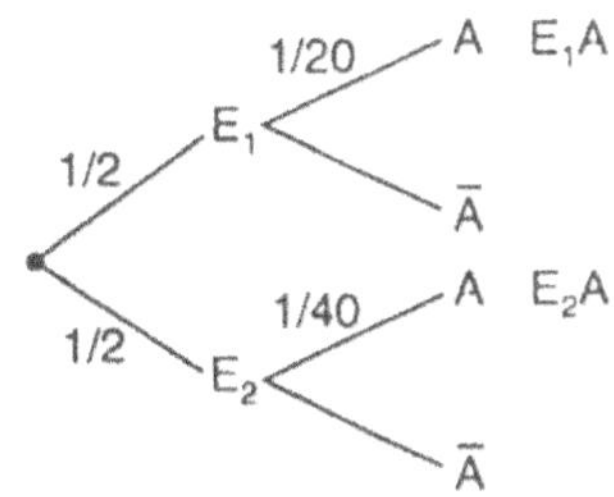

$\therefore \quad P(E_1) = \frac{1}{2}$ and $P(E_2) = \frac{1}{2}$, because there are equal number of men and women.

Let A be the event of choosing an oretor

$\therefore \quad P(A/E_1)$ = probability that a man is an oretor

$$= \frac{5}{100} = \frac{1}{20}$$

$P(A/E_2)$ = probability that a woman is an oretor

$$= \frac{25}{1000} = \frac{1}{40}$$

$\therefore \quad P(\text{oretor is chosen}) = P(A) = P(E_1A \text{ or } E_2A)$

$= P(E_1A) + P(E_2A) = P(E_1)P(A/E_1) + P(E_2)P(A/E_2)$

$$= \frac{1}{2} \times \frac{1}{20} + \frac{1}{2} \times \frac{1}{40} = \frac{3}{80}.$$

Example 5. *One bag contains 5 white and 4 black balls. Another bag contains 7 white and 9 black balls. A ball is transferred from the first bag to the second bag and then a ball is drawn from the second. Find the probability that this ball is white.*

Sol. Let W_1 and B_1 be the events of transferring a white ball and a black ball from bag I to bag II respectively.

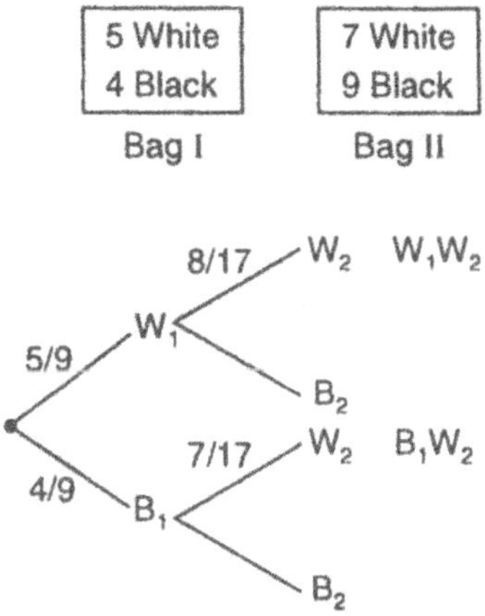

Let W_2 be the event of drawing a white ball from the bag II after the transfer of ball.

$\therefore$ Required probability

$$= P(W_2) = P(W_1W_2 \text{ or } B_1W_2)$$
$$= P(W_1W_2) + P(B_1W_2)$$
$$= P(W_1)P(W_2/W_1) + P(B_1)P(W_2/B_1)$$
$$= \frac{5}{5+4} \times \frac{7+1}{(7+1)+9} + \frac{4}{5+4} \times \frac{7}{7+(9+1)}$$
$$= \frac{5}{9} \times \frac{8}{17} + \frac{4}{9} \times \frac{7}{17} = \frac{4}{9}.$$

Example 6. *There are two bags. The first bag contains 5 white and 3 black balls and the second bag contains 3 white and 5 black balls. Two balls are drawn at random from the first bag and are put into the second bag, without noticing their colours. Then two balls are drawn from the second bag. Find the probability that these balls are white and black.*

Sol. Let E_1, E_2 and E_3 be the events of transferring 2 white, 1 white and 1 black, 2 black balls respectively from the first bag to second bag.

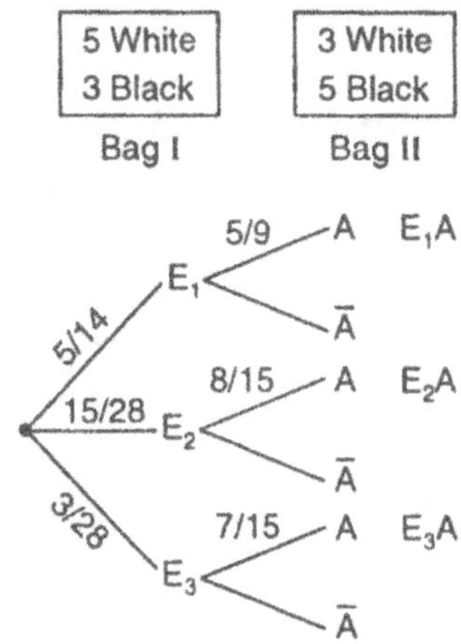

$$P(E_1) = \frac{{}^5C_2}{{}^8C_2} = \frac{10}{28} = \frac{5}{14}$$

$$P(E_2) = \frac{{}^5C_1 \times {}^3C_1}{{}^8C_2} = \frac{5 \times 3}{28} = \frac{15}{28}$$

$$P(E_3) = \frac{{}^3C_2}{{}^8C_2} = \frac{3}{28}$$

Let A be the event of drawing one white and one black ball from the second bag.

$$\therefore \quad P(A) = P(E_1A \text{ or } E_2A \text{ or } E_3A)$$
$$= P(E_1A) + P(E_2A) + P(E_3A)$$
$$= P(E_1)P(A/E_1) + P(E_2)P(A/E_2) + P(E_3)P(A/E_3)$$
$$= \frac{5}{14} \times \frac{{}^5C_1 \times {}^5C_1}{{}^{10}C_2} + \frac{15}{28} \times \frac{{}^4C_1 \times {}^6C_1}{{}^{10}C_2} + \frac{3}{28} \times \frac{{}^3C_1 \times {}^7C_1}{{}^{10}C_2}$$
$$= \frac{5}{14} \times \frac{5}{9} + \frac{15}{28} \times \frac{8}{15} + \frac{3}{28} \times \frac{7}{15} = \mathbf{\frac{673}{1260}}.$$

Example 7. *Two cards are drawn from a pack of 52 cards at random and kept out. Then one card is drawn from the remaining 50 cards. Find the probability that it is an ace.*

Sol. Let E_1, E_2 and E_3 be the events of drawing 2 aces, 1 ace and 1 non ace, 2 non aces in the first draw of 2 cards.

$$\therefore \quad P(E_1) = \frac{{}^4C_2}{{}^{52}C_2} = \frac{1}{221}, \qquad P(E_2) = \frac{{}^4C_1 \times {}^{48}C_1}{{}^{52}C_2} = \frac{32}{221}$$

and $$P(E_3) = \frac{^{48}C_2}{^{52}C_2} = \frac{188}{221}.$$

Let E be the event of drawing an ace in the second draw of one card.

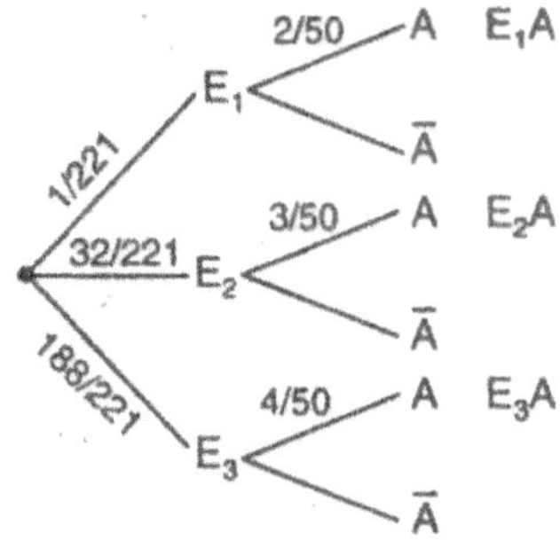

Required probability

$= P(A) = P(E_1A \text{ or } E_2A \text{ or } E_3A)$

$= P(E_1)P(A/E_1) + P(E_2)P(A/E_2) + P(E_3)P(A/E_3)$

$$= \frac{1}{221} \times \frac{4-2}{52-2} + \frac{32}{221} \times \frac{4-1}{52-2} + \frac{188}{221} \times \frac{4-0}{52-2}$$

$$= \frac{2+96+752}{221 \times 50} = \frac{850}{11050} = \frac{1}{13}.$$

Example 8. *Two machines A and B produce respectively 60% and 40% of the total numbers of a items of a factory. The percentages of defective output of these machines are respectively 2% and 5%. If an item is selected at random, what is the probability that the item is (i) defective (ii) non-defective ?*

Sol. Let E_1, E_2 be the events of drawing an item produced by machine A and machine B respectively. Let A be the event of selecting a defective item.

$\therefore$ $\bar{A}$ represent the event of selecting a non-defective item. We have

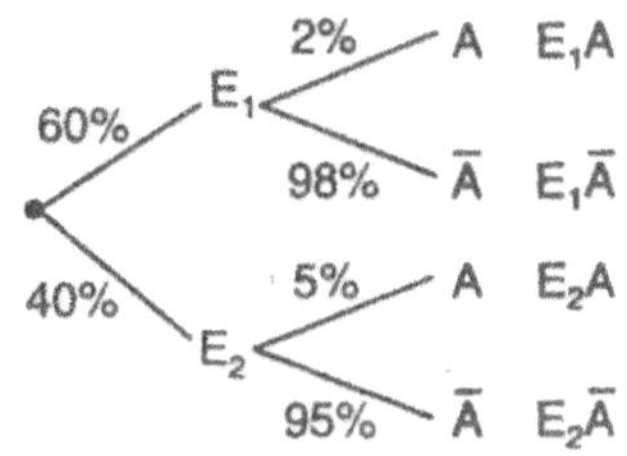

$P(E_1) = 60\%$, $P(E_2) = 40\%$.

$P(A/E_1)$ = probability that a defective item is produced by $A = 2\%$.

$P(A/E_2)$ = probability that a defective item is produced by $B = 5\%$.

(*i*) P(selected item is defective)

$= P(A) = P(E_1A \text{ or } E_2A) = P(E_1A) + P(E_2A)$

$= P(E_1)P(A/E_1) + P(E_2)P(A/E_2) = (60\%)(2\%) + (40\%)(5\%)$

$$= \frac{60}{100} \times \frac{2}{100} + \frac{40}{100} \times \frac{5}{100} = \frac{320}{10000} = \mathbf{0.032}$$

(*ii*) P(selected item is non-defective)

$= P(\bar{A}) = P(E_1\bar{A} \text{ or } E_2\bar{A}) = P(E_1\bar{A}) + P(E_2\bar{A})$

$= P(E_1)P(\bar{A}/E_1) + P(E_2)P(\bar{A}/E_2) = (60\%)(98\%) + (40\%)(95\%)$

$$= \frac{60}{100} \times \frac{98}{100} + \frac{40}{100} \times \frac{95}{100} = \frac{9680}{10000} = \mathbf{0.968}.$$

EXERCISE 5.7

1. A bag contains 3 white and 2 black balls and another bag contains 2 white and 4 black balls. One bag is chosen at random. From the selected bag, one ball is drawn. Find the probability that the ball drawn is white.

2. A factory has two machines A and B. Past records show that the machine A produced 60% of the items of output and machine B produced 40% of the items. Further 2% of the items produced by machine A were defective and 1% produced by machine B were defective. If an item is drawn at random, what is the probability that it is defective ?

3. Find the probability of drawing a one-rupee coin from a purse with two compartments one of which contains 3 fifty-paise coins and 2 one-rupee coins and other contains 2 fifty-paise coins and 3 one-rupee coins.

4. An unbiased coin is tossed. If the result is a head, a pair of unbiased dice is rolled and the sum of the numbers obtained is noted. If the result is a tail, a card from a well shuffled pack of eleven cards numbered 2, 3, 4, ..., 12 is picked and the number on the card is noted. What is the probability that the noted number is either 7 or 8 ?

5. Two third of the students in a class are boys and the rest girls. It is known that the probability of a girl getting a first class is 0.25 and that of a boy getting first class is 0.28. Find the probability that a student chosen at random will get first class marks in the examination.

6. In a bolt factory, machines A, B and C manufacture 25%, 35% and 40% of the total bolts. Of their outputs 5%, 4% and 2% are respectively defective bolts. A bolt is drawn at random from the output. What is the probability that the bolt drawn is defective ?

7. We are given three boxes as follows :

   Box I has 10 light bulbs of which 4 are defective.

   Box II has 6 light bulbs of which 1 is defective.

   Box III has 8 light bulbs of which 3 are defective.

   We select a box at random and then draw a bulb at random. What is the probability that the bulb is defective ?

8. The contents of three bags I, II and III are as follows :

   Bag I : 1 white, 2 black and 3 red balls,

   Bag II : 2 white, 1 black and 1 red balls, and

   Bag III : 4 white, 5 black and 3 red balls.

   A bag is chosen at random and two balls are drawn. What is the probability that the balls are white and red ?

9. A bag contains 6 white and 7 black balls, and another bag contains 4 white and 5 black balls. A ball is taken from the first bag and without seeing its colour is put in the second bag. A ball is taken from the latter. Find the probability that the ball drawn is white.

10. An urn contains 10 white and 3 black balls, while another urn contains 3 white and 5 black balls. Two balls are drawn from the first urn and put in to the second urn and then a ball is drawn from the latter. What is the probability of drawing a white ball ?

Answers

1. $\frac{7}{15}$ 2. $\frac{2}{125}$ 3. $\frac{1}{2}$ 4. $\frac{193}{792}$

5. 0.27 6. 0.345 7. $\frac{113}{360}$ 8. $\frac{118}{495}$

9. $\frac{29}{65}$ 10. $\frac{59}{130}$.

II. BAYES' THEOREM

5.20. MOTIVATION

Let there be two or more urns, each containing some white balls and red balls. Suppose an urn is chosen at random and a ball is drawn from that chosen urn. By using *addition theorem* and *multiplication theorem*, we can find the probability of drawing a white ball (or red ball) from the urn chosen.

But in case, we are given that the ball drawn is white and we are interested in finding the probability of the event that the ball was drawn from the Ist urn or IInd urn etc., then the situation is not the same as in the previous case. Now the probability of the drawn urn will depend upon the colour of the drawn ball.

To tackle this type of problems, Bayes' theorem is used. This theorem was enunciated by British mathematician **Thomos Bayes** in 1763.

5.21. BAYES' THEOREM

Let $E_1, E_2, \ldots\ldots, E_n$ be n mutually exclusive and exhaustive events, with non-zero probabilities, of a random experiment. If A be any arbitrary event of the sample space of the above experiment with P(A) > 0, then prove that

$$P(E_i/A) = \frac{P(E_i)P(A/E_i)}{\sum_{j=1}^{n} P(E_j)P(A/E_j)}, \quad 1 \le i \le n. \qquad \textit{(KU 2006, 2007)}$$

Proof. Let S be the sample space of the random experiment.

$\therefore \quad S = E_1 \cup E_2 \cup \ldots\ldots \cup E_n \qquad (\because \; E_1, E_2, \ldots\ldots, E_n$ are exhaustive)

Now $\quad A = S \cap A = (E_1 \cup E_2 \cup \ldots\ldots \cup E_n) \cap A$

$= (E_1 \cap A) \cup (E_2 \cap A) \cup \ldots\ldots \cup (E_n \cap A)$

$\therefore \quad P(A) = P(E_1 \cap A) + P(E_2 \cap A) + \ldots\ldots + P(E_n \cap A)$*

$= P(E_1)P(A/E_1) + P(E_2)P(A/E_2) + \ldots\ldots + P(E_n)P(A/E_n)$

or
$$P(A) = \sum_{j=1}^{n} P(E_j)P(A/E_j) \qquad \ldots(1)$$

Now,
$$P(E_i/A) = \frac{P(E_i \cap A)}{P(A)}, \; 1 \le i \le n$$

$\therefore$
$$P(E_i/A) = \frac{P(E_i)P(A/E_i)}{\sum_{j=1}^{n} P(E_j)P(A/E_j)}, \; 1 \le i \le n. \qquad \text{[Using (1)]}$$

---

* **Why this step.** Since $E_1, E_2, \ldots, E_n$ are mutually exclusive, we have $E_l \cap E_m = \phi$ for $l \neq m$.
Now $(E_l \cap A) \cap (E_m \cap A) = (E_l \cap E_m) \cap A = \phi \cap A = \phi$

$\therefore$ $E_1 \cap A, E_2 \cap A, \ldots\ldots, E_n \cap A$ are also *pairwise m.e.* and *extended addition theorem* is applicable.

Remark. If $n = 2$, then

$$P(E_1/A) = \frac{P(E_1)P(A/E_1)}{P(E_1)P(A/E_1) + P(E_2)P(A/E_2)}$$

and
$$P(E_2/A) = \frac{P(E_2)P(A/E_2)}{P(E_1)P(A/E_1) + P(E_2)P(A/E_2)}.$$

Example 1. *Two urns contain 1 white, 6 red and 4 white, 3 red balls. One of the urns is selected at random and a ball is drawn from it. Find :*

(*i*) *the probability of drawing a white ball.*

(*ii*) *the probability of drawing the ball from first urn, if the ball drawn is white.*

Sol. Let E_1, E_2 be the events of drawing the first urn and the second urn respectively.

$$\therefore \quad P(E_1) = \tfrac{1}{2},\ P(E_2) = \tfrac{1}{2}.$$

Urn I	Urn II
1 White 6 Red	4 White 3 Red

Let A be the event of drawing a white ball.

$$\therefore \quad P(A/E_1) = \frac{1}{1+6} = \frac{1}{7} \text{ and } P(A/E_2) = \frac{4}{4+3} = \frac{4}{7}$$

(*i*) P(getting white ball)

$$= P(A) = P(E_1A \text{ or } E_2A) = P(E_1A) + P(E_2A)$$
$$= P(E_1)P(A/E_1) + P(E_2)P(A/E_2)$$
$$= \left(\frac{1}{2}\times\frac{1}{7}\right) + \left(\frac{1}{2}\times\frac{4}{7}\right) = \mathbf{\frac{5}{14}}$$

(*ii*) P(White ball drawn is from the first urn)

$$= P(E_1/A) = \frac{P(E_1)P(A/E_1)}{P(E_1)P(A/E_1) + P(E_2)P(A/E_2)}$$

$$= \frac{\frac{1}{2}\times\frac{1}{7}}{\left(\frac{1}{2}\times\frac{1}{7}\right)+\left(\frac{1}{2}\times\frac{4}{7}\right)} = \frac{\frac{1}{14}}{\frac{5}{14}} = \mathbf{\frac{1}{5}}.$$

Note. In finding the probability of drawing a white ball in part (*i*), we do not require the use of **Bayes' theorem.**

Example 2. *In 1988, there will be three candidates for the position of principal – A, B and C. The chances of their selection are in the proportion 4 : 2 : 3 respectively. The probability that A, if selected, will introduce co-education in the college is 0.3. The probability of B and C doing the same are respectively 0.5 and 0.8. What is the probability that there will be co-education in the college in 1988 ? Also find the probability that the co-education in the college was introduced by the principal B.*

Sol. Let E_1, E_2, E_3 be the events of selection of A, B, C as principal respectively. Let A be the event of introduction of co-education in the college.

$$\therefore \quad P(E_1) = \frac{4}{4+2+3} = \frac{4}{9},\quad P(E_2) = \frac{2}{4+2+3} = \frac{2}{9} \text{ and } P(E_3) = \frac{3}{4+2+3} = \frac{3}{9}$$

Also, $$P(A/E_1) = \frac{3}{10},\quad P(A/E_2) = \frac{5}{10},\quad P(A/E_3) = \frac{8}{10}$$

Now, P(co-education is introduced in the college)

$$= P(A) = P(E_1A \text{ or } E_2A \text{ or } E_3A) = P(E_1A) + P(E_2A) + P(E_3A)$$

$$= P(E_1)P(A/E_1) + P(E_2)P(A/E_2) + P(E_3)P(A/E_3)$$

$$= \left(\frac{4}{9} \times \frac{3}{10}\right) + \left(\frac{2}{9} \times \frac{5}{10}\right) + \left(\frac{3}{9} \times \frac{8}{10}\right) = \frac{46}{90} = \frac{\mathbf{23}}{\mathbf{45}}.$$

By *Bayes' theorem*,

P(Co-education was introduced by the principal B)

$$= P(E_2/A) = \frac{P(E_2)P(A/E_2)}{P(A)} = \frac{\frac{2}{9} \times \frac{5}{10}}{\frac{23}{45}} \;\frac{\mathbf{5}}{\mathbf{23}}.$$

Example 3. *Assume that a factory has two machines. Past records shows that machine I produces 30% of the items of output and machine II produces 70% of the items. Further, 5% of the items produced by machine I were defective and only 1% produced by machines II were defective. An item is drawn at random and found to be defective. What is the probability that it was produced by machine I or machine II ?*

Sol. Let E_1, E_2 be the events of drawing an item produced by machine I and II respectively. Let A be the event of drawing a defective item.

$$\therefore \qquad P(E_1) = \frac{30}{100} = \frac{3}{10}, \qquad P(E_2) = \frac{70}{100} = \frac{7}{10}$$

$$\text{Also,} \qquad P(A/E_1) = \frac{5}{100}, \qquad P(A/E_2) = \frac{1}{100}$$

The events E_1 and E_2 are mutually exclusive and exhaustive.

By *Bayes' theorem*,

P(machine I produced the defective item) $= P(E_1/A)$

$$= \frac{P(E_1)P(A/E_1)}{P(E_1)P(A/E_1) + P(E_2)P(A/E_2)}$$

$$= \frac{\frac{3}{10} \times \frac{5}{100}}{\left(\frac{3}{10} \times \frac{5}{100}\right) + \left(\frac{7}{10} \times \frac{1}{100}\right)} = \frac{15}{15+7} = \frac{\mathbf{15}}{\mathbf{22}}$$

Also, P(machine II produced the defective item) $= (P(E_2/A)$

$$= \frac{P(E_2)P(A/E_2)}{P(E_1)P(A/E_1) + P(E_2)P(A/E_2)}$$

$$= \frac{\frac{7}{10} \times \frac{1}{100}}{\left(\frac{3}{10} \times \frac{5}{100}\right) + \left(\frac{7}{10} \times \frac{1}{100}\right)} = \frac{7}{15+7} = \frac{\mathbf{7}}{\mathbf{22}}.$$

Example 4. *In a bolt factory, machines A, B and C manufacture 60%, 25% and 15% respectively. Of the total of their output, 1%, 2% and 1% are defective bolts. A bolt is drawn at random from the total production and found to be defective. From which machine, the defective bolt is expected to have been manufactured ?*

Sol. Let E_1, E_2, E_3 be the events of drawing a bolt produced by machine A, B and C respectively. Let A be the event of drawing a defective bolt.

$$\therefore \quad P(E_1) = \frac{60}{100}, \quad P(E_2) = \frac{25}{100}, \quad P(E_3) = \frac{15}{100}$$

$$P(A/E_1) = \frac{1}{100}, \quad P(A/E_2) = \frac{2}{100}, \quad P(A/E_3) = \frac{1}{100}$$

The events E_1, E_2 and E_3 are mutually exclusive and exhaustive.

By *Bayes' theorem.*

P(defective bolt is produced by machine A) $= P(E_1/A)$

$$= \frac{P(E_1)P(A/E_1)}{P(E_1)P(A/E_1) + P(E_2)P(A/E_2) + P(E_3)P(A/E_3)}$$

$$= \frac{\frac{60}{100} \times \frac{1}{100}}{\left(\frac{60}{100} \times \frac{1}{100}\right) + \left(\frac{25}{100} \times \frac{2}{100}\right) + \left(\frac{15}{100} \times \frac{1}{100}\right)} = \frac{60}{60 + 50 + 15} = \frac{60}{125} = \mathbf{\frac{12}{25}}$$

P(defective bolt is produced by machine B) $= P(E_2/A)$

$$= \frac{P(E_2)P(A/E_2)}{P(E_1)P(A/E_1) + P(E_2)P(A/E_2) + P(E_3)P(A/E_3)}$$

$$= \frac{\frac{25}{100} \times \frac{2}{100}}{\left(\frac{60}{100} \times \frac{1}{100}\right) + \left(\frac{25}{100} \times \frac{2}{100}\right) + \left(\frac{15}{100} \times \frac{1}{100}\right)} = \frac{50}{60 + 50 + 15} = \frac{50}{125} = \mathbf{\frac{10}{25}}$$

P(defective bolt is produced by machine C) $= P(E_3/A)$

$$= \frac{P(E_3)P(A/E_3)}{P(E_1)P(A/E_1) + P(E_2)P(A/E_2) + P(E_3)P(A/E_3)}$$

$$= \frac{\frac{15}{100} \times \frac{1}{100}}{\left(\frac{60}{100} \times \frac{1}{100}\right) + \left(\frac{25}{100} \times \frac{2}{100}\right) + \left(\frac{15}{100} \times \frac{1}{100}\right)} = \frac{15}{60 + 50 + 15} = \frac{15}{125} = \mathbf{\frac{3}{25}}.$$

$\therefore$ Most probably, the defective bolt has been drawn from the output of machine A.

$$\left(\because \frac{12}{25} > \frac{10}{25}, \frac{3}{25}\right)$$

Example 5. *An insurance company insured 2000 scooter drivers, 4000 car drivers and 6000 truck drivers. The probability of an accident involving a scooter driver, car driver and a truck driver is 0.01, 0.03 and 0.15 respectively. One of the insured drivers meets with an accident. What is the probability that he is a car driver ?*

Sol. Let E_1, E_2, E_3 be the events of drawing scooter driver, car driver, truck driver respectively.

Total number of drivers = 2000 + 4000 + 6000 = 12000

$$\therefore \quad P(E_1) = \frac{2000}{12000} = \frac{1}{6}, \quad P(E_2) = \frac{4000}{12000} = \frac{1}{3} \text{ and } P(E_3) = \frac{6000}{12000} = \frac{1}{2}$$

Let A be the event of getting an accident.

$\therefore \quad P(A/E_1) = 0.01, P(A/E_2) = 0.03$ and $P(A/E_3) = 0.15$

The events E_1, E_2 and E_3 are mutually exclusive and exhaustive.

$\therefore$ By **Bayes' theorem**

P(accident involved car driver) $= P(E_2/A)$

$$= \frac{P(E_2)P(A/E_2)}{P(E_1)P(A/E_1) + P(E_2)P(A/E_2) + P(E_3)P(A/E_3)}$$

$$= \frac{\frac{1}{3} \times 0.03}{\frac{1}{6} + 0.01 + \frac{1}{3} \times 0.03 + \frac{1}{2} \times 0.15} = \frac{0.01}{0.52/6} = \frac{0.06}{0.52} = \mathbf{\frac{3}{26}}.$$

Example 6. *The following observations were made at a clinic, where HIV virus test is performed :*

(i) 15% of the patients at the clinic have HIV virus

(ii) among those that have HIV virus, 95% test positive on the ELISA test

(iii) among those that do not have HIV virus, 2% test positive on the ELISA test.

Find the probability that a patient has the HIV virus if the ELISA test is positive.

Sol. Let E_1 and E_2 be the events that 'a patient has HIV virus' and 'a patient does not have HIV virus' respectively. Let A be the event that the ELISA test is positive

$$\therefore \quad P(E_1) = \frac{15}{100} = 0.15$$

and $$P(E_2) = P(\overline{E}_1) = 1 - 0.15 = 0.85$$

Also, $$P(A/E_1) = \frac{95}{100} = 0.95$$

and $$P(A/E_2) = \frac{2}{100} = 0.02$$

By **Bayes's Theorem,**

P(patient has HIV virus if the ELISA test is positive)

$$= P(E_1/A) = \frac{P(E_1)\,P(A/E_1)}{P(E_1)\,P(A/E_1) + P(E_2)\,P(A/E_2)}$$

$$= \frac{(0.15)(0.95)}{(0.15)(0.95) + (0.85)(0.02)} = \mathbf{0.893}.$$

Example 7. *In a test, an examinee either guesses or copies or knows the answer to a multiple choice question with four choices. The probability that he makes a guess is 1/3 and the probability that he copies the answer is 1/6. The probability that his answer is correct, given that he copied it, is 1/8. Find the probability that he knew the answer to the question, given that he correctly answered it.*

Sol. Let E_1, E_2, E_3 be the events that he 'guesses the answer', 'copies the answer', 'knows the answer' respectively.

$$\therefore \quad P(E_1) = \frac{1}{3}, \quad P(E_2) = \frac{1}{6}, \quad P(E_3) = 1 - \frac{1}{3} - \frac{1}{6} = \frac{1}{2}$$

($\because$ E_1, E_2, E_3 are *m.e.* and exhaustive)

Let A be the event that he answers correctly.

$\therefore$ $P(A/E_1)$ = Probability that he answers correctly on the basis of guess

$= \frac{1}{4}$ ($\because$ There are 4 choices to answer)

$P(A/E_2)$ = Probability that he answers correctly on the basis of copying

$= \frac{1}{8}$ (given)

$P(A/E_3)$ = Probability that he answers correctly when he knows the answer

$= 1.$ (A/E_3 is a sure event)

Required probability

= Probability that he knows the answer given that he has correctly answered it

$= P(E_3/A)$

By **Bayes' theorem**, we have $P(E_3/A) = \dfrac{P(E_3)P(A/E_3)}{P(E_1)P(A/E_1) + P(E_2)P(A/E_2) + P(E_3)P(A/E_3)}$

$$= \frac{\frac{1}{2}\times 1}{\left(\frac{1}{3}\times\frac{1}{4}\right)+\left(\frac{1}{6}\times\frac{1}{8}\right)+\left(\frac{1}{2}\times 1\right)} = \frac{\frac{1}{2}}{\frac{1}{12}+\frac{1}{48}+\frac{1}{2}} = \frac{\frac{1}{2}}{\frac{4+1+24}{48}} = \frac{1}{2}\times\frac{48}{29} = \mathbf{\frac{24}{29}}.$$

Example 8. *By examining the chest X-ray, the probability that T.B. is detected when a person is actually suffering is 0.99. The probability that the doctor diagnoses incorrectly that a person has T.B. on the basis of X-ray is 0.001. In a certain city, 1 in 1,000 persons suffers from T.B. A person is selected at random and is diagnosed to have T.B. What is the chance that he actually has T.B. ?*

Sol. Let E_1 and E_2 be the events that a person selected is "suffering" and "not suffering" from T.B. respectively.

$$\therefore \quad P(E_1) = \frac{1}{1000} = 0.001 \quad\text{and}\quad P(E_2) = 1 - \frac{1}{1000} = \frac{999}{1000} = 0.999.$$

Let A be the event that the doctor diagnoses a person to have T.B.

$\therefore$ $P(A/E_1)$ = Probability that the T.B. is detected by the doctor when the person is actually suffering

$= 0.99$

$P(A/E_2)$ = Probability that the T.B. is detected by the doctor when the person is actually not suffering

$= 0.001$

Required probability

= Probability that a person has T.B. who has already been declared to have T.B.

$= P(E_1/A).$

By **Bayes' theorem,** we have $P(E_1/A) = \dfrac{P(E_1)P(A/E_1)}{P(E_1)P(A/E_1) + P(E_2)P(A/E_2)}$

$$= \frac{(0.001)(0.99)}{(0.001)(0.99)+(0.999)(0.001)} = \frac{0.00099}{0.00099+0.000999} = \frac{0.00099}{0.001989} = \frac{990}{1989} = \mathbf{\frac{110}{221}}.$$

Example 9. *A doctor is to visit a patient. From the past experience, it is known that the probabilities that he will come by train, bus, scooter or by other means of transport are respectively $\frac{3}{10}, \frac{1}{5}, \frac{1}{10}$ and $\frac{2}{5}$. The probabilities that he will be late are $\frac{1}{4}, \frac{1}{3}$ and $\frac{1}{12}$, if he comes by train, bus and scooter respectively, but if he comes by any other means of transport, then he will not be late. When he arrives, he is late. What is the probability that he comes by train ?*

Sol. Let E_1, E_2, E_3, E_4 be the events that the doctor comes by train, bus, scooter, other means respectively.

$$\therefore \quad P(E_1) = \frac{3}{10}, \quad P(E_2) = \frac{1}{5}, \quad P(E_3) = \frac{1}{10}, \quad P(E_4) = \frac{2}{5}.$$

Let A be the event that the doctor is late.

$\therefore \quad P(A/E_1)$ = Probability that the doctor is late when he comes by train

$$= \frac{1}{4}$$

Similarly, $P(A/E_2) = \frac{1}{3}, \quad P(A/E_3) = \frac{1}{12}, \quad P(A/E_4) = 0.$

Required probability = Probability that he comes by train when he is late

$= P(E_1/A)$

By **Bayes' theorem,** we have

$$P(E_1/A) = \frac{P(E_1)P(A/E_1)}{P(E_1)P(A/E_1) + P(E_2)P(A/E_2) + P(E_3)P(A/E_3) + P(E_4)P(A/E_4)}$$

$$= \frac{\frac{3}{10} \times \frac{1}{4}}{\left(\frac{3}{10} \times \frac{1}{4}\right) + \left(\frac{1}{5} \times \frac{1}{3}\right) + \left(\frac{1}{10} \times \frac{1}{12}\right) + \left(\frac{2}{5} \times 0\right)}$$

$$= \frac{\frac{3}{40}}{\frac{3}{40} + \frac{1}{15} + \frac{1}{120} + 0} = \frac{\frac{3}{40}}{\frac{9+8+1}{120}} = \frac{3}{40} \times \frac{120}{18} = \frac{1}{2}.$$

EXERCISE 5.8

1. Two urns contain 4 white, 6 blue and 4 white, 5 blue balls. One of the urns is selected at random and a ball is drawn. If the ball drawn is white, find the probability that it is drawn from the second urn.
2. Three urns contain 6 red, 4 black ; 4 red, 6 black and 5 red, 5 black balls respectively. One of the urns is selected at random and a ball is drawn. If ball drawn is red, find the probability that it is drawn from the first urn.
3. Assume that a factory has two machines. Past records shows that machine I produces 60% of the items of output and machine II produces 40% of the items. Further, 2% of the items produced by machine I were defective and only 1% produced by machine II were defective. If a defective item is drawn at random, what is the probability that it was produced by machine I ?
4. In a bolt factory, machines A and B manufacture equal number of bolts. Of the total of their output, 1% and 2% are defective. A bolt is drawn at random from the total production and found to be defective. From which machine, the defective bolt is expected to have been produced ?

5. State and prove Bayes' theorem. In a bolt factory, machines A, B and C manufacture 25%, 35% and 40% of the total of output. Of their output, 5, 4 and 2 percent are defective. A bolt is drawn at random and is found to be defective. What is the probability that it was manufactured by the machine A ?

6. A manufacturing firm produces pipes in two plants with daily production volume of 5000 and 7000 units respectively. According to past experience, it is known that the fraction of defective outputs produced by the plants are 0.01 and 0.02 respectively. If a pipe is selected at random from a day's total production and found to be defective, find out the probability that it came from the second plant.

7. In a Post Office, two clerks are engaged in checking incoming mails. On an average, the first clerk (A_1) checks 60 percent mail, while the second (A_2) does the remaining. A_1 has an error rate of 0.03 and A_2 has an error rate of 0.02. A mail is selected at random from incoming mails of a day and is discovered to have an error, find the probability that it was checked by clerk A_2.

Answers

1. 10/19	2. 2/5	3. 3/4	4. B
5. 25/69	6. 14/19	7. 4/13.	

III. PROBABILITY DISTRIBUTION

5.22. VARIABLE

We have already studied a lot about frequency distributions. These distributions are based upon observations, *i.e.*, the frequencies for different values of the variable, under consideration, are based on actual observation. For example, if an unbiased coin is tossed 100 times, we may get head 57 times. Here, 57 is the observed frequency but theoretically we shall expect head 50 times. In this section, we shall study *probability distributions and frequency distributions* which are based upon theoretical considerations.

A symbol which can assume any of a prescribed set of values is called a **variable.** A variable is generally denoted by x, y, z, ...

A real valued function defined on the sample space of a random experiment is called a **random variable.** Thus, the values of a random variable are real numbers, connected with the outcomes of the experiment. In the experiment of toss of two coins, if we define the random variable (x) as the number of heads, then the values of the random variable x are 0, 1, 1, 2 corresponding to the outcomes *TT, TH, HT, HH* respectively. In case there are three coins, then the values of this random variable are 0, 1, 1, 1, 2, 2, 2, 3 corresponding to the outcomes *TTT, TTH, THT, HTT, HHT, HTH, THH, HHH* respectively.

We can define any number of random variables on the same sample space.

If x denote the random variable, defined as the cube of the number of tails, in the random experiment of toss of two coins, than we have :

Sample points	*HH*	*HT*	*TH*	*TT*
x	$(0)^3 = 0$	$(1)^3 = 1$	$(1)^3 = 1$	$(2)^3 = 8$

A **discrete random variable** is a random variable which can take only finite number of values. For example, in the experiment of drawing three cards from a pack of playing cards, the random variable "no. of kings drawn" is a discrete random variable, taking 4 values 0, 1, 2 and 3.

A **continuous random variable** is a random variable which can assume any value between certain limits.

5.23. PROBABILITY DISTRIBUTION OF A DISCRETE RANDOM VARIABLE

Let x be a discrete random variable assuming values $x_1, x_2, x_3, \ldots, x_n$ corresponding to various outcomes of a random experiment. If the probability of occurrence of $x = x_i$ is $P(x_i) = p_i$, $1 \le i \le n$ such that $p_1 + p_2 + p_3 + \ldots + p_n = 1$, then the function, $P(x_i) = p_i$ is called the **probability function** of the random variable x and the set $\{P(x_1), P(x_2), P(x_3), \ldots, P(x_n)\}$ is called the **probability distribution** of x.

Example 1. *Find the probability distribution of the discrete random variable "the number of sixes" in two tosses of a die.*

Sol. Let E be the event of getting 'six' in a throw of die.

$\therefore$ $P(E) = 1/6.$

Let x denote the discrete random variable *number of sixes*.

$\therefore$ x can take values 0, 1 and 2.

Now $$P(x = 0) = P(\overline{E}_1 \overline{E}_2) = P(\overline{E}_1)P(\overline{E}_2) = \frac{5}{6} \times \frac{5}{6} = \frac{25}{36}$$

$$P(x = 1) = P(E_1 \overline{E}_2 \text{ or } \overline{E}_1 E_2) = P(E_1 \overline{E}_2) + P(\overline{E}_1 E_2)$$

$$= P(E_1)\, P(\overline{E}_2) + P(\overline{E}_1)\, P(E_2)$$

$$= \left(\frac{1}{6} \times \frac{5}{6}\right) + \left(\frac{5}{6} \times \frac{1}{6}\right) = \frac{10}{36} = \frac{5}{18}$$

$$P(x = 2) = P(E_1 E_2) = P(E_1)P(E_2) = \frac{1}{6} \times \frac{1}{6} = \frac{1}{36}.$$

$\therefore$ The required probability distribution is

x	0	1	2
$P(x)$	$\frac{25}{36}$	$\frac{5}{18}$	$\frac{1}{36}$

Example 2. *3 bad articles are mixed with 7 good ones. Find the probability distribution of the number of bad articles if 3 articles are drawn at random without replacement.*

Sol. Let x denote the discrete random variable *number of bad articles*.

$\therefore$ x can take values, 0, 1, 2, 3.

Now $$P(x = 0) = P(\text{no bad article}) = \frac{^7C_3}{^{10}C_3} = \frac{7}{24}$$

$$P(x = 1) = P(\text{1 bad article}) = \frac{^3C_1 . ^7C_3}{^{10}C_3} = \frac{21}{40}$$

$$P(x = 2) = P(\text{2 bad articles}) = \frac{^3C_2 . ^7C_1}{^{10}C_3} = \frac{7}{40}$$

$$P(x = 3) = P(\text{all bad articles}) = \frac{^3C_3}{^{10}C_3} = \frac{1}{120}$$

∴ The required probability distribution (P.D.) is

x	0	1	2	3
$P(x)$	$\frac{7}{24}$	$\frac{21}{40}$	$\frac{7}{40}$	$\frac{1}{120}$

Example 3. *A pair of unbiased dice is tossed. If the random variable x is the sum of two numbers obtained on two dice, find the probability distribution of x.*

Sol. Let S be the sample space of the experiment.

∴ $S = \{(1, 1), (1, 2), ..., (6, 5), (6, 6)\}$

The possible values of x are 2, 3, 4, ..., 12.

Now,
$$P(x = 2) = P\{(1, 1)\} = 1/36$$
$$P(x = 3) = P\{(1, 2), (2, 1)\} = 2/36$$
$$P(x = 4) = P\{(1, 3), (2, 2), (3, 1)\} = 3/36$$
$$P(x = 5) = P\{(1, 4), (2, 3), (3, 2), (4, 1)\} = 4/36$$
$$P(x = 6) = P\{(1, 5), (2, 4), (3, 3), (4, 2), (5, 1)\} = 5/36$$
$$P(x = 7) = P\{(1, 6), (2, 5), (3, 4), (4, 3), (5, 2), (6, 1)\} = 6/36$$
$$P(x = 8) = P\{(2, 6), (3, 5), (4, 4), (5, 3), (6, 2)\} = 5/36$$
$$P(x = 9) = P\{(3, 6), (4, 5), (5, 4), (6, 3)\} = 4/36$$
$$P(x = 10) = P\{(4, 6), (5, 5), (6, 4)\} = 3/36$$
$$P(x = 11) = P\{(5, 6), (6, 5)\} = 2/36$$
$$P(x = 12) = P\{(6, 6)\} = 1/36.$$

Also, $$\sum_{i=2}^{12} P(x = x_i) = \frac{1}{36} + \frac{2}{36} + \frac{3}{36} + \frac{4}{36} + \frac{5}{36} + \frac{6}{36} + \frac{5}{36} + \frac{4}{36} + \frac{3}{36} + \frac{2}{36} + \frac{1}{36} = \frac{36}{36} = 1.$$

The required P.D. is

x	2	3	4	5	6	7	8	9	10	11	12
$P(x)$	$\frac{1}{36}$	$\frac{2}{36}$	$\frac{3}{36}$	$\frac{4}{36}$	$\frac{5}{36}$	$\frac{6}{36}$	$\frac{5}{36}$	$\frac{4}{36}$	$\frac{3}{36}$	$\frac{2}{36}$	$\frac{1}{36}$

Example 4. *Obtain the probability distribution of the total number of heads occurring in three tosses of an unbiased coin.*

Sol. Let x denote the discrete random variable *number of heads*.

∴ The possible values of x are 0, 1, 2, 3.

Now
$$P(x = 0) = P(TTT) = P(T)\,P(T)\,P(T) = \frac{1}{2} \times \frac{1}{2} \times \frac{1}{2} = \frac{1}{8}$$
$$P(x = 1) = P(HTT \text{ or } THT \text{ or } TTH) = P(HTT) + P(THT) + P(TTH)$$
$$= P(H)\,P(T)\,P(T) + P(T)\,P(H)\,P(T) + P(T)\,P(T)\,P(H)$$
$$= \left(\frac{1}{2} \times \frac{1}{2} \times \frac{1}{2}\right) + \left(\frac{1}{2} \times \frac{1}{2} \times \frac{1}{2}\right) + \left(\frac{1}{2} \times \frac{1}{2} \times \frac{1}{2}\right) = \frac{3}{8}$$
$$P(x = 2) = P(HHT \text{ or } HTH \text{ or } THH) = P(HHT) + P(HTH) + P(THH)$$
$$= P(H)\,P(H)\,P(T) + P(H)\,P(T)\,P(H) + P(T)\,P(H)\,P(H)$$

$$= (\tfrac{1}{2} \times \tfrac{1}{2} \times \tfrac{1}{2}) + (\tfrac{1}{2} \times \tfrac{1}{2} \times \tfrac{1}{2}) + (\tfrac{1}{2} \times \tfrac{1}{2} \times \tfrac{1}{2}) = \tfrac{3}{8}.$$

$$P(x = 3) = P(HHH) = P(H)\,P(H)\,P(H) = \tfrac{1}{2} \times \tfrac{1}{2} \times \tfrac{1}{2} = \tfrac{1}{8}.$$

Also $\displaystyle\sum_{i=0}^{3} P(x = i) = \frac{1}{8} + \frac{3}{8} + \frac{3}{8} + \frac{1}{8} = 1.$

$\therefore$ The required P.D. is

x	0	1	2	3
$P(x)$	1/8	3/8	3/8	1/8

Example 5. *Find the probability distribution of the number of doublets in four throws of a pair of dice.*

Sol. Let S be the sample space of the experiment.

$\therefore$ $S = \{(1, 1), (1, 2), ..., (6, 5), (6, 6)\}$

Let E be the event of getting a doublet in a throw of two dice

$\therefore$ $E = \{(1, 1), (2, 2), ..., (6, 6)\}$

$\therefore$ $P(E) = 6/36 = 1/6$

Let x denote the discrete random variable '*no. of doublets*' in four throws.

$\therefore$ The possible values of x are 0, 1, 2, 3, 4.

Now $P(x = 0) = P(\bar{E}_1 \bar{E}_2 \bar{E}_3 \bar{E}_4) = P(\bar{E}_1)\,P(\bar{E}_2)\,P(\bar{E}_3)\,P(\bar{E}_4)$

$$= \frac{5}{6} \times \frac{5}{6} \times \frac{5}{6} \times \frac{5}{6} = \frac{625}{1296}$$

$$P(x = 1) = P(E_1 \bar{E}_2 \bar{E}_3 \bar{E}_4 \text{ or } \bar{E}_1 E_2 \bar{E}_3 \bar{E}_4 \text{ or } \bar{E}_1 \bar{E}_2 E_3 \bar{E}_4 \text{ or } \bar{E}_1 \bar{E}_2 \bar{E}_3 E_4)$$

$$= \left(\frac{1}{6} \times \frac{5}{6} \times \frac{5}{6} \times \frac{5}{6}\right) + \left(\frac{5}{6} \times \frac{1}{6} \times \frac{5}{6} \times \frac{5}{6}\right) + \left(\frac{5}{6} \times \frac{5}{6} \times \frac{1}{6} \times \frac{5}{6}\right) + \left(\frac{5}{6} \times \frac{5}{6} \times \frac{5}{6} \times \frac{1}{6}\right)$$

$$= \frac{500}{1296}$$

$$P(x = 2) = P(E_1 E_2 \bar{E}_3 \bar{E}_4 \text{ or } \bar{E}_1 \bar{E}_2 E_3 E_4 \text{ or } \bar{E}_1 E_2 E_3 \bar{E}_4 \text{ or } E_1 \bar{E}_2 \bar{E}_3 E_4 \text{ or } \bar{E}_1 E_2 \bar{E}_3 E_4 \text{ or } E_1 \bar{E}_2 E_3 \bar{E}_4)$$

$$= \left(\frac{1}{6} \times \frac{1}{6} \times \frac{5}{6} \times \frac{5}{6}\right) + \left(\frac{5}{6} \times \frac{5}{6} \times \frac{1}{6} \times \frac{1}{6}\right) + \left(\frac{5}{6} \times \frac{1}{6} \times \frac{1}{6} \times \frac{5}{6}\right) + \left(\frac{1}{6} \times \frac{5}{6} \times \frac{5}{6} \times \frac{1}{6}\right) + \left(\frac{5}{6} \times \frac{1}{6} \times \frac{5}{6} \times \frac{1}{6}\right) + \left(\frac{1}{6} \times \frac{5}{6} \times \frac{1}{6} \times \frac{5}{6}\right)$$

$$= \frac{150}{1296}$$

$$P(x = 3) = P(E_1 E_2 E_3 \bar{E}_4 \text{ or } E_1 E_2 \bar{E}_3 E_4 \text{ or } E_1 \bar{E}_2 E_3 E_4 \text{ or } \bar{E}_1 E_2 E_3 E_4)$$

$$= \left(\frac{1}{6} \times \frac{1}{6} \times \frac{1}{6} \times \frac{5}{6}\right) + \left(\frac{1}{6} \times \frac{1}{6} \times \frac{5}{6} \times \frac{1}{6}\right) + \left(\frac{1}{6} \times \frac{5}{6} \times \frac{1}{6} \times \frac{1}{6}\right) + \left(\frac{5}{6} \times \frac{1}{6} \times \frac{1}{6} \times \frac{1}{6}\right)$$

$$= \frac{20}{1296}$$

$$P(x = 4) = P(E_1E_2E_3E_4) = P(E_1)P(E_2)P(E_3)P(E_4)$$
$$= \frac{1}{6} \times \frac{1}{6} \times \frac{1}{6} \times \frac{1}{6} = \frac{1}{1296}.$$

Also $\sum_{i=0}^{4} P(x = i) = \frac{625}{1296} + \frac{500}{1296} + \frac{150}{1296} + \frac{20}{1296} + \frac{1}{1296} = 1.$

$\therefore$ The required P.D. is

x	0	1	2	3	4
$P(x)$	$\frac{625}{1296}$	$\frac{500}{1296}$	$\frac{150}{1296}$	$\frac{20}{1296}$	$\frac{1}{1296}$

EXERCISE 5.9

1. Find the probability distribution of the number of heads in two throws of an unbiased coin.
2. Find the probability distribution of the number of tails in three throws of an unbiased coin.
3. Find the probability distribution of the number of ones in a single throw of two dice.
4. Find the probability distribution of the number of sixes in a single throw of three dice.
5. Find the probability distribution of the number of sixes in a single throw of four dice.
6. Three cards are drawn with replacements from a pack of playing cards. Find the probability distribution of the number of spades in the cards drawn.
7. Three cards are drawn without replacements from a pack of playing cards. Find the probability distribution of the number of diamonds in the cards drawn.

Answers

1.

x (no. of heads)	0	1	2
$P(x)$	1/4	1/2	1/4

2.

x (no. of tails)	0	1	2	3
$P(x)$	1/8	3/8	3/8	1/8

3.

x (no. of ones)	0	1	2
$P(x)$	$\frac{25}{36}$	$\frac{10}{36}$	$\frac{1}{36}$

4.

x (no. of sixes)	0	1	2	3
$P(x)$	$\frac{125}{216}$	$\frac{75}{216}$	$\frac{15}{216}$	$\frac{1}{216}$

5.

x (no. of sixes)	0	1	2	3	4
P(x)	$\frac{625}{1296}$	$\frac{500}{1296}$	$\frac{150}{1296}$	$\frac{20}{1296}$	$\frac{1}{1296}$

6.

x (no. of spades)	0	1	2	3
P(x)	$\frac{27}{64}$	$\frac{27}{64}$	$\frac{9}{64}$	$\frac{1}{64}$

7.

x (no. of diamonds)	0	1	2	3
P(x)	$\frac{703}{1700}$	$\frac{741}{1700}$	$\frac{234}{1700}$	$\frac{22}{1700}$

5.24. MEAN AND VARIANCE OF A RANDOM VARIABLE

We know the method of finding the mean and variance of frequency distributions. In a frequency distribution, we have frequencies corresponding to different values of the variable. Similarly in a probability distribution, we have probabilities corresponding to different admissible values of the discrete random variable.

Now, we shall extend the idea of mean and variance for probability distributions.

Let x be a discrete random variable assuming values $x_1, x_2, \ldots, x_n$ with respective probabilities $p_1, p_2, \ldots, p_n$ with $p_1 + p_2 + \ldots + p_n = 1$.

We define, mean (μ) $$= \frac{\sum_{i=1}^{n} p_i x_i}{\sum_{i=1}^{n} p_i} = \sum_{i=1}^{n} p_i x_i \qquad \left(\because \sum_{i=1}^{n} p_i = 1\right)$$

and variance $$= \frac{\sum_{i=1}^{n} p_i (x_i - \mu)^2}{\sum_{i=1}^{n} p_i} = \sum_{i=1}^{n} p_i (x_i - \mu)^2 .$$

We have $$\sum_{i=1}^{n} p_i (x_i - \mu)^2 = \sum_{i=1}^{n} p_i (x_i^2 + \mu^2 - 2\mu x_i)$$

$$= \sum_{i=1}^{n} p_i x_i^2 + \mu^2 \sum_{i=1}^{n} p_i - 2\mu \sum_{i=1}^{n} p_i x_i$$

$$= \sum_{i=1}^{n} p_i x_i^2 + \mu^2 . 1 - 2\mu.\mu = \sum_{i=1}^{n} p_i x_i^2 - \mu^2 .$$

$\therefore$ **Mean** $(\mu) = \sum_{i=1}^{n} p_i x_i$ and **variance** $(\sigma^2) = \sum_{i=1}^{n} p_i x_i - \mu^2$.

In short, we write, $\mu = \Sigma px$ and variance $= \Sigma px^2 - \mu^2$.

The mean of random variable x is called the **expected value** of x and is denoted by $E(x)$. The mean and variance of a random variable are also referred to as the mean and variance of the corresponding P.D.

Remark 1. $E(x) = p_1x_1 + p_2x_2 + ... + p_nx_n$.

Remark 2. For the sake of simplicity, $\sum_{i=1}^{n} p_i x_i$ and $\sum_{i=1}^{n} p_i x_i^2$ are written in Σpx and Σpx^2 respectively.

Remark 3. S.D. of probability distribution $= \sqrt{\text{variance}} = \sqrt{\Sigma px^2 - \mu^2}$.

Example 1. *A die is tossed twice. A 'success' is 'getting an odd number' in a random toss. Find the expected value and variance of the number of successes.*

Sol. Let E be the event of getting a success *i.e.*, of getting an odd number. On a die, odd numbers are 1, 3, 5.

$$P(E) = \frac{3}{6} = \frac{1}{2} \quad \text{and} \quad P(\overline{E}) = 1 - \frac{1}{2} = \frac{1}{2}.$$

Let x denote the random variable 'number of successes'

$\therefore$ The possible values of x are 0, 1, 2.

$$P(x = 0) = P(\overline{E}_1 \overline{E}_2) = P(\overline{E}_1)\, P(\overline{E}_2) = \frac{1}{2} \times \frac{1}{2} = \frac{1}{4}$$

$$P(x = 1) = P(E_1\overline{E}_2 \text{ or } \overline{E}_1 E_2) = P(E_1\overline{E}_2) + P(\overline{E}_1 E_2)$$

$$= P(E_1)\, P(\overline{E}_2) + P(\overline{E}_1)\, P(E_2) = \left(\frac{1}{2} \times \frac{1}{2}\right) + \left(\frac{1}{2} \times \frac{1}{2}\right) = \frac{1}{2}$$

$$P(x = 2) = P(E_1E_2) = P(E_1)\, P(E_2) = \frac{1}{2} \times \frac{1}{2} = \frac{1}{4}.$$

Calculation of Mean and Variance

x	p	px	px^2
0	$\frac{1}{4}$	0	0
1	$\frac{1}{2}$	$\frac{1}{2}$	$\frac{1}{2}$
2	$\frac{1}{4}$	$\frac{1}{2}$	1
		$\Sigma px = 1$	$\Sigma px^2 = \frac{3}{2}$

$$E(x) = \text{Mean } (\mu) = \Sigma px = \mathbf{1}$$

$$\text{Variance} = \Sigma px^2 - \mu^2 = \frac{3}{2} - (1)^2 = \mathbf{0.5}.$$

Example 2. *Three urns contain respectively 3 green and 2 white balls, 5 green and 6 white balls and 2 green and 4 white balls. One ball is drawn from each urn. Find the expected value and variance of the probability distribution of the random variable "no. of white balls drawn".*

Sol. Let x denote the discrete random variable, '*no. of white ball drawn*'.

Urn I	Urn II	Urn III
3 Green, 2 White	5 Green, 6 White	2 Green, 4 White

$\therefore$ The possible values of x are 0, 1, 2, 3. Let G_i, W_i be the events of drawing a green ball, white ball respectively from the urn i $(1 \le i \le 3)$.

Now
$$P(x=0) = P(G_1G_2G_3) = P(G_1)P(G_2)P(G_3) = \frac{3}{5}\times\frac{5}{11}\times\frac{2}{6} = \frac{30}{330} = \frac{1}{11}$$

$$P(x=1) = P(W_1G_2G_3 \text{ or } G_1W_2G_3 \text{ or } G_1G_2W_3)$$
$$= \left(\frac{2}{5}\times\frac{5}{11}\times\frac{2}{6}\right)+\left(\frac{3}{5}\times\frac{6}{11}\times\frac{2}{11}\right)+\left(\frac{3}{5}\times\frac{5}{11}\times\frac{4}{6}\right)$$
$$= \frac{20+36+36}{330} = \frac{116}{330} = \frac{58}{165}$$

$$P(x=2) = P(W_1W_2G_3 \text{ or } W_1G_2W_3 \text{ or } G_1W_2W_3)$$
$$= \left(\frac{2}{5}\times\frac{6}{11}\times\frac{2}{6}\right)+\left(\frac{2}{5}\times\frac{5}{11}\times\frac{4}{6}\right)+\left(\frac{3}{5}+\frac{6}{11}+\frac{4}{6}\right)$$
$$= \frac{24+40+720}{330} = \frac{136}{330} = \frac{68}{165}$$

$$P(x=3) = P(W_1W_2W_3) = \frac{2}{5}\times\frac{6}{11}\times\frac{4}{6} = \frac{48}{330} = \frac{8}{55}.$$

$\therefore$ The P.D. is

x	0	1	2	3
$P(x)$	$\frac{1}{11}$	$\frac{58}{165}$	$\frac{68}{165}$	$\frac{8}{55}$

Calculation of Mean and Variance

x	$p = P(x)$	px	px^2
0	$\frac{30}{300}$	0	0
1	$\frac{116}{330}$	$\frac{116}{330}$	$\frac{116}{330}$
2	$\frac{136}{330}$	$\frac{272}{330}$	$\frac{544}{330}$
3	$\frac{48}{330}$	$\frac{144}{330}$	$\frac{432}{330}$
Total		$\frac{532}{330}$	$\frac{1092}{330}$

Now, expected value of x $= E(x) = \Sigma px = \dfrac{532}{330} = \mathbf{1.6121}$

Variance $= \Sigma px^2 - (\text{mean})^2$

$= \dfrac{1092}{330} - (1.6121)^2 = \mathbf{0.7102}.$

Example 3. *Two unbiased dice are thrown. Find the expected value and S.D. of the probability distribution of the random variable "sum of numbers on the dice".*

Sol. Let x denote the random variable '*sum of numbers on the dice*'.

We have already studied the probability distribution of the discrete random variable x.

Calculate of Mean and S.D.

x	p	px	px^2
2	$\frac{1}{36}$	$\frac{2}{36}$	$\frac{4}{36}$
3	$\frac{2}{36}$	$\frac{6}{36}$	$\frac{18}{36}$
4	$\frac{3}{36}$	$\frac{12}{36}$	$\frac{48}{36}$
5	$\frac{4}{36}$	$\frac{20}{36}$	$\frac{100}{36}$
6	$\frac{5}{36}$	$\frac{30}{36}$	$\frac{180}{36}$
7	$\frac{6}{36}$	$\frac{42}{36}$	$\frac{294}{36}$
8	$\frac{5}{36}$	$\frac{40}{36}$	$\frac{320}{36}$
9	$\frac{4}{36}$	$\frac{36}{36}$	$\frac{324}{36}$
10	$\frac{3}{36}$	$\frac{30}{36}$	$\frac{300}{36}$
11	$\frac{2}{36}$	$\frac{22}{36}$	$\frac{242}{36}$
12	$\frac{1}{36}$	$\frac{12}{36}$	$\frac{144}{36}$
Total		$\frac{252}{36}$	$\frac{1974}{36}$

Now, expected value of $x = \Sigma px = \frac{252}{36} = \mathbf{7}$

$$\text{S.D.} = \sqrt{\Sigma px^2 - (\text{mean})^2} = \sqrt{\frac{1974}{36} - (7)^2} = \mathbf{2.4152.}$$

Example 4. *Five defective articles are mixed with twenty good ones. Four articles are drawn simultaneously at random. Find the mean and S.D. of the defective articles drawn.*

Sol. Let x denote the discrete random variable *no. of defective articles drawn.*

$\therefore$ The possible values of x are 0, 1, 2, 3, 4.

Now

$$P(x = 0) = \frac{^{20}C_4}{^{15}C_4} = 0.3830$$

$$P(x = 1) = \frac{^5C_1 \times {^{20}C_3}}{^{25}C_4} = 0.4506$$

$$P(x=2) = \frac{{}^5C_2 \times {}^{20}C_2}{{}^{25}C_4} = 0.1502$$

$$P(x=3) = \frac{{}^5C_3 \times {}^{20}C_1}{{}^{25}C_4} = 0.0158$$

$$P(x=4) = \frac{{}^5C_4}{{}^{25}C_4} = 0.0004.$$

∴ The P.D. is

x	0	1	2	3	4
$P(x)$	0.3830	0.4506	0.1502	0.0158	0.0004

Calculation of Mean and S.D.

x	p	px	px^2
0	0.3830	0	0
1	0.4506	0.4506	0.4506
2	0.1502	0.3004	0.6008
3	0.0158	0.0474	0.1422
4	0.0004	0.0016	0.0064
Total		0.8	1.2

Now, mean $= \Sigma px = \mathbf{0.8}$

$$\text{S.D.} = \sqrt{\Sigma px^2 - (\text{mean})^2}$$

$$= \sqrt{1.2 - (0.8)^2} = \mathbf{0.7483.}$$

EXERCISE 5.10

1. Two urns contain 5 black and 4 white balls and 4 black and 5 white balls. One ball is drawn from each urn. Find the expected value and variance of the probability distribution of the random variable "no. of white balls drawn".
2. Find the S.D. of probability distribution of the number of aces drawn, when a draw of 2 cards, one at a time, is made with replacement.
3. Find the S.D. of probability distribution of the number of aces drawn, when a draw of 2 cards, one at a time, is made without replacement.
4. Three defective pencils are mixed with ten good ones. Two pencils are drawn simultaneously at random. Find the expected value of defective pencils drawn.
5. A draw of 3 balls is made from an urn containing 7 red and 6 black balls. Find the S.D. of the probability distribution of the discrete random variable "no.of black balls drawn".

Answers

1. 1, 0.4938 **2.** 0.3768 **3.** 0.3730 **4.** 0.4616
5. 0.7882.

IV. BINOMIAL DISTRIBUTION

5.25. INTRODUCTION OF BINOMIAL DISTRIBUTION

The binomial distribution is a particular type of probability distribution. This was discovered by **James Bernoulli (1654—1705)** in the year 1700. This distribution mainly deals with attributes. An attribute is either present or absent with respect to elements of a population. For example, if a coin is tossed, we get either *head* or *tail*. The workers of a factory may be classified as *skilled* and *unskilled*. An item of a population of articles produced in a firm may be either defective or non-defective.

5.26. CONDITIONS

The following conditions are essential for the applicability of binomial distribution :

(*i*) **The random experiment is performed for a finite and fixed number of trials.** If in an experiment, a coin is tossed repeatedly or a ball is drawn from an urn repeatedly, then each toss or draw is called a **trial.** For example, if a coin is tossed 6 times, then this experiment has 6 trials. The number of trials is an experiment is generally denoted by 'n'.

(*ii*) **The trials are independent.** By this we mean that the result of a particular trial is not going to effect the result of any other trial. For example, if a coin is tossed or a die is thrown, the trials would be independent. If from a pack of playing cards, some draws of one card are made without replacing the cards, then the trials would not be independent. But if the card drawn is replaced before the next draw, then the trials would be independent.

(*iii*) **Each trial must result in either "success" or "failure".** In other words, in every trial, there should be only two possible outcomes *i.e., success* or *failure*. For example, if a coin is tossed, then either *head* or *tail* is observed. Similarly, if an item is examined, it is either *defective* or *non-defective*.

(*iv*) **The probability of success in each trial is same.** In other words, this condition requires that the probability of *success* should not change in different trials. For example, if a sample of two items is drawn, then the probability of exactly one being defective will be constant in different trials provided the items are replaced before the next draw.

5.27. BINOMIAL VARIABLE

A random variable which counts the number of successes in a random experiment with trials satisfying above four conditions is called a **binomial variable**.

For example, if a coin is tossed 5 times and the event of getting head is *success*, then the possible values of the binomial variable are 0, 1, 2, 3, 4, 5. This is so, because, the minimum number of successes is 0 and maximum number is 5.

5.28. BINOMIAL PROBABILITY FUNCTION

When a fair coin is tossed, we have only two possibilities : head and tail. Let us call the occurrence of head as 'success'. Therefore, the occurrence of tail would be a 'failure'. Let this coin be tossed 5 times. Suppose we are interested in finding the probability of getting 4 heads and 1 tail *i.e.*, of getting 4 successes. If S and F denote 'success' and 'failure' in a trial respectively, then there are ${}^5C_4 = 5$ ways of having 4 successes.

These are : *SSSSF, SSSFS, SSFSS, SFSSS, FSSSS.*

The probability of getting 4 successes in each case is $\left(\frac{1}{2}\right)^4\left(\frac{1}{2}\right)$, because the trials are independent.

$\therefore$ By using *addition theorem,* the required probability of having 4 successes is ${}^5C_4\left(\frac{1}{2}\right)^4\left(\frac{1}{2}\right)$, which is equal to $\frac{5}{32}$. Now we shall generalise this method of finding the probabilities for different values of *binomial variables.*

Let a random experiment satisfying the conditions of **binomial distribution** be performed. Let the number of trials in the experiment be n. Let p denote the probability of *success* in any trial.

$\therefore$ Probability of failure, $q = 1 - p$

Let x denote the binomial variable corresponding to this experiment.

$\therefore$ The possible values of x are $0, 1, 2, \ldots\ldots, n$.

If there are r successes in n trials, then there would be $n - r$ failures. One of the ways in which r successes may occur is

$$\underbrace{SS \ldots\ldots\ldots SS}_{r \text{ times}} \quad \underbrace{FF \ldots\ldots\ldots F}_{n-r \text{ times}}$$

where S and F denote success and failure in trials.

Now, $P(SS \ldots\ldots SFF \ldots\ldots F) = P(S)P(S) \ldots\ldots P(S)P(F)P(F) \ldots\ldots P(F)$

($\because$ the trials are independent)

$$= p.p \ldots\ldots p.q.q \ldots\ldots q = p^r q^{n-r}.$$

We know that ${}^nC_r$ is the number of combinations of n things taking r at a time. Therefore, the number of ways in which r successes can occur in n trials is equal to the number of ways of choosing r trials (for successes) out of total n trials *i.e.*, it is ${}^nC_r$. Therefore, there are ${}^nC_r$ ways in which, we get r successes and $n - r$ failures and the probability of occurrence of each of these ways is $p^r q^{n-r}$. Hence the probability of r successes in n trials in any order is

$$P(x = r) = p^r q^{n-r} + p^r q^{n-r} + \ldots\ldots {}^nC_r \text{ terms} \qquad \text{(By addition theorem)}$$

or $$\mathbf{P(x = r) = {}^nC_r p^r q^{n-r},\ 0 \le r \le n.}$$

This is called the **binomial probability function.** The corresponding **binomial distribution** is

x	0	1	2	....	n
$P(x)$	${}^nC_0 p^0 q^n$	${}^nC_1 p^1 q^{n-1}$	${}^nC_2 p^2 q^{n-2}$	....	${}^nC_n p^n q^0$

The probabilities of 0 success, 1 success, 2 successes, ..., n successes are respectively the 1st, 2nd, 3rd, ..., $(n + 1)$th terms in the binomial expansion of $(q + p)^n$. This is why, it is called **binomial distribution.**

5.29. BINOMIAL FREQUENCY DISTRIBUTION

If a random experiment, satisfying the requirements of binomial distribution, is repeated N times, then the expected frequency of getting r $(0 \le r \le n)$ successes is given by

$$\mathbf{N.P(x = r) = N.{}^nC_r p^r q^{n-r},\ \ 0 \le r \le n.}$$

The *frequencies* of getting 0 success, 1 success, 2 successes, ..., n successes are respectively the 1st, 2nd, 3rd, ..., $(n + 1)$th terms in the expansion of $N(q + p)^n$.

Example 1. *An unbiased coin is tossed six times. Find the probability of obtaining :*

(*i*) *no head* (*ii*) *all heads*

(*iii*) *at least one head i.e., one or more heads*

(*iv*) *at least 4 heads* (KU 2007) (*v*) *more than 4 heads*

(*vi*) *more than 4 and less than 6 heads* (*vii*) *more than 6 heads.*

Sol. Let p be the probability of success *i.e.*, of getting head in the toss of the coin.

$$\therefore \qquad n = 6, p = \frac{1}{2} \quad \text{and} \quad q = 1 - p = 1 - \frac{1}{2} = \frac{1}{2}.$$

Let x be the binomial variable **'no. of successes'**.

By **Binomial distribution**, $P(x = r) = {}^nC_r p^r q^{n-r}, \quad 0 \le r \le n.$

$$\therefore \qquad P(x = r) = {}^6C_r \left(\frac{1}{2}\right)^r \left(\frac{1}{2}\right)^{6-r} = {}^6C_r \left(\frac{1}{2}\right)^6 = {}^6C_r \frac{1}{64}, \quad 0 \le r \le n.$$

(*i*) $P(\text{no head}) = P(x = 0) = {}^6C_0 \dfrac{1}{64} = \mathbf{\dfrac{1}{64}}$

(*ii*) $P(\text{all heads}) = P(x = 6) = {}^6C_6 \dfrac{1}{64} = \mathbf{\dfrac{1}{64}}$

(*iii*) $P(\text{at least one head}) = 1 - P(\text{no head}) = 1 - \dfrac{1}{64} = \mathbf{\dfrac{63}{64}}$ [Using part (*i*)]

(*iv*) $P(\text{at least 4 heads}) = P(x \ge 4) = P(x = 4 \text{ or } 5 \text{ or } 6)$

$$= P(x = 4) + P(x = 5) + P(x = 6)$$

$$= {}^6C_4 \frac{1}{64} + {}^6C_5 \frac{1}{64} + {}^6C_6 \frac{1}{64}$$

$$= (15 + 6 + 1) \frac{1}{64} = \frac{22}{64} = \mathbf{\frac{11}{32}}$$

(*v*) $P(\text{more than 4 heads})$

$$= P(x > 4) = P(x = 5 \text{ or } 6) = P(x = 5) + P(x = 6)$$

$$= {}^6C_5 \frac{1}{64} + {}^6C_6 \frac{1}{64} = (6 + 1) \frac{1}{64} = \mathbf{\frac{7}{64}}$$

(*vi*) $P(\text{more than 4 heads and less than 6 heads})$

$$= P(4 < x < 6) = P(x = 5) = {}^6C_5 \frac{1}{64} = \frac{6}{64} = \mathbf{\frac{3}{32}}.$$

(*vii*) $P(\text{more than 6 heads}) = P(x > 6) = \mathbf{0}.$ ($\because$ The event is impossible)

Example 2. *A die is thrown 4 times. Getting a number greater than 2 is a success. Find the probability of getting (i) exactly 1 success (ii) less than 3 successes (iii) more than 3 successes.*

Sol. Let p be the probability of success *i.e.*, of getting number greater than 2 in the throw of a die.

$$\therefore \qquad n = 4, p = \frac{4}{6} = \frac{2}{3} \quad \text{and} \quad q = 1 - p = 1 - \frac{2}{3} = \frac{1}{3}.$$

Let x be the binomial variable "*no. of successes*".

By **Binomial distribution**, $P(x = r) = {}^nC_r\, p^r q^{n-r}, 0 \le r \le n.$

$\therefore \qquad P(x = r) = {}^4C_r \left(\frac{2}{3}\right)^r \left(\frac{1}{3}\right)^{4-r}, \; 0 \le r \le 4.$

(*i*) $P(\text{exactly 1 success}) = P(x = 1)$

$$= {}^4C_1 \left(\frac{2}{3}\right)^1 \left(\frac{1}{3}\right)^{4-1} = 4 \times \frac{2}{3} \times \frac{1}{27} = \mathbf{\frac{8}{81}}.$$

(*ii*) $P(\text{less than 3 successes}) = P(x < 3)$

$$= P(x = 0 \text{ or } x = 1 \text{ or } x = 2)$$

$$= P(x = 0) + P(x = 1) + P(x = 2) \qquad \text{(Addition theorem for m.e. events)}$$

$$= {}^4C_0 \left(\frac{2}{3}\right)^0 \left(\frac{1}{3}\right)^{4-0} + {}^4C_1 \left(\frac{2}{3}\right)^1 \left(\frac{1}{3}\right)^{4-1} + {}^4C_2 \left(\frac{2}{3}\right)^2 \left(\frac{1}{3}\right)^{4-2}$$

$$= \left(1 \times 1 \times \frac{1}{81}\right) + \left(4 \times \frac{2}{3} \times \frac{1}{27}\right) + \left(6 \times \frac{4}{9} \times \frac{1}{9}\right)$$

$$= \frac{1 + 8 + 24}{81} = \mathbf{\frac{33}{81}}.$$

(*iii*) $P(\text{more than 3 successes}) = P(x > 3)$

$$= P(x = 4) = {}^4C_4 (\tfrac{2}{3})^4 (\tfrac{1}{3})^0 = 1 \times \frac{16}{81} \times 1 = \mathbf{\frac{16}{81}}.$$

Example 3. *The probability of a bomb hitting a target is 1/5. It is known that the attack of 2 bombs is sufficient to destroy a bridge. If 6 bombs are targetted at the bridge, find the probability of the bridge being destroyed.*

Sol. Let p be the probability of success *i.e.*, a bomb hitting the bridge.

$\therefore \qquad n = 6, p = \frac{1}{5} \quad \text{and} \quad q = 1 - p = 1 - \frac{1}{5} = \frac{4}{5}.$

Let x be the binomial variable "*no. of successes*".

By **Binomial distribution,** $P(x = r) = {}^nC_r p^r q^{n-r}, \; 0 \le r \le n.$

$\therefore \qquad P(x = r) = {}^6C_r \left(\frac{1}{5}\right)^r \left(\frac{4}{5}\right)^{6-r}, \; 0 \le r \le 6.$

Now, $P(\text{the bridge is destroyed}) = P(x \ge 2) = 1 - P(x < 2)$

$$= 1 - \{P(x = 0) + P(x = 1)\}$$

$$= 1 - \left\{ {}^6C_0 \left(\frac{1}{5}\right)^0 \left(\frac{4}{5}\right)^6 + {}^6C_1 \left(\frac{1}{5}\right)^1 \left(\frac{4}{5}\right)^5 \right\}$$

$$= 1 - \left\{ \frac{4096}{15625} + \frac{6144}{15625} \right\} = \frac{5385}{15625} = \mathbf{0.3446.}$$

Example 4. *There are 20% chances for a worker of an industry to suffer from an occupational disease. 50 workers were selected at random and examined for the occupational disease. Find the probability that (i) only one worker is found suffering from the disease ; (ii) more than 3 are suffering from the disease ; (iii) none is suffering from the disease.*

Sol. Let p be the probability of success *i.e.*, a worker is suffering from disease.

$\therefore \qquad n = 50, p = \frac{20}{100} = \frac{1}{5} \quad \text{and} \quad q = 1 - p = 1 - \frac{1}{5} = \frac{1}{4}.$

Let x be the binomial variable, "*no. of successes*".

By **Binomial distribution,** $P(x=r) = {}^nC_r p^r q^{n-r},\ 0 \le r \le n.$

$$\therefore \quad P(x=r) = {}^{50}C_r \left(\frac{1}{5}\right)^r \left(\frac{4}{5}\right)^{50-r},\ 0 \le r \le 50.$$

(*i*) P(only one is suffering) $= P(x=1)$

$$= {}^{50}C_1 \left(\frac{1}{5}\right)^1 \left(\frac{4}{5}\right)^{49} = 50 \times \frac{1}{5} \times \left(\frac{4}{5}\right)^{49} = \mathbf{10}\left(\frac{\mathbf{4}}{\mathbf{5}}\right)^{\mathbf{49}}.$$

(*ii*) P(more than 3 are suffering) $= P(x>3) = 1 - P(x \le 3)$

$= 1 - P(x=0 \text{ or } x=1 \text{ or } x=2 \text{ or } x=3)$

$= 1 - \{P(x=0) + P(x=1) + P(x=2) + P(x=3)\}$

$$= 1 - \left\{ {}^{50}C_0 \left(\frac{1}{5}\right)^0 \left(\frac{4}{5}\right)^{50} + {}^{50}C_1 \left(\frac{1}{5}\right)^1 \left(\frac{4}{5}\right)^{49} + {}^{50}C_2 \left(\frac{1}{5}\right)^2 \left(\frac{4}{5}\right)^{48} + {}^{50}C_3 \left(\frac{1}{5}\right)^3 \left(\frac{4}{5}\right)^{47} \right\}$$

$$= 1 - \left\{ 1 \times 1 \times \left(\frac{4}{5}\right)^{50} + 50 \times \frac{1}{5} \times \left(\frac{4}{5}\right)^{49} + \left(\frac{50 \times 49}{1 \times 2}\right) \times \left(\frac{1}{5}\right)^2 \times \left(\frac{4}{5}\right)^{48} + \left(\frac{50 \times 49 \times 48}{1 \times 2 \times 3}\right) \times \left(\frac{1}{5}\right)^3 \times \left(\frac{4}{5}\right)^{47} \right\}$$

$$= 1 - \frac{4^{47}}{5^{50}} \{64 + (50 \times 16) + (1225 \times 4) + (19600 \times 1)\} = \mathbf{1} - \left(\frac{\mathbf{4^{47}}}{\mathbf{5^{50}}} \times \mathbf{25364}\right).$$

(*iii*) P(none is suffering) $= P(x=0)$

$$= {}^{50}C_0 \left(\frac{1}{5}\right)^0 \left(\frac{4}{5}\right)^{50} = 1 \times 1 \times \left(\frac{4}{5}\right)^{50} = \left(\frac{\mathbf{4}}{\mathbf{5}}\right)^{\mathbf{50}}.$$

Example 5. *Assuming that half the population is vegetarian so that the chance of an individual being a vegetarian is $\frac{1}{2}$ and assuming that 100 investigators can take samples of 10 individuals to see whether they are vegetarian, how many investigators would you expect to report that three persons or less were vegetarian.*

Sol. Let p be the probability of success *i.e.*, an individual is vegetarian.

$$\therefore \quad n = 10, N = 100, p = \frac{1}{2} \text{ and } q = 1 - p = 1 - \frac{1}{2} = \frac{1}{2}.$$

Let x be the binomial variable, '*no. of successes*'.

By **Binomial distribution,** $P(x=r) = {}^nC_r p^r q^{n-r},\ 0 \le r \le n.$

$$\therefore \quad P(x=r) = {}^{10}C_r \left(\frac{1}{2}\right)^r \left(\frac{1}{2}\right)^{10-r} = {}^{10}C_r \left(\frac{1}{2}\right)^{10} = {}^{10}C_r \frac{1}{1024},\ 0 \le r \le 10.$$

Now, P(3 or less vegetarian) $= P(x \le 3)$

$$= P(x=0) + P(x=1) + P(x=2) + P(x=3)$$

$$= {}^{10}C_0 \frac{1}{1024} + {}^{10}C_1 \frac{1}{1024} + {}^{10}C_2 \frac{1}{1024} + {}^{10}C_3 \frac{1}{1024}$$

$$= (1 + 10 + 45 + 120) \frac{1}{1024} = \frac{176}{1024} = 0.1719.$$

$\therefore$ No. of investigators expected to report 3 or less vegetarians in their samples

$$= 100\,(0.1719) = 17.19 \simeq \mathbf{17}.$$

Example 6. *There are 64 beds in a garden and 3 seeds of a particular type of flower are sown in each bed. The probability of a flower being white is 1/4. Find the number of beds with 3, 2, 1 and 0 white flowers.*

Sol. Let p be the probability of success *i.e.*, the flower is white.

$$\therefore \qquad n = 3,\ N = 64,\ p = \frac{1}{4},\ q = 1 - p = 1 - \frac{1}{4} = \frac{3}{4}.$$

Let x be the binomial variable, '*no. of successes*'.

By **Binomial distribution,** $P(x = r) = {}^nC_r p^r q^{n-r}$, $0 \le r \le n$.

$$\therefore \qquad P(x = r) = {}^3C_r \left(\frac{1}{4}\right)^r \left(\frac{3}{4}\right)^{3-r}, \ 0 \le r \le 3.$$

$$\therefore \quad P(3 \text{ white flowers}) = P(x = 3) = {}^3C_3 \left(\frac{1}{4}\right)^3 \left(\frac{3}{4}\right)^0 = \frac{1}{64}$$

$$P(2 \text{ white flowers}) = P(x = 2) = {}^3C_2 \left(\frac{1}{4}\right)^2 \left(\frac{3}{4}\right)^1 = \frac{9}{64}$$

$$P(1 \text{ white flower}) = P(x = 1) = {}^3C_1 \left(\frac{1}{4}\right)^1 \left(\frac{3}{4}\right)^2 = \frac{27}{64}$$

$$P(\text{no white flower}) = P(x = 0) = {}^3C_0 \left(\frac{1}{4}\right)^0 \left(\frac{3}{4}\right)^3 = \frac{27}{64}.$$

$\therefore$ Number of beds with 3 white flowers

$$= 64 \times P(x = 3) = 64 \times \frac{1}{64} = \mathbf{1}$$

Number of beds with 2 white flowers

$$= 64 \times P(x = 2) = 64 \times \frac{9}{64} = \mathbf{9}$$

Number of beds with 1 white flower

$$= 64 \times P(x = 1) = 64 \times \frac{27}{64} = \mathbf{27}$$

Number of beds with no white flower

$$= 64 \times P(x = 0) = 64 \times \frac{27}{64} = \mathbf{27}.$$

Example 7. *An irregular six faced die is thrown and the probability that in 10 throws it will give five even numbers is twice the probability that it will give four even numbers. How many times in 10,000 sets of 10 throws, would you expect to give no even number.*

Sol. Let the event of getting an even number in a single throw of die be the 'success' of the experiment.

Let probability of success $= p$

$\therefore$ Probability of failure, $q = 1 - p$

Let x denote the number of successes in the experiment. The possible values of x are 0, 1, 2, ..., 10.

By **Binomial distribution,** $P(x = r) = {}^{10}C_r p^r q^{10-r}$, $0 \le r \le 10$.

By the given condition,

$$P(x = 5) = 2.\ P(x = 4)$$

$$\therefore \quad {}^{10}C_5 p^5 q^5 = 2 \times {}^{10}C_4 p^4 q^6$$

$$\therefore \quad \frac{10.9.8.7.6}{1.2.3.4.5}\ p = 2 \times \frac{10.9.8.7}{1.2.3.4}\ q$$

$$\Rightarrow \quad p = \frac{5}{3} q \ \Rightarrow \ p = \frac{5}{3}(1-p)$$

$$\Rightarrow \quad p = \frac{5}{8} \text{ and } q = 1 - \frac{5}{8} = \frac{3}{8}.$$

$\therefore$ No. of times no even number is excepted out of 10,000 sets of 10 throws

$$= 10{,}000 \times P(x = 0)$$

$$= 10{,}000\ {}^{10}C_0 \left(\frac{5}{8}\right)^0 \left(\frac{3}{8}\right)^{10} = 0.549 \simeq \mathbf{1}.$$

Example 8. *In an experiment, a fair die is thrown 6 times. The event of occurring number greater than 4 is the 'success' of the experiment. Find the Binomial distribution of the experiment. If the same experiment is repeated 3645 times, find also the Binomial frequency distribution.*

Sol. Let p be the probability of success *i.e.*, of getting number greater than 4

$$\therefore \quad n = 6, p = \frac{2}{6} = \frac{1}{3} \text{ and } q = 1 - p = 1 - \frac{1}{3} = \frac{2}{3}.$$

Let x be the binomial variable, "*no. of successes*".

By **Binomial distribution,** $P(x = r) = {}^nC_r p^r q^{n-r}$, $0 \le r \le n$.

$$\therefore \quad P(x = r) = {}^6C_r \left(\frac{1}{3}\right)^r \left(\frac{2}{3}\right)^{6-r}, \ 0 \le r \le 6.$$

Now

$$P(x = 0) = {}^6C_0 \left(\frac{1}{3}\right)^0 \left(\frac{2}{3}\right)^6 = 1 \times 1 \times \frac{64}{729} = \frac{64}{729}$$

$$P(x = 1) = {}^6C_1 \left(\frac{1}{3}\right)^1 \left(\frac{2}{3}\right)^5 = 6 \times \frac{1}{3} \times \frac{32}{243} = \frac{192}{729}$$

$$P(x = 2) = {}^6C_2 \left(\frac{1}{3}\right)^2 \left(\frac{2}{3}\right)^4 = 15 \times \frac{1}{9} \times \frac{16}{81} = \frac{240}{729}$$

$$P(x = 3) = {}^6C_3 \left(\frac{1}{3}\right)^3 \left(\frac{2}{3}\right)^3 = 20 \times \frac{1}{27} \times \frac{8}{27} = \frac{160}{729}$$

$$P(x = 4) = {}^6C_4 \left(\frac{1}{3}\right)^4 \left(\frac{2}{3}\right)^2 = 15 \times \frac{1}{81} \times \frac{4}{9} = \frac{60}{729}$$

$$P(x = 5) = {}^6C_5 \left(\frac{1}{3}\right)^5 \left(\frac{2}{3}\right)^1 = 6 \times \frac{1}{243} \times \frac{2}{3} = \frac{12}{729}$$

$$P(x = 6) = {}^6C_6 \left(\frac{1}{3}\right)^6 \left(\frac{2}{3}\right)^0 = 1 \times \frac{1}{729} \times 1 = \frac{1}{729}.$$

∴ The required distributions are as follows :

Binomial Distribution		*Binomial Frequency Distribution*	
x	$P(x)$	x	$N. P(x) = 3645 \times P(x)$
0	$\frac{64}{729}$	0	$3645 \times \frac{64}{729} = 320$
1	$\frac{192}{729}$	1	$3645 \times \frac{192}{729} = 960$
2	$\frac{240}{729}$	2	$3645 \times \frac{240}{729} = 1200$
3	$\frac{160}{729}$	3	$3645 \times \frac{160}{729} = 800$
4	$\frac{60}{729}$	4	$3645 \times \frac{60}{729} = 300$
5	$\frac{12}{729}$	5	$3645 \times \frac{12}{729} = 60$
6	$\frac{1}{729}$	6	$3645 \times \frac{1}{729} = 5$

Example 9. *The probability of a man hitting a target is* $\frac{1}{4}$. *If he fires 7 times, what is the probability of his hitting the target at least twice ? How many least number of times must he fire so that the probability of his hitting at least once is greater than* $\frac{2}{3}$.

Sol. Let p be the probability of success *i.e.*, the man hitting the target.

$$\therefore \quad n = 7, p = \frac{1}{4} \quad \text{and} \quad q = 1 - p = 1 - \frac{1}{4} = \frac{3}{4}$$

Let x be the binomial variable, "*no. of successes*".

By **Binomial distribution,** $P(x = r) = {}^nC_r p^r q^{n-r}$, $0 \le r \le n$

$$\therefore \quad P(x = r) = {}^7C_r \left(\frac{1}{4}\right)^r \left(\frac{3}{4}\right)^{7-r}, \; 0 \le r \le 7$$

(*i*) P(hitting at least twice) $= P(x \ge 2) = 1 - P(x < 2)$

$$= 1 - P(x = 0 \text{ or } x = 1) = 1 - \{P(x = 0) + P(x = 1)\}$$

$$= 1 - \left\{{}^7C_0\left(\frac{1}{4}\right)^0\left(\frac{3}{4}\right)^7 + {}^7C_1\left(\frac{1}{4}\right)^1\left(\frac{3}{4}\right)^6\right\}$$

$$= 1 - \left\{1 \times 1 \times \left(\frac{3}{4}\right)^7 + 7 \times \frac{1}{4} \times \left(\frac{3}{4}\right)^6\right\} = \mathbf{0.5551.}$$

(*ii*) Let n be the number of times, he should fire so that the probability of hitting the target at least once is greater than $\frac{2}{3}$.

In this case, $P(x = r) = {}^nC_r \left(\frac{1}{4}\right)^r \left(\frac{3}{4}\right)^{n-r}$, $0 \le r \le n$

By the given condition,

$$P(x \ge 1) > \tfrac{2}{3} \quad \Rightarrow \quad 1 - P(x = 0) > \tfrac{2}{3}$$

$$\Rightarrow \quad 1 - {}^nC_0\left(\frac{1}{4}\right)^0\left(\frac{3}{4}\right)^n > \frac{2}{3} \qquad \Rightarrow \quad 1 - \left(\frac{3}{4}\right)^n > \frac{2}{3} \text{ or } \left(\frac{3}{4}\right)^n < \frac{1}{3}$$

This inequility holds if $n \geq 4$.

$\therefore \qquad n = 4.$

EXERCISE 5.11

1. If a die is thrown 6 times and getting either 5 or 6 is considered a success. Obtain the probability of getting 0, 1, 2, 3, 4, 5, 6 successes.
2. An unbiased coin is tossed 8 times. Find the probability of obtaining (*i*) exactly 2 heads (*ii*) more than 2 heads (*iii*) all heads.
3. The incidence of an occupational disease in a factory is such that the workers have a 25% chances of suffering from it. What is the probability that out of 6 workmen, 4 or more contact the disease ?
4. If on an average 1 vessel in every 10 is wrecked, find the probability that out of 5 vessels expected to arrive, at least 4 will arrive safely.
5. Out of 800 families with 4 children each, what percentage would be expected to have (*a*) 2 boys and 2 girls (*b*) at least one boy (*c*) no girl (*d*) at most 2 girls ? Assume equal probabilities for boys and girls.
6. Six dice are thrown 729 times. How many times do you expect at least 3 dice to show either five or six ?
7. In a certain town, 20% of population is literate, and assume that 200 investigators takes a sample of 10 individuals to see whether they are literate. How many investigators would you expect to report that 3 persons or less are literate in their samples ?
8. Five coins are tossed 3,200 times. Find the frequencies of getting x heads $(0 \leq x \leq 5)$.
9. An insurance agent sells policies to 6 men, all of identical age. According to the acturial tables, the probability that a man of this particular age will be alive 20 years hence is 3/4. Find the probability that in 20 years (*i*) at least two men (*ii*) at most two men, will be alive.
10. During war, 1 ship out of 9 was sunk on an average in making a certain voyage. What was the probability that exactly 3 out of a convoy of 6 ships would arrive safely ?
11. A machine manufacturing screws is known to produce 5% defectives. In a random sample of 15 screws, what is the probability that there are :

    (*i*) exactly 3 defectives and (*ii*) not more than 3 defectives ?
12. The probability of a man hitting a target is $\frac{1}{3}$. How many least number of times, must he fire so that the probability of hitting the target at least once is more than 90% ?
13. 10 students are chosen at random from a college in which 20% students do not reside in the college compound. Find the probability that exactly 8 of the students do not reside in the college compound.
14. The probability that an evening student will be graduate is 0.8. Determine the probability that out of 5 students :

    (*i*) none (*ii*) one (*iii*) at least one, will be graduate.
15. The probability that a student will be graduate is 0.4. Determine the probability that out of 5 students (*i*) none (*ii*) one (*iii*) at least one (*iv*) all will be graduate.
16. Out of 10 watches, 4 are defective. Three watches are selected at random. Find the probability that :

    (*i*) no watch is defective (*ii*) one watch is defective.
17. A coin is tossed 5 times. What is the probability of getting at least 3 heads ? (*KU 2006 C*)

Answers

1. $\frac{64}{729}, \frac{192}{729}, \frac{240}{729}, \frac{160}{729}, \frac{60}{729}, \frac{12}{729}, \frac{1}{729}$
2. $\frac{7}{64}, \frac{219}{256}, \frac{1}{256}$
3. $\frac{77}{2048}$
4. $\frac{45927}{50000}$
5. 37.5%, 93.75%, 6.25%, 68.75%
6. 233
7. 176
8. 100, 500, 1000, 1000, 500, 100
9. 0.9954, 0.0376
10. = 0.0193
11. (*i*) 0.0307 (*ii*) 0.9945
12. $n = 6$
13. 0.000074
14. (*i*) 0.00032 (*ii*) 0.0064 (*iii*) 0.99968
15. (*i*) 0.07776 (*ii*) 0.2592 (*iii*) 0.92224 (*iv*) 0.01024
16. (*i*) $\frac{27}{125}$ (*ii*) $\frac{54}{125}$
17. $\frac{1}{2}$.

V. PROPERTIES OF BINOMIAL DISTRIBUTION

5.30. MEAN OF B.D.

Let x be a binomial random variable and

$$P(x = r) = {}^nC_r p^r q^{n-r},\ 0 \le r \le n.$$

The *mean* of x is the average number of successes.

$$\therefore \quad \text{Mean,} \quad \mu = \sum_{r=0}^{n} r.P(x=r) = \sum_{r=0}^{n} r.{}^nC_r p^r q^{n-r}$$

$$= 0.\,{}^nC_0 p^0 q^n + 1.{}^nC_1 p^1 q^{n-1} + 2.\,{}^nC_2 p^2 q^{n-2} + \ldots\ldots + n.\,{}^nC_n p^n.q^0$$

$$= 0 + n.pq^{n-1} + 2.\frac{n(n-1)}{1.2} p^2 q^{n-2} + \ldots\ldots. + n.1.p^n$$

$$= np\left\{q^{n-1} + \frac{n-1}{1} pq^{n-2} + \ldots. + p^{n-1}\right\}$$

$$= np\,\{{}^{n-1}C_0 p^0 q^{n-1} + {}^{n-1}C_1 p^1 q^{n-2} + \ldots. + {}^{n-1}C_{n-1} p^{n-1} q^0\}$$

$$= np\,(q+p)^{n-1} = np(1)^{n-1} = np.$$

$\therefore$ Mean of x = **np.**

5.31. VARIANCE AND S.D. OF B.D. *(KU 2006 C)*

Let x be a binomial random variable and

$$P(x = r) = {}^nC_r p^r q^{n-r},\ 0 \le r \le n.$$

The variance and standard deviation of x measures the dispersion of the binomial distribution and are given by

$$\text{Variance} = \sum_{r=0}^{n} r^2.P(x=r) - \mu^2$$

and

$$\text{S.D} = \sqrt{\sum_{r=0}^{n} r^2.P(x=r) - \mu^2}.$$

Now $\sum_{r=0}^{n} r^2.P(x=r) = \sum_{r=0}^{n} r^2.{}^nC_r p^r q^{n-r}$

$$= 0.\,{}^nC_0 p^0 q^n + 1^2.\,{}^nC_1 p^1 q^{n-1} + 2^2.\,{}^nC_2 p^2 q^{n-2} + 3^2.{}^nC_3 p^n q^{n-3} + ... + n^2.\,{}^nC_n p^n q^0$$

$$= 0 + 1.\frac{n}{1} pq^{n-1} + 2^2.\frac{n(n-1)}{1.2} p^2 q^{n-2} + \frac{3^2.\,n(n-1)(n-2)}{1.2.3} p^3 q^{n-3} + ... + n^2.1.p^n.1$$

$$= np\left\{q^{n-1} + \frac{2(n-1)}{1} pq^{n-2} + \frac{3(n-1)(n-2)}{1\times 2} p^2 q^{n-3} + ... + np^{n-1}\right\}$$

$$= np\left\{\left(q^{n-1} + \frac{n-1}{1} pq^{n-2} + \frac{(n-1)(n-2)}{1\times 2} p^2 q^{n-3} + ... + p^{n-1}\right)\right.$$

$$\left. + \left(\frac{n-1}{1} pq^{n-2} + \frac{2(n-1)(n-2)}{1\times 2} p^2 q^{n-3} + ... + (n-1)p^{n-1}\right)\right\}$$

$$= np\{(q+p)^{n-1} + (n-1)\,p(q^{n-2} + (n-2)\,pq^{n-3} + ... + p^{n-2})\}$$

$$= np\{1 + (n-1)\,p(q+p)^{n-2}\} = np\,\{1 + (n-1)\,p.1\}$$

$$= np\{1 + np - p\} = np + n^2p^2 - np^2.$$

$\therefore$ Variance $= \sum_{r=0}^{n} r^2.P(x=r) - \mu^2 = (np + n^2p^2 - np^2) - (np)^2 = np - np^2$

$$= np(1-p) = \mathbf{npq}.$$

Also, **S.D.** $= \sqrt{\textbf{variance}} = \sqrt{\mathbf{npq}}$.

5.32. RECURRENCE FORMULA FOR B.D.

Let x be a binomial random variable and

$$P(x=r) = {}^nC_r p^r q^{n-r} \quad 0 \le r \le n.$$

For $0 \le k < n$, $P(k) = {}^nC_k\, p^k q^{n-k}$ and $P(k+1) = {}^nC_{k+1}\, p^{k+1}\, q^{n-(k+1)}$

Dividing, we get $\dfrac{P(k+1)}{P(k)} = \dfrac{{}^nC_{k+1} p^{k+1} q^{n-k-1}}{{}^nC_k p^k q^{n-k}}$

$$= \frac{n!}{(k+1)!\,(n-(k+1))!}\cdot\frac{k!\,(n-k)!}{n!}\cdot\frac{p}{q} = \frac{n-k}{k+1}\,\frac{p}{q}.$$

$\therefore$ $$\mathbf{P(k+1) = \frac{n-k}{k+1}.\frac{p}{q}\,P(k)\ \text{for}\ \ 0 \le k < n.}$$

This the required **recurrence formula.**

Example 1. *The mean and S.D. of a binomial distribution are 20 and 4 respectively. Calculate n, p and q.*

Sol. Let the binomial distribution be

$$P(x=r) = {}^nC_r p^r q^{n-r},\ 0 \le r \le n.$$

$\therefore$ Mean $= np$ and S.D. $= \sqrt{npq}$.

We are given mean $= 20$, S.D. $= 4$.

$\therefore \qquad np = 20, \sqrt{npq} = 4$

$\Rightarrow \qquad \sqrt{20q} = 4 \quad \Rightarrow \quad 20q = 16 \quad \Rightarrow \quad q = \frac{4}{5}$

$\therefore \qquad p = 1 - q = 1 - \frac{4}{5} = \frac{1}{5}.$

$np = 20$ implies $n \times \frac{1}{5} = 20$ *i.e.*, $n = 100$

$\therefore \quad \mathbf{n = 100, p = \frac{1}{5}, q = \frac{4}{5}}.$

Example 2. *If the sum of the mean and the variance of a binomial distribution of 5 trials is 9/5, then find the binomial distribution.*

Sol. Let the binomial distribution be

$$P(x = r) = {}^nC_r\, p^r q^{n-r},\ 0 \le r \le n.$$

$\therefore \qquad$ Mean $= np$ and variance $= npq$.

By the given condition,

$$np + npq = \frac{9}{5} \quad \text{and} \quad n = 5.$$

$\Rightarrow \qquad 5p + 5p(1-p) = \frac{9}{5} \quad \Rightarrow \quad 5p + 5p - 5p^2 = \frac{9}{5}$

$\Rightarrow \qquad 25p^2 - 50p + 9 = 0 \quad \therefore \quad p = \frac{1}{5}$

$\therefore \qquad q = 1 - p = 1 - \frac{1}{5} = \frac{4}{5}.$

$\therefore$ The binomial distribution is

$$\mathbf{P(x = r) = {}^5C_r \left(\frac{1}{5}\right)^r \left(\frac{4}{5}\right)^{5-r},\ 0 \le r \le 5.}$$

Example 3. *Is the following statement correct ? "The mean and variance of a binomial distribution are respectively 6 and 9".*

Sol. Let the binomial distribution be

$$P(x = r) = {}^nC_r\, p^r q^{n-r}\ 0, \le r \le n.$$

Now $\qquad$ mean $= 6 \quad \Rightarrow \quad np = 6$

$\qquad$ variance $= 9 \quad \Rightarrow \quad npq = 9$

$\therefore \qquad 6q = 9 \quad \Rightarrow \quad q = \frac{3}{2}.$

This is impossible, because probability of an event can never be greater than 1.

$\therefore$ The given statement is not correct.

EXERCISE 5.12

1. Determine the binomial distribution whose mean is 5 and standard deviation is $\sqrt{2.5}$.
2. Determine the probability of 3 successes in a binomial distribution whose mean and variance are respectively 2 and $\frac{3}{2}$.

3. For a binomial distribution, the mean is 6 and the standard deviation is $\sqrt{2}$. Find the probability of getting 7 successes.
4. Is there any inconsistency in the statement. "The mean of a Binomial Distribution is 80 and S.D. is 8." If no inconsistency is found, what shall be the values of p, q and n ?

Answers

1. $P(x = r) = {}^{10}C_r \left(\frac{1}{2}\right)^r \left(\frac{1}{2}\right)^{10-r}$, $0 \le r \le 10$

2. 0.2076 3. 0.2341 4. $\frac{1}{5}, \frac{4}{5}$, 400

SECTION II

6

Discrete Numeric Functions

6.1. INTRODUCTION

We know that a function from a set A to a set B is a rule or a set of rules which assigns to each element in A a unique element in B. In this chapter, we shall consider only those functions which are defined from the set of whole numbers* to the set of real numbers.

6.2. DISCRETE NUMERIC FUNCTION

A function defined from the set of whole numbers to the set of real numbers is called a **discrete numeric function** or simply a **numeric function.**

Thus the domain of a numeric function is **W** and its codomain is **R**.

Numeric functions are denoted by bold face lower case letters **a, b, c,** .... or by lower case letters a, b, c, ...

Let **a** be a numeric function. The values of the function **a** at the points 0, 1, 2, ..., n, ... are **a**(0), **a**(1), **a**(2), ..., **a**(n), ... respectively. For the sake of simplicity, these values are written as $a_0, a_1, a_2, \ldots\ldots, a_n, \ldots\ldots$ respectively.

Thus $a_0, a_1, a_2, \ldots\ldots, a_n, \ldots\ldots$ denote the values of the numeric function **a** at 0, 1, 2, ..., n, ... respectively.

The numeric function **a** is also written as $(a_0, a_1, a_2, \ldots\ldots, a_n, \ldots\ldots)$.

Remark. Let **a** be a numeric function defined by $a_r = \dfrac{4r+3}{r^2+1}$, $r \geq 0$. Here the numeric function '**a**' has only one simple expression to find its value for every value of r. In such a case, we use the notation

$$\mathbf{a} = \frac{4r+3}{r^2+1}, \; r \geq 0.$$

Example 1. *Let* **a** *be the numeric function defined as follows :*

$$a_r = \begin{cases} 2+r & 0 \leq r \leq 5 \\ 2-r & r > 5, r \text{ is odd} \\ 2/r & r > 5, \ r \text{ is even} \end{cases}.$$

Find the values of a_2, a_8, a_{17}.

---

*The set of whole numbers contains the numbers 0, 1, 2, 3, 4, ... and denoted by **W**. Thus, **W** is the set of all non-negative integers.

Sol. We have $a_r = \begin{cases} 2+r & 0 \le r \le 5 \\ 2-r & r > 5, r \text{ is odd} \\ 2/r & r > 5, r \text{ is even} \end{cases}$.

$$0 \le 2 \le 5 \Rightarrow a_2 = 2 + 2 = 4$$

$$8 > 5 \text{ and } 8 \text{ is even} \Rightarrow a_8 = \frac{2}{8} = \frac{1}{4},$$

$$17 > 5 \text{ and } 17 \text{ is odd} \Rightarrow a_{17} = 2 - 17 = -15.$$

Example 2. *Let* **a** *be a numeric function such that a_r is equal to the remainder when the integer r is divided by 19. Find a_0, a_{15}, a_{37}, a_{38}.*

Sol.

$$0 = 0.(19) + 0 \Rightarrow a_0 = 0$$
$$15 = 0.(19) + 15 \Rightarrow a_{15} = 15$$
$$37 = 1.(19) + 18 \Rightarrow a_{37} = 18$$
$$38 = 2.(19) + 0 \Rightarrow a_{38} = 0.$$

Example 3. *A ball is dropped to the floor from a height of 20 m. Suppose that the ball always rebounds to reach half of the height from which it falls.*

(i) *Let a_r denote the height it reaches in the rth rebound. Find the numeric function* **a.**

(ii) *Let b_r denote the loss in height during the rth rebound. Express b_r in terms of a_r. Also, find the numeric function* **b.**

Sol. (i) a_1 = height attained by the ball in the Ist rebound

$$= \frac{20}{2} \text{ m}$$

a_2 = height attained by the ball in the IInd rebound

$$= \frac{1}{2}\left(\frac{20}{2}\right) = \frac{20}{2^2} \text{ m}$$

....

a_r = height attained by the ball in the rth rebound $= \frac{20}{2^r}$ m

....

$$\therefore \quad \mathbf{a} = \frac{20}{2^r} \text{ m}, r \ge 0.$$

(ii) b_1 = loss in height during the Ist rebound

$$= \left(20 - \frac{20}{2}\right) \text{ m}$$

b_2 = loss in height during the IInd rebound

$$= \left(20 - \frac{20}{2^2}\right) \text{ m}$$

....

b_r = loss in height during the rth rebound

$$= \left(20 - \frac{20}{2^r}\right) \text{ m}$$

....

$\therefore \qquad \mathbf{b_r = 20 - a_r,\ r \geq 0}$

Also, $\qquad \mathbf{b = 20 - \dfrac{20}{2^r},\ r \geq 0.}$

Example 4. *Let* **a** *be a numeric function, where* a_r *is the altitude of an aircraft, in thousand of feet at the rth minute. Let the aircraft take off after spending 10 minutes on the ground, climbs up at a uniform speed to a cruising altitude of 30,000 feet in 10 minutes, starts to descend uniformly after 110 minutes of flying time and lands 10 minutes later. Find the numeric function* **a.**

Sol. $0 \leq r \leq 10 \Rightarrow a_r = 0$ ($\because$ The aircraft is on the ground)

For $10 \leq r \leq 20$, let $a_r = lr + m$.

$a_{10} = 0 \Rightarrow 10l + m = 0$ ...(1)

$a_{20} = 30^* \Rightarrow 20l + m = 30$ ...(2)

Solving (1) and (2), we get $l = 3$, $m = -30$.

$\therefore \quad a_r = 3r - 30 = 3(r - 10),\ 10 \leq r \leq 20$

$r = 10 \Rightarrow a_{10} = 3(10 - 10) = 0$, which is already known.

$\therefore \quad a_r = 3(r - 10),\ 11 \leq r \leq 20$

$21 \leq r \leq 120 \Rightarrow a_r = 30$ ($\because$ $10 + 110 = 120$)

For $120 \leq r \leq 130$, let $a_r = ur + v$. ($\because$ Lending time is 10 minutes)

$a_{120} = 30 \Rightarrow 120u + v = 30$ ...(3)

$a_{130} = 0 \Rightarrow 130u + v = 0$ ...(4)

Solving (3) and (4), we get $u = -3$ and $v = 390$.

$\therefore \quad a_r = -3r + 390 = 3(130 - r),\ 120 \leq r \leq 130$

$r = 120 \Rightarrow a_{120} = 3(130 - 120) = 30$, which is already known.

$\therefore \quad a_r = 3(130 - r),\ 121 \leq r \leq 130.$

Also, for $r \geq 131$, we have $a_r = 0$.

$\therefore$ The numeric function **a** is given below :

$$\mathbf{a_r = \begin{cases} 0 & 0 \leq r \leq 10 \\ 3(r-10) & 11 \leq r \leq 20 \\ 30 & 21 \leq r \leq 120 \\ 3(130 - r) & 121 \leq r \leq 130 \\ 0 & r \geq 131 \end{cases}}$$

EXERCISE 6.1

1. Let a be a numeric function defined as follows :

$$a_r = \begin{cases} 1 + 2r & 0 \leq r \leq 4 \\ \sqrt{r} & 5 \leq r \leq 10. \\ r^2 + 1 & r \geq 11 \end{cases}$$

Find the values of a_2, a_{15}, a_8.

$^*a_{20} = 30$ means the height of the aircraft is 30 thousand feet at the end of 20th minute.

2. Let **a** be a numeric function such that a_r is equal to 0 if the integer r is divisible by 7, and is equal to 1 otherwise. Find a_0, a_2, a_7, a_{15}.
3. Let **a** be a numeric function such that a_r is equal to r^2 if the integer r is a prime number, and is equal to $r - 4$ otherwise. Find the value of $a_4 - 2a_3 + a_{17} + a_{20}$.
4. A person deposits Rs. 5000 in a saving account at an interest rate of 6% *p.a.* compounded annually. Find the numeric function **a**, where a_r is the amount of Rs. 5000 at the end of the rth year, $r \geq 0$.

Answers

1. 5, 226, $\sqrt{8}$ 2. 0, 1, 0, 1 3. 287 4. $\mathbf{a} = 5000\,(1.06)^r,\ r \geq 0$

6.3. ALGEBRA OF NUMERIC FUNCTIONS

I. Scaled version of a numeric function. Let **a** be a numeric function and α, any real number. The numeric function whose value at r is α times a_r is called the **scaled version** of **a** with **scaling factor** α. This scaled version of **a** is denoted by α**a**.

Illustration. Let **a** be a numeric function, where

$$a_r = \begin{cases} 4 & 0 \leq r \leq 3 \\ r+7 & 4 \leq r \leq 9 \\ 10 & r \geq 10 \end{cases}.$$

The scaled version 9**a** or **a** with scaling factor 9 is given by

$$(9\mathbf{a})(r) = \begin{cases} 36 & 0 \leq r \leq 3 \\ 9(r+7) & 4 \leq r \leq 9 \\ 90 & r \geq 10 \end{cases}.$$

II. Sum of two numeric functions. Let **a** and **b** be two numeric functions. The **sum** of **a** and **b** is a numeric function whose value at r is equal to the sum of the values of **a** and ***b*** at r. Thus if **c** be the sum of numeric functions **a** and **b** then

$$c_r = a_r + b_r,\ r \geq 0.$$

Illustration. Let **a** and **b** be numeric functions, where

$$a_r = \begin{cases} 2r+1 & 0 \leq r \leq 2 \\ 4 & 3 \leq r \leq 7 \\ r^2 & r \geq 8 \end{cases}$$

and
$$b_r = \begin{cases} 1-r & 0 \leq r \leq 3 \\ 4r & 4 \leq r \leq 17 \\ 15 & r \geq 18 \end{cases}.$$

Let **c** be the sum of numeric functions **a** and **b.**

By definitions, $c_r = a_r + b_r,\ r \geq 0$.

We shall consider the cases :

$$0 \leq r \leq 2,\ r = 3,\ 4 \leq r \leq 7,\ 8 \leq r \leq 17,\ r \geq 18.$$

$$\begin{aligned}
0 \leq r \leq 2 &\Rightarrow c_r = a_r + b_r = (2r + 1) + (1 - r) = r + 2 \\
r = 3 &\Rightarrow c_3 = a_3 + b_3 = 4 + (1 - 3) = 2 \\
4 \leq r \leq 7 &\Rightarrow c_r = a_r + b_r = 4 + 4r = 4(r + 1) \\
8 \leq r \leq 17 &\Rightarrow c_r = a_r + b_r = r^2 + 4r \\
r \geq 18 &\Rightarrow c_r = a_r + b_r = r^2 + 15
\end{aligned}$$

$\therefore$ The sum **c** of numeric functions **a** and **b** is given by

$$c_r = \begin{cases} r+2 & 0 \le r \le 2 \\ 2 & r = 3 \\ 4(r+1) & 4 \le r \le 7 \\ r^2 + 4r & 8 \le r \le 17 \\ r+15 & r \ge 18 \end{cases}.$$

III. Product of two numeric functions. Let **a** and **b** be two numeric functions. The **product** of **a** and **b** is a numeric function whose value at r is equal to the product of the values of **a** and **b** at r. Thus if **c** be the product of numeric functions **a** and **b** then

$$c_r = a_r\, b_r,\ r \ge 0.$$

Illustration. Let **d** be the product of the numeric functions **a** and **b** defined in the above illustration.

$$\begin{aligned} \therefore \quad 0 \le r \le 2 &\Rightarrow d_r = a_r\, b_r = (2r+1)(1-r) = -2r^2 + r + 1 \\ r = 3 &\Rightarrow d_r = a_3 b_3 = 4(1-3) = -8 \\ 4 \le r \le 7 &\Rightarrow d_r = a_r\, b_r = 4(4r) = 16r \\ 8 \le r \le 17 &\Rightarrow d_r = a_r\, b_r = r^2(4r) = 4r^3 \\ r \ge 18 &\Rightarrow d_r = a_r\, b_r = r^2 . 15 = 15r^2 \end{aligned}$$

$\therefore$ The product **d** of numeric functions **a** and **b** is given by

$$d_r = \begin{cases} -2r^2 + r + 1 & 0 \le r \le 2 \\ -8 & r = 3 \\ 16r & 4 \le r \le 7 \\ 4r^3 & 8 \le r \le 17 \\ 15r^2 & r \ge 18 \end{cases}.$$

Remark. The sum and product of numeric functions **a** and **b** are denoted by **a + b** and **ab** respectively.

Example 1. *Let* **a** *be a numeric function such that a_r is equal to the remainder when the integer r is divided by 17. Let* **b** *be a numeric function such that b_r is equal to 0 if the integer r is divisible by 3, and is equal to 1 otherwise.*

(i) Let $c_r = a_r + b_r$. For what value of r is $c_r = 0$? For what value of r is $c_r = 1$?

(ii) Let $d_r = a_r\, b_r$. For what value of r is $d_r = 0$? For what value of r is $d_r = 1$?

Sol. We have $a = (0, 1, 2, 3, 4, 5, 6, 7, 8, 9, 10, 11, 12, 13, 14, 15, 16, 0, 1, 2, ...)$

and $b = (0, 1, 1, 0, 1, 1, 0, 1, 1, 0, 1, 1, 0, 1, 1, 0, 1, 1, 0, 1 ......)$

(*i*) $c_r = 0$ if $a_r + b_r = 0$ or if $a_r = 0$ and $b_r = 0$

or if r is divisible by 17 and 3 both or if r is divisible by 51.

$\therefore$ $c_r = 0$ if, $r = 0, 51, 102, 153, ...$

$c_r = 1$ if, $a_r + b_r = 1$ if $a_r = 0, b_r = 1$ or $a_r = 1, b_r = 0$

$a_r = 0, b_r = 1$ hold only when r is divisible by 17 and not divisible by 3.

$\therefore$ $c_r = 1$ if, $r = 17, 34, 68, 85, ...$

$a_r = 1, b_r = 0$ hold only when r is of the form $17k + 1$ and divisible by 3.

$\therefore$ $c_r = 1$ if, $r = 18, 69, 120, 171, ...$

(*ii*) $d_r = 0$ if. $a_r\ b_r = 0$ if $a_r = 0$ or $b_r = 0$ or if r is a divisible by 17 or by 3.

$\therefore$ $d_r = 0$ if, $r = 0, 3, 6, 9, 12, 15, 17, 18, ...$

$d_r = 1$ if $a_r\ b_r = 1$ if $a_r = 1$, $b_r = 1$ if r is of the form $17k + 1$ and not divisible by 3.

$\therefore$ $d_r = 0$ if $r = 1, 35, 52, 86, ...$

6.4. ACCUMULATED SUM OF A NUMERIC FUNCTION

If **a** is a numeric function then the numeric function whose value at r is equal to $\sum_{i=0}^{r} a_i$ is called the **accumulated sum** of the numeric function **a**.

Illustration. Let **a** be a numeric function, where $a_r = r$, $r \geq 0$.

Let **b** be the accumulated sum of the numeric function **a**.

$$\therefore \quad b_r = \sum_{i=0}^{r} a_i = a_0 + a_1 + a_2 + ... + a_r$$

$$= 0 + 1 + 2 + ... + r$$

$$= \frac{r(r+1)}{2}, \ r \geq 0.$$

Example 2. *Nitin deposits Rs. 12000, in a saving account at an interest rate of 7% p.a. compounded annually. Find the numeric function* **a**, *where a_r is the amount of Rs. 12000 at the end of the rth year, $r \geq 0$. Also find the accumulated sum of the numeric function* **a** *and interpret it.*

Sol. Rate of interest = 7% p.a.

$\therefore$ a_r = amount of Rs. 12000 at the end of r years

$$= \text{Rs. } 12000\left(1+\frac{7}{100}\right)^r = \text{Rs. } 12000\ (1.07)^r$$

$$\therefore \quad \mathbf{a = Rs.\ 12000\ (1.07)^r, \quad r \geq 0.}$$

Let **b** be the accumulated sum of the numeric function **a**.

$$\therefore \quad b_r = \sum_{i=0}^{r} a_i = a_0 + a_1 + a_2 + ... + a_r$$

$$\therefore \quad b_r = \text{Rs. } 12000 + \text{Rs. } 12000\ (1.07) + \text{Rs. } 12000\ (1.07)^2 + ... + \text{Rs. } 12000\ (1.07)^r \quad ...(1)$$

$$= \text{Rs. } \frac{12000\ ((1.07)^{r+1} - 1)}{1.07 - 1} = \text{Rs. } \frac{1200000}{7}\ ((1.07)^{r+1} - 1)$$

$$\therefore \quad \mathbf{b = Rs.\ \frac{1200000}{7}\ ((1.07)^{r+1} - 1), \quad r \geq 0.}$$

(1) $\Rightarrow$ $b_r = \text{Rs. } 12000\ (1.07)^r + \text{Rs. } 12000\ (1.07)^{r-1} + ... + \text{Rs. } 12000\ (1.07) + \text{Rs. } 12000$

∴ If Nitin deposits Rs. 12,000 per year and the rate of interest is 7% p.a. then b_r is equal to the total amount of an annuity of $r + 1$ instalments of Rs. 12000 each at the end of $r + 1$ years when each instalment is paid at the end of the year.

6.5. SHIFTING OF A NUMERIC FUNCTION

Let **a** be a numeric function and i, a positive integer.

(*i*) The symbol $S^i\mathbf{a}$ is used to denote the numeric function defined below :

$$(S^i\mathbf{a})(r) = \begin{cases} 0 & r = 0, 1, 2, \ldots\ldots, i-1 \\ a_{r-i} & r \geq i. \end{cases}$$

(*ii*) The symbol $S^{-i}\mathbf{a}$ is used to denote the numeric function defined below :

$$(S^{-i}\mathbf{a})(r) = a_{r+i} \qquad r \geq 0.$$

Example 3. *Let* **a** *be a numeric function defined as :*

$$a_r = \begin{cases} 1 & 0 \leq r \leq 10 \\ 2 & r \geq 11 \end{cases}.$$

Find the numeric functions :

(*i*) S^2 **a** (*ii*) S^5 **a** (*iii*) S^{-4} **a** (*iv*) S^{-7} **a**.

Sol. We have $a_r = \begin{cases} 1 & 0 \leq r \leq 10 \\ 2 & r \geq 11 \end{cases}.$

(*i*) Let the numeric function $S^2\mathbf{a}$ be denoted by **b**.

$$\therefore \qquad b_r = \begin{cases} 0 & r = 0, 1 \\ a_{r-2} & r \geq 2 \end{cases}$$

By definition of **a**, $a_{r-2} = 1$ if $0 \leq r-2 \leq 10$ if $2 \leq r \leq 12$

$a_{r-2} = 2$ if $r - 2 \geq 11$ if $r \geq 13$

$$\therefore \qquad b_r = \begin{cases} 0 & r = 0, 1 \\ 1 & 2 \leq r \leq 12 \\ 2 & r \geq 13 \end{cases}.$$

(*ii*) Let the numeric function $S^5\mathbf{a}$ be denoted by **c**.

$$\therefore \qquad c_r = \begin{cases} 0 & r = 0, 1, 2, 3, 4 \\ a_{r-5} & r \geq 5 \end{cases}$$

By definition of **a**, $a_{r-5} = 1$ if $0 \leq r-5 \leq 10$ if $5 \leq r \leq 15$

$a_{r-5} = 2$ if $r - 5 \geq 11$ if $r \geq 16$

$$\therefore \qquad c_r = \begin{cases} 0 & r = 0, 1, 2, 3, 4 \\ 1 & 5 \leq r \leq 15 \\ 2 & r \geq 16 \end{cases}.$$

(*iii*) Let the numeric function $S^{-4}\mathbf{a}$ be denoted by **d**.

$$\therefore \qquad d_r = a_{r+4} \quad r \geq 0$$

By definition of **a**, $a_{r+4} = 1$ if $0 \leq r+4 \leq 10$ if $-4 \leq r \leq 6$ if $0 \leq r \leq 6$

$a_{r+4} = 2$ if $r + 4 \geq 11$ if $r \geq 7$

$$\therefore \qquad d_r = \begin{cases} 1 & 0 \leq r \leq 6 \\ 2 & r \geq 7 \end{cases}.$$

(*iv*) Let the numeric function $S^{-7}\mathbf{a}$ be denoted by **e**.

$\therefore \qquad e_r = a_{r+7} \quad r \geq 0$

By definition of **a**, $a_{r+7} = 1$ if $0 \leq r + 7 \leq 10$ if $0 \leq r \leq 3$

$a_{r+7} = 2$ if $r + 7 \geq 11$ if $r \geq 4$

$$\therefore \qquad e_r = \begin{cases} 1 & 0 \leq r \leq 3 \\ 2 & r \geq 4 \end{cases}.$$

Example 4. *Let* **a** *be a numeric function defined as*

$$a_r = \begin{cases} 5 & 0 \leq r \leq 3 \\ 2^r + 3 & r \geq 4 \end{cases}.$$

Find the numeric functions (i) $S^2\mathbf{a}$ *(ii)* $S^{-2}\mathbf{a}$.

Sol. We have $\qquad a_r = \begin{cases} 5 & 0 \leq r \leq 3 \\ 2^r + 3 & r \geq 4 \end{cases}$

(*i*) Let the numeric function $S^2\ \mathbf{a}$ be denoted by **b**.

$$\therefore \qquad b_r = \begin{cases} 0 & r = 0, 1 \\ a_{r-2} & r \geq 2 \end{cases}$$

By definition of **a**, $a_{r-2} = 5$ if $0 \leq r - 2 \leq 3$ if $2 \leq r \leq 5$

$a_{r-2} = 2^{r-2} + 3$ if $r - 2 \geq 4$ if $r \geq 6$

$$\therefore \qquad b_r = \begin{cases} 0 & r = 0, 1 \\ 5 & 2 \leq r \leq 5. \\ 2^{r-2} + 3 & r \geq 6 \end{cases}$$

(*ii*) Let the numeric function $S^{-2}\mathbf{a}$ be denoted by **c**.

$\therefore \qquad c_r = a_{r+2} \quad r \geq 0$

By definition of **a**, $a_{r+2} = 5$ if $0 \leq r + 2 \leq 3$ if $0 \leq r \leq 1$

$a_{r+2} = 2^{r+2} + 3$ if $r + 2 \geq 4$ if $r \geq 2$

$$\therefore \qquad c_r = \begin{cases} 5 & r = 0, 1 \\ 2^{r+2} + 3 & r \geq 2 \end{cases}.$$

Remark. It is interesting to observe the numeric functions **a**, $S^2\mathbf{a}$ and $S^{-2}\mathbf{a}$:

a : (5, 5, 5, 5, 19, 35, 67, 131, ...)

$S^2\mathbf{a}$: (0, 0, 5, 5, 5, 5, 19, 35, ...)

$S^{-2}\mathbf{a}$: (5, 5, 19, 35, 67, 131, 259, 515, ...).

6.6. FORWARD DIFFERENCE OF A NUMERIC FUNCTION

Let **a** be a numeric function. The **forward difference** of **a** is a numeric function whose value at r is equal to $a_{r+1} - a_r$ for $r \geq 0$. The forward difference of **a** is denoted by $\Delta\mathbf{a}$.

$\therefore \qquad (\Delta\mathbf{a})(r) = a_{r+1} - a_r \quad r \geq 0$

Similarly, we can find $\Delta(\Delta\mathbf{a})$. This numeric function is denoted by $\Delta^2\mathbf{a}$.

6.7. BACKWARD DIFFERENCE OF A NUMERIC FUNCTION

Let **a** be a numeric function. The **backward difference** of **a** is a numeric function whose value is a_0 at $r = 0$ and is equal to $a_r - a_{r-1}$ for $r \geq 1$. The backward difference of **a** is denoted by $\nabla\mathbf{a}$.

$$\therefore \qquad (\nabla \mathbf{a})(r) = \begin{cases} a_0 & r = 0 \\ a_r - a_{r-1} & r \geq 1 \end{cases}$$

Similarly, we can find $\nabla(\nabla \mathbf{a})$. This numeric function is denoted by $\nabla^2 \mathbf{a}$.

For example, if $\quad \mathbf{a} = (1, 4, 9, 16, 25, 36, ...)$

then $\quad \Delta \mathbf{a} = (4 - 1, 9 - 4, 16 - 9, 25 - 16, 36 - 25, 49 - 36, ...)$

$\qquad = (3, 5, 7, 9, 11, 13, ...)$

and $\quad \nabla \mathbf{a} = (1, 4 - 1, 9 - 4, 16 - 9, 25 - 16, 36 - 25, ...)$

$\qquad = (1, 3, 5, 7, 9, 11, ...)$.

It may be noted that

$$S^{-1}(\nabla a) = \Delta a.$$

Example 5. *Let* **a** *be a numeric function defined as*

$$a_r = r^3 - 2r^2 + 3r + 2, \; r \geq 0.$$

Find the numeric functions $\Delta \mathbf{a}$, $\Delta^2 \mathbf{a}$ *and* $\Delta^3 \mathbf{a}$.

Sol. We have $\quad a_r = r^3 - 2r^2 + 3r + 2, \; r \geq 0$

Let **b** denote the numeric function $\Delta \mathbf{a}$.

$\therefore \quad b_r = a_{r+1} - a_r, \; r \geq 0$

$\therefore \quad b_r = [(r + 1)^3 - 2(r + 1)^2 + 3(r + 1) + 2] - [r^3 - 2r^2 + 3r + 2]$

or $\quad \mathbf{b_r = 3r^2 - r + 2, \quad r \geq 0.}$

Let **c** denote the numeric function $\Delta^2 \mathbf{a}$.

$\therefore \quad c_r = b_{r+1} - b_r, \; r \geq 0$

$\therefore \quad c_r = [3(r + 1)^2 - (r + 1) + 2] - [3r^2 - r + 2]$

or $\quad \mathbf{c_r = 6r + 2. \quad r \geq 0.}$

Let **d** represent the numeric function $\Delta^3 a$.

$\therefore \quad d_r = c_{r+1} - c_r, \; r \geq 0$

$\therefore \quad d_r = [6(r + 1) + 2] - [6r + 2]$

or $\quad \mathbf{d_r = 6, \quad r \geq 0.}$

Alternatively, we can write

$$\Delta \mathbf{a} = 3r^2 - r + 2, \quad r \geq 0$$

$$\Delta^2 \mathbf{a} = 6r + 2, \quad r \geq 0$$

$$\Delta^3 \mathbf{a} = 6, \quad r \geq 0.$$

Example 6. *Let* **a** *be a numeric function given by*

$$a_r = \begin{cases} 0 & 0 \leq r \leq 3 \\ 2^r + 5 & r \geq 4 \end{cases}.$$

Find the numeric functions $\Delta \mathbf{a}$ *and* $\nabla \mathbf{a}$.

Sol. We have $\quad a_r = \begin{cases} 0 & 0 \leq r \leq 3 \\ 2^r + 5 & r \geq 4 \end{cases}.$

Let **b** denote the forward difference numeric function $\Delta \mathbf{a}$ of **a**.

$\therefore \quad b_r = a_{r+1} - a_r, \; r \geq 0$

$$0 \le r \le 2 \quad \Rightarrow \quad b_r = 0 - 0 = 0$$
$$r = 3 \quad \Rightarrow \quad b_3 = a_4 - a_3 = (2^4 + 5) - 0 = 21$$
$$r \ge 4 \quad \Rightarrow \quad b_r = a_{r+1} - a_r = (2^{r+1} + 5) - (2^r + 5)$$
$$= 2^{r+1} - 2^r = 2^r (2 - 1) = 2^r.$$

$$\therefore \qquad b_r = \begin{Bmatrix} \mathbf{0} & \mathbf{0 \le r \le 2} \\ \mathbf{21} & \mathbf{r = 3} \\ \mathbf{2^r} & \mathbf{r \ge 4} \end{Bmatrix}$$

Let **c** denote the backward difference numeric function $\nabla\mathbf{a}$ of **a**.

$$\therefore \qquad c_0 = 0, c_r = a_r - a_{r-1}, \quad r \ge 1$$
$$1 \le r \le 3 \quad \Rightarrow \quad c_r = 0 - 0 = 0$$
$$r = 4 \quad \Rightarrow \quad c_r = a_4 - a_3 = (2^4 + 5) - 0 = 21$$
$$r \ge 5 \quad \Rightarrow \quad c_r = a_r - a_{r-1} = (2^r + 5) - (2^{r-1} + 5)$$
$$= 2^r - 2^{r-1} = 2^{r-1} (2 - 1) = 2^{r-1}$$

$$\therefore \qquad c_r = \begin{Bmatrix} \mathbf{0} & \mathbf{0 \le r \le 3} \\ \mathbf{21} & \mathbf{r = 4} \\ \mathbf{2^{r-1}} & \mathbf{r \ge 5} \end{Bmatrix}.$$

Example 7. *If* **a, b, c** *be numeric functions such that* **c = ab,** *show that*

$$(\Delta\mathbf{c})(r) = a_{r+1} \cdot (\Delta\mathbf{b})(r) + b_r \cdot (\Delta\mathbf{a})(r).$$

Sol. We have
$$(\Delta\mathbf{a})(r) = a_{r+1} - a_r$$
$$(\Delta\mathbf{b})(r) = b_{r+1} - b_r$$
$$(\Delta\mathbf{c})(r) = c_{r+1} - c_r$$
$$\therefore \qquad \text{L.H.S.} = (\Delta\mathbf{c})(r) = c_{r+1} - c_r = a_{r+1}\, b_{r+1} - a_r\, b_r$$
$$\text{R.H.S.} = a_{r+1} \cdot (\Delta\mathbf{b})(r) + b_r \cdot (\Delta\mathbf{a})(r)$$
$$= a_{r+1} (b_{r+1} - b_r) + b_r (a_{r+1} - a_r)$$
$$= a_{r+1}\, b_{r+1} - a_{r+1} b_r + b_r a_{r+1} - b_r a_r$$
$$= a_{r+1}\, b_{r+1} - a_r\, b_r$$
$$\therefore \qquad (\Delta\mathbf{c})(r) = a_{r+1} \cdot (\Delta\mathbf{b})(r) + b_r \cdot (\Delta\mathbf{a})(r).$$

6.8. CONVOLUTION OF NUMERIC FUNCTIONS

Let **a** and **b** be two numeric functions. The **convolution** of **a** and **b** is a numeric function whose value at r is

$$a_0 b_r + a_1 b_{r-1} + \ldots\ldots + a_i\, b_{r-i} + \ldots\ldots + a_r b_0, \quad r \ge 0.$$

Thus if **c** be the convolution of numeric functions **a** and **b** then

$$c_r = a_0 b_r + a_1 b_{r-1} + \ldots\ldots + a_i\, b_{r-i} + \ldots\ldots + a_r\, b_0, \quad r \ge 0$$

The convolution of **a** and **b** is denoted by **a ∗ b**.

Equivalently, we can write

$$(\mathbf{a} * \mathbf{b})(r) = a_0\, b_r + a_1 b_{r-1} + \ldots\ldots + a_i\, b_{r-i} + \ldots\ldots + a_r\, b_0, \ r \ge 0.$$

In particular,

$$(\mathbf{a} * \mathbf{b})(0) = a_0 b_0$$
$$(\mathbf{a} * \mathbf{b})(1) = a_0 b_1 + a_1 b_0$$
$$(\mathbf{a} * \mathbf{b})(2) = a_0 b_2 + a_1 b_1 + a_2 b_0$$
$$(\mathbf{a} * \mathbf{b})(3) = a_0 b_3 + a_1 b_2 + a_2 b_1 + a_3 b_0$$

....

Remark. If **a** and **b** are numeric functions then **a** $*$ **b** and **b** $*$ **a** are equal numeric functions.

Example 8. *If* **a** *and* **b** *are numeric functions given by* $a_r = 5^r,\ r \geq 0,\ b_r = 7^r,\ r \geq 0$, *then show that* $(\mathbf{a} * \mathbf{b})(r) = \sum_{i=0}^{r} 5^i\, 7^{r-i},\ r \geq 0.$

Sol. We have $\quad a_r = 5^r,\quad r \geq 0,\quad b_r = 7^r,\quad r \geq 0.$

By definition, $(\mathbf{a} * \mathbf{b})(r) = a_0 b_r + a_1 b_{r-1} + ...+ a_i\, b_{r-1} + ... + a_r\, b_0$

$$= \sum_{i=0}^{r} a_i b_{r-i} = \sum_{i=0}^{r} 5^i 7^{r-i}$$

$$\therefore \qquad (\mathbf{a} * \mathbf{b})(\mathbf{r}) = \sum_{\mathbf{i}=0}^{\mathbf{r}} \mathbf{5^i 7^{r-i}},\quad \mathbf{r} \geq \mathbf{0}.$$

Example 9. *Determine the numeric function* **a** $*$ **a**, *where the numeric function* **a** *is defined as*

$$a_r = \begin{cases} 1 & 0 \leq r \leq 2 \\ 0 & r \geq 3 \end{cases}.$$

Sol. We have
$$a_r = \begin{cases} 1 & 0 \leq r \leq 2 \\ 0 & r \geq 3 \end{cases}.$$

$$\therefore \qquad (\mathbf{a} * \mathbf{a})(0) = a_0 a_0 = 1(1) = 1$$
$$(\mathbf{a} * \mathbf{a})(1) = a_0 a_1 + a_1 a_0 = 1(1) + 1(1) = 2$$
$$(\mathbf{a} * \mathbf{a})(2) = a_0 a_2 + a_1 a_1 + a_2 a_0 = 1(1) + 1(1) + 1(1) = 3$$
$$(\mathbf{a} * \mathbf{a})(3) = a_0 a_3 + a_1 a_2 + a_2 a_1 + a_3 a_0 = 1(0) + 1(1) + 1(1) + 0(1) = 2$$
$$(\mathbf{a} * \mathbf{a})(4) = a_0 a_4 + a_1 a_3 + a_2 a_2 + a_3 a_1 + a_4 a_0$$
$$= 1(0) + 1(0) + 1(1) + 0(1) + 0(1) = 1$$
$$(\mathbf{a} * \mathbf{a})(5) = a_0 a_5 + a_1 a_4 + a_2 a_3 + a_3 a_2 + a_4 a_1 + a_5 a_0$$
$$= 1(0) + 1(0) + 1(0) + 0(1) + 0(1) + 0(1) = 0$$

Also $\qquad (\mathbf{a} * \mathbf{a})(r) = 0,\quad r \geq 6$

$$(\mathbf{a} * \mathbf{a})(r) = \begin{cases} \mathbf{1} & \mathbf{r = 0} \\ \mathbf{2} & \mathbf{r = 1} \\ \mathbf{3} & \mathbf{r = 2} \\ \mathbf{2} & \mathbf{r = 3} \\ \mathbf{1} & \mathbf{r = 4} \\ \mathbf{0} & \mathbf{r \geq 5} \end{cases}.$$

Example 10. *Determine the numeric function* **a** $*$ **b**, *where the numeric functions* **a** *and* **b** *are defined as*

$$a_r = \begin{cases} 0 & 0 \le r \le 2 \\ 2^r & r \ge 3 \end{cases}, \quad b_r = 2^r, \quad r \ge 0.$$

Sol. We have $\quad \mathbf{a} = (0, \ 0, \ 0, \ 2^3, \ 2^4, \ 2^5, \ \ldots\ldots)$

and $\quad \mathbf{b} = (1, \ 2, \ 2^2, \ 2^3, \ 2^4, \ 2^5, \ \ldots\ldots)$

Let **c** denote the numeric function **a** $*$ **b**.

$$\therefore \quad c_0 = a_0 b_0 = 0(1) = 0$$

$$c_1 = a_0 b_1 + a_1 b_0 = 0(2) + 0(1) = 0$$

$$c_2 = a_0 b_2 + a_1 b_1 + a_2 b_0 = 0(2^2) + 0(2) + 0(1) = 0$$

$$c_3 = a_0 b_3 + a_1 b_2 + a_2 b_1 + a_3 b_0$$
$$= 0(2^3) + 0(2^2) + 0(2) + 2^3(1) = 2^3$$

$$c_4 = a_0 b_4 + a_1 b_3 + a_2 b_2 + a_3 b_1 + a_4 b_0$$
$$= 0(2^4) + 0(2^3) + 0(2^2) + 2^3.2^1 + 2^4.1 = 2.2^4$$

$$c_r = a_0 b_r + a_1 \, b_{r-1} + \ldots\ldots + a_i \, b_{r-i} + \ldots\ldots + a_r b_0$$
$$= 0(2^r) + 0(2^{r-1}) + 0(2^{r-2}) + 2^3.2^{r-3} + \ldots\ldots + 2^i.2^{r-i} + \ldots\ldots + 2^r.1$$
$$= 2^r + \ldots\ldots + 2^r + \ldots\ldots + 2^r \ (r-2 \text{ terms})$$
$$= (r-2)2^r$$

....

$$\therefore \quad (\mathbf{a} * \mathbf{b})(r) = \begin{cases} 0 & 0 \le \mathrm{r} \le 2 \\ (\mathrm{r}-2)2^{\mathrm{r}} & \mathrm{r} \ge 3 \end{cases}.$$

Example 11. *Let* $S = \{x, y, z, \alpha, \beta\}$. *Let* **a** *and* **b** *be numeric functions defined as follows* :

a_r = *no. of sequences of length r using English letters x, y and z,* $r \ge 0$,

b_r = *no. of sequences of length r using Greek letters* α *and* β, $r \ge 0$.

If **c** *be the numeric function, where* c_r *is the number of sequences of length r using letters of the set S, then show that* **c** = **a** $*$ **b**.

and $$c_r = \sum_{i=0}^{r} 3^i \, 2^{r-i} \quad r \ge 0.$$

Sol. c_r = no. of sequences of length r using letters of the set S.

$\therefore$ In each sequence in c_r, there are r letters in total and there may be any number of English and Greek letters.

$\therefore$ In a sequence, if there are i English letters then there must be $r - i$ Greek letters, $0 \le i \le r$.

$$\therefore \quad c_r = a_0 \, b_r + a_1 b_{r-1} + \ldots\ldots + a_i \, b_{r-i} + \ldots\ldots + a_r \, b_0$$

(Using principles of counting)

$$= (\mathbf{a} * \mathbf{b})(r)$$

$$\therefore \quad \mathbf{c} = \mathbf{a} * \mathbf{b}.$$

Also $a_r = 3 \,.\, 3 \,.\, 3. \ldots\ldots r$ times $= 3^r$

$b_r = 2 \,.\, 2 \,.\, 2. \ldots\ldots r$ times $= 2^r$

$$\therefore \quad c_r = 3^0 2^r + 3^1 2^{r-1} + \ldots\ldots + 3^i \, 2^{r-i} + \ldots\ldots + 3^r \, 2^0$$

$$\therefore \qquad c_r = \sum_{i=0}^{r} 3^i \, 2^{r-i}, \; r \ge 0.$$

EXERCISE 6.2

1. Let **a** and **b** be numeric functions defined as follows :

$$a_r = \begin{cases} 0 & 0 \le r \le 2 \\ 2^{-r} + 5 & r \ge 3 \end{cases} \quad \text{and} \quad b_r = \begin{cases} 3 - 2^r & 0 \le r \le 1 \\ r + 2 & r \ge 2 \end{cases}.$$

Find the sum and product of numeric functions **a** and **b**.

2. Let **a** and **b** be numeric functions defined as follows :

$$a_r = \begin{cases} 0 & 0 \le r \le 3 \\ 2^{-r} + 3 & r \ge 4 \end{cases} \quad \text{and} \quad b_r = \begin{cases} 1 - 2^r & 0 \le r \le 2 \\ r + 4 & r \ge 3 \end{cases}.$$

Find the sum and product of numeric functions **a** and **b**.

3. Let **a** be a numeric function, where $a_r = 5^r$, $r \ge 0$. Find the accumulated sum of the numeric function **a**.

4. Let **a** be a numeric function given by

$$a_r = \begin{cases} 2 & 0 \le r \le 20 \\ 7 & r \ge 21 \end{cases}.$$

Find the numeric functions S^4**a** and S^7**a**.

5. Let **a** be a numeric function given by

$$a_r = \begin{cases} 9 & 0 \le r \le 20 \\ 14 & r \ge 21 \end{cases}.$$

Find the numeric functions S^{-4}**a** and S^{-7}**a**.

6. Let **a** be a numeric function defined as :

$$a_r = \begin{cases} -8 & 0 \le r \le 10 \\ 2^{-r} + 7r + 9 & r \ge 11 \end{cases}.$$

Find the numeric functions :

(*i*) S^3**a** (*ii*) S^{-3}**a**.

7. Let **a** be a numeric function defined as :

$$a_r = 4r^2 - r + 7, \quad r \ge 0.$$

Find the numeric functions Δ**a**, Δ^2**a**, Δ^3**a**.

8. Let **a** and **b** be numeric functions defined as :

$$a_r = r + 1, \; r \ge 0$$
$$b_r = 2^r, \; r \ge 0.$$

Find the numeric function Δ(**ab**).

9. Let **a** be a numeric function given by

$$a_r = \begin{cases} 0 & 0 \le r \le 2 \\ 2^{-r} + 5 & r \ge 3 \end{cases}.$$

Find the numeric functions Δ**a** and ∇**a**.

10. Let **a** and **b** be two numeric functions. The quotient of **a** and **b** denoted by **a/b** is a numeric function whose value at r is equal to a_r/b_r. Let **d** = **a/b**. Show that :

$$(\Delta\mathbf{d})(r) = \frac{b_r \,.\, (\Delta\mathbf{a})(r) - a_r \,.\, (\Delta\mathbf{b})(r)}{b_r\, b_{r+1}}.$$

11. Determine the numeric function **a** * **b**, where the numeric functions **a** and **b** are defined as :

$$a_r = \begin{cases} 4 & 0 \le r \le 2 \\ 0 & r \ge 3 \end{cases}$$

and

$$b_r = \begin{cases} 17 & 0 \le r \le 1 \\ 0 & r \ge 2 \end{cases}.$$

12. Determine the numeric function **a** * **b**, where the numeric functions **a** and **b** are defined as :

$$a_r = 5^r, \qquad r \ge 0.$$

and

$$b_r = \begin{cases} 0 & 0 \le r \le 1 \\ 5^r & r \ge 2 \end{cases}.$$

Answers

1. $(\mathbf{a} + \mathbf{b})(r) = \begin{cases} 3 - 2^r & 0 \le r \le 1 \\ 4 & r = 2 \\ 2^{-r} + r + 7 & r \ge 3 \end{cases}$

$(\mathbf{ab})(r) = \begin{cases} 0 & 0 \le r \le 2 \\ r \,.\, 2^{-r} + 2^{-r+1} + 5r + 10 & r \ge 3 \end{cases}$

2. $(\mathbf{a} + \mathbf{b})(r) = \begin{cases} 1 - 2^r & 0 \le r \le 2 \\ 7 & r = 3 \\ 2^{-r} + r + 7 & r \ge 4 \end{cases}$

$(\mathbf{ab})(r) = \begin{cases} 0 & 0 \le r \le 3 \\ r \,.\, 2^{-r} + 2^{-r+2} + 3r + 12 & r \ge 4 \end{cases}$

3. (Accumulated sum) $(r) = \dfrac{5^{r+1} - 1}{4}, r \ge 0$

4. If $S^4\mathbf{a}$ is denoted by **b** then $b_r = \begin{matrix} 0 & 0 \le r \le 3 \\ 2 & 4 \le r \le 24 \\ 7 & r \ge 25 \end{matrix}$

If $S^7\mathbf{a}$ is denoted by **c** then $c_r = \begin{cases} 0 & 0 \le r \le 6 \\ 2 & 7 \le r \le 27 \\ 7 & r \ge 28 \end{cases}$

5. If $S^{-4}\mathbf{a}$ is denoted by **b** then $b_r = \begin{cases} 9 & 0 \le r \le 16 \\ 14 & r \ge 17 \end{cases}$

If $S^{-7}\mathbf{a}$ is denoted by **c** then $c_r = \begin{cases} 9 & 0 \le r \le 13 \\ 14 & r \ge 14 \end{cases}$

6. (*i*) If $S^3\mathbf{a}$ is denoted by $\mathbf{b}$ then $b_r = \begin{cases} 0 & r = 0, 1, 2 \\ -8 & 3 \le r \le 13 \\ 2^{-r+3} + 7r - 12 & r \ge 14 \end{cases}$

(*ii*) If $S^{-3}\mathbf{a}$ is denoted by $\mathbf{c}$ then $c_r = \begin{cases} -8 & 0 \le r \le 7 \\ 2^{-r-3} + 7r + 30 & r \ge 8 \end{cases}$

7.
$$(\Delta\mathbf{a})(r) = 8r + 3, r \ge 0$$
$$(\Delta^2\mathbf{a})(r) = 8, r \ge 0$$
$$(\Delta^3\mathbf{a})(r) = 0, r \ge 0$$

8. $(\Delta(\mathbf{ab}))(r) = 2^r (r + 3), r \ge 0$

9. $(\Delta\mathbf{a})(r) = \begin{cases} 0 & 0 \le r \le 1 \\ 41/8 & r = 2 \\ -2^{-(r+1)} & r \ge 3 \end{cases}$ $(\nabla\mathbf{a})(r) = \begin{cases} 0 & 0 \le r \le 2 \\ 41/8 & r = 3 \\ -2^{-r} & r \ge 4 \end{cases}$

11. $(\mathbf{a} * \mathbf{b})_r = \begin{cases} 68 & r = 0 \\ 136 & r = 1 \\ 136 & r = 2 \\ 68 & r = 3 \\ 0 & r \ge 4 \end{cases}$ 12. $(\mathbf{a} * \mathbf{b})_r = \begin{cases} 0 & 0 \le r \le 1 \\ (r - 1)5^r & r \ge 2 \end{cases}$

6.9. GENERATING FUNCTIONS

If $\mathbf{a} = (a_0, a_1, a_2, \ldots\ldots, a_r, \ldots\ldots)$ is a numeric function then the infinite series

$a_0 + a_1 z + a_2 z^2 + \ldots\ldots + a_r z^r + \ldots\ldots$ is called the **generating function** of the numeric function $\mathbf{a}$ and is denoted by $A(z)$.

For example, the generating function $A(z)$ of the numeric function $\mathbf{a} = (1, 2, 2^2, \ldots\ldots, 2^r, \ldots\ldots)$ is $1 + 2z + 2^2z^2 + \ldots\ldots + 2^rz^r + \ldots\ldots$

$$\therefore \quad A(z) = 1 + 2z + (2z)^2 + \ldots\ldots + (2z)^r + \ldots\ldots$$

$$= \frac{1}{1-2z} \qquad \text{(Assuming } |z| < 1/2\text{)}$$

$\frac{1}{1-2z}$ is called the *closed form* of the generating function $A(z)$.

Example 1. *Find the generating functions in their closed form of the following numeric functions :*

(*i*) $\mathbf{a}$, *where* $a_r = r + 1, r \ge 0$ (*ii*) $\mathbf{b}$, *where* $b_r = r, r \ge 0$

(*iii*) $\mathbf{c}$, *where* $c_r = \dfrac{1}{r!}, r \ge 0$ (*iv*) $\mathbf{d}$, *where* $d_r = \begin{cases} 2 & r = 0 \\ 7 & r = 1 \\ 2^{r+1} & r \ge 2 \end{cases}$

(*v*) $\mathbf{e}$, *where* $e_r = 3^r + 4^r, r \ge 0.$

Sol. (*i*) We have $a_r = r + 1, \ r \geq 0$

$\therefore$ $\mathbf{a} = (1, 2, 3, 4, \ldots\ldots, r + 1, \ldots\ldots)$

$\therefore$ Generating function of **a**

$$= A(z) = 1 + 2z + 3z^2 + 4z^3 + \ldots\ldots + (r + 1)z^r + \ldots\ldots$$

$$= (1 - z)^{-2} = \frac{1}{(1-\mathbf{z})^2}.$$

(*ii*) We have $b_r = r, \ r \geq 0$

$\therefore$ $\mathbf{b} = (0, 1, 2, 3, \ldots\ldots, r, \ldots\ldots).$

$\therefore$ Generating function of **b**

$$= B(z) = 0 + 1.z + 2z^2 + 3z^3 + \ldots\ldots + rz^r + \ldots\ldots$$

$$= z(1 + 2z + 3z^2 + \ldots\ldots + rz^{r-1} + \ldots\ldots)$$

$$= z(1 - z)^{-2} = \frac{\mathbf{z}}{(1-\mathbf{z})^2}.$$

(*iii*) We have $c_r = \dfrac{1}{r!}, \quad r \geq 0$

$$\therefore \qquad \mathbf{c} = \left(1, \frac{1}{1!}, \frac{1}{2!}, \frac{1}{3!}, \ldots\ldots, \frac{1}{r!}, \ldots\ldots\right).$$

$\therefore$ Generating function of **c**

$$= C(z) = 1 + \frac{z}{1!} + \frac{z^2}{2!} + \frac{z^3}{3!} + \ldots\ldots + \frac{z^r}{r!} + \ldots\ldots = \mathbf{e}^z.$$

(*iv*) We have
$$d_r = \begin{cases} 2 & r = 0 \\ 7 & r = 1 \\ 2^{r+1} & r \geq 2 \end{cases}$$

$\therefore$ $\mathbf{d} = (2, 7, 2^3, 2^4, \ldots\ldots, 2^{r+1}, \ldots\ldots).$

$\therefore$ Generating function of **d**

$$= D(z) = 2 + 7z + 2^3z^2 + 2^4z^3 + \ldots\ldots + 2^{r+1}\,z^r + \ldots\ldots$$

$$= 2 + 7z + 2((2z)^2 + (2z)^3 + \ldots\ldots + (2z)^r + \ldots\ldots)$$

$$= 2 + 7z + 2[(1 + 2z + (2z)^2 + (2z)^3 + \ldots\ldots + (2z)^r + \ldots\ldots) - 1 - 2z]$$

$$= 3z + 2\left(\frac{1}{1-2z}\right) = 3z + \frac{2}{1-2z} = \frac{\mathbf{2 + 3z - 6z^2}}{\mathbf{1 - 2z}}.$$

(*v*) We have $e_r = 3^r + 4^r, \quad r \geq 0$

$\therefore$ $e = (1 + 1, 3 + 4, 3^2 + 4^2, 3^3 + 4^3, \ldots\ldots, 3^r + 4^r, \ldots\ldots).$

$\therefore$ Generating function of **e**

$$= E(z) = (1 + 1) + (3 + 4)z + (3^2 + 4^2)z^2 + (3^3 + 4^3)z^3 + \ldots\ldots + (3^r + 4^r)z^r + \ldots\ldots$$

$$= (1 + 3z + 3^2z^2 + 3^3z^3 + \ldots\ldots + 3^rz^r + \ldots\ldots) + (1 + 4z + 4^2z^2 + 4^3z^3 + \ldots\ldots + 4^rz^r + \ldots\ldots)$$

$$= (1 + 3z + (3z)^2 + (3z)^3 + \ldots\ldots + (3z)^r + \ldots\ldots)$$
$$+ (1 + 4z + (4z)^2 + (4z)^3 + \ldots\ldots + (4z)^r + \ldots\ldots)$$
$$= \frac{1}{1-3z} + \frac{1}{1-4z} = \frac{\mathbf{2-7z}}{\mathbf{(1-3z)(1-4z)}}.$$

Example 2. *Find the numeric functions whose generating functions are given below :*

(*i*) $\dfrac{3+2z-6z^2}{1-3z}$ (*ii*) $\dfrac{3}{(1-2z)^2}$

(*iii*) $\dfrac{3z}{(1-z)(1+2z)}$ (*iv*) $\dfrac{z}{(1-z)^2}$.

Sol. (*i*) Let $A(z) = \dfrac{3+2z-6z^2}{1-3z}$

Dividing, we get $A(z) = 2z + \dfrac{3}{1-3z}$

$\therefore$ $A(z) = 2z + 3(1 + (-3z))^{-1}$

$$= 2z + 3\left[1 + (-1)(-3z) + \frac{(-1)(-2)}{2!}(-3z)^2 + \frac{(-1)(-2)(-3)}{3!}(-3z)^3 + \ldots\ldots\right]$$
$$= 2z + 3[1 + 3z + 9z^2 + 27z^3 + \ldots\ldots]$$
$$= 3 + 11z + 3^3z^2 + 3^4z^3 + \ldots\ldots$$
$$= a_0 + a_1z + a_2z^2 + a_3z^3 + \ldots\ldots$$

where $a_0 = 3$, $a_1 = 11$, $a_r = 3^{r+1}$, $r \geq 2$.

$\therefore$ The numeric function of $A(z)$ is **a**, where

$$a_r = \begin{cases} 3 & r = 0 \\ 11 & r = 1. \\ 3^{r+1} & r \geq 2 \end{cases}$$

(*ii*) Let $A(z) = \dfrac{3}{(1-2z)^2}$

$\therefore$ $A(z) = 3(1 + (-2z))^{-2}$

$$= 3\left[1 + (-2)(-2z) + \frac{(-2)(-3)}{2!}(-2z)^2 + \frac{(-2)(-3)(-4)}{3!}(-2z)^3 + \ldots\ldots\right]$$
$$= 3[1 + 2.2z + 3.2^2z^2 + 4.2^3z^3 + \ldots\ldots]$$
$$= 3 + 3.2.2z + 3.3.2^2z^2 + 3.4.2^3z^3 + \ldots\ldots$$
$$= a_0 + a_1z + a_2z^2 + a_3z^3 + \ldots\ldots,$$

where $a_r = 3(r + 1)2^r$, $r \geq 0$.

$\therefore$ The numeric function of $A(z)$ is **a**,

where $a_r = 3(r+1)2^r$, $r \geq 0$.

(*iii*) Let $$A(z) = \frac{3z}{(1-z)(1+2z)}$$

Breaking $A(z)$ into partial fractions, we have

$$A(z) = \frac{3(1)}{(1-z)(3)} + \frac{3(-1/2)}{(3/2)(1+2z)} = \frac{1}{1-z} - \frac{1}{1+2z} = \frac{1}{1-z} - \frac{1}{1-(-2z)}$$

$$= (1 + z + z^2 + z^3 + ...) - (1 + (-2z) + (-2z)^2 + (-2z)^3 + ...)$$

$$= (1-1) + (1-(-2))z + (1-(-2)^2)z^2 + (1-(-2)^3)z^3 + ...$$

$$= a_0 + a_1 z + a_2 z^2 + a_3 z^3 + ..., \text{ where } a_r = 1-(-2)^r, r \geq 0$$

$\therefore$ The numeric function of $A(z)$ is **a**, where, $a_r = 1-(-2)^r, r \geq 0$.

(*iv*) Let $$A(z) = \frac{z}{(1-z)^2}$$

$\therefore$ $$A(z) = z(1-z)^{-2}$$

$$= z\left(1 + (-2)(-z) + \frac{(-2)(-3)}{2!}(-z)^2 + \frac{(-2)(-3)(-4)}{3!}(-z)^3 + ...\right)$$

$$= z(1 + 2z + 3z^2 + 4z^3 + ...) = z + 2z^2 + 3z^3 + 4z^4 + ...$$

$$= 0 + 1.z + 2z^2 + 3z^3 + 4z^4 + ...$$

$$= a_0 + a_1 z + a_2 z^2 + a_3 z^3 + a_4 z^4 + ..., \text{ where } a_r = r, r \geq 0.$$

$\therefore$ The numeric function of $A(z)$ is **a**, where $a_r = r, r \geq 0$.

The following table shows the generating functions of some important numeric functions :

S.No.	*Numeric function* $\mathbf{a} = (a_0, a_1, ..., a_n, ...)$	*Generating function* $A(z) = \sum_{n=0}^{\infty} a_n z^n$
1.	$a_n = 1$	$\frac{1}{1-z}$
2.	$a_n = \lambda^n$	$\frac{1}{1-\lambda z}$
3.	$a_n = (-1)^n$	$\frac{1}{1+z}$
4.	$a_n = n+1$	$\frac{1}{(1-z)^2}$
5.	$a_n = n$	$\frac{z}{(1-z)^2}$
6.	$a_n = \frac{1}{n!}$	e^z.

Theorem I. Let a = $(a_1, a_2, \ldots\ldots)$, b = $(b_1, b_2, \ldots\ldots)$, c = $(c_1, c_2, \ldots\ldots)$ be numeric functions and A(z), B(z), C(z) their corresponding generating functions. Prove the following :

(*i*) $b_r = \alpha a_r$, $r \geq 0 \Rightarrow B(z) = \alpha A(z)$

(*ii*) $b_r = \alpha^r a_r$, $r \geq 0 \Rightarrow B(z) = A(\alpha z)$

(*iii*) $c_r = a_r + b_r$, $r \geq 0 \Rightarrow C(z) = A(z) + B(z)$.

Proof. (*i*) We have $A(z) = a_0 + a_1 z + a_2 z^2 + \ldots\ldots$ ...(1)

and $B(z) = b_0 + b_1 z + b_2 z^2 + \ldots\ldots$

$\therefore$ $B(z) = \alpha a_0 + (\alpha a_1)z + (\alpha a_2)z^2 + \ldots\ldots$

$= \alpha(a_0 + a_1 z + a_2 z^2 + \ldots\ldots)$

$= \alpha A(z)$ [Using (1)]

$\therefore$ $\mathbf{B(z) = \alpha A(z).}$

(*ii*) We have $A(z) = a_0 + a_1 z + a_2 z^2 + \ldots\ldots$ ...(1)

and $B(z) = b_0 + b_1 z + b_2 z^2 + \ldots\ldots$

$\therefore$ $B(z) = \alpha^0 a_0 + (\alpha^1 a_1)z + (\alpha^2 a_2)z^2 + \ldots\ldots$

$= a_0 + a_1(\alpha z) + a_2(\alpha z)^2 + \ldots\ldots$

$= A(\alpha z)$ [Using (1)]

$\therefore$ $\mathbf{B(z) = A(\alpha z)}$

(*iii*) We have $A(z) = a_0 + a_1 z + a_2 z^2 + \ldots\ldots$ ...(1)

$B(z) = b_0 + b_1 z + b_2 z^2 + \ldots\ldots$ ...(2)

and $C(z) = c_0 + c_1 z + c_2 z^2 + \ldots\ldots$

$\therefore$ $C(z) = (a_0 + b_0) + (a_1 + b_1)z + (a_2 + b_2)z^2 + \ldots\ldots$

$= (a_0 + a_1 z + a_2 z^2 + \ldots\ldots) + (b_0 + b_1 z + b_2 z^2 + \ldots\ldots)$

$= A(z) + B(z)$ [Using (1) and (2)]

$\therefore$ $\mathbf{C(z) = A(z) + B(z).}$

Remarks. (*i*) If **b** = α**a** then we have $B(z) = \alpha A(z)$.

(*ii*) If **c** = **a** + **b** then we have $C(z) = A(z) + B(z)$.

(*iii*) If **c** = **ab** then there is no simple form in which $C(z)$ can be expressed in terms of $A(z)$ and $B(z)$.

Theorem II. Let a, b, c be numeric functions and A(z), B(z), C(z) be their corresponding generating functions. If c is the convolution of numeric functions a and b then prove that

$$\mathbf{C(z) = A(z)\ B(z).}$$

Proof. Let $A(z) = a_0 + a_1 z + a_2 z^2 + \ldots\ldots$

$B(z) = b_0 + b_1 z + b_2 z^2 + \ldots\ldots$

and $C(z) = c_0 + c_1 z + c_2 z^2 + \ldots\ldots$

Since **c** = **a** $*$ **b**, we have

$c_0 = a_0 b_0$

$c_1 = a_0 b_1 + a_1 b_0$

....

$c_r = a_0 b_r + a_1 b_{r-1} + \ldots\ldots + a_{r-1} b_1 + a_r b_0$

....

Now $A(z)\ B(z) = (a_0 + a_1 z + a_2 z^2 + \ldots\ldots)(b_0 + b_1 z + b_2 z^2 + \ldots\ldots)$

$= a_0 b_0 + (a_0 b_1 + a_1 b_0)z + (a_0 b_2 + a_1 b_1 + a_2 b_0)z^2 + \ldots\ldots\ldots$

$= c_0 + c_1 z + c_2 z^2 + \ldots\ldots = C(z).$

$\therefore$ $\mathbf{C(z) = A(z)\ B(z).}$ ($\because$ c is convolution of **a** and **b**)

Example 3. *Find the generating function of the numeric function* **a**, *where*

$$a_r = \begin{cases} 1 & 0 \le r \le 3 \\ 5 & 4 \le r \le 7. \\ 0 & r \ge 8 \end{cases}$$

Sol. Let $A(z)$ be the generating function of the numeric function **a**.

$$\therefore \quad A(z) = a_0 + a_1 z + a_2 z^2 + \ldots\ldots$$
$$= 1 + 1.z + 1.z^2 + 1.z^3 + 5.z^4 + 5.z^5 + 5.z^6 + 5.z^7 + 0.z^8 + 0.z^9 + \ldots\ldots$$
$$= \mathbf{1 + z + z^2 + z^3 + 5z^4 + 5z^5 + 5z^6 + 5z^7.}$$

Example 4. *Let* **a** *and* **b** *be numeric functions, where* $a_r = 2^r, r \ge 0$ *and* $b_r = 3^r, r \ge 0$. *Find the numeric function* **a** $*$ **b**.

Sol. We have $\quad a_r = 2^r, \quad b_r = 3^r, \quad r \ge 0$

$$\therefore \quad A(z) = a_0 + a_1 z + a_2 z^2 + \ldots\ldots$$
$$= 2^0 + 2^1 z + 2^2 z^2 + \ldots\ldots$$
$$= 1 + 2z + (2z)^2 + \ldots\ldots = \frac{1}{1-2z}$$

Similarly, $\quad B(z) = \dfrac{1}{1-3z}$

Let **c** denote the numeric function **a** $*$ **b**.

$$\therefore \quad C(z) = A(z)\, B(z)$$
$$= \frac{1}{1-2z} \cdot \frac{1}{1-3z} = \frac{1}{(1-2z)(1-3z)}$$
$$= \frac{1}{(1-2z)(1-3(1/2))} + \frac{1}{(1-2(1/3))(1-3z)}$$
$$= -\frac{2}{1-2z} + \frac{3}{1-3z} = \frac{3}{1-3z} - \frac{2}{1-2z}$$
$$= 3[1 + (3z) + (3z)^2 + \ldots\ldots] - 2[1 + (2z) + (2z)^2 + \ldots\ldots]$$
$$= (3 - 2) + (3^2 - 2^2)z + (3^3 - 2^3)z^2 + \ldots\ldots$$

$$\therefore \quad c_r = 3^{r+1} - 2^{r+1}, \quad r \ge 0$$

$$\therefore \quad (\mathbf{a} * \mathbf{b})(r) = \mathbf{3^{r+1} - 2^{r+1}}, \quad \mathbf{r \ge 0.}$$

$\therefore$ The numeric function **a** $*$ **b** is known.

Example 5. *If $A(z)$ be the generating function of a numeric function* **a**, *then show that the generating function of the accumulated sum of* **a** *is* $\dfrac{A(z)}{1-z}$.

Sol. Let $\quad \mathbf{a} = (a_0, a_1, a_2, \ldots\ldots)$

$$\therefore \quad A(z) = a_0 + a_1 z + a_2 z^2 + \ldots\ldots$$

Let $\quad \mathbf{b} = (1, 1, 1, \ldots\ldots)$ be a numeric function

$$\therefore \quad B(z) = 1 + 1.z + 1.z^2 + \ldots\ldots = \frac{1}{1-z}$$

Let **c** denote the numeric function **a** * **b**.

$$\therefore \quad c_r = a_0 b_r + a_1 b_{r-1} + \ldots\ldots + a_{r-1} b_1 + a_r b_0, \; r \geq 0$$

$$= a_0.1 + a_1.1 + \ldots\ldots + a_{r-1}.1 + a_r.1 = \sum_{i=0}^{r} a_i$$

$$\therefore \quad c_r = \sum_{i=0}^{r} a_i, \; r \geq 0$$

$\therefore$ **c** *i.e.*, **a** * **b** is the accumulated sum of **a**.

$\therefore$ Generating function of accumulated sum of **a**

$$= \text{Generating function of } \mathbf{a} * \mathbf{b}$$

$$= A(z)\, B(z)$$

$$= A(z) \cdot \frac{1}{1-z} = \frac{\mathbf{A(z)}}{\mathbf{1-z}}.$$

Example 6. *Let A(z) be the generating function of a numeric function* **a**. *For any positive integer i, show that :*

(*i*) *the generating function of* S^i**a** *is* $z^i A(z)$.

(*ii*) *the generating function of* S^{-i}**a** *is* $z^{-i}[A(z) - a_0 - a_1 z - a_2 z^2 - \ldots\ldots - a_{i-1} z^{i-1}]$.

Sol. Let $\quad \mathbf{a} = (a_0, a_1, a_2, \ldots\ldots)$

$$\therefore \quad A(z) = a_0 + a_1 z + a_2 z^2 + \ldots\ldots$$

(*i*) Let **b** denote the numeric function S^i**a**.

$$\therefore \quad b_r = \begin{cases} 0 & 0 \leq r \leq i-1 \\ a_{r-i} & r \geq i \end{cases}$$

$\therefore$ Generating function of S^i**a**

$$= b_0 + b_1 z + b_2 z^2 + \ldots\ldots + b_{i-1} z^{i-1} + b_i z^i + b_{i+1} z^{i+1} + \ldots\ldots$$

$$= 0 + 0.z + 0.z^2 + \ldots\ldots + 0.z^{i-1} + a_0 z^i + a_1 z^{i+1} + a_2 z^{i+2} + \ldots\ldots$$

$$= z^i\, [a_0 + a_1 z + a_2 z^2 + \ldots\ldots] = z^i A(z).$$

$\therefore$ **Generating function of** $\mathbf{S^i a = z^i A(z)}$.

(*ii*) Let **c** denote the numeric function S^{-i}**a**.

$$\therefore \quad c_r = a_{r+i}, \; r \geq 0$$

$\therefore$ Generating function of S^{-i}**a**

$$= c_0 + c_1 z + c_2 z^2 + \ldots\ldots$$

$$= a_i + a_{i+1} z + a_{i+2} z^2 + \ldots\ldots$$

$$= z^{-i}[a_i z^i + a_{i+1} z^{i+1} + a_{i+2} z^{i+2} + \ldots\ldots]$$

$$= z^{-i}\, [A(z) - a_0 - a_1 z - a_2 z^2 - \ldots\ldots - a_{i-1} z^{i-1}]$$

$\therefore$ **Generating function of** $\mathbf{S^{-i} a}$

$$= \mathbf{z^{-i}[A(z) - a_0 - a_1 z - a_2 z^2 - \ldots\ldots - a_{i-1} z^{i-1}]}.$$

Example 7. *Let A(z) and B(z) be the generating functions of two numeric functions* **a** *and* **b** *respectively. Show that :*

(*i*) *if* **b** = Δ**a** *then* $B(z) = \dfrac{A(z) - a_0}{z} - A(z)$ (*ii*) *if* **b** = ∇**a** *then* $B(z) = A(z) - z\,A(z)$.

Sol. Let $\mathbf{a} = (a_0, a_1, a_2, a_3, \ldots\ldots)$

and $\mathbf{b} = (b_0, b_1, b_2, b_3, \ldots\ldots)$

$\therefore$ $A(z) = a_0 + a_1 z + a_2 z^2 + a_3 z^3 + \ldots\ldots$

and $B(z) = b_0 + b_1 z + b_2 z^2 + b_3 z^3 + \ldots\ldots$

(*i*) We have $\mathbf{b} = \Delta\mathbf{a}$.

$\therefore$ $b_r = a_{r+1} - a_r \quad r \geq 0$

Now,
$$\frac{A(z) - a_0}{z} - A(z) = \frac{1}{z}[(a_0 + a_1 z + a_2 z^2 + a_3 z^3 + \ldots\ldots) - a_0] - (a_0 + a_1 z + a_2 z^2 + a_3 z^3 + \ldots\ldots)$$
$$= (a_1 + a_2 z + a_3 z^2 + \ldots\ldots) - (a_0 + a_1 z + a_2 z^2 + a_3 z^3 + \ldots\ldots)$$
$$= (a_1 - a_0) + (a_2 - a_1)z + (a_3 - a_2)z^2 + \ldots\ldots$$
$$= b_0 + b_1 z + b_2 z^2 + \ldots\ldots = B(z)$$

$\therefore$
$$\mathbf{B(z) = \frac{A(z) - a_0}{z} - A(z).}$$

(*ii*) We have $\mathbf{b} = \nabla\mathbf{a}$.

$\therefore$ $b_0 = a_0, \; b_r = a_r - a_{r-1}, \quad r \geq 1$

Now,
$$A(z) - zA(z) = (a_0 + a_1 z + a_2 z^2 + a_3 z^3 + \ldots\ldots) - z(a_0 + a_1 z + a_2 z^2 + a_3 z^4 + \ldots\ldots)$$
$$= a_0 + (a_1 - a_0)z + (a_2 - a_1)z^2 + (a_3 - a_2)z^3 + \ldots\ldots$$
$$= b_0 + b_1 z + b_2 z^2 + b_3 z^3 + \ldots\ldots = B(z)$$

$\therefore$
$$\mathbf{B(z) = A(z) - zA(z).}$$

EXERCISE 6.3

1. Find the generating functions in their closed form of the following numeric functions :

   (*i*) $\mathbf{a} = (1, 1, 1, 1, \ldots\ldots, 1, \ldots\ldots)$ (*ii*) $\mathbf{b} = (1, 3, 3^2, 3^3, \ldots\ldots, 3^r, \ldots\ldots)$

   (*iii*) $\mathbf{c} = (5, 5(4), 5(4)^2, 5(4)^3, \ldots\ldots, 5(4)^r, \ldots\ldots)$.

2. Find the generating functions in their closed form of the following numeric functions :

   (*i*) **a**, where $a_r = \alpha^r, \quad r \geq 0$ (*ii*) **b**, where $b_r = \begin{cases} 3 & r = 0 \\ 11 & r = 1 \\ 3^{r+1} & r \geq 2 \end{cases}$

   (*iii*) **c**, where $c_r = 2^r + 5^r, \quad r \geq 0$ (*iv*) **d**, where $d_r = \begin{cases} 2^r & r \text{ is even} \\ -2^r & r \text{ is odd.} \end{cases}$

3. Find the numeric functions whose generating functions are given below :

   (*i*) e^z (*ii*) $\dfrac{2 + 3z - 6z^2}{1 - 2z}$

   (*iii*) $\dfrac{2 - 7z}{(1 - 3z)(1 - 4z)}$ (*iv*) $(1 - 4z)^{-2}$. *(KU 2007)*

4. The generating function of a numeric function **a** is $\frac{2}{1-4z^2}$. Show that $a_r = \begin{cases} 0 & r \text{ is odd} \\ 2^{r+1} & r \text{ is even} \end{cases}$.

5. Find the generating function of the numeric function **a**, where $a_r = 1^r + 2^r + 3^r$, $r \geq 0$.

6. Let **a** and **b** be numeric functions, where $a_r = 2^r$, $r \geq 0$, $b_r = 4^r$, $r \geq 0$. Find the numeric function **a** $*$ **b**.

Answers

1. (*i*) $A(z) = \frac{1}{1-z}$ (*ii*) $B(z) = \frac{1}{1-3z}$ (*iii*) $C(z) = \frac{5}{1-4z}$

2. (*i*) $A(z) = \frac{1}{1-\alpha z}$ (*ii*) $B(z) = \frac{3+2z-6z^2}{1-3z}$ (*iii*) $C(z) = \frac{2-7z}{(1-2z)(1-5z)}$

   (*iv*) $D(z) = \frac{1}{1+2z}$

3. (*i*) **a**, where $a_r = \frac{1}{r!}$, $r \geq 0$ (*ii*) **a**, where $a_r = \begin{cases} 2 & r = 0 \\ 7 & r = 1 \\ 2^{r+1} & r \geq 2 \end{cases}$

   (*iii*) **a**, where $a_r = 3^r + 4^r$, $r \geq 0$ (*iv*) **a**, where $a_r = (r+1)4^r$, $r \geq 0$

5. $\frac{1}{1-z} + \frac{1}{1-2z} + \frac{1}{1-3z}$

6. $(\mathbf{a} * \mathbf{b})(r) = 2^{2r+1}\,! - 2^r$, $r \geq 0$.

7

Recurrence Relations

7.1. INTRODUCTION

We know that a sequence is a list of objects arranged in a definite order. A sequence can be represented either *recursively* or *explicitly*. A relation used to express a sequence recursively is called a *recurrence relation.*

In this chapter, we shall learn the methods of solving recurrence relations. This knowledge would help us to express a sequence explicitly which is given in recursive form. A recurrence relation is also called a *difference equation.*

7.2. RECURSION

A computation (or object) is said to be defined **recursively** if it is defined in terms of itself.

For example, we know that

$$n\,! = n(n-1) \ldots\ldots 3.2.1, \; n \in \mathrm{N} \quad \text{and} \quad 0\,! = 1.$$

By using recursion, $n\,!$ can also be defined as follows :

$$n\,! = n\,.\,(n-1)!, \; n \geq 1 \quad \text{and} \quad 0\,! = 1.$$

Recursion is a very powerful technique of solving problems in discrete mathematics.

7.3. SEQUENCE

A list of objects arranged in a definite order is called a **sequence.** In a sequence, the objects occurring at first place, second place, third place, ... are respectively called the first term, second term, third term, ... of the sequence.

The *n*th term of a sequence is generally written as a_n or b_n or c_n etc.

The sequence $a_1, a_2, a_3, \ldots\ldots, a_n, \ldots\ldots$ is symbollically written as $\{a_n\}$. The terms of a sequence may all be different, or, some may be repeated.

Sometimes a sequence is also written as $a_0, a_1, a_2, a_3, \ldots\ldots$

For this sequence, the first term, second term, third term, ... are respectively a_0, a_1, a_2, ...

The following are some of the sequences :

(*i*) 1, 2, 1, 2, ... (*ii*) 1, 2, 3, 4.

(*iii*) 4, 9, 16, 25, ... (*iv*) 1, 3, 5, 7, 9, 11.

A sequence containing only finitely many terms is called a **finite sequence** and a sequence not containing finitely many terms is called an **infinite sequence.**

The sequences (*ii*) and (*iv*) given above are finite sequences and sequences (*i*) and (*iii*) are infinite sequences.

7.4. SEQUENCE AS A DISCRETE NUMERIC FUNCTION

Let $a_1, a_2, a_3, \ldots\ldots$ be an infinite real sequence. Define a function $f : \mathbf{W} \to \mathbf{R}$ by $f(n) = a_{n+1}, n \in \mathbf{W}$. The images of the elements 0, 1, 2, 3, ... under f are $a_1, a_2, a_3, a_4, \ldots\ldots$ respectively. Thus a sequence can be considered as a discrete numeric function.

Conversely, let $\phi : \mathbf{W} \to \mathbf{R}$ by a discrete numeric function. The images $\phi(0), \phi(1), \phi(2), \phi(3), \ldots\ldots$ form an infinite real sequence with nth term equal to $\phi(n-1)$. Thus, a discrete numeric function gives an infinite real sequence.

Therefore an infinite real sequence can be viewed as a discrete numeric function and conversely.

7.5. REPRESENTATION OF A SEQUENCE

We know that the infinite sequence $a_1, a_2, a_3, \ldots\ldots$ is represented as $\{a_n\}$. It is possible to represent some of the sequences in a compact form.

There are two kinds of formulas to represent a sequence in a compact form.

(*i*) **Recursive formula.** A formula is called a **recursive formula** if a term of the sequence is defined by a formula involving its preceding terms and the formula also include information about the beginning of the sequence. Such informations are called the **initial conditions** for the sequence.

Illustrations 1. The recursive formula $a_n = 2a_{n-1} + 1, 2 \le n \le 6, a_1 = 5$ defines the finite sequence :

$$5,\ 2(5)+1 = 11,\quad 2(11)+1 = 23,\quad 2(23)+1 = 47,$$
$$2(47)+1 = 95,\quad 2(95)+1 = 191 \quad i.e.,\quad 5, 11, 23, 47, 95, 191.$$

2. The recursive formula $b_n = 4b_{n-1}, n > 1, b_1 = 2$ defines the infinite sequence :

$$2, 8, 32, 128, \ldots\ldots$$

(*ii*) **Explicit formula.** A formula is called an **explicit formula** if a term of the sequence is defined using only its position number.

Illustrations 1. The explicit formula $a_n = (-2)^n, 1 \le n \le 5$ defines the finite sequence $-2, 4, -8, 16, -32$.

2. The explicit formula $s_n = 2n + 7, n \ge 1$ defines the infinite sequence 9, 11, 13, 15, ...

7.6. FIBONACCI SEQUENCE

The **Fibonacci sequence** starts with two numbers 1, 1 and contains numbers that are equal to the sum of their two immediate predecessors.

$\therefore$ If $\{f_n\}$ denotes the Fibonacci sequence then

$$f_1 = 1,\quad f_2 = 1,\quad f_3 = f_1 + f_2 = 1 + 1 = 2,$$
$$f_4 = f_2 + f_3 = 1 + 2 = 3,\quad f_5 = f_3 + f_4 = 2 + 3 = 5, \ldots\ldots$$

$\therefore$ The Fibonacci sequence is **1, 1, 2, 3, 5, 8, 13, 21, ...**

This sequence can be represented by recursive formula as follows :

$$f_n = f_{n-1} + f_{n-2}, n \ge 3, f_1 = 1, f_2 = 1.$$

Later on we would be able to represent the Fibonacci sequence by an explicit formula.

Remark. The Fibonacci sequence can also be expressed as follows :

$$f_n = f_{n-1} + f_{n-2}, n \ge 2, f_0 = 1, f_1 = 1.$$

Example 1: *Find the first five terms of the following sequences given by recursive formulas :*

(*i*) $a_n = 2a_{n-1} + 1$ *for* $n \geq 2$; $a_1 = 7$

(*ii*) $b_n = 3b_{n-1} + 4b_{n-2}$ *for* $n \geq 3$; $b_1 = 0$, $b_2 = 5$.

Sol. (*i*) We have $\quad a_n = 2a_{n-1} + 1, n \geq 2 \quad$ ...(1)

We are given that $\quad a_1 = 7$.

For $n = 2$, (1) $\Rightarrow$ $a_2 = 2a_1 + 1 = 2(7) + 1 = 15$

For $n = 3$, (1) $\Rightarrow$ $a_3 = 2a_2 + 1 = 2(15) + 1 = 31$

For $n = 4$, (1) $\Rightarrow$ $a_4 = 2a_3 + 1 = 2(31) + 1 = 63$

For $n = 5$, (1) $\Rightarrow$ $a_5 = 2a_4 + 1 = 2(63) + 1 = 127$

$\therefore$ The first five terms of the given sequence are **7, 15, 31, 63, 127.**

(*ii*) We have $\quad b_n = 3b_{n-1} + 4b_{n-2}, n \geq 3 \quad$ ...(1)

Also, $\quad b_1 = 0, b_2 = 5$

For $n = 3$, (1) $\Rightarrow$ $b_3 = 3b_2 + 4b_1 = 3(5) + 4(0) = 15$

For $n = 4$, (1) $\Rightarrow$ $b_4 = 3b_3 + 4b_2 = 3(15) + 4(5) = 65$

For $n = 5$, (1) $\Rightarrow$ $b_5 = 3b_4 + 4b_3 = 3(65) + 4(15) = 255$

$\therefore$ The first five terms of the given sequence are **0, 5, 15, 65, 255.**

Example 2. *Find the first five terms of the following sequences given by explicit formulas :*

(*i*) $t_n = 2^n - n(2)^{n-1}, n \geq 1$ $\qquad$ (*ii*) $a_n = 5(2)^n + 2(3)^n, n \geq 1$.

Sol. (*i*) We have $\quad t_n = 2^n - n(2)^{n-1}, n \geq 1 \quad$ ...(1)

For $n = 1$, (1) $\Rightarrow$ $t_1 = 2^1 - 1(2)^{1-1} = 2 - 1 = 1$

For $n = 2$, (1) $\Rightarrow$ $t_2 = 2^2 - 2(2)^{2-1} = 4 - 4 = 0$

For $n = 3$, (1) $\Rightarrow$ $t_3 = 2^3 - 3(2)^{3-1} = 8 - 12 = -4$

For $n = 4$, (1) $\Rightarrow$ $t_4 = 2^4 - 4(2)^{4-1} = 16 - 32 = -16$

For $n = 5$, (1) $\Rightarrow$ $t_5 = 2^5 - 5(2)^{5-1} = 32 - 80 = -48$

$\therefore$ The first five terms of the given sequence are **1, 0, – 4, – 16, – 48.**

(*ii*) We have $\quad a_n = 5(2)^n + 2(3)^n, n \geq 1 \quad$ ...(1)

For $n = 1$, (1) $\Rightarrow$ $a_1 = 5(2)^1 + 2(3)^1 = 10 + 6 = 16$

For $n = 2$, (1) $\Rightarrow$ $a_2 = 5(2)^2 + 2(3)^2 = 20 + 18 = 38$

For $n = 3$, (1) $\Rightarrow$ $a_3 = 5(2)^3 + 2(3)^3 = 40 + 54 = 94$

For $n = 4$, (1) $\Rightarrow$ $a_4 = 5(2)^4 + 2(3)^4 = 80 + 162 = 242$

For $n = 5$, (1) $\Rightarrow$ $a_5 = 5(2)^5 + 2(3)^5 = 160 + 486 = 646$

$\therefore$ The first five terms of the given sequence are **16, 38, 94, 242, 646.**

7.7. RECURRENCE RELATION

In the recursive representation of a sequence, a term of the sequence is defined by a relation involving its preceding terms and this representation also include initial conditions.

A relation is called a **recurrence relation** for a sequence if a term of the sequence is defined by a relation involving its preceding terms.

In other words, a **recurrence relation** for a sequence $\{a_n\}$ is a relation that expresses a_n in terms of one or more preceding terms of the sequence, *viz*, $a_0, a_1, \ldots\ldots, a_{n-1}$.

For the infinite sequence 2, 8, 32, 128, ..., the recursive formula is

$$a_n = 4a_{n-1} \text{ for } n \geq 2, \quad a_1 = 2.$$

Thus $a_n = 4a_{n-1}$ for $n \geq 2$ is a recurrence relation for the sequence 2, 8, 32, 128, ...

Example 3. *Ram deposits Rs. 5000 in a saving account at a Bank. The rate of interest is 6% p.a. compounded annually. Find a recurrence relation to compute the amount A_n in the account at the end of n years.*

Sol. We have A_n = amount at the end of n years

$\therefore$ A_0 = Rs. 5,000

For $n \geq 1$, A_n = amount at the end of $(n-1)$ years + interest for the nth year

$$= A_{n-1} + \frac{A_{n-1} \times 6 \times 1}{100} = (1.06)\,A_{n-1}.$$

$\therefore$ The required recurrence relation is $\mathbf{A_n = (1.06)\ A_{n-1}, n \geq 1}$. Here initial condition is A_0 = Rs. 5,000.

Example 4. *There are n guests in a meeting hall. Each person shakes hand with every body else exactly once. Define the number of handshakes that occur, recursively.*

Sol. Let H_n denote the number of handshakes when there are n persons.

If $n = 1$, there is no handshake.

$\therefore$ $H_1 = 0$

Let $n > 1$. Let Mr. X be any arbitrary but fixed guest. There are $n - 1$ guests excluding Mr. X and H_{n-1} handshakes occur among these $n - 1$ guests.

Since Mr. X is to shake hand with each of $n - 1$ guests, the additional handshakes are $n - 1$.

$\therefore$ $\mathbf{H_n = H_{n-1} + (n - 1) \text{ for } n \geq 2, H_1 = 0.}$

Example 5. *Let a_n be the number of n-element sequences without repeats that can be made from the set {1, 2, 3, ..., n}. Find a recurrence relation to compute a_n.*

Sol. For $n = 1$, there is only one 1-element sequence from the set {1}.

$\therefore$ $a_1 = 1$

Let $n \geq 2$. By definition, a_{n-1} is the number of $(n - 1)$ element sequences without repeats that can be made from the set {1, 2, 3, ..., $n - 1$}.

For each $(n - 1)$ element sequence, we insert n in turn in each of the n available places (at the beginning, at the end and between any pair of numbers in the sequence).

$\therefore$ Each $(n - 1)$ element sequence gives n sequences of n elements.

$\therefore$ $a_n = na_{n-1}$

$\therefore$ The required recurrence relation is $\mathbf{a_n = na_{n-1}}$ **for n ≥ 2.** Here initial condition is $a_1 = 1$.

Example 6. *Let A = {0, 1}. Find a recurrence relation for a_n, the number of strings of length n in A*, the set of all finite sequences of elements of A, that do not contain adjacent 0's.*

Sol. Since A = {0, 1}, the strings in A^* contain only 0 and 1. The elements of A^* are also called **bit strings**.

We have a_n = number of n-bit strings not containing 00

For $n = 1$, the required bit strings are 0, 1. $\therefore$ $a_1 = 2$

For $n = 2$, the required bit strings are 01, 10, 11. $\therefore$ $a_2 = 3$

Since we are to consider only those strings which do not contain 00, the string 00 is not to be counted in a_2.

Let $n \geq 3$. We are to count all n-bit strings not containing 00. An n-bit string not containing 00 begins with either 10 or 11 or 01.

Any $(n - 1)$ bit string not containing 00 can be catenated with 1 from left to get a n-bit string not containing 00 and beginning with either 10 or 11.

Also, any $(n-2)$-bit string not containing 00 can be catenated with 01 from left to get a n-bit string not containing 00 and beginning with 01.

$\therefore$ $$a_n = a_{n-1} + a_{n-2}$$

$\therefore$ The required recurrence relation is $\mathbf{a_n = a_{n-1} + a_{n-2}}$ **for** $\mathbf{n \geq 3}$.

Here initial conditions are $a_1 = 2, a_2 = 3$.

Remarks. Let A^* be the set of all finite sequences of the elements of a set A. The elements of A^* are also called **strings.** In a string, we avoid writing commas between the terms. If $s_1 = \alpha_1\alpha_2 \ldots\ldots \alpha_l$ and $s_2 = \beta_1\beta_2 \ldots\ldots \beta_n$ be two strings in A^* then the string $\alpha_1\alpha_2 \ldots\ldots \alpha_l\beta_1\beta_2 \ldots\ldots \beta_n$ is called a **catenation** of s_1 and s_2 and written as s_1s_2.

Example 7. *Let A = {0, 1}. Find a recurrence relation for a_n, the number of strings of length n in A*, the set of all finite sequences of elements of A, that do not contain 111.*

Sol. We have a_n = number of n-bit strings of length n not containing 111

For $n = 1$, the required bit strings are 0, 1 $\therefore$ $a_1 = 2$

For $n = 2$, the required bit strings are 00, 01, 10, 11. $\therefore$ $a_2 = 4$

For $n = 3$, the required bit strings are 000, 100, 010, 001, 110, 101, 011. $\therefore$ $a_3 = 7$

Let $n \geq 4$. We are to count all n-bit strings not containing 111. An n-bit string not containing 111 begin with either 00 or 01 or 10 or 11.

Any $(n-1)$ bit string not containing 111 can be catenated with 0 from left to get a n-bit string not containing 111 and beginning with either 00 or 01.

Any $(n-2)$ bit string not containing 111 can be catenated with 10 from left to get a n-bit string not containing 111 and beginning with 10.

Also, any $(n-3)$-bit string not containing 111 can be catenated with 110 from left to get a n-bit string not containing 111 and beginning with 11.

$\therefore$ $$a_n = a_{n-1} + a_{n-2} + a_{n-3}$$

$\therefore$ The required recurrence relation is $\mathbf{a_n = a_{n-1} + a_{n-2} + a_{n-3}}$ **for** $\mathbf{n \geq 4}$.

Here initial conditions are $a_1 = 2, a_2 = 4, a_3 = 7$.

7.8. SOLUTION OF RECURRENCE RELATION

A sequence satisfying a given recurrence relation is called its **solution.**

A recurrence relation may be satisfied by more than one sequence.

For example, let $\{a_n\}$ be a sequence satisfying the recurrence relation

$$a_n = 2a_{n-1} + 1, \quad n \geq 2.$$

The sequence 5, 11, 23, 47, ... satisfies this recurrence relation.

The above recurrence relation is also satisfied by the sequence 3, 7, 15, 31, ...

$\therefore$ The solution of a recurrence relation is not unique. In addition, if some initial condition(s) are also given then there may exist a unique sequence satisfying that recurrence relation.

For example, the solution of the recurrence relation :

$a_n = a_{n-1} + 3, n \geq 2$ with $a_1 = 2$ is the sequence 2, 5, 8, 11, ...

The explicit formula for this sequence is $a_n = 3n - 1, n \geq 1$.

Example 8. *Find the recurrence relation for the sequence $\{a_n\}$, given by explicit formula $a_n = (3 + 7n)4^n, n \geq 1$.*

Sol. We have $$a_n = (3 + 7n)4^n \quad \ldots(1)$$

$\Rightarrow$ $$a_n = 3(4)^n + 7n(4)^n \quad \ldots(2)$$

Replacing n by $n - 1$, we get

$$a_{n-1} = 3(4)^{n-1} + 7(n-1)4^{n-1} \quad ...(3)$$

$4 \times (3) \Rightarrow \quad 4a_{n-1} = 3(4)^n + 7(n-1)\,4^n \quad ...(4)$

$(2) - (4) \Rightarrow \quad a_n - 4a_{n-1} = 7(4)^n \quad ...(5)$

$(5) \Rightarrow \quad a_{n-1} - 4a_{n-2} = 7(4)^{n-1} \quad ...(6)$

$4 \times (6) \Rightarrow \quad 4a_{n-1} - 16a_{n-2} = 7(4)^n \quad ...(7)$

$(5) - (7) \Rightarrow \quad a_n - 8a_{n-1} + 16a_{n-2} = 0$

∴ The recurrence relation for the given sequence is

$$\mathbf{a_n - 8a_{n-1} + 16a_{n-2} = 0,\ n \ge 3.}$$

Remark. For the above sequence :

$$a_1 = (3 + 7)(4) = 40 \text{ and } a_2 = (3 + 14)(16) = 272.$$

∴ By definition, the solution of the recurrence relation

$a_n - 8a_{n-1} + 16a_{n-2} = 0$, $n \ge 3$ with initial conditions $a_1 = 40$, $a_2 = 272$ is

$$\mathbf{a_n = (3 + 7n)4^n,\ n \ge 1.}$$

EXERCISE 7.1

1. Find the first five terms of the following sequences given by recursive formulas :
   (*i*) $a_n = a_{n-1} + 3$ for $n \ge 2$, $a_1 = 12$ (*ii*) $b_n = b_{n-1}$ for $n \ge 2$, $b_1 = 4$
   (*iii*) $c_n = nc_{n-1} + n^2c_{n-2}$ for $n \ge 3$; $c_1 = c_2 = 1$ (*iv*) $d_n = d_{n-1} + d_{n-3}$ for $n \ge 4$, $d_1 = 1, d_2 = 2, d_3 = 0$.
2. Find the first five terms of the following sequences given by explicit formulas :
   (*i*) $a_n = 28 + 2(2)^n$, $n \ge 0$ (*ii*) $b_n = 3(4)^n + (-2)^n$, $n \ge 1$
   (*iii*) $c_n = (1 + 3n - 2n^2)(-1)^n$, $n \ge 0$ (*iv*) $d_n = -9(2)^n + 2(5)^n + 2n + 8$, $n \ge 0$.
3. Mahesh deposits Rs. 17,000 in a saving account which yields interest at the rate of $5\frac{1}{2}$ % p.a. compounded annually. Find a recurrence relation to compute the amount a_n in the account at the end of n years.
4. Derive a recurrence relation for obtaining the amount A_n at the end of n years on an investment of Rs. 10,000 at 5% p.a. compounded annually.
5. If the annual interest rate is i per rupee and it is compounded m times per year, find a recurrence relation for obtaining the amount A_n at the end of n periods on an investment of Rs. 2,000.
6. Find the recurrence relation for the sequence $\{b_n\}$, given by explicit formula :

$$b_n = 2(3)^n + 5(-4)^n,\ n \ge 1.$$

Answers

1. (*i*) 12, 15, 18, 21, 24 (*ii*) 4, 4, 4, 4, 4
   (*iii*) 1, 1, 12, 64, 620 (*iv*) 1, 2, 0, 1, 3
2. (*i*) 30, 32, 36, 44, 60 (*ii*) 10, 52, 184, 784, 3040
   (*iii*) 1, – 2, – 1, 8, – 19 (*iv*) 1, 2, 26, 192, 1122
3. $a_n = (1.055)a_{n-1}$, $n \ge 1$
4. $A_n = (1.05)\,A_{n-1}$, $n \ge 1$
5. $A_n = \left(1 + \frac{i}{m}\right) A_{n-1}$, $n \ge 1$
6. $b_n + b_{n-1} - 12b_{n-2} = 0$, $n \ge 3$.

7.9. SOLUTION OF RECURRENCE RELATION USING BACKTRACKING

The backtracking method of solving a recurrence relation supplemented with initial condition(s) is used to solve simple type of recurrence relations. This method is also known as **substitution method** and also as **iterative method.**

In the **backtracking** method, the recurrence relation is applied iteratively and a pattern is identified for its solution *i.e.*, for the explicit formula for the given recurrence relation.

Example 1. *By using backtracking, solve the following recurrence relations :*

(i) $a_n = a_{n-1} + 5,\ n \geq 2$ *with* $a_1 = 7$ *(ii)* $b_n = 3b_{n-1} + 1,\ n \geq 2$ *with* $b_1 = 4$

(iii) $c_n = c_{n-1} + n,\ n \geq 1$ *with* $c_0 = 1.$

Sol. *(i)* We have

$$a_n = a_{n-1} + 5, n \geq 2 \text{ with } a_1 = 7 \qquad \text{...(1)}$$

We backtrack the value of a_n by substituting the definition of a_{n-1}, a_{n-2} and so on until we get a pattern.

$$(1) \Rightarrow \quad a_n = a_{n-1} + 5$$
$$= (a_{n-2} + 5) + 5 = a_{n-2} + 2(5)$$
$$= (a_{n-3} + 5) + 2(5) = a_{n-3} + 3(5)$$

$\therefore$ After k iterations, we get

$$a_n = a_{n-k} + k(5) \qquad \text{...(2)}$$

Putting $n - k = 1$, we get $k = n - 1$.

$\therefore$ Letting $k = n - 1$ in (2), we get

$$a_n = a_1 + (n - 1)5$$

$$\Rightarrow \quad a_n = 7 + (n - 1)5 \ \Rightarrow\ a_n = 5n + 2$$

$\therefore$ The solution of (1) is $\mathbf{a_n = 5n + 2,\ n \geq 1}$.

(ii) We have $b_n = 3b_{n-1} + 1, n \geq 2$ with $b_1 = 4$ ...(1)

We backtrack the value of b_n by substituting the definition of b_{n-1}, b_{n-2} and so on until we get a pattern.

$$(1) \Rightarrow \quad b_n = 3b_{n-1} + 1$$
$$= 3(3b_{n-2} + 1) + 1 = 3^2\, b_{n-2} + 3 + 1$$
$$= 3^2\,(3b_{n-3} + 1) + 3 + 1 = 3^3\, b_{n-3} + 3^2 + 3 + 1$$

$\therefore$ After k iterations, we get

$$b_n = 3^k b_{n-k} + 3^{k-1} + 3^{k-2} + \ldots\ldots + 3 + 1 \qquad \text{...(2)}$$

Putting $n - k = 1$, we get $k = n - 1$.

$\therefore$ Letting $k = n - 1$ in (2), we get

$$b_n = 3^{n-1}\, b_1 + 3^{n-1-1} + 3^{n-1-2} + \ldots\ldots + 3 + 1$$
$$= 3^{n-1} \,.\, 4 + 3^{n-2} + 3^{n-3} + \ldots\ldots + 3 + 1$$
$$= 4\,.3^{n-1} + (1 + 3 + \ldots\ldots + 3^{n-3} + 3^{n-2})$$
$$= 4\,.\,3^{n-1} + \frac{1.(1-3^{n-1})}{1-3} = 4\,.\,3^{n-1} + \frac{1}{2}\,(3^{n-1} - 1)$$

$$= \frac{9}{2} \cdot 3^{n-1} - \frac{1}{2} = \frac{1}{2} \cdot 3^{n+1} - \frac{1}{2} = \frac{3^{n+1} - 1}{2}$$

$\therefore$ The solution of (1) is

$$\mathbf{b_n = \frac{3^{n+1} - 1}{2}, n \geq 1.}$$

(*iii*) We have $c_n = c_{n-1} + n, n \geq 1$ with $c_0 = 1$ ...(1)

We backtrack the value of c_n by substituting the definition of c_{n-1}, c_{n-2} and so on until we get a pattern.

(1) $\Rightarrow$

$$\begin{aligned} c_n &= c_{n-1} + n \\ &= c_{n-2} + (n-1) + n \\ &= c_{n-3} + (n-2) + (n-1) + n \\ &= c_{n-4} + (n-3) + (n-2) + (n-1) + n \end{aligned}$$

$\therefore$ After k iterations, we get

$$c_n = c_{n-k} + (n-k+1) + (n-k+2) + \ldots\ldots + (n-1) + n \quad \ldots(2)$$

Putting $n - k = 0$, we get $k = n$.

$\therefore$ Letting $k = n$ in (2), we get

$$\begin{aligned} c_n &= c_0 + 1 + 2 + \ldots\ldots + (n-1) + n \\ &= 1 + (1 + 2 + \ldots\ldots + (n-1) + n) \\ &= 1 + \frac{n(n+1)}{2} \end{aligned}$$

$\therefore$ The solution of (1) is $\mathbf{c_n = \frac{n(n+1)}{2} + 1, n \geq 0.}$

Example 2. *Find an explicit formula for the* recurrence relation :

$a_n = 3a_{n-1} + 1, n \geq 2,$ *where* $a_0 = 0, a_1 = 1.$ (K U 2006)

Sol. We have $a_n = 3a_{n-1} + 1, n \geq 2$ with $a_0 = 0, a_1 = 1$ ...(1)

(1) $\Rightarrow$

$$\begin{aligned} a_n &= 3a_{n-1} + 1 \\ &= 3(3a_{n-2} + 1) + 1 = 3^2 a_{n-2} + 3 + 1 \\ &= 3^2 (3a_{n-3} + 1) + 3 + 1 = 3^3 a_{n-3} + 3^2 + 3 + 1 \end{aligned}$$

$\therefore$ After k iterations, we get

$$a_n = 3^k a_{n-k} + (3^{k-1} + 3^{k-2} + \ldots\ldots + 3 + 1) \quad \ldots(2)$$

Putting $n - k = 1$, we get $k = n - 1$.

Letting $k = n - 1$ in (2), we get

$$\begin{aligned} a_n &= 3^{n-1} a_1 + (3^{n-2} + 3^{n-3} + \ldots\ldots + 3 + 1) \\ &= 3^{n-1}(1) + 3^{n-2} + 3^{n-3} + \ldots\ldots + 3 + 1 \\ &= 1 + 3 + \ldots\ldots + 3^{n-2} + 3^{n-1} \\ &= \frac{1(1 - 3^n)}{1 - 3} = \frac{3^n - 1}{2} \end{aligned}$$

$$\therefore \quad a_n = \frac{3^n - 1}{2}, n \geq 1$$

Also, $$\frac{(3)^0 - 1}{2} = \frac{1-1}{2} = 0$$

$\therefore$ $$a_n = \frac{3^n - 1}{2}\ , n \geq 1 \text{ also satisfies } a_0 = 0.$$

The required explicit formula is

$$\mathbf{a_n = \frac{3^n - 1}{2}\ , n \geq 0.}$$

Example 3. *Solve the recurrence relation*

$$p_n = a - \frac{b}{k} p_{n-1}$$

for the price in an economic model, where a, b, k are positive parameters and p_0 is the initial price.

Sol. We have $$p_n = a - \frac{b}{k} p_{n-1}$$

$\Rightarrow$ $$p_n = a + \lambda p_{n-1} \qquad ...(1) \qquad \left(\text{Putting } \lambda = -\frac{b}{k}\right)$$

$$= a + \lambda(a + \lambda p_{n-2}) = a + \lambda a + \lambda^2 p_{n-2}$$

$$= a + \lambda a + \lambda^2 (a + \lambda p_{n-3}) = a + \lambda a + \lambda^2 a + \lambda^3 p_{n-3}$$

$\therefore$ After k iterations, we get

$$p_n = a + \lambda a + \lambda^2 a + \ldots\ldots + \lambda^{k-1} a + \lambda^k p_{n-k} \qquad ...(2)$$

Putting $n - k = 0$, we get $k = n$.

Letting $k = n$ in (2), we get

$$p_n = a + \lambda a + \lambda^2 a + \ldots\ldots + \lambda^{n-1} a + \lambda^n p_0$$

$$= \frac{a(1-\lambda^n)}{1-\lambda} + \lambda^n p_0 = \frac{a}{1-\lambda} - \frac{a\lambda^n}{1-\lambda} + \lambda^n p_0$$

$$= \lambda^n \left(p_0 - \frac{a}{1-\lambda}\right) + \frac{a}{1-\lambda} = \left(-\frac{b}{k}\right)^n \left(p_0 - \frac{a}{1+\frac{b}{k}}\right) + \frac{a}{1+\frac{b}{k}}$$

$\therefore$ $$\mathbf{p_n = \left(-\frac{b}{k}\right)^n \left(p_0 - \frac{ak}{k+b}\right) + \frac{ak}{k+b}\ , n \geq 0.}$$

Example 4. *Let K_n be the picture obtained by drawing n non-collinear dots and joining each pair of vertices by an edge. Find the recurrence relation for the number of edges of K_n and solve it.*

Sol. Let a_n denote the number of edges of K_n, $n \geq 1$.

If $n = 1$, there is no edge. $\therefore$ $a_1 = 0$

Let $n > 1$. Let A be any arbitrary but fixed dot. There are $n - 1$ dots in K_n excluding A and a_{n-1} edges are obtained by joining these $n - 1$ dots. Since the dot A is to be joined by each of $n - 1$ dots by an edge, the additional edges are $n - 1$.

$\therefore \quad a_n = a_{n-1} + (n - 1), n > 1$

$\therefore$ The required recurrence relation is

$$\mathbf{a_n = a_{n-1} + (n - 1), n > 1.}$$

Here, the initial condition is $a_1 = 0$.

We have $\quad a_n = a_{n-1} + (n - 1) \quad$ ...(1)

$$\therefore \quad a_n = a_{n-2} + (n - 2) + (n - 1)$$
$$= a_{n-3} + (n - 3) + (n - 2) + (n - 1)$$

$\therefore$ After k iterations, we get

$$a_n = a_{n-k} + (n - k) + (n - k + 1) + \ldots\ldots + (n - 2) + (n - 1)$$

Putting $n - k = 1$, we get $k = n - 1$.

$\therefore$ $k = n - 1$ implies

$$a_n = a_{n-n+1} + (n - n + 1) + (n - n + 1 + 1) + \ldots\ldots + (n - 2) + (n - 1)$$

$$\Rightarrow \quad a_n = a_1 + (1 + 2 + \ldots\ldots + (n - 2) + (n - 1))$$
$$= 0 + \frac{(n-1)(n-1+1)}{2} = \frac{n(n-1)}{2}$$

$\therefore$ The required solution is

$$\mathbf{a_n = \frac{n(n-1)}{2}, n \geq 1.}$$

EXERCISE 7.2

By using backtracking, solve the following recurrence relations (Q. No. 1—8)

1. $a_n = a_{n-1} + 5, n \geq 2$ with $a_1 = 3$
2. $b_n = b_{n-1} - 7, n \geq 2$ with $b_1 = 10$
3. $c_n = 2c_{n-1} + 1, n \geq 2$ with $c_1 = 7$
4. $d_n = 5d_{n-1} + 3, n \geq 2$ with $d_1 = 3$
5. $e_n = 2e_{n-1}, n \geq 1$ with $e_0 = 1$
6. $f_n = f_{n-1} + 4n, n \geq 1$ with $f_0 = 0$
7. $g_n = ng_{n-1}$ with $g_1 = 7$
8. $a_n = 3a_{n-1}, n \geq 2$ with $a_1 = 1$. *(KU 2006 C)*
9. Solve the recurrence relation of the Handshake problem :

$$H_n = H_{n-1} + (n - 1), n > 1 \quad \text{with } H_1 = 0.$$

Answers

1. $a_n = 5n - 2, n \geq 1$
2. $b_n = -7n + 17, n \geq 1$
3. $c_n = 2^{n+2} - 1, n \geq 1$
4. $d_n = \frac{15}{4} \cdot 5^{n-1} - \frac{3}{4}, n \geq 1$
5. $e_n = 2^n, n \geq 0$
6. $f_n = 2n(n + 1), n \geq 0$
7. $g_n = 7(n!), n \geq 1$
8. $a_n = 3^{n-1}, n \geq 1$
9. $H_n = \frac{n(n-1)}{2}, n \geq 1.$

7.10. LINEAR RECURRENCE RELATION WITH CONSTANT COEFFICIENTS

A recurrence relation of the form

$$a_n = c_1a_{n-1} + c_2a_{n-2} + \ldots\ldots + c_ka_{n-k} + f(n) \qquad \ldots(1)$$

where c_1, c_2, ..., c_k are constants and $f(n)$, a function of n, is called a **linear recurrence relation with constant coefficients.** The recurrence relation (1) is called linear, because the expression $c_1a_{n-1} + c_2a_{n-2} + \ldots\ldots + c_ka_{n-k}$ is sum of multiples of the previous terms of the sequence. The recurrence relation (1) is called with constant coefficients, because c_1, c_2, ..., c_k are all constants, rather than functions depending upon n. In the recurrence relation (1), if $c_k \neq 0$ then a_n depends on its k immediate predecessors : $a_{n-1}, a_{n-2}, \ldots\ldots, a_{n-k}$. We say that (1) is a recurrence relation of **order** k.

In (1), if $f(n) = 0$, then (1) is called a **homogeneous recurrence relation** and if $f(n) \neq 0$ then it is called a **non-homogeneous recurrence relation.**

Illustrations. (*i*) The recurrence relation $a_n = a_{n-1} + a_{n-2}^2$ is not linear, because a_{n-2} is multiplied by itself.

(*ii*) The recurrence relation $a_n = 2a_{n-1} + 3na_{n-2} + n^2$ is not with constant coefficients, because a_{n-2} is multiplied by a term involving n.

(*iii*) The recurrence relation $a_n = a_{n-1} + a_{n-3} + 2n$ is not homogeneous, because the right side constains the term $2n$, which is non-zero. This recurrence relation is a non-homogeneous recurrence relation.

(*iv*) The recurrence relation $a_n = 2a_{n-2} + 7a_{n-3}$ is a linear homogeneous recurrence relation with constant coefficients and its order is 3.

(*v*) The recurrence relation $a_n = a_{n-1} + 2$ is a linear non-homogeneous recurrence relation with constant coefficients and its order is 1.

7.11. SOLUTION OF A LINEAR HOMOGENEOUS RECURRENCE RELATION WITH CONSTANT COEFFICIENTS

We know that a linear recurrence relation with constant coefficients is either homogeneous or non-homogeneous. In the present section, we shall learn the method of solving a linear homogeneous recurrence relation with constant coefficients. This knowledge would also be used in solving linear non-homogeneous recurrence relations with constant coefficients.

Let $$a_n = c_1a_{n-1} + c_2a_{n-2} + \ldots\ldots + c_ka_{n-k} \qquad \ldots(1)$$

be a linear homogeneous recurrence relation with constant coefficients and its order be k.

$\therefore \qquad c_k \neq 0$

Let $a_n = Ar^n$ be a solution of (1).

$\therefore$ (1) $\Rightarrow \qquad Ar^n = c_1Ar^{n-1} + c_2Ar^{n-2} + \ldots\ldots + c_kAr^{n-k}$

Dividing by Ar^{n-k}, we get

$$r^k = c_1r^{k-1} + c_2r^{k-2} + \ldots\ldots + c_k$$

$\therefore$ r is a root of the polynomial equation

$$x^k = c_1x^{k-1} + c_2x^{k-2} + \ldots\ldots + c_k \qquad \ldots(2)$$

of degree k.

The polynomial equation (2) is called the **characteristic equation** of the recurrence relation (1).

The characteristic equation (2) will have k roots and these roots may or may not be distinct.

The roots of the characteristic equation are called the **characteristic roots** of the given recurrence relation.

Let the characteristic roots of (2) be $r_1, r_2, \ldots\ldots, r_k$.

Case I. The roots are distinct.

In this case, the solution of the recurrence relation is

$$a_n = A_1 r_1^n + A_2 r_2^n + \ldots\ldots + A_k r_k^n$$

This solution involves k arbitrary constants $A_1, A_2, \ldots\ldots, A_k$ and their values may be selected so that this solution may satisfy k initial conditions.

Case II. The roots are not distinct.

Let $r_1 = r_2 = r_3$ and all other roots be distinct.

In this case, the solution of the recurrence relation (1) is

$$a_n = (A_1 + nA_2 + n^2A_3)r_1^n + A_4 r_4^n + \ldots\ldots + A_n r_k^n$$

This solution involves k arbitrary constants $A_1, A_2 \ldots\ldots, A_k$ and their values may be selected so that this solution may satisfy k initial conditions.

We accept this general result and prove this result for the case, $k = 2$.

Theorem. Let r and s be the characteristic roots of the linear homogeneous recurrence relation

$$\mathbf{a_n = ca_{n-1} + da_{n-2}}$$

with constant coefficients and of order 2. Prove the following :

(*i*) **If $\mathbf{r \neq s}$, the solution of the given relation is $\mathbf{a_n = Ar^n + Bs^n}$.**

(*ii*) **If $\mathbf{r = s}$, the solution of the given relation is $\mathbf{a_n = (A + nB)r^n}$.**

Here A and B are arbitrary constants.

Proof. Given recurrence relation is

$$a_n = ca_{n-1} + da_{n-2} \qquad \ldots(1)$$

The characteristic equation of (1) is

$$x^2 = cx + d \quad i.e., \quad x^2 - cx - d = 0$$

The characteristic roots of (1) are r and s.

$$\therefore \qquad r^2 - cr - d = 0 \qquad \ldots(2)$$

$$\text{and} \qquad s^2 - cs - d = 0 \qquad \ldots(3)$$

(*i*) Let $r \neq s$. We shall show that $a_n = Ar^n + Bs^n$ and $a_n = ca_{n-1} + da_{n-2}$ represent the same sequence.

We have $\qquad a_n = Ar^n + Bs^n$

$$\therefore \quad a_n - ca_{n-1} - da_{n-2} = (Ar^n + Bs^n) - c(Ar^{n-1} + Bs^{n-1}) - d(Ar^{n-2} + Bs^{n-2})$$

$$= Ar^{n-2}(r^2 - cr - d) + Bs^{n-2}(s^2 - cs - d)$$

$$= Ar^{n-2}(0) + Bs^{n-2}(0) = 0$$

$$\therefore \qquad a_n = ca_{n-1} + da_{n-2}.$$

(*ii*) Let $r = s$.

$$\therefore \qquad \text{Sum of roots} = r + s = -\frac{-c}{1} = c$$

$$\Rightarrow \qquad 2r = c \quad \Rightarrow \quad r = \frac{c}{2}$$

Also, disc. of $x^2 - cx - d = 0$ is zero.

$$\therefore \qquad (-c)^2 - 4(1)(-d) = 0 \quad \text{or} \quad c^2 + 4d = 0$$

We shall show that $a_n = (A + nB)r^n$ and $a_n = ca_{n-1} + da_{n-2}$ represent the same sequence.

We have $\qquad a_n = (A + nB)r^n$

$$\therefore \quad a_n - ca_{n-1} - da_{n-2} = (A + nB)r^n - c(A + (n-1)B)r^{n-1} - d(A + (n-2)B)r^{n-2}$$

$$= Ar^{n-2}(r^2 - cr - d) + Br^{n-2}(nr^2 - c(n-1)r - d(n-2))$$

$$= Ar^{n-2}(0) + Br^{n-2}[n(r^2 - cr - d) + cr + 2d]$$

$$= Br^{n-2}\left[n(0) + c\left(\frac{c}{2}\right) + 2d\right]$$

$$= Br^{n-2}\left[\frac{c^2 + 4d}{2}\right] = Br^{n-2}\left[\frac{0}{2}\right] = 0$$

$$\therefore \qquad a_n = ca_{n-1} + da_{n-2}.$$

$\therefore$ The theorem holds.

Example 1. *Solve the following recurrence relations :*

(*i*) $a_n = 7a_{n-1} - 12a_{n-2}$ *for* $n \geq 3$

(*ii*) $a_n = 6a_{n-1} - 9a_{n-2}$ *for* $n \geq 3$

(*iii*) $a_n - 9a_{n-1} + 26a_{n-2} - 24a_{n-3} = 0$ *for* $n \geq 4$

(*iv*) $a_n + 6a_{n-1} + 12a_{n-2} + 8a_{n-3} = 0$ *for* $n \geq 4$

(*v*) $a_n - 6a_{n-1} + 9a_{n-2} - 4a_{n-3} = 0$ *for* $n \geq 4$

(*vi*) $a_{r+4} + 2a_{r+3} + 3a_{r+2} + 2a_{r+1} + a_r = 0$ *for* $r \geq 1$.

Sol. (*i*) We have $\qquad a_n = 7a_{n-1} - 12a_{n-2}$ ...(1)

The order of (1) is 2.

The characteristic equation of (1) is

$$x^2 = 7x - 12 \quad i.e., \quad x^2 - 7x + 12 = 0$$

$\therefore \qquad x = 3, 4$. The roots are distinct.

$\therefore$ The solution of (1) is

$$\mathbf{a_n = A(3)^n + B(4)^n \text{ for } n \geq 1,}$$

where A and B are arbitrary constants.

(*ii*) We have $\qquad a_n = 6a_{n-1} - 9a_{n-2}$ ...(1)

The order of (1) is 2.

The characteristic equation of (1) is

$$x^2 = 6x - 9 \quad i.e., \quad x^2 - 6x + 9 = 0.$$

$\therefore \qquad x = 3, 3$. Here the roots are repeated.

$\therefore$ The solution of (1) is

$$\mathbf{a_n = (A + nB)(3)^n \text{ for } n \geq 1,}$$

where A and B are arbitrary constants.

(*iii*) We have $\quad a_n - 9a_{n-1} + 26a_{n-2} - 24a_{n-3} = 0$ ...(1)

The order of (1) is 3.

The characteristic equation of (1) is

$$x^3 - 9x^2 + 26x - 24 = 0$$

$x = 2$ is a root, because $8 - 36 + 52 - 24 = 0$.

Dividing $x^3 - 9x^2 + 26x - 24$ by $x - 2$, we get $x^2 - 7x + 12$.

$$x^2 - 7x + 12 = 0 \quad \Rightarrow \quad x = 3, 4$$

$\therefore$ Characteristic roots of (1) are 2, 3, 4. These roots are distinct.

$\therefore$ The solution of (1) is

$$\mathbf{a_n = A(2)^n + B(3)^n + C(4)^n \text{ for } n \geq 1,}$$

where A, B, C are arbitrary constants.

(*iv*) We have $a_n + 6a_{n-1} + 12a_{n-2} + 8a_{n-3} = 0$ ...(1)

The order of (1) is 3.

The characteristic equation of (1) is

$$x^3 + 6x^2 + 12x + 8 = 0.$$

$\Rightarrow$ $(x + 2)^3 = 0 \quad \therefore \quad x = -2, -2, -2 \quad \therefore$ The roots are repeated.

$\therefore$ The solution of (1) is

$$\mathbf{a_n = (A + nB + n^2C)(-2)^n \text{ for } n \geq 1,}$$

where A, B, C are arbitrary constants.

(*v*) We have $a_n - 6a_{n-1} + 9a_{n-2} - 4a_{n-3} = 0$ ...(1)

The order of (1) is 3.

The characteristic equation of (1) is

$$x^3 - 6x^2 + 9x - 4 = 0$$

$x = 1$ is a root, because $1 - 6 + 9 - 4 = 0$.

Dividing $x^3 - 6x^2 + 9x - 4$ by $x - 1$, we get $x^2 - 5x + 4$.

$$x^2 - 5x + 4 = 0 \quad \Rightarrow \quad x = 1, 4$$

$\therefore$ The characteristic roots of (1) are 1, 1, 4. The root 1 is repeated twice.

$\therefore$ The solution of (1) is

$$a_n = (A + nB)(1)^n + C(4)^n$$

i.e., $$\mathbf{a_n = A + nB + C(4)^n \text{ for } n \geq 1,}$$

where A, B, C are arbitrary constants.

(*vi*) We have $a_{r+4} + 2a_{r+3} + 3a_{r+2} + 2a_{r+1} + a_r = 0$ ...(1)

The order of (1) is 4.

The characteristic equation of (1) is

$$x^4 + 2x^3 + 3x^2 + 2x + 1 = 0.$$

$$\Rightarrow \quad x^4 + 2x^3 = -3x^2 - 2x - 1$$

$$\Rightarrow \quad x^4 + 2x^3 + x^2 = -2x^2 - 2x - 1$$

$$\Rightarrow \quad (x^2 + x)^2 = -2x^2 - 2x - 1 \qquad ...(2)$$

$$\therefore \quad (x^2 + x + \lambda)^2 = (x^2 + x)^2 + 2\lambda(x^2 + x) + \lambda^2$$

$$= -2x^2 - 2x - 1 + 2\lambda(x^2 + x) + \lambda^2$$

$$= 2(\lambda - 1)x^2 + 2(\lambda - 1)x + \lambda^2 - 1$$

$$= (\lambda - 1)[2x^2 + x + \lambda + 1]$$

Let $\lambda = 1$.

$\therefore \quad (x^2 + x + 1)^2 = (1 - 1)(2x^2 + x + 1 + 1) = 0$

$\therefore \quad (x^2 + x + 1)^2 = 0$

$$x^2 + x + 1 = 0 \Rightarrow x = \frac{-1 \pm \sqrt{3}i}{2}$$

$\therefore$ The characteristic roots if (1) are $\frac{-1 \pm \sqrt{3}i}{2}, \frac{-1 \pm \sqrt{3}i}{2}$

$\therefore$ The roots $\frac{-1 + \sqrt{3}i}{2}$ and $\frac{-1 - \sqrt{3}i}{2}$ are repeated twice each.

$\therefore$ The solution of (1) is

$$\mathbf{a_r = (A + rB)\left(\frac{-1+\sqrt{3}i}{2}\right)^r + (C + rD)\left(\frac{-1-\sqrt{3}i}{2}\right)^r \text{ for } r \geq 1,}$$

where A, B, C, D are arbitrary constants.

Example 2. *Solve the following recurrence relations subject to given initial conditions :*

(*i*) $a_n = 5a_{n-1} - 6a_{n-2}$ for $n \geq 2$ with $a_0 = 7$, $a_1 = 16$

(*ii*) $b_n = b_{n-1} + 2b_{n-2}$ for $n \geq 2$ with $b_0 = 1$, $b_1 = 8$

(*iii*) $c_n = -8c_{n-1} - c_{n-2}$ for $n \geq 2$ with $c_0 = 0$, $c_1 = 1$

(*iv*) $d_n = 6d_{n-1} - 11d_{n-2} + 6d_{n-3}$ for $n \geq 3$ with $d_0 = 1$, $d_1 = 5$, $d_2 = 15$.

Sol. (*i*) We have $a_n = 5a_{n-1} - 6a_{n-2}$ ...(1)

The order of (1) is 2.

The characteristic equation of (1) is

$$x^2 = 5x - 6 \text{ i.e., } x^2 - 5x + 6 = 0.$$

$$\therefore \quad x = 2, 3$$

$\therefore$ The solution of (1) is

$$a_n = A2^n + B3^n, n \geq 0$$

$$a_0 = 7 \Rightarrow A2^0 + B3^0 = 7 \Rightarrow A + B = 7 \quad ...(2)$$

$$a_1 = 16 \Rightarrow A2^1 + B3^1 = 16 \Rightarrow 2A + 3B = 16 \quad ...(3)$$

Solving (2) and (3), we get $A = 5$, $B = 2$

$\therefore$ The required solution is $\mathbf{a_n = 5(2)^n + 2(3)^n, n \geq 0}$.

(*ii*) We have $b_n = b_{n-1} + 2b_{n-2}$ ...(1)

The order of (1) is 2.

The charateristic equation of (1) is

$$x^2 = x + 2 \text{ i.e., } x^2 - x - 2 = 0.$$

$$\therefore \quad x = -1, 2$$

$\therefore$ The solution of (1) is

$$b_n = A(-1)^n + B(2)^n, \quad n \geq 0$$

$$b_0 = 1 \Rightarrow A(-1)^0 + B(2)^0 = 1 \Rightarrow A + B = 1 \quad ...(2)$$

$$b_1 = 8 \Rightarrow A(-1)^1 + B(2)^1 = 8 \Rightarrow -A + 2B = 8 \quad ...(3)$$

Solving (2) and (3), we get

$$A = -2, B = 3.$$

$\therefore$ The required solution is

$$\mathbf{b_n = -2(-1)^n + 3(2)^n, n \geq 0.}$$

(*iii*) We have $\quad c_n = -8c_{n-1} - c_{n-2}$ ...(1)

The order of (1) is 2.

The characteristic equation of (1) is

$$x^2 = -8x - 1 \text{ i.e., } x^2 + 8x + 1 = 0.$$

$$\therefore \quad x = \frac{-8 \pm \sqrt{64-4}}{2} = -4 \pm \sqrt{15}$$

$\therefore$ The solution of (1) is

$$c_n = A(-4+\sqrt{15})^n + B(-4-\sqrt{15})^n, \quad n \geq 0$$

$$c_0 = 0 \Rightarrow A(-4+\sqrt{15})^0 + B(-4-\sqrt{15})^0 = 0$$

$$\Rightarrow \quad A + B = 0 \qquad \text{...(2)}$$

$$c_1 = 1 \Rightarrow A(-4+\sqrt{15})^1 + B(-4-\sqrt{15})^1 = 1$$

$$\Rightarrow -4(A+B) + \sqrt{15}(A-B) = 1 \Rightarrow -4(0) + \sqrt{15}(A-B) = 1$$

$$\Rightarrow A - B = \frac{1}{\sqrt{15}} \qquad \text{...(3)}$$

Solving (2) and (3), we get

$$A = \frac{1}{2\sqrt{15}}, B = -\frac{1}{2\sqrt{15}}.$$

$\therefore$ The required solution is

$$\mathbf{c_n = \frac{1}{2\sqrt{15}}(-4+\sqrt{15})^n - \frac{1}{2\sqrt{15}}(-4-\sqrt{15})^n, n \geq 0.}$$

(*iv*) We have $\quad d_n = 6d_{n-1} - 11d_{n-2} + 6d_{n-3}$ ...(1)

$\therefore$ The order of (1) is 3.

The characteristic equation of (1) is

$$x^3 = 6x^2 - 11x + 6 \quad \text{i.e.,} \quad x^3 - 6x^2 + 11x - 6 = 0 \qquad \text{...(2)}$$

$x = 1$ is a root of (2), because $1 - 6 + 11 - 6 = 0$

Dividing $x^3 - 6x^2 + 11x - 6$ by $x - 1$, we get $x^2 - 5x + 6$.

$\therefore$ The depressed equation is $x^2 - 5x + 6 = 0$

$\Rightarrow \quad x = 2, 3$

$\therefore$ The characteristic roots of (1) are 1, 2, 3.

$\therefore$ The solution of (1) is

$$d_n = A(1)^n + B(2)^n + C(3)^n$$

i.e., $$d_n = A + B(2)^n + C(3)^n, \quad n \geq 0$$

$$d_0 = 1 \Rightarrow A + B(2)^0 + C(3)^0 = 1 \Rightarrow A + B + C = 1 \qquad \text{...(3)}$$

$$d_1 = 5 \Rightarrow A + B(2)^1 + C(3)^1 = 5 \Rightarrow A + 2B + 3C = 5 \qquad \text{...(4)}$$

$d_2 = 15 \Rightarrow A + B(2)^2 + C(3)^2 = 15 \Rightarrow A + 4B + 9C = 15$ ...(5)

(4) – (3) $\Rightarrow B + 2C = 4$ ...(6)

(5) – (4) $\Rightarrow 2B + 6C = 10 \Rightarrow B + 3C = 5$ ...(7)

(7) – (6) $\Rightarrow C = 1$

$\therefore$ (6) $\Rightarrow B + 2(1) = 4 \Rightarrow B = 2$

(3) $\Rightarrow A + 2 + 1 = 1 \Rightarrow A = -2$

$\therefore$ The required solution is

$$d_n = -2 + 2(2)^n + 1(3)^n$$

or

$$\mathbf{d_n = 2^{n+1} + 3^n - 2, n \geq 0.}$$

Example 3. *It is given that white tiger population of a place is 30 at time n = 0 and 32 at time n = 1. Also the increase from time n – 1 to time n is twice the increase from time n – 2 to time n – 1. Write recurrence relation for growth rate and solve it.*

Sol. Let a_n be the tiger population at time n.

$$\therefore \quad a_0 = 30, \quad a_1 = 32$$

The increase from time $n - 1$ to time n is $a_n - a_{n-1}$ and the increase from time $n - 2$ to time $n - 1$ is $a_{n-1} - a_{n-2}$.

$\therefore$ We have $a_n - a_{n-1} = 2(a_{n-1} - a_{n-2})$

$$\Rightarrow \quad a_n - 3a_{n-1} + 2a_{n-2} = 0 \qquad \text{...(1)}$$

$\therefore$ The recurrence relation for growth rate is $\mathbf{a_n - 3a_{n-1} + 2a_{n-2} = 0, n \geq 2.}$

The characteristic equation of (1) is

$$x^2 - 3x + 2 = 0. \quad \therefore \quad x = 1, 2$$

$\therefore$ The solution of (1) is

$$a_n = A(1)^n + B(2)^n$$

i.e.,

$$a_n = A + B(2)^n, n \geq 0$$

$a_0 = 30 \Rightarrow A + B = 30$ ...(2)

$a_1 = 32 \Rightarrow A + 2B = 32$ ...(3)

(3) – (2) $\Rightarrow B = 2$

$\therefore$ (2) $\Rightarrow A = 30 - 2 = 28$

$$\therefore \quad a_n = 28 + 2(2)^n$$

i.e.,

$$\mathbf{a_n = 28 + 2^{n+1}, n \geq 0.}$$

Example 4. *Find an explicit formula for the Fibonacci sequence.* (K U 2006)

Sol. The Fibonacci sequence is 1, 1, 2, 3, 5, 8, 13, 21, ...

The recursive formula for this sequence is

$$f_n = f_{n-1} + f_{n-2} \text{ for } n \geq 2 \text{ with } f_0 = 1, f_1 = 1.$$

The recurrence relation is

$$f_n = f_{n-1} + f_{n-2} \text{ for } n \geq 2 \qquad \text{...(1)}$$

The order of (1) is 2.

The characteristic equation of (1) is

$$x^2 = x + 1 \text{ i.e., } x^2 - x - 1 = 0$$

$$\therefore \quad x = \frac{1 \pm \sqrt{1+4}}{2} = \frac{1+\sqrt{5}}{2}, \frac{1-\sqrt{5}}{2}$$

Let $\alpha = \dfrac{1+\sqrt{5}}{2}$ and $\beta = \dfrac{1-\sqrt{5}}{2}$

$\therefore$ The solution of (1) is

$$f_n = A\alpha^n + B\beta^n, \; n \geq 0, \quad \text{...(2)}$$

where A and B are arbitrary constants.

$$f_0 = 1 \;\Rightarrow\; A + B = 1 \quad \text{...(3)}$$

$$f_1 = 1 \;\Rightarrow\; A\alpha + B\beta = 1 \quad \text{...(4)}$$

(4) $\Rightarrow$ $A\alpha + (1-A)\beta = 1 \;\Rightarrow\; (\alpha-\beta)A = 1-\beta$

$$\Rightarrow \quad A = \frac{1-\beta}{\alpha-\beta} = \frac{1-\dfrac{1-\sqrt{5}}{2}}{\dfrac{1+\sqrt{5}}{2}-\dfrac{1-\sqrt{5}}{2}} = \frac{1+\sqrt{5}}{2\sqrt{5}}$$

$$(3) \Rightarrow \quad B = 1 - A = 1 - \frac{1+\sqrt{5}}{2\sqrt{5}} = -\frac{1-\sqrt{5}}{2\sqrt{5}}$$

$$\therefore \; (2) \Rightarrow \quad f_n = \left(\frac{1+\sqrt{5}}{2\sqrt{5}}\right)\left(\frac{1+\sqrt{5}}{2}\right)^n + \left(-\frac{1-\sqrt{5}}{2\sqrt{5}}\right)\left(\frac{1-\sqrt{5}}{2}\right)^n$$

or

$$\mathbf{f_n = \frac{1}{\sqrt{5}}\left[\left(\frac{1+\sqrt{5}}{2}\right)^{n+1} - \left(\frac{1-\sqrt{5}}{2}\right)^{n+1}\right], \; n \geq 0.}$$

This is the required explicit formula for the Fibonacci sequence.

Example 5. *Solve the following recurrence relations subjected to given initial conditions :*

(i) $a_n = 2a_{n-1} - a_{n-2}$ *for* $n \geq 3$, $a_1 = 1.5$, $a_2 = 3$

(ii) $a_r - 4a_{r-1} + 4a_{r-2} = 0$ *for* $r \geq 2$, $a_0 = 1$, $a_1 = 6$.

Sol. *(i)* We have $a_n = 2a_{n-1} - a_{n-2}$ ...(1)

The order of (1) is 2.

The characteristic equation of (1) is

$$x^2 = 2x - 1 \text{ i.e., } x^2 - 2x + 1 = 0.$$

$\therefore$ $x = 1, 1$. Here the roots are repeated.

$\therefore$ The solution of (1) is

$$a_n = (A + nB)(1)^n \quad \text{i.e.,} \quad a_n = A + nB, \; n \geq 1$$

$$a_1 = 1.5 \;\Rightarrow\; A + B = 1.5 \quad \text{...(2)}$$

$$a_2 = 3 \;\Rightarrow\; A + 2B = 3 \quad \text{...(3)}$$

Solving (2) and (3), we get $A = 0, B = 1.5$.

$\therefore$ The required solution is

$$a_n = 0 + 1.5\,n \quad \text{i.e.,} \quad \mathbf{a_n = 1.5\,n, \; n \geq 1.}$$

(ii) We have $a_r - 4a_{r-1} + 4a_{r-2} = 0$ ...(1)

The order of (1) is 2.

The characteristic equation of (1) is

$$x^2 - 4x + 4 = 0$$

$\therefore$ $x = 2, 2$ Here the roots are repeated.

$\therefore$ The solution of (1) is

$$a_r = (A + rB)(2)^r, \quad r \geq 0$$

i.e.,

$$a_r = A(2)^r + rB(2)^r, \quad r \geq 0$$

$$a_0 = 1 \Rightarrow A(2)^0 + 0.B(2)^0 = 1 \Rightarrow A = 1$$

$$a_1 = 6 \Rightarrow A(2)^1 + 1.B(2)^1 = 6$$

$$\Rightarrow 2A + 2B = 6 \Rightarrow A + B = 3 \Rightarrow B = 3 - 1 = 2$$

$\therefore$ The required solution is

$$\mathbf{a_r = (1 + 2r)(2)^r, r \geq 0.}$$

EXERCISE 7.3

1. Check whether the following recurrence relations are linear homogeneous relations with constant coefficients or not. If yes, write the order of the relation :
   (*i*) $a_n = 2a_{n-1} + 4a_{n-2} + 7a_{n-3}$ (*ii*) $b_n = 5b_{n-1}\, b_{n-3}$
   (*iii*) $c_n = c_{n-1} + 7$ (*iv*) $d_n = nd_{n-1} + 3d_{n-2}$.
2. Solve the following recurrence relations :
   (*i*) $a_n = 5a_{n-1} - 4a_{n-2}$ for $n \geq 2$ (*ii*) $b_n + b_{n-1} - 12b_{n-2} = 0$ for $n \geq 3$
   (*iii*) $a_r - 6a_{r-1} + 8a_{r-2} = 0$ for $r \geq 3$ (*iv*) $a_r + a_{r-1} + a_{r-2} = 0$ for $r \geq 2$.
3. Solve the following recurrence relations :
   (*i*) $a_n = 14a_{n-1} - 49a_{n-2}$ for $n \geq 3$ (*ii*) $b_n + 2b_{n-1} + b_{n-2} = 0$ for $n \geq 2$
   (*iii*) $9y_{k+2} - 6y_{k+1} + y_k = 0$ for $k \geq 0$ (*iv*) $a_r - 4a_{r-1} + 4a_{r-2} = 0$ for $r \geq 2$.
4. Solve the following recurrence relations :
   (*i*) $R(k) - 4R(k-1) - 11R(k-2) + 30R(k-3) = 0$ for $k \geq 3$
   (*ii*) $4a_n - 20a_{n-1} + 17a_{n-2} - 4a_{n-3} = 0$ for $n \geq 3$
   (*iii*) $y_{k+4} + 4y_{k+3} + 8y_{k+2} + 8y_{k+1} + 4y_k = 0$ for $k \geq 0$.
5. Solve the following recurrence relations subject to the given initial conditions :
   (*i*) $a_n = 2a_{n-1} - a_{n-2}$ for $n \geq 2$ with $a_0 = 1, a_1 = 4$ (*KU 2006 C, 2007*)
   (*ii*) $b_n = 2b_{n-1} + 8b_{n-2}$ for $n \geq 2$ with $b_0 = 4, b_1 = 10$
   (*iii*) $c_n = 2c_{n-1} + 15c_{n-2}$ for $n \geq 2$ with $c_0 = 0, c_1 = -1$
   (*iv*) $t_n = -2t_{n-1} + 15t_{n-2}$ for $n \geq 2$ with $t_0 = 0, t_1 = 1$
   (*v*) $t_n = 2t_{n-1} - 2t_{n-2}$ for $n \geq 3$ with $t_1 = 1, t_2 = 4$
6. Solve the following recurrence relations subject to the given initial conditions :
   (*i*) $a_n = 6a_{n-1} - 11a_{n-2} + 6a_{n-3}$ for $n \geq 3$ with $a_0 = 2, a_1 = 5, a_2 = 15$
   (*ii*) $b_n = -3b_{n-1} - 3b_{n-2} - b_{n-3}$ for $n > 3$ with $b_0 = 1, b_1 = -2, b_2 = -1$.
7. Solve the following recurrence relations subjected to the given initial conditions :
   (*i*) $9a_r - 6a_{r-1} + a_{r-2} = 0$ for $r \geq 2$ with $a_0 = 0, a_1 = 2$
   (*ii*) $a_n = 10a_{n-1} - 25\, a_{n-2}$ for $n \geq 2$ with $a_0 = 1, a_1 = 3$
   (*iii*) $t_n = 6t_{n-1} - 9t_{n-2}$ for $n \geq 2$ with $a_0 = 1, a_1 = 6$
   (*iv*) $a_n = 5a_{n-1} - 8a_{n-2} + 4a_{n-3}$ for $n \geq 3$ with $a_0 = 0, a_1 = 1, a_2 = 2$.
8. Suppose that the population of a village is 100 at time $n = 0$ and 110 at time $n = 1$. The population increase from time $n - 1$ to time n is double the increase from time $n - 2$ to time $n - 1$. Find a recurrence relation for the population at time n and hence solve it.

Answers

1. (*i*) Yes, 3 (*ii*) No, because the relation is not linear
(*iii*) No, because the relation is not homogeneous
(*iv*) No, because the coefficients of d_{n-1} is not a constant
2. (*i*) $a_n = A + B(4)^n, n \geq 0$ (*ii*) $b_n = A(3)^n + B(-4)^n, n \geq 1$
(*iii*) $a_r = A(2)^r + B(4)^r, r \geq 1$ (*iv*) $a_r = A\left(\frac{-1+\sqrt{3}i}{2}\right)^n + B\left(\frac{-1-\sqrt{3}i}{2}\right)^n, r \geq 0$
3. (*i*) $a_n = (A + nB)(7)^n, n \geq 1$ (*ii*) $b_n = (A + nB)(-1)^n, n \geq 0$
(*iii*) $y_k = (A + kB)\left(\frac{1}{3}\right)^k, k \geq 0$ (*iv*) $a_r = (A + rB)(2)^r, r \geq 0$
4. (*i*) $R(k) = A(2)^k + B(-3)^k + C(5)^k, k \geq 0$ (*ii*) $a_n = (A + nB)\left(\frac{1}{2}\right)^n + C(4)^n, n \geq 0$
(*iii*) $y_k = (A + kB)(-1 + i)^k + (C + kD)(-1 - i)^k, k \geq 0$
5. (*i*) $a_n = 3n + 1, n \geq 0$ (*ii*) $b_n = 3(4)^n + (-2)^n, n \geq 0$
(*iii*) $c_n = -\frac{1}{8}(5)^n + \frac{1}{8}(-3)^n, n \geq 0$ (*iv*) $t_n = -\frac{1}{8}(-5)^n + \frac{1}{8}(3)^n, n \geq 0$
(*v*) $t_n = \frac{-1-2i}{2}(1+i)^n + \frac{-1+2i}{2}(1-i)^n, n \geq 1$
6. (*i*) $a_n = 1 - 2^n + 2(3)^n, n \geq 0$ (*ii*) $b_n = (1 + 3n - 2n^2)(-1)^n, n \geq 0$
7. (*i*) $a_r = 2r\left(\frac{1}{3}\right)^{r-1}, r \geq 0$ (*ii*) $a_n = (5 - 2n)(5)^{n-1}, n \geq 0$
(*iii*) $t_n = (1 + n)(3)^n, n \geq 0$ (*iv*) $a_n = 2^{n+1} - n2^{n-1} - 2, n \geq 0$
8. $a_n = 90 + 10(2)^n, n \geq 0.$

7.12. PARTICULAR SOLUTION OF A LINEAR NON-HOMOGENEOUS RECURRENCE RELATION WITH CONSTANT COEFFICIENTS

Let $$a_n = c_1 a_{n-1} + c_2 a_{n-2} + \ldots\ldots + c_k a_{n-k} + f(n) \quad \ldots(1)$$
be a linear non-homogeneous recurrence relation with constant coefficients and its order be k.

$$\therefore \quad f(n) \neq 0, c_k \neq 0$$

In this section, we shall find a solution of the given recurrence relation which is free from arbitrary constants. Such a solution is called the **particular solution** of the given recurrence relation. If some initial conditions are also given with the relation then its particular solution may or may not satisfy those initial conditions. The problem of finding the solution of recurrence relation (1) with given initial conditions will be taken up later on.

There is no general method of finding the particular solution of recurrence relation (1). The method depends upon the form of the function $f(n)$. We shall consider the method of finding the particular solution (*PS*) of (1) in the following four cases :

Case I. If $f(n)$ is a polynomial of the form $\alpha_0 n^m + \alpha_1 n^{m-1} + \ldots\ldots + \alpha_m$, $m \in \mathbf{N}$ and the solution of the recurrence relation $a_n = c_1 a_{n-1} + c_2 a_{n-2} + \ldots\ldots + c_k\, a_{n-k}$ is not a polynomial of degree m then the particular solution of (1) is assumed to be $P_0 n^m + P_1 n^{m-1} + \ldots\ldots + P_m$, where

$P_0, P_1, ..., P_m$ are constants. This expression for particular solution of (1) is substituted in the recurrence relation (1) and the values of constants $P_0, P_1, ..., P_m$ are calculated by comparing the terms.

Example 1. *Find the particular solution of the following recurrence relations :*

(i) $a_{n+2} - 4a_n = n^2 + n - 1$ *for* $n \geq 0$ *(ii)* $a_n + 5a_{n-1} + 6a_{n-2} = 3n^2 - 2n + 1$ *for* $n \geq 2$

Sol. *(i)* We have $a_{n+2} - 4a_n = n^2 + n - 1$ ...(1)

Here $f(n) = n^2 + n - 1$. This is a polynomial of degree 2.

Let *PS* of (1) be $a_n = P_0n^2 + P_1n + P_2$

$\therefore$ $a_{n+2} = P_0(n + 2)^2 + P_1(n + 2) + P_2$

Substituting these values in (1), we get

$$P_0(n + 2)^2 + P_1(n + 2) + P_2 - 4[P_0n^2 + P_1n + P_2] = n^2 + n - 1$$

$$\Rightarrow \quad (P_0 - 4P_0)n^2 + (4P_0 + P_1 - 4P_1)n + 4P_0 + 2P_1 + P_2 - 4P_2 = n^2 + n - 1$$

$$\Rightarrow \quad -3P_0n^2 + (4P_0 - 3P_1)n + (4P_0 + 2P_1 - 3P_2) = n^2 + n - 1$$

Comparing the terms, we get

$$-3P_0 = 1,\ 4P_0 - 3P_1 = 1 \quad \text{and} \quad 4P_0 + 2P_1 - 3P_2 = -1.$$

Solving these equations, we get

$$P_0 = -\frac{1}{3},\ P_1 = -\frac{7}{9},\ P_2 = -\frac{17}{27}$$

$\therefore$ The particular solution of (1) is

$$\mathbf{a_n = -\frac{1}{3}\,n^2 - \frac{7}{9}\,n - \frac{17}{27},\ n \geq 0.}$$

(ii) We have $a_n + 5a_{n-1} + 6a_{n-2} = 3n^2 - 2n + 1$ ...(1)

Here $f(n) = 3n^2 - 2n + 1$. This is a polynomial of degree 2.

Let *PS* of (1) be $a_n = P_0n^2 + P_1n + P_2$

$\therefore$ $a_{n-1} = P_0(n - 1)^2 + P_1(n - 1) + P_2$

$a_{n-2} = P_0(n - 2)^2 + P_1(n - 2) + P_2$

Substituting these values in (1), we get

$$P_0n^2 + P_1n + P_2 + 5[P_0(n - 1)^2 + P_1(n - 1) + P_2] + 6[P_0(n - 2)^2 + P_1(n - 2) + P_2] = 3n^2 - 2n + 1$$

$$\Rightarrow \quad (P_0 + 5P_0 + 6P_0)n^2 + (P_1 - 10P_0 + 5P_1 - 24P_0 + 6P_1)n + P_2 + 5P_0 - 5P_1 + 5P_2 + 24P_0 - 12P_1 + 6P_2 = 3n^2 - 2n + 1$$

$$\Rightarrow \quad 12P_0n^2 + (-34P_0 + 12P_1)n + (29P_0 - 17P_1 + 12P_2) = 3n^2 - 2n + 1$$

Comparing the terms, we get

$$12P_0 = 3 \quad ...(2)$$

$$-34P_0 + 12P_1 = -2 \quad ...(3)$$

$$29P_0 - 17P_1 + 12P_2 = 1 \quad ...(4)$$

Solving (2), (3), (4), we get

$$P_0 = \frac{1}{4},\ P_1 = \frac{13}{24},\ P_2 = \frac{71}{288}$$

$\therefore$ The particular solution of (1) is

$$\mathbf{a_n = \frac{1}{4}\,n^2 + \frac{13}{24}\,n + \frac{71}{288},\ n \geq 0.}$$

Case II. If $f(n)$ is a constant then the particular solution is assumed to be P, a constant, provided 1 is not a characteristic root of the relation (1).

In case 1 is a characteristic root of multiplicity $\lambda(\geq 1)$ then the particular solution is assumed to be $n^\lambda P$, where P is a constant. The expression for particular solution is substituted in the relation (1) and the value of P is calculated by comparing the terms.

Example 2. *Find the particular solution of the following recurrence relations :*

(*i*) $a_n - 5a_{n-1} + 6a_{n-2} = 4$, *for* $n \geq 2$ (*ii*) $a_n - 2a_{n-1} + a_{n-2} = 11$, *for* $n \geq 2$

Sol. (*i*) We have $a_n - 5a_{n-1} + 6\,a_{n-2} = 4$ ...(1)

Here $f(n) = 4$. This is a constant.

The characteristic equation of (1) is

$$x^2 - 5x + 6 = 0 \quad \therefore \quad x = 2, 3$$

$\therefore$ 1 is not a characteristic root of (1).

$\therefore$ Let PS of (1) be $a_n = P$, a constant

$\Rightarrow$ $a_{n-1} = P$ and $a_{n-2} = P$

Substituting these values in (1), we get

$$P - 5P + 6P = 4$$

$\Rightarrow$ $2P = 4 \Rightarrow P = 2$

$\therefore$ The particular solution of (1) is $a_n = \mathbf{2}$, $\mathbf{n \geq 0}$.

(*ii*) We have $a_n - 2a_{n-1} + a_{n-2} = 11$ ...(1)

Here $f(n) = 11$. This is a constant.

The characteristic equation (1) is

$$x^2 - 2x + 1 = 0. \quad \therefore \quad x = 1, 1$$

$\therefore$ 1 is a characteristic root of (1) of multiplicity 2.

Let PS of (1) be $a_n = n^2P$

$\Rightarrow$ $a_{n-1} = (n-1)^2\,P$ and $a_{n-2} = (n-2)^2\,P$

$\therefore$ Substituting these values in (1), we get

$$n^2P - 2(n-1)^2P + (n-2)^2P = 11$$

$\Rightarrow$ $(P - 2P + P)n^2 + (4P - 4P)n + (-2P + 4P) = 11$

$\Rightarrow$ $2P = 11 \Rightarrow P = 11/2$

$\therefore$ The particular solution of (1) is $a_n = \mathbf{\frac{11}{2}\,n^2,\ n \geq 0}$.

Case III. If $f(n)$ is of the form β^n then the particular solution is assumed to be $P\beta^n$ provided β is not a characteristic root of the relation (1).

In case β is a characteristic root of multiplicity $\lambda(\geq 1)$ then the particular solution is assumed to be $n^\lambda P\beta^n$. The expression for particular solution is substituted in the relation (1) and the value of P is calculated by comparing the terms.

Example 3. *Find the particular solution of the following recurrence relations :*

(*i*) $a_{r+2} - 5a_{r+1} + 6a_r = 5^r$, *for* $r \geq 0$ (*ii*) $a_n - 5a_{n-1} + 6a_{n-2} = 3^n + n$, *for* $r \geq 2$

Sol. (*i*) We have $a_{r+2} - 5a_{r+1} + 6a_r = 5^r$ ...(1)

Here $f(r) = 5^r$. This is of the form β^r with $\beta = 5$.

The characteristic equation of (1) is

$$x^2 - 5x + 6 = 0 \quad \therefore \quad x = 2, 3$$

$\therefore$ 5 is not a characteristic root of (1).

Let PS of (1) be $a_r = P(5)^r$.

$\Rightarrow \qquad a_{r+2} = P(5)^{r+2}$ and $a_{r+1} = P(5)^{r+1}$

Substituting these values in (1), we get

$$P(5)^{r+2} - 5P(5)^{r+1} + 6P(5)^r = 5^r$$

$$\Rightarrow \qquad 25P - 25P + 6P = 1 \;\Rightarrow\; P = 1/6$$

$\therefore$ The particular solution of (1) is

$$a_r = \frac{1}{6}(5)^r, \quad r \geq 0.$$

(*ii*) We have $a_n - 5a_{n-1} + 6a_{n-2} = 3^n + n$ ...(1)

Here $f(n) = 3^n + n$. 3^n is of the form β^n, where $\beta = 3$ and n is a polynomial of degree 1.

The characteristic equation of (1) is $x^2 - 5x + 6 = 0 \quad \therefore \quad x = 2, 3$

$\therefore$ 3 is a characteristic root of (1) of multiplicity 1.

Let the part of PS of (1) corresponding to 3^n be $nP(3)^n$.

Let the part of PS of (1) corresponding to n be $P_0n + P_1$.

$\therefore$ PS of (1) is $\quad a_n = nP(3)^n + P_0n + P_1$

$$\Rightarrow \qquad a_{n-1} = (n-1)P(3)^{n-1} + P_0(n-1) + P_1$$

and $$a_{n-2} = (n-2)P(3)^{n-2} + P_0(n-2) + P_2$$

Substituting these values in (1), we get

$$nP(3)^n + P_0n + P_1 - 5\,[(n-1)P(3)^{n-1} + P_0(n-1) + P_1] + 6[(n-2)P(3)^{n-2} + P_0(n-2) + P_1] = 3^n + n$$

$$\Rightarrow \left(nP - \frac{5}{3}(n-1)P + \frac{6}{9}(n-2)P\right)3^n + (P_0 - 5P_0 + 6P_0)n + (P_1 + 5P_0 - 5P_1 - 12P_0 + 6P_1) = 3^n + n$$

$$\Rightarrow \frac{1}{3}(3nP - 5nP + 5P + 2nP - 4P)3^n + 2P_0n + (-7P_0 + 2P_1) = 3^n + n$$

$$\frac{1}{3}P(3)^n + 2P_0n + (-7P_0 + 2P_1) = 3^n + n$$

Comparing the terms, we get

$$\frac{1}{3}P = 1, \; 2P_0 = 1, \; -7P_0 + 2P_1 = 0$$

$$\therefore \qquad P = 3, \; P_0 = \frac{1}{2}, \; P_1 = \frac{7}{4}$$

$\therefore$ The particular solution of (1) is

$$a_n = n(3)(3)^n + \frac{1}{2}n + \frac{7}{4} \quad i.e., \quad \mathbf{a_n = n(3)^{n+1} + \frac{1}{2}n + \frac{7}{4}, n \geq 0.}$$

Case IV. If $f(n)$ is of the form $(\alpha_0 n^m + \alpha_1 n^{m-1} + \ldots\ldots + \alpha_m)\beta^n$, $m \in \mathbf{W}$, then the particular solution is assumed to be $(P_0n^m + P_1n^{m-1} + \ldots\ldots + P_m)\beta^n$, provided β is not a characteristic root of the relation (1).

In case β is a characteristic root of multiplicity λ (≥ 1) then the particular solution is assumed to be

$$n^\lambda(P_0n^m + P_1n^{m-1} + \ldots\ldots + P_m)\beta^n.$$

The expression for particular solution is substituted in the relation (1) and the values of constants $P_0, P_1, \ldots\ldots, P_m$ are calculated by comparing the terms.

Note. Case II is a particular case of **case IV.** because if $f(n) = P$, a constant, then we can write $f(n)$ as $P(1)^n$. Here the polynomial is the constant polynomial P and β is equal to 1.

Example 4. *Find the particular solution of the following recurrence relations :*

(*i*) $a_n + 5a_{n-1} + 4a_{n-2} = 56(3)^n$, *for* $n \geq 2$ (K U 2006)

(*ii*) $a_n - 6a_{n-1} + 9a_{n-2} = n(3)^n$, *for* $n \geq 2$.

Sol. (*i*) We have $a_n + 5a_{n-1} + 4a_{n-2} = 56(3)^n$ ...(1)

Here $f(n) = 56(3)^n$. This is the product of a constant polynomial '56' with β^n, where $\beta = 3$.

The characteristic equation of (1) is

$$x^2 + 5x + 4 = 0 \quad \therefore \quad x = -1, -4$$

$\therefore$ 3 is not a characteristic root of (1).

Let *PS* of (1) be $a_n = P(3)^n$

$\Rightarrow$ $a_{n-1} = P(3)^{n-1}$ and $a_{n-2} = P(3)^{n-2}$

Substituting the values of a_{n-1} and a_{n-2} in (1), we get

$$P(3)^n + 5P(3)^{n-1} + 4P(3)^{n-2} = 56(3)^n$$

$$\Rightarrow \quad \left(P + \frac{5}{3}P + \frac{4}{9}P\right) 3^n = 56(3)^n$$

$$\Rightarrow \quad \frac{28}{9}P = 56 \quad \Rightarrow \quad P = 18$$

$\therefore$ The particular solution is

$$a_n = 18(3)^n \quad i.e., \quad \mathbf{a_n = 2(3)^{n+2}, n \geq 0}$$

(*ii*) We have $a_n - 6a_{n-1} + 9a_{n-2} = n(3)^n$ ...(1)

Here $f(n) = n(3)^n$. This is the product of polynomial 'n' with β^n, where $\beta = 3$.

The characteristic equation of (1) is

$$x^2 - 6x + 9 = 0 \quad \therefore \quad x = 3, 3$$

$\therefore$ 3 is a characteristic root of (1) of multiplicity 2.

Let *PS* of (1) be $a_n = n^2(P_0 n + P_1)\, 3^n$

$\therefore$ $a_{n-1} = (n-1)^2\, (P_0\,(n-1) + P_0)\, 3^{n-1}$

and $a_{n-2} = (n-2)^2\, (P_0(n-2) + P_1)\, 3^{n-2}$

Substituting the values in (1), we get

$$n^2(P_0 n + P_1)3^n - 6(n-1)^2(P_0(n-1) + P_1)\, 3^{n-1} + 9(n-2)^2(P_0\,(n-2) + P_1)\, 3^{n-2} = n(3)^n$$

$$\Rightarrow \quad n^2(P_0 n + P_1) - 2(n-1)^2\, (P_0\,(n-1) + P_1) + (n-2)^2\, (P_0(n-2) + P_1) = n$$

$$\Rightarrow \quad n^2\,(P_0 n + P_1) - 2(n^2 - 2n + 1)\,(nP_0 - P_0 + P_1) + (n^2 - 4n + 4)(4P_0 - 2P_0 + P_1) = n$$

$$\Rightarrow \quad (P_0 - 2P_0 + P_0)n^3 + (P_1 + 2P_0 - 2P_1 + 4P_0 - 2P_0 + P_1 - 4P_0)n^2$$
$$+ (-4P_0 + 4P_1 - 2P_0 + 8P_0 - 4P_1 + 4P_0)n + (nP_0 - 2P_1 - 8P_0 + 4P_1) = n$$

$$\Rightarrow \quad 6P_0 n + (-6P_0 + 2P_1) = n$$

Comparing the terms, we get $6P_0 = 1$ and $-6P_0 + 2P_1 = 0$

$$\therefore \quad P_0 = \frac{1}{6}, P = \frac{1}{2}$$

$\therefore$ The particular solution of (1) is

$$a_n = n^2\left(\frac{1}{6}n + \frac{1}{2}\right)3^n \quad i.e., \quad a_n = \frac{n^2(n+3)}{2}\,3^{n-1}, n \geq 0.$$

EXERCISE 7.4

Find the particular solution of the following recurrence relations :

1. (i) $a_n - 5a_{n-1} + 6a_{n-2} = n + 1, n \geq 2$ (ii) $a_r + 5a_{r-1} + 6a_{r-2} = 3r^2, r \geq 2$
(iii) $a_n - a_{n-1} - 2a_{n-2} = 2n^2, n \geq 2$ (iv) $a_{n+2} - 2a_{n+1} + a_n = 3n + 5, n \geq 0.$
2. (i) $a_n - 4a_{n-1} + 5a_{n-2} = 2, n \geq 2$ (ii) $a_n + a_{n-1} = 4, n \geq 1$
(iii) $a_n - a_{n-1} = 4, n \geq 1$ (iv) $a_n - 2a_{n-1} + a_{n-2} = 18, n \geq 2.$
3. (i) $a_n - 2a_{n-1} = 4^n, n \geq 1$ (ii) $a_n - 2a_{n-1} = 2^n, n \geq 1$
(iii) $a_{r+2} - 3a_{r+1} + 2a_r = 5^r, r \geq 0$ (iv) $a_{r+2} - 3a_{r+1} + 2a_r = 2^r, r \geq 0.$
4. (i) $a_r + a_{r-1} = 3r(2)^r, r \geq 1$ (ii) $a_n - 4a_{n-1} = 6(4^n), n \geq 1$ *(KU 2006 C)*
(iii) $a_n - a_{n-1} - 2a_{n-2} = 3n(4)^n, n \geq 2$ (iv) $a_r + 5a_{r-1} + 6a_{r-2} = 42(4)^r, r \geq 2.$

Answers

1. (i) $a_n = \frac{1}{2}n + \frac{9}{4}, n \geq 0$ (ii) $a_r = \frac{1}{4}r^2 + \frac{17}{24}r + \frac{115}{288}, r \geq 0$
(iii) $a_n = -n^2 - 5n - 8, n \geq 0$ (iv) $a_n = \frac{1}{2}n^3 + n^2, n \geq 0$
2. (i) $a_n = 1, n \geq 0$ (ii) $a_n = 2, n \geq 0$
(iii) $a_n = 4n, n \geq 0$ (iv) $a_n = 9n^2, n \geq 0$
3. (i) $a_n = 2^{2n+1}, n \geq 0$ (ii) $a_n = n(2)^n, n \geq 0$
(iii) $a_r = \frac{1}{12}(5)^r, r \geq 0$ (iv) $a_r = r(2)^{r-1}, r \geq 0$
4. (i) $a_r = \left(2r + \frac{2}{3}\right)2^r, r \geq 0$ (ii) $a_n = 3n(2)^{2n+1}, n \geq 0$
(iii) $a_n = \left(\frac{24}{5}n - \frac{96}{25}\right)4^n, n \geq 0$ (iv) $a_r = 4^{r+2}, r \geq 0.$

Hint

1. (iv) The solution of $a_{n+2} - 2a_{n+1} + a_n = 0$ is $(A + nB)(1)^n$ *i.e.*, $A + nB$, which is a polynomial of degree same as that of $3n + 5$. Try $P_0n^3 + P_1n^2$ as the particular solution.

7.13. TOTAL SOLUTION OF A LINEAR NON-HOMOGENEOUS RECURRENCE RELATION WITH CONSTANT COEFFICIENTS

Let
$$a_n = c_1a_{n-1} + c_2a_{n-2} + \ldots\ldots + c_ka_{n-k} + f(n) \quad \ldots(1)$$
be a linear non-homogeneous recurrence relation with constant coefficients and its order be k.

$$\therefore \quad f(n) \neq 0, c_k \neq 0$$

In this section, we shall find a solution of the given recurrence relation which will also satisfy the initial conditions, if given with the recurrence relation. Such a solution is called the **total solution** of the given recurrence relation.

The solution of the corresponding homogeneous recurrence relation

$$a_n = c_1 a_{n-1} + c_2 a_{n-2} + \ldots\ldots + c_k\, a_{n-k} \qquad \ldots(2)$$

is called the **homogeneous solution** (HS) of the given recurrence relation (1).

The homogeneous solution of (1) *i.e.*, the solution of (2) will contain k arbitrary constants. The particular solution of (1) will not contain any arbitrary constant. We shall show that the sum of the homogeneous solution of (1) and the particular solution of (1) will also be a solution of the given recurrence relation (1).

Let the sequences $a_1^{(h)}, a_2^{(h)}, \ldots\ldots, a_n^{(h)}, \ldots$ and $a_1^{(p)}, a_2^{(p)}, \ldots\ldots, a_n^{(p)}, \ldots\ldots$ be respectively the homogeneous solution and the particular solution of (1).

$\therefore$ (2) $\Rightarrow$ $$a_n^{(h)} = c_1 a_{n-1}^{(h)} + c_2 a_{n-2}^{(h)} + \ldots\ldots + c_k a_{n-k}^{(h)}$$

(1) $\Rightarrow$ $$a_n^{(p)} = c_1 a_{n-1}^{(p)} + c_2 a_{n-2}^{(p)} + \ldots\ldots + c_k a_{n-k}^{(p)} + f(n)$$

Adding, we get $a_n^{(h)} + a_n^{(p)} = c_1 (a_{n-1}^{(h)} + a_{n-1}^{(p)}) + c_2 (a_{n-2}^{(h)} + a_{n-2}^{(p)})$

$$+ \ldots\ldots + c_k (a_{n-k}^{(h)} + a_{n-k}^{(p)}) + f(n)$$

$\therefore$ The sequence $a_1^{(h)} + a_1^{(p)}, a_2^{(h)} + a_2^{(p)}, \ldots\ldots, a_n^{(h)} + a_n^{(p)}, \ldots\ldots$

satisfies the given recurrence relation (1). This solution will contain k arbitrary constants. These arbitrary constants can be chosen so that this solution may satisfy the given initial conditions. Thus, the total solution of the given recurrence relation is the sum of the homogeneous solution and the particular solution of the given relation.

The following are the working rules to find the total solution of a linear non-homogeneous recurrence relation with constant coefficients.

Step I. *Find the homogeneous solution of the given recurrence relation which is nothing but the general solution of the corresponding homogeneous recurrence relation.*

Step II. *Find the particular solution of the given recurrence relation.*

Step III. *Find the sum of these two solutions and call it the total solution of the given recurrence relation.*

Step IV. *Find the arbitrary constants involved in the total solution so that this solution may satisfy the given initial conditions.*

Example 1. *Find the total solution of the following recurrence relations.*

(*i*) $y_{k+1} - 5y_k = 5^k, k \geq 0$ *with* $y_0 = 7$ (*ii*) $a_n + 4a_{n-1} = 7, n \geq 1$ *with* $a_0 = 3$.

Sol. (*i*) We have $y_{k+1} - 5y_k = 5^k$ ...(1)

The characteristic equation of (1) is $x - 5 = 0$. $\therefore$ $x = 5$

$\therefore$ *HS* of (1) $= A5^k$

Here $f(k) = 5^k$. This is of the form β^k, where $\beta = 5$.

Here 5 is a characteristic root of (1).

$\therefore$ PS of (1) = $kP5^k$, where P is a constant.

$\therefore$ (1) $\Rightarrow$ $(k+1)P5^{k+1} - 5kP5^k = 5^k$

$\Rightarrow$ $5(k+1)P - 5kP = 1$

$\Rightarrow$ $5P = 1 \Rightarrow P = 1/5$

$\therefore$ PS of (1) $= k\left(\frac{1}{5}\right)5^k = k5^{k-1}$

$\therefore$ Total solution (TS) of (1) is $y_k = A5^k + k5^{k-1}$ (Using $TS = HS + PS$)

or $$y_k = \left(A + \frac{k}{5}\right)5^k,\ k \geq 0$$

The initial condition is $y_0 = 7$.

$$y_0 = 7 \Rightarrow \left(A + \frac{0}{5}\right)(5)^0 = 7 \Rightarrow A = 7$$

$\therefore$ The required total solution of (1) is

$$\mathbf{y_k = \left(7 + \frac{k}{5}\right)5^k,\ k \geq 0.}$$

(*ii*) We have $a_n + 4a_{n-1} = 7$ ...(1)

The characteristic equation of (1) is $x + 4 = 0$. $\therefore$ $x = -4$

$\therefore$ HS of (1) $= A(-4)^n$

Here $f(n) = 7$. This is a constant. 1 is not a characteristic root of (1).

$\therefore$ PS of (1) = P, a constant

$\therefore$ (1) $\Rightarrow$ $P + 4P = 7 \Rightarrow P = 7/5$

$\therefore$ PS of (1) $= \frac{7}{5}$

$\therefore$ Total solution (TS) of (1) is

$$a_n = A(-4)^n + \frac{7}{5},\ n \geq 0$$ (Using $TS = HS + PS$)

$$a_0 = 3 \Rightarrow A(-4)^0 + \frac{7}{5} = 3 \Rightarrow A = 3 - \frac{7}{5} = \frac{8}{5}$$

$\therefore$ The required total solution of (1) is

$$\mathbf{a_n = \frac{8}{5}(-4)^n + \frac{7}{5},\ n \geq 0.}$$

Example 2. *Find the total solution of the following recurrence relations :*

(*i*) $S(n) + 2S(n-1) + S(n-2) = 20,\ n \geq 2$ *with* $S(0) = 2,\ S(1) = 3$

(*ii*) $a_r + a_{r-1} + a_{r-2} = r2^r,\ r \geq 2$

(*iii*) $a_n - 4a_{n-1} + 4a_{n-2} = 3n + 2^n,\ n \geq 2$ *with* $a_0 = 13,\ a_1 = 22$

(*iv*) $a_{r+2} - 3a_{r+1} + 2a_r = 5^r,\ r \geq 0$ *with* $a_0 = \frac{1}{12},\ a_1 = \frac{7}{12}$.

Sol. (*i*) We have $S(n) + 2S(n-1) + S(n-2) = 20$ ...(1)

The characteristic equation of (1) is

$$x^2 + 2x + 1 = 0.\quad \therefore\quad x = -1, -1$$

$\therefore$ HS of (1) $= (A + nB)(-1)^n$

Here $f(n) = 20$. This is a constant. Also, 1 is not a characteristic root of (1).

$\therefore$ PS of (1) $= P$, a constant

$\therefore$ (1) $\Rightarrow$ $P + 2P + P = 20$ $\Rightarrow$ $4P = 20$ $\Rightarrow$ $P = 5$

$\therefore$ PS of (1) $= 5$

$\therefore$ Total solution (TS) of (1) is

$$S(n) = (A + nB)(-1)^n + 5, \; n \geq 0$$

The initial conditions are $S(0) = 9$, $S(1) = 3$.

$$\begin{aligned} S(0) = 2 \;&\Rightarrow\; (A + 0.B)(-1) + 5 = 2 \\ &\Rightarrow\; A + 5 = 2 \;\Rightarrow\; A = -3 \\ S(1) = 3 \;&\Rightarrow\; (A + 1.B)(-1)^1 + 5 = 3 \\ &\Rightarrow\; -A - B + 5 = 3 \\ &\Rightarrow\; B = 2 - A = 2 - (-3) = 5 \end{aligned}$$

$\therefore$ The required total solution (TS) of (1) is

$$\mathbf{S(n) = (-3 + 5n)(-1)^n + 5, \; n \geq 0.}$$

(*ii*) We have $a_r + a_{r-1} + a_{r-2} = r2^r$ ...(1)

The characteristic equation of (1) is

$$x^2 + x + 1 = 0.$$

$$\therefore \qquad x = \frac{-1 \pm \sqrt{3}i}{2}$$

$$\therefore \quad HS \text{ of } (1) = A\left(\frac{-1 + \sqrt{3}i}{2}\right)^r + B\left(\frac{-1 - \sqrt{3}i}{2}\right)^r$$

Here $f(r) = r2^r$. This is the product of the polynomial r with β^r, where $\beta = 2$. Here 2 is not a characteristic root of (1).

$\therefore$ PS of (1) $= (P_0 r + P_1)2^r$, where P_0 and P_1 are constants.

(1) $\Rightarrow$ $(P_0 r + P_1)\,2^r + (P_0(r-1) + P_1)2^{r-1} + (P_0(r-2) + P_1)\,2^{r-2} = r\,2^r$

$$\Rightarrow \quad 2^r\left[P_0 r + P_1 + \frac{1}{2}(P_0 r - P_0 + P_1) + \frac{1}{4}(P_0 r - 2P_0 + P_1)\right] = r2^r$$

$$\Rightarrow \quad \left(P_0 + \frac{P_0}{2} + \frac{P_0}{4}\right)r + P_1 - \frac{1}{2}P_0 + \frac{1}{2}P_1 - \frac{1}{2}P_0 + \frac{1}{4}P_1 = r$$

$$\Rightarrow \quad \frac{7}{4}P_0 r + \left(\frac{7}{4}P_1 - P_0\right) = r$$

Comparing the terms, we get $\quad \dfrac{7}{4}P_0 = 1, \; \dfrac{7}{4}P_1 - P_0 = 0$

Solving, we get $\quad P_0 = \dfrac{4}{7}, \; P_1 = \dfrac{16}{49}$

$$\therefore \quad PS \text{ of } (1) = \left(\frac{4}{7}r + \frac{16}{49}\right)2^r = \left(\frac{28r + 16}{49}\right)2^r$$

$\therefore$ Total solution (TS) of (1) is

$$\mathbf{a_r = A\left(\frac{-1+\sqrt{3}i}{2}\right)^r + B\left(\frac{-1-\sqrt{3}i}{2}\right)^r + \left(\frac{28r+16}{49}\right)2^r, r \geq 0.}$$

Remark. In this problem, the initial conditions are not given. If we are given initial conditions then we choose those values of A and B for which the total solution satisfies the given initial conditions.

(*iii*) We have $a_n - 4a_{n-1} + 4a_{n-2} = 3n + 2^n$ ...(1)

The characteristic equation of (1) is $x^2 - 4x + 4 = 0$. $\therefore$ $x = 2, 2$

$\therefore$ *HS* of (1) $= (A + nB)2^n$

Here $f(n) = 3n + 2^n$. This is the sum of the polynomial $3n$ and β^n, where $\beta = 2$.

$\therefore$ Part of *PS* of (1) corresponding to $3n$ is $P_0n + P_1$ and part of *PS* of (1) corresponding to 2^n is $n^2P_22^n$. ($\because$ 2 is a ch. root of multiplicity 2)

$\therefore$ *PS* of (1) $= P_0n + P_1 + n^2P_22^n$

$$\therefore \quad (1) \Rightarrow P_0n + P_1 + n^2P_22^n - 4[P_0(n-1) + P_1 + (n-1)^2P_2 2^{n-1}] + 4[P_0(n-2) + P_1 + (n-2)^2 P_2 2^{n-2}] = 3n + 2^n$$

$$\Rightarrow (n^2P_2 - 2(n-1)^2P_2 + (n-2)^2P_2]2^n + (P_0 - 4P_0 + 4P_0)n + (P_1 + 4P_0 - 4P_1 - 8P_0 + 4P_1) = 3n + 2^n$$

$$\Rightarrow \quad 2P_22^n + P_0n + (-4P_0 + P_1) = 3n + 2^n$$

Comparing the terms, we get

$$2P_2 = 1,\ P_0 = 3,\ -4P_0 + P_1 = 0$$

$$\therefore \quad P_0 = 3,\ P_1 = 12,\ P_2 = \frac{1}{2}$$

$$\therefore \quad PS \text{ of } (1) = 3n + 12 + n^2\left(\frac{1}{2}\right)2^n$$

$$= 3(n+4) + n^2 2^{n-1}$$

$\therefore$ Total solution (*TS*) of (1) is

$$a_n = (A + nB)2^n + 3(n+4) + n^2 2^{n-1},\ n \geq 0$$

The initial conditions are $a_0 = 13$, $a_1 = 22$.

$$a_0 = 13 \Rightarrow (A+0).1 + 3(4) + 0 = 13 \Rightarrow A = 1$$

$$a_1 = 22 \Rightarrow (A+B).2 + 3(5) + 1 = 22$$

$$\Rightarrow 2A + 2B = 6 \Rightarrow A + B = 3 \Rightarrow B = 3 - A = 3 - 1 = 2$$

$\therefore$ The required total solution of (1) is

$$\mathbf{a_n = (1 + 2n)2^n + 3(n+4) + n^2 2^{n-1}, n \geq 0.}$$

(*iv*) We have $a_{r+2} - 3a_{r+1} + 2a_r = 5^r$ ...(1)

The characteristic equation of (1) is

$$x^2 - 3x + 2 = 0 \quad \therefore \quad x = 1, 2$$

$\therefore$ *HS* of (1) $= A(1)^r + B(2)^r = A + B2^r$

Here $f(r) = 5^r$. This is of the form β^r, where $\beta = 5$. Also, 5 is not a characteristic root of (1).

$\therefore$ *PS* of (1) $= P5^r$

$$\therefore \quad (1) \Rightarrow P5^{r+2} - 3P5^{r+1} + 2P5^r = 5^r$$

$$\Rightarrow \quad 25P - 15P + 2P = 1$$

$$\Rightarrow \quad 12P = 1 \Rightarrow P = \frac{1}{12}$$

$\therefore$ PS of (1) $= \frac{1}{12}(5)^r$

$\therefore$ Total solution of (1) is

$$a_r = A + B2^r + \frac{1}{12}(5)^r,\ r \geq 0$$

The initial conditions are $a_0 = \frac{1}{12}$, $a_1 = -\frac{7}{12}$

$$a_0 = \frac{1}{12} \quad \Rightarrow \quad A + B + \frac{1}{12}(1) = \frac{1}{12} \quad \Rightarrow \quad A + B = 0$$

$$a_1 = -\frac{7}{12} \quad \Rightarrow \quad A + 2B + \frac{5}{12} = -\frac{7}{12} \quad \Rightarrow \quad (A + B) + B = -1 \quad \Rightarrow \quad B = -1$$

$\therefore$ $A = -B = -(-1) = 1$

$\therefore$ The required total solution of (1) is

$$a_r = 1 + (-1)2^r + \frac{1}{12}(5)^r$$

or $$\mathbf{a_r = 1 - 2^r + \frac{1}{12}(5)^r,\ r \geq 0.}$$

EXERCISE 7.5

Find the total solution of the following recurrence relations :

1. (*i*) $a_n - 3a_{n-1} = n,\ n \geq 1$ with $a_0 = 3$ (*ii*) $a_n + a_{n-1} = 4,\ n \geq 1$ with $a_0 = 12$
   (*iii*) $a_r - a_{r-1} = 4,\ r \geq 1$ with $a_0 = 5$ (*iv*) $a_r - 2a_{r-1} = 3(2)^r,\ r \geq 1$ with $a_0 = 1$.
2. (*i*) $S(k) - S(k-1) - 6S(k-2) = -30k \geq 2$ with, $S(0) = 20$, $S(1) = -5$
   (*ii*) $a_n + 6a_{n-1} + 9a_{n-2} = 3,\ n \geq 2$ with $a_0 = 0$, $a_1 = 1$
   (*iii*) $a_n - 4a_{n-1} + 5a_{n-2} = 2,\ n \geq 2$ (*iv*) $a_{r+2} - 5a_{r+1} + 6a_r = 5^r,\ r \geq 0$ with $a_0 = \frac{49}{12}$, $a_1 = \frac{77}{12}$.
3. (*i*) $a_r + 5a_{r-1} + 6a_{r-2} = 3r^2,\ r \geq 2$ with $a_0 = \frac{691}{288}$, $a_1 = -\frac{1049}{288}$
   (*ii*) $a_n - a_{n-1} - 2a_{n-2} = 3n(4)^n,\ r \geq 2$ with $a_0 = -\frac{21}{25}$, $a_1 = \frac{171}{25}$
   (*iii*) $a_n - 6a_{n-1} + 9a_{n-2} = n(3)^n,\ n \geq 2$ with $a_0 = 0$, $a_1 = 5$
   (*iv*) $a_n + 5a_{n-1} + 4a_{n-2} = 56(3)^n,\ n \geq 2$ with $a_0 = 22$, $a_1 = 47$
4. $a_n - 5a_{n-1} + 6a_{n-2} = 1.$ *(KU 2007)*

Answers

1. (*i*) $a_n = \frac{1}{4}[5(3)^{n+1} - 2n - 3],\ n \geq 0$ (*ii*) $a_n = 10(-1)^n + 2,\ n \geq 0$
   (*iii*) $a_r = 4r + 5,\ r \geq 0$ (*iv*) $a_r = 2^r(3r + 1),\ r \geq 0$

2. (*i*) $S(k) = 11(-2)^k + 4(3)^k + 5, k \geq 0$ (*ii*) $a_n = -\left(\frac{4n+9}{48}\right)(-3)^n + \frac{3}{16}, n \geq 0$

(*iii*) $a_n = A(2+i)^n + B(2-i)^n + 1, n \geq 0$ (*iv*) $a_r = \frac{37}{3}(2)^{r-1} - \frac{9}{4}(3)^r + \frac{1}{6}(5)^r, r \geq 0$

3. (*i*) $a_r = (-2)^r + (-3)^r + \frac{1}{4}r^2 + \frac{17}{24}r + \frac{115}{288}, r \geq 0$

(*ii*) $a_n = 2^{n+1} + (-1)^n + \frac{6(5n-4)}{25}(4)^{n+1}, n \geq 0$

(*iii*) $a_n = n(3)^n + \frac{n^2(n+3)(3)^{n-1}}{2}, n \geq 0$ (*iv*) $a_n = (-4)^n + 3(-1)^n + 18(3)^n, n \geq 0$

4. $a_n = A(2)^n + B(3)^n + \frac{1}{2}, n \geq 0.$

7.14. SOLUTION OF A RECURRENCE RELATION BY THE METHOD OF GENERATING FUNCTIONS

Let $a_0, a_1, a_2, ...$ be an infinite sequence. The generating function of this sequence *i.e.*, of the numeric function $(a_0, a_1, a_2, ...)$ is the infinite series $a_0 + a_1z + a_2z^2 + ...$ and this sum is denoted by $A(z)$.

In this section, we shall involve the use of generating functions in finding the solution of recurrence relations. For a given recurrence relation, we shall first find the generating function of the required solution and using this generating function, we shall find the desired solution of the given recurrence relation.

Example 1. *Find an explicit formula for the recurrence relation* $a_n = 3a_{n-1} + 1, n \geq 2$, *where* $a_0 = 0, a_1 = 1$. (KU 2006)

Sol. We have $a_n = 3a_{n-1} + 1, n \geq 2$ ...(1)

Multiplying (1) by z^n, we get

$$a_nz^n = 3\,a_{n-1}\,z^n + z^n, n \geq 2 \quad ...(2)$$

Summing (2) for all n, $n \geq 2$, we get

$$\sum_{n=2}^{\infty} a_nz^n = 3\sum_{n=2}^{\infty} a_{n-1}z^n + \sum_{n=2}^{\infty} z^n \quad ...(3)$$

Let $A(z)$ be the generating function of the sequence $a_0, a_1, a_2, ...$

$$\therefore \quad A(z) = a_0 + a_1z + a_2z^2 + ... = \sum_{n=0}^{\infty} a_nz^n$$

Now
$$\sum_{n=2}^{\infty} a_nz^n = a_2z^2 + a_3z^3 + a_4z^4 + ... = A(z) - a_0 - a_1z$$

$$= A(z) - 0 - 1.z = A(z) - z \quad (\because \; a_0 = 0, a_1 = 1)$$

$$\sum_{n=2}^{\infty} a_{n-1}z^n = a_1z^2 + a_2z^3 + a_3z^4 + ...$$

$$= z(a_1z + a_2z^2 + a_3z^3 + ...)$$

$$= z(A(z) - a_0) = z(A(z) - 0) = zA(z)$$

$$\sum_{n=2}^{\infty} z^n = z^2 + z^3 + z^4 + \ldots\ldots = \frac{z^2}{1-z}$$

$\therefore$ (3) becomes

$$A(z) - z = 3.z\, A(z) + \frac{z^2}{1-z}$$

$$\Rightarrow \qquad (1 - 3z)\, A(z) = z + \frac{z^2}{1-z} = \frac{z}{1-z}$$

$$\Rightarrow \qquad A(z) = \frac{z}{(1-3z)(1-z)} = \frac{1/3}{(1-3z)\left(1-\frac{1}{3}\right)} + \frac{1}{(1-3(1))(1-z)}$$

$$= \frac{1}{2(1-3z)} - \frac{1}{2(1-z)} = \frac{1}{2}\left[\frac{1}{1-3z} - \frac{1}{1-z}\right]$$

$$= \frac{1}{2}\,[(1 + 3z + 3^2z^2 + \ldots\ldots + 3^n z^n + \ldots\ldots) - (1 + z + z^2 + \ldots\ldots + z^n + \ldots\ldots)]$$

$$= \frac{1}{2}\,[(1-1) + (3-1)z + (3^2-1)z^2 + \ldots\ldots + (3^n - 1)z^n + \ldots\ldots]$$

$$\therefore \qquad \mathbf{a_n = \frac{1}{2}\,(3^n - 1),\ n \geq 0.}$$

This is the required solution of the given recurrence relation.

Example 2. *Find the explicit formula for the Fibonacci sequence by the method of generating functions.*

Sol. The Fibonacci sequence is 1, 1, 2, 3, 5, 8, 13, 21, ...

The recursive formula for this sequence is

$$f_n = f_{n-1} + f_{n-2},\ n \geq 2 \qquad \ldots(1)$$

with initial conditions $f_0 = f_1 = 1$.

Multiplying (1) by z^n, we get

$$f_n z^n = f_{n-1} z^n + f_{n-2} z^n,\ n \geq 2 \qquad \ldots(2)$$

Summing (2) for all n, $n \geq 2$, we get

$$\sum_{n=2}^{\infty} f_n z^n = \sum_{n=2}^{\infty} f_{n-1} z^n + \sum_{n=2}^{\infty} f_{n-2} z^n \qquad \ldots(3)$$

Let $F(z)$ be the generating function of the Fibonacci sequence $f_0, f_1, f_2, \ldots\ldots$

$$\therefore \qquad F(z) = f_0 + f_1 z + f_2 z^2 + \ldots\ldots = \sum_{n=0}^{\infty} f_n z^n$$

Now $\sum_{n=2}^{\infty} f_n z^n = f_2 z^2 + f_3 z^3 + f_4 z^4 + \ldots\ldots = F(z) - f_0 - f_1 z$

$$= F(z) - 1 - (1)z = F(z) - 1 - z$$

$$\sum_{n=2}^{\infty} f_{n-1} z^n = f_1 z^2 + f_2 z^3 + f_3 z^4 + \ldots\ldots = z(f_1 z + f_2 z^2 + f_3 z^3 + \ldots\ldots)$$

$$= z(F(z) - f_0) = z(F(z) - 1)$$

$$\sum_{n=2}^{\infty} f_{n-2} z^n = f_0 z^2 + f_1 z^3 + f_2 z^4 + \ldots\ldots = z^2(f_0 + f_1 z + f_2 z^2 + \ldots\ldots)$$

$$= z^2 F(z)$$

$\therefore$ (3) becomes

$$F(z) - 1 - z = z(F(z) - 1) + z^2 F(z)$$

$\Rightarrow$ $(1 - z - z^2)\, F(z) = 1 + z - z$

$$\Rightarrow F(z) = \frac{1}{1-z-z^2} = \frac{1}{1-(z^2+z)} = \frac{1}{\frac{5}{4} - \left(z+\frac{1}{2}\right)^2} = \frac{1}{\left(\frac{\sqrt{5}}{2} - z - \frac{1}{2}\right)\left(\frac{\sqrt{5}}{2} + z + \frac{1}{2}\right)}$$

$$= \frac{1}{-\left(\frac{1-\sqrt{5}}{2} + z\right)\left(\frac{1+\sqrt{5}}{2} + z\right)} = \frac{1}{-\frac{1-5}{4}\left(1 + \frac{2}{1-\sqrt{5}} z\right)\left(1 + \frac{2}{1+\sqrt{5}} z\right)}$$

$$= \frac{1}{\left(1 + \frac{2(1+\sqrt{5})}{1-5} z\right)\left(1 + \frac{2(1-\sqrt{5})}{1-5} z\right)} = \frac{1}{\left(1 - \frac{1+\sqrt{5}}{2} z\right)\left(1 - \frac{1-\sqrt{5}}{2} z\right)}$$

$$= \frac{1}{\left(1 - \frac{1+\sqrt{5}}{2} z\right)\left(1 - \left(\frac{1-\sqrt{5}}{2}\right)\left(\frac{2}{1+\sqrt{5}}\right)\right)} + \frac{1}{\left(1 - \left(\frac{1+\sqrt{5}}{2}\right)\left(\frac{2}{1-\sqrt{5}}\right)\right)\left(1 - \frac{1-\sqrt{5}}{2} z\right)}$$

$$= \frac{1}{\frac{2\sqrt{5}}{1+\sqrt{5}}\left(1 - \frac{1+\sqrt{5}}{2} z\right)} - \frac{1}{\frac{2\sqrt{5}}{1-\sqrt{5}}\left(1 - \frac{1-\sqrt{5}}{2} z\right)}$$

$$= \frac{\frac{1+\sqrt{5}}{2}}{\sqrt{5}\left(1 - \frac{1+\sqrt{5}}{2} z\right)} - \frac{\frac{1-\sqrt{5}}{2}}{\sqrt{5}\left(1 - \frac{1-\sqrt{5}}{2} z\right)}$$

$$= \frac{1}{\sqrt{5}}\left[\sum_{n=0}^{\infty}\left(\frac{1+\sqrt{5}}{2}\right)^{n+1} z^n - \sum_{n=0}^{\infty}\left(\frac{1-\sqrt{5}}{2}\right)^{n+1} z^n\right]$$

$$= \sum_{n=0}^{\infty} \frac{1}{\sqrt{5}} \left(\left(\frac{1+\sqrt{5}}{2} \right)^{n+1} - \left(\frac{1-\sqrt{5}}{2} \right)^{n+1} \right) z^n$$

$$\therefore \qquad \mathbf{f_n} = \frac{1}{\sqrt{5}} \left(\left(\frac{1+\sqrt{5}}{2} \right)^{n+1} - \left(\frac{1-\sqrt{5}}{2} \right)^{n+1} \right), \ n \geq 0.$$

This is the required explicit formula for the Fibonacci sequence.

Example 3. *Solve the recurrence relation $a_{n+2} - 3a_{n+1} + 2a_n = 0$, $n \geq 0$ with the initial conditions $a_0 = 2$ and $a_1 = 3$ by the method of generating functions.*

Sol. We have $a_{n+2} - 3a_{n+1} + 2a_n = 0, n \geq 0$ ...(1)

Multiplying (1) by z^{n+2}, we get

$$a_{n+2}\, z^{n+2} - 3a_{n+1}\, z^{n+2} + 2a_n z^{n+2} = 0, \quad n \geq 0 \qquad ...(2)$$

Summing (2) for all n, $n \geq 0$, we get

$$\sum_{n=0}^{\infty} a_{n+2} z^{n+2} - 3 \sum_{n=0}^{\infty} a_{n+1} z^{n+2} + 2 \sum_{n=0}^{\infty} a_n z^{n+2} = 0 \qquad ...(3)$$

Let $A(z)$ be the generating function of the sequence $a_0, a_1, a_2, \ldots\ldots$

$$\therefore \qquad A(z) = a_0 + a_1 z + a_2 z^2 + \ldots\ldots = \sum_{n=0}^{\infty} a_n z^n$$

Now
$$\sum_{n=0}^{\infty} a_{n+2} z^{n+2} = a_2 z^2 + a_3 z^3 + a_4 z^4 + \ldots\ldots = A(z) - a_0 - a_1 z$$

$$= A(z) - 2 - 3z$$

$$\sum_{n=0}^{\infty} a_{n+1} z^{n+2} = a_1 z^2 + a_2 z^3 + a_3 z^4 + \ldots\ldots = z(a_1 z + a_2 z^2 + a_3 z^3 + \ldots\ldots)$$

$$= z(A(z) - a_0) = z(A(z) - 2)$$

$$\sum_{n=0}^{\infty} a_n z^{n+2} = a_0 z^2 + a_1 z^3 + a_2 z^4 + \ldots\ldots = z^2(a_0 + a_1 z + a_2 z^2 + \ldots\ldots)$$

$$= z^2 A(z)$$

$\therefore$ (3) becomes

$$(A(z) - 2 - 3z) - 3z(A(z) - 2) + 2z^2 A(z) = 0$$

$$\Rightarrow \qquad (1 - 3z + 2z^2)\, A(z) = 2 + 3z - 6z$$

$$\Rightarrow \qquad A(z) = \frac{2 - 3z}{1 - 3z + 2z^2}$$

$$\Rightarrow \qquad A(z) = \frac{2-3z}{(1-z)(1-2z)} = \frac{2-3(1)}{(1-z)(1-2(1))} + \frac{2-3\left(\frac{1}{2}\right)}{\left(1-\frac{1}{2}\right)(1-2z)}$$

$$= \frac{1}{1-z} + \frac{1}{1-2z} = \sum_{n=0}^{\infty} z^n + \sum_{n=0}^{\infty} (2^n z^n) = \sum_{n=0}^{\infty} (1+2^n) z^n$$

$\therefore$ $\mathbf{a_n = 1 + 2^n, n \geq 0.}$

This is the required solution.

Example 4. *Solve the recurrence relation $a_{r+2} - 2a_{r+1} + a_r = 2^r$ for $r \geq 0$ $a_0 = 2, a_1 = 1$ by the method of generating functions.*

Sol. We have $a_{r+2} - 2a_{r+1} + a_r = 2^r, r \geq 0$ ...(1)

Multiplying (1) by z^r, we get

$$a_{r+2} z^r - 2a_{r+1} z^r + a_r z^r = 2^r z^r, r \geq 0 \qquad ...(2)$$

Summing (2) for all r, $r \geq 0$, we get

$$\sum_{r=0}^{\infty} a_{r+2} z^r - 2\sum_{r=0}^{\infty} a_{r+1} z^r + \sum_{r=0}^{\infty} a_r z^r = \sum_{r=0}^{\infty} 2^r z^r \qquad ...(3)$$

Let $A(z)$ be the generating function of the sequence $a_0, a_1, a_2, ...$

$$\therefore \qquad A(z) = a_0 + a_1 z + a_2 z^2 + ... = \sum_{r=0}^{\infty} a_r z^r$$

Now
$$\sum_{r=0}^{\infty} a_{r+2} z^r = a_2 + a_3 z + a_4 z^2 + ... = \frac{1}{z^2}(a_2 z^2 + a_3 z^3 + a_4 z^4 + ...)$$

$$= \frac{1}{z^2}(A(z) - a_0 - a_1 z) = \frac{1}{z^2}(A(z) - 2 - 1.z) = \frac{1}{z^2}(A(z) - 2 - z)$$

$$\sum_{r=0}^{\infty} a_{r+1} z^r = a_1 + a_2 z + a_3 z^2 + ... = \frac{1}{z}(a_1 z + a_2 z^2 + a_3 z^3 + ...)$$

$$= \frac{1}{z}(A(z) - a_0) = \frac{1}{z}(A(z) - 2)$$

$$\sum_{r=0}^{\infty} a_r z^r = a_0 + a_1 z + a_2 z^2 + ... = A(z)$$

$$\sum_{r=0}^{\infty} 2^r z^r = 1 + 2z + 2^2 z^2 + ... = \frac{1}{1-2z}$$

$\therefore$ (3) becomes

$$\frac{1}{z^2}(A(z) - 2 - z) - 2 \cdot \frac{1}{z}(A(z) - 2) + A(z) = \frac{1}{1-2z}$$

$$\Rightarrow \qquad A(z) - 2 - z - 2z(A(z) - 2) + z^2 A(z) = \frac{z^2}{1-2z}$$

$$\Rightarrow \quad (1 - 2z + z^2)\,A(z) = \frac{z^2}{1-2z} + 2 + z - 4z$$

$$\Rightarrow \quad (1 - z)^2 A(z) = \frac{z^2}{1-2z} + 2 - 3z$$

$$\Rightarrow \quad A(z) = \frac{z^2}{(1-2z)(1-z)^2} + \frac{2}{(1-z)^2} - \frac{3z}{(1-z)^2}$$

Using partial fractions, we have

$$\frac{z^2}{(1-2z)(1-z)^2} = \frac{1}{1-2z} - \frac{1}{(1-z)^2}$$

$$\therefore \quad A(z) = \left(\frac{1}{1-2z} - \frac{1}{(1-z)^2}\right) + \frac{2}{(1-z)^2} - \frac{3z}{(1-z)^2}$$

$$= \frac{1}{1-2z} + \frac{1}{(1-z)^2} - \frac{3z}{(1-z)^2}$$

$$= \sum_{r=0}^{\infty} (2z)^r + \sum_{r=0}^{\infty} (r+1)z^r - 3\sum_{r=0}^{\infty} rz^r \qquad \text{(Note this step)}$$

$$= \sum_{r=0}^{\infty} (2^r + r + 1 - 3r)\, z^r = \sum_{r=0}^{\infty} (2^r - 2r + 1)z^r$$

$\therefore$ $\mathbf{a_r = 2^r - 2r + 1,\ r \geq 0.}$

This is the required solution of the given recurrence relation.

EXERCISE 7.6

Solve the following recurrence relations by the method of generating functions :

1. $a_n = 3a_{n-1} + 2, n \geq 1$ with $a_0 = 1$
2. $a_n = 4a_{n-1} + 6(4)^n, n \geq 1$ with $a_0 = 1$
3. $a_r - 7a_{r-1} + 10a_{r-2} = 0, r \geq 2$ with $a_0 = 3, a_1 = 3$
4. $a_{r+2} - 5a_{r+1} + 6a_r = 2, r \geq 0$ with $a_0 = 1, a_2 = 2$
5. $a_{n+2} - 5a_{n+1} + 6a_n = 5^n, n \geq 0.$

Answers

1. $a_n = 2(3)^n - 1, n \geq 0$
2. $a_n = (1 + 6n)4^n, n \geq 0$
3. $a_r = 4(2)^r - 5^r, r \geq 0$
4. $a_r = 1 - 2^r + 3^r, r \geq 0$
5. $a_n = A(2)^n + B(3)^n + \dfrac{1}{6}(5)^n, n \geq 0.$

8

Mathematical Logic

8.1. INTRODUCTION

The dictionary meaning of *'Logic'* is the *'science of reasoning'*. The language of mathematics is very neat and concise and surpasses every other language in its precision and bravity. The study of logic through the use of mathematical symbols is called **mathematical logic** or **mathematical reasoning.** The mathematical logic is also called *'symbolic logic'*. Since symbols are abstract and neutral, they give clear expression to our thoughts. The mathematical approach to logic was first propounded by British mathematician **George Boole.** On this account, the mathematical logic is also called **Boolean logic.**

8.2. STATEMENT

A sentence is called a **statement** or a **proposition** if it is either true or false but not both.

Statements are denoted by letters p, q, r ....

Illustrations. (*i*) '2 + 6 = 8' is a statement because it is true.

(*ii*) 'Calcutta is in England' is a statement because it is false.

(*iii*) 'Where are you going ?' is not a statement because it is neither true nor false.

(*iv*) '7 divides 92' is a statement because it is false.

(*v*) 'Two individuals are always related' is a statement because it is false.

(*vi*) 'Today is sunday' is not a statement because it is neither true nor false. On the other hand, the sentence, 'On monday, it can be said that it is sunday' is a statement because it is a false sentence.

(*vii*) 'The equation $ax^2 + bx + c = 0$ always has real roots' is a statement because it is false.

(*viii*) 'The equation $ax^2 + bx + c = 0$, where $a, b, c \in \mathbf{R}$, $b^2 - 4ac \geq 0$ has real roots' is a statement because it is true.

8.3. TRUTH VALUE OF A STATEMENT

We know that a statement is either true or false.

The truth or falsity of a statement is called its **truth value.**

If a statement is true then its truth value is denoted by '*T*' and if a statement is false then its truth value is denoted by '*F*'.

Illustrations. (*i*) The truth value of the statement '2 + 3 = 6' is *F*, because this statement is false.

(*ii*) The truth value of the statement '64 is the square of 8' is T, because this statement is true.

EXERCISE 8.1

1. Which of the following sentences are statements ?
   (*i*) 10 divided by 2 gives 5.
   (*ii*) It may rain today.
   (*iii*) London is in America.
   (*iv*) He is not honest.
   (*v*) The positive square root of 16 is 4.
   (*vi*) $x^2 - 5x + 6 = 0$.
   (*vii*) $x^2 - 5x + 6 = 0$ when $x = 6$.
   (*viii*) $x^2 - 5x + 6 = 0$ when $x = 2$.
   (*ix*) 4 is a prime number.
   (*x*) Come here !
   (*xi*) The sum of 5 and 7 is greater than 10.
   (*xii*) The sum of all interior angles of a triangle is 180°.
   (*xiii*) All real numbers are complex numbers.
   (*xiv*) The sides of a quadrilateral have equal length.
2. Write the truth values of the following statements :
   (*i*) $ax^2 + bx + c = 0$ may have non-real roots.
   (*ii*) There are only finite number of integers.
   (*iii*) The intersection of two non-empty sets is always non-empty.
   (*iv*) The capital of America is New York.
   (*v*) Two individuals may be relatives.

Answers

1. (*i*), (*iii*), (*v*), (*vii*), (*viii*), (*ix*), (*xi*), (*xii*), (*xiii*)
2. (*i*) T (*ii*) F (*iii*) F (*iv*) T (*v*) T.

8.4. TRUTH TABLE

A table indicating the truth values of one or more statements is called a **truth table.**

The truth tables for one statement 'p', two statements 'p, q', three statements 'p, q, r' are shown below in figure (*i*), (*ii*), (*iii*) respectively :

p
T
F

(*i*)

p	q
T	T
T	F
F	T
F	F

(*ii*)

p	q	r
T	T	T
T	T	F
T	F	T
F	T	T
T	F	F
F	T	F
F	F	T
F	F	F

(*iii*)

The second row of figure (*ii*) represent the case when p is true and q is false. Similarly, the fourth row of figure (*iii*) represent the case when p is false, q is true and r is true.

In case of n statements, there are 2^n distinct possible arrangements of truth values of the statements.

8.5. PHRASES 'THERE EXISTS' AND 'FOR EVERY'

The phrases *there exists* and *for every* appear frequently in statements. The phrase 'there exists' means there is at least one. The phrase 'for every' means for all. These phrases are called **quantifiers.**

8.6. BASIC LOGICAL OPERATIONS

There are three **basic logical operations** called negation, conjunction and disjunction. These basic logical operations correspond to English words "not", "and" and "or" respectively. These operations are used to form new statements from given statements.

A statement whose truth value does not explicitly depend on another statement is called a **simple statement.** If two or more simple statements are combined by using basic logical operations then the resulting statement is called a **compound statement.**

Simple statements which on combining form a compound statement are called the **component statements** of the compound statement under considerations. The compound statement S consisting of component statements $p, q, r, ...$ is written as $S(p, q, r, ...)$. The truth value of a compound statement depends upon the truth values of its component statements.

8.7. NEGATION

If p is any statement, then the denial of statement p is called the **negation** of the statement p and written as $\sim p$.

The negation of statement p is formed by inserting the word 'not' in p or by writing 'It is false that ......' before p.

Illustrations. (*i*) Let p be the statement : 4 is a factor of 12.

$\therefore$ $\sim p$ can be written as : '4 is not a factor of 12' or as 'it is false that 4 is a factor of 12. Here truth value of p is T and that of $\sim p$ is F.

(*ii*) Let q be the statement : 'Jaipur is in Bangla Desh'.

$\therefore$ $\sim q$ can be written as : 'Jaipur is not in Bangla Desh' or as 'it is false that Jaipur is in Bangla Desh'.

Here truth value of q is F and that of $\sim q$ is T.

(*iii*) Let r be the statement : 'All teachers are male'.

$\therefore$ $\sim r$ is the statement : 'Not all teachers are male'.

$\sim r$ is not the statement : 'No teacher is male'.

$\sim r$ can also be expressed as : 'There exists a teacher who is not male'.

p	$\sim p$
T	F
F	T

The truth value of negation of a statement is always opposite to the truth value of the original statement.

Let p be any statement. The truth values of p and $\sim p$ can also be shown in the form of a table, called **truth table.** In the truth table, the first line states that if p is true then $\sim p$ is false and the second line states that if p is false then $\sim p$ is true.

Example 1. *Write the negation of the following statements :*

(*i*) $3 + 7 = 10$ (*ii*) $8 \leq 15$

(*iii*) *All doctors are men* (*iv*) *Shimla is in H.P.*

(*v*) *All triangle are not equilateral triangle.*

(*vi*) *Every natural number is an integer.*

Sol. (*i*) Let $p : 3 + 7 = 10.$

$\therefore$ $\sim p$ is **$3 + 7 \neq 10$.**

(*ii*) Let $p : 8 \leq 15.$

$\therefore$ $\sim p$ is **$8 > 15$.**

(*iii*) Let p : All doctors are men.

$\therefore$ $\sim p$ is **'There exists a doctor, who is not a man'**.

(*iv*) Let p : Shimla is in H.P.

$\therefore$ $\sim p$: **Shimla is not in H.P.**

(*v*) Let p : All triangles are not equilateral triangle.

$\therefore$ $\sim p$ is **'There exists a triangle which is an equilateral triangle'.**

(*vi*) Let p : Every natural number is an integer

$\therefore$ $\sim p$ is **"There exists a natural number which is not an integer"**.

Example 2. *Write the negation of the following statements :*

(*i*) *p : For every positive real number x, the number x – 1 is also positive.*

(*ii*) *q : All cats scratch.*

(*iii*) *r : For every real number x, either x > 1 or x < 1.*

(*iv*) *s : There exists a number x such that 0 < x < 1.*

Sol. (*i*) We have

p : For every positive real number x, the number $x - 1$ is also positive.

$\therefore$ $\sim p$ is the statement :

'There exists a positive real number x, for which x – 1 is not positive'.

(*ii*) We have

q : All cats scratch.

$\therefore$ $\sim q$ is the statement :

'There exists a cat who does not scratch'.

(*iii*) We have

r : For every real number x, either $x > 1$ or $x < 1$

$\therefore$ $\sim r$ is the statement :

'There exists a real number x for which neither x > 1 nor x < 1'.

(*iv*) We have

s : There exists a number x such that $0 < x < 1$.

$\therefore$ $\sim s$ is the statement :

'For every number x either $x \leq 0$ or $x \geq 1$'.

Example 3. *Identify the quantifier in the following statements and write the negation of the statements :*

(*i*) *There exists a number which is equal to its square.*

(*ii*) *For every real number x, x is less than x + 1.*

(*iii*) *There exists a capital for every state in India.*

Sol. (*i*) Let p : There exists a number which is equal to its square.

In p, the quantifier is 'there exists'.

$\sim p$: **Every number is not equal to its square.**

(*ii*) Let p : For every real number x, x is less than $x + 1$.

In p, the quantifier is 'for every'.

$\sim p$: **There exists a real number x such that x is not less than x + 1.**

(*iii*) Let p : There exists a capital for every state in India.

In p, the quantifier is 'for every'.

$\sim p$: **There exists a state in India having no capital.**

Example 4. *Find the truth values of the statement $\sim p$ if the statement p is :*

(*i*) $\log_a mn = \log_a m - \log_a n$ (*ii*) $(3 + 9) - 7 = 4$

(*iii*) $2^{(3^2)} \neq (2^3)^2$ (*iv*) *7.3 is an irrational number.*

Sol. (*i*) We have $p : \log_a mn = \log_a m - \log_a n$.

$\therefore$ $\sim p$ is $\log_a mn \neq \log_a m - \log_a n$.

Since $\log_a mn = \log_a m + \log_a n$, truth value of $\sim p$ is **T.**

Remark. It would be wrong to write $\sim p$ as $\log_a mn = \log_a m + \log_a n$.

(*ii*) We have $p : (3 + 9) - 7 = 4$.

$\therefore$ $\sim p$ is $(3 + 9) - 7 \neq 4$

Since $(3 + 9) - 7 = 12 - 7 = 5$ and $5 \neq 4$, the truth value of $\sim p$ is **T.**

(*iii*) We have $p : 2^{(3^2)} \neq (2^3)^2$.

$\therefore$ $\sim p$ is $2^{(3^2)} = (2^3)^2$

Since $2^{(3^2)} = 2^9 = 512$ and $(2^3)^2 = (8)^2 = 64$ and $512 \neq 64$, the truth value of $\sim p$ is **F.**

(*iv*) We have p : 7.3 is an irrational number.

$\therefore$ $\sim p$ is 7.3 is not an irrational number.

Since $7.3 \in \mathbf{Q}$, the truth value of $\sim p$ is **T.**

Example 5. *Are the following pairs of statements negations of each other ?*

(*i*) *The number x is not a rational number.*

The number x is not an irrational number.

(*ii*) *The number x is not a rational number.*

The number x is an irrational number.

(*iii*) *$x + y = y + x$ is true for every real numbers x and y.*

There exists real numbers x and y for which $x + y = y + x$.

Sol. (*i*) Let p : The number x is not a rational number.

and q : The number x is not an irrational number.

$\therefore$ $\sim p$: The number x is a rational number.

$\sim p$ is not same as q because if $x = 2 + 3i$, then x is neither an irrational number nor a rational number.

$\therefore$ Given pair of statements is not negation of each other.

(*ii*) Let p : The number x is not a rational number

and q : The number x is an irrational number.

$\therefore$ $\sim p$: The number x is a rational number.

$\sim p$ is not same as q because if $x = 2$, then x is a rational number but it is not an irrational number.

$\therefore$ Given pair of statements is not negation of each other.

(*iii*) Let $p : x + y = y + x$ is true for every real numbers x and y.

and q : There exists real numbers x and y for which $x + y = y + x$.

$\therefore$ $\sim p$: There exists real numbers x and y for which $x + y \neq y + x$.

$\therefore$ $\sim p$ is not same as q.

$\therefore$ Given pair of statements is not negation of each other.

8.8. CONJUNCTION

If two statements are combined by using the connecting word 'and ', then the resulting statement is called a **conjunction.** The conjunction of statements p and q is denoted by $p \wedge q$.

For example, let

p : Monsoon is very good this year

and q : The rivers are rising,

then their conjunction $p \wedge q$ denotes the statement :

'Monsoon is very good this year and the rivers are rising.

p	q	$p \wedge q$
T	T	T
T	F	F
F	T	F
F	F	F

The conjunction $p \wedge q$ is defined to be true when p and q are both true, otherwise it is false. The adjoining truth table represents the truth values of the conjunction $p \wedge q$. In the truth table, the first line say that if p is true, q is true then $p \wedge q$ is true. The other lines have analogous meaning.

Example 6. *Let p and q stand for the statements : 'Nitin is intelligent' and 'Nitin is hardworking' respectively. Describe the following statements :*

(*i*) $p \wedge q$, (*ii*) $\sim p \wedge q$, (*iii*) $p \wedge \sim q$, (*iv*) $\sim p \wedge \sim q$.

Sol. We have p : Nitin is intelligent

and q : Nitin is hardworking.

(*i*) $p \wedge q$: Nitin is intelligent and Nitin is hardworking.

(*ii*) $\sim p \wedge q$: Nitin is not intelligent and Nitin is hardworking.

(*iii*) $p \wedge \sim q$: Nitin is intelligent and Nitin is not hardworking.

(*iv*) $\sim p \wedge \sim q$: Nitin is not intelligent and Nitin is not hardworking.

Example 7. *Find the truth values of the following statements :*

(*i*) *2 divides 4 and 3 + 7 = 10* (*ii*) *2 divides 7 and 8 + 10 = 18*

(*iii*) *7 divides 14 and 8 + 2 = 12* (*iv*) *3 divides 16 and 2 + 5 = 8.*

Sol. We know that the conjunction $p \wedge q$ of p and q is true only when p and q are both true.

(*i*) Truth value of '2 divides 4' is T.

Truth value of 3 + 7 = 10 is T.

$\therefore$ Truth value of '2 divides 4 and 3 + 7 = 10' is **T.**

(*ii*) Truth value of '2 divides 7' is F.

Truth value of '8 + 10 = 18' is T.

$\therefore$ Truth value of '2 divides 7 and 8 + 10 = 18' is **F.**

(*iii*) Truth value of '7 divides 14' is T.

Truth value of '8 + 2 = 12' is F.

$\therefore$ Truth value of '7 divides 14 and 8 + 2 = 12' is **F.**

(*iv*) Truth value of '3 divides 16' is F.

Truth value of '2 + 5 = 8' is F.

$\therefore$ Truth value of '3 divides 16 and 2 + 5 = 8' is **F.**

Example 8. *Find the truth values of :*

(*i*) $\sim p \wedge q$ (*ii*) $\sim (p \wedge q)$.

Sol. (*i*) **Truth values of $\sim p \wedge q$**

p	q	$\sim p$	$\sim p \wedge q$
T	T	F	F
T	F	F	F
F	T	T	T
F	F	T	F

(*ii*) **Truth values of $\sim (p \wedge q)$**

p	q	$p \wedge q$	$\sim (p \wedge q)$
T	T	T	F
T	F	F	T
F	T	F	T
F	F	F	T

8.9. DISJUNCTION

If two statements are combined by using the connecting word 'or', then the resulting statement is called a **disjunction.**

The disjunction of two statements p and q is denoted by $p \vee q$.

For example, let $p : 8 \leq 10$ and $q : 4$ is an integer, then their disjunction $p \vee q$ denotes the statement : '$8 \leq 10$ or 4 is an integer'.

The disjunction $p \vee q$ is defined to be true if at least one of p and q is true, otherwise it is false. The adjoining truth table represents the truth values of the disjunction $p \vee q$. In the truth table, the first line say that if p is true, q is true then $p \vee q$ is true. The other lines have analogous meaning.

p	q	$p \vee q$
T	T	T
T	F	T
F	T	T
F	F	F

Remark. The disjunction $p \vee q$ is false only when p and q are both false.

Example 9. *Let p and q stand for the statements 'Kamla is tall' and 'Bimla is beautiful' respectively. Describe the following statements :*

(*i*) $p \vee q$ (*ii*) $\sim p \vee q$ (*iii*) $p \vee \sim p$ (*iv*) $\sim p \vee \sim q$.

Sol. We have p : Kamla is tall

and q : Bimla is beautiful.

(*i*) $p \vee q$: Kamla is tall or Bimla is beautiful.

(*ii*) $\sim p \vee q$: Kamla is not tall or Bimla is beautiful.

(*iii*) $p \vee \sim q$: Kamla is tall or Bimla is not beautiful.

(*iv*) $\sim p \vee \sim q$: Kamla is not tall or Bimla is not beautiful.

Example 10. *For each of the following compound statements, first identify the connecting word and then break it into component statements.*

(*i*) *All rational numbers are real and all real numbers are not complex.*

(*ii*) *Square of an integer is positive or negative.*

(*iii*) *The sand heats up quickly in the sun and does not cool down fast at night.*

(*iv*) $x = 2$ *and* $x = 3$ *are the roots of the equation* $3x^2 - x - 10 = 0$.

Sol. (*i*) Given compound statement is : 'All rational numbers are real and all real numbers are not complex'.

Here the connecting word is 'and'.

The component statements are :

(*a*) All rational numbers are real.

(*b*) All real numbers are not complex.

(*ii*) Given compound statement is : 'Square of an integer is positive or negative'.

Here the connecting word is 'or'.

The component statements are :

(*a*) Square of an integer is positive.

(*b*) Square of an integer is negative.

(*iii*) Given compound statement is : 'The sand heats up quickly in the sun and does not cool down fast at night'.

Here the connecting word is 'and'.

The component statement are :

(*a*) The sand heats up quickly in the sun.

(*b*) The sand does not cool down fast at night.

(*iv*) Given compound statement is :

'$x = 2$ and $x = 3$ are the roots of the equation $3x^2 - x - 10 = 0$'.

Here the connecting word is 'and'.

The component statement are :

(*a*) $x = 2$ is a root of the equation $3x^2 - x - 10 = 0$.

(*b*) $x = 3$ is a root of the equation $3x^2 - x - 10 = 0$.

Example 11. *Find the truth values of :*

(*i*) $\sim p \vee q$ (*ii*) $\sim (p \vee q)$.

Sol. (*i*) **Truth values of ~ p ∨ q**

p	q	$\sim p$	$\sim p \vee q$
T	T	F	T
T	F	F	F
F	T	T	T
F	F	T	T

(*ii*) **Truth values of ~ (p ∨ q)**

p	q	$p \vee q$	$\sim (p \vee q)$
T	T	T	F
T	F	F	T
F	T	F	T
F	F	F	T

Example 12. *Let p and q stand for the statements : 'It is hot' and 'It is humid' respectively. Describe the following statements :*

(*i*) $\sim p$ (*ii*) $\sim q$ (*iii*) $p \wedge q$ (*iv*) $p \vee q$ (*v*) $\sim p \wedge q$ (*vi*) $p \vee \sim q$ (*vii*) $\sim p \vee \sim q$ (*viii*) $\sim p \wedge \sim q$.

Sol. We have p : It is hot

and q : It is humid.

(*i*) $\sim p$: It is not hot or It is false that it is hot.

(*ii*) $\sim q$: It is not humid.

(*iii*) $p \wedge q$: It is hot and humid .

(*iv*) $p \vee q$: It is hot or it is humid.

(*v*) $\sim p \wedge q$: It is not hot and it is humid.

(*vi*) $p \vee \sim q$: It is hot or it is not humid.

(*vii*) $\sim p \vee \sim q$: It is not hot or it is not humid.

(*viii*) $\sim p \wedge \sim q$: It is not hot and it is not humid.

Example 13. *Find the truth values of the following compound statements :*

(*i*) *Honesty is the best policy or* $3 < 7$ (*ii*) *Honesty is the best policy or* $4 > 7$

(*iii*) *Honesty is the worst policy or* $5 \geq 3$ (*iv*) *Honesty is the worst policy or* $11 < 9$.

Sol. We know that the disjunction $p \vee q$ of p and q is true only when at least one of p and q is true.

(*i*) Truth value of 'Honesty is the best policy' is *T*.

Truth value of '$3 < 7$' is *T*.

∴ Truth value of 'Honesty is the best policy or $3 < 7$' is **T.**

(*ii*) Truth value of 'Honesty is the best policy' is *T*.

Truth value of '$4 > 7$' is *F*.

∴ Truth value of 'Honesty is the best policy or $4 > 7$' is **T.**

(*iii*) Truth value of 'Honesty is the worst policy' is *F*.

Truth value of '$5 \geq 3$' is *T*.

∴ Truth value of 'Honesty is the worst policy or $5 \geq 3$' is **T.**

(*iv*) Truth value of 'Honesty is the worst policy' is *F* .

Truth value of '$11 < 9$' is *F*.

∴ Truth value of 'Honesty is the worst policy or $11 < 9$' is **F.**

Example 14. *Find the truth values of the following compound statements :*

(*i*) $4 + 2 = 6$ *and* $9 + 7 = 15$ (*ii*) *3 divides 9 and Ch. of log 273.5 is 2*

(*iii*) $5 + 3 = 2$ *or* $5 \times 3 = 15$ (*iv*) *4 divides 17 or* $3 + 4 = 7$.

Sol. (*i*) Truth value of '$4 + 2 = 6$' is *T*.

Truth value of '$9 + 7 = 15$' is *F*.

∴ Truth value of '$4 + 2 = 6$ and $9 + 7 = 15$' is **F.**

(*ii*) Truth value of '3 divides 9' is *T*.

Truth value of 'Ch. of log 273.5 is 2' is *T*.

∴ Truth value of '3 divides 9 and Ch. of log 273.5 is 2' is **T.**

(*iii*) Truth value of '$5 + 3 = 2$' is *F*.

Truth value of '$5 \times 3 = 15$' is *T*.

∴ Truth value of '$5 + 3 = 2$ or $5 \times 3 = 15$' is **T.**

(*iv*) Truth value of '4 divides 17' is *F*.

Truth value of '$3 + 4 = 7$' is *T*

∴ Truth value of '4 divides 17 or $3 + 4 = 7$' is **T.**

Example 15. *Find the truth values of :*

(*i*) $\sim (p \vee \sim q)$ (*ii*) $\sim (\sim p \wedge \sim q)$.

Sol. (*i*) **Truth values of ~ (p ∨ ~ q)**

p	q	$\sim q$	$p \vee \sim q$	$\sim (p \vee \sim q)$
T	T	F	T	F
T	F	T	T	F
F	T	F	F	T
F	F	T	T	F

(*ii*) **Truth values of ~ (~ p ∧ ~ q)**

p	q	$\sim p$	$\sim q$	$\sim p \wedge \sim q$	$\sim (\sim p \wedge \sim q)$
T	T	F	F	F	T
T	F	F	T	F	T
F	T	T	F	F	T
F	F	T	T	T	F

Example 16. *Write down the truth table for the compound statement :*

$$(\sim p \vee q) \wedge (\sim p \wedge \sim q).$$

Sol. **Truth table for (~ p ∨ q) ∧ (~ p ∧ ~ q)**

p	q	$\sim p$	$\sim q$	$\sim p \vee q$	$\sim p \wedge \sim q$	$(\sim p \vee q) \wedge (\sim p \wedge \sim q)$
T	T	F	F	T	F	F
T	F	F	T	F	F	F
F	T	T	F	T	F	F
F	F	T	T	T	T	T

Example 17. *Find the truth values of the following compound statements :*

(*i*) $(p \vee \sim r) \wedge (q \vee \sim r)$ (*ii*) $\sim (p \vee \sim q) \wedge (\sim p \vee r)$.

Sol. (*i*) **Truth values of (p ∨ ~ r) ∧ (q ∨ ~ r)**

p	q	r	$\sim r$	$p \vee \sim r$	$q \vee \sim r$	$(p \vee \sim r) \wedge (q \vee \sim r)$
T	T	T	F	T	T	T
T	T	F	T	T	T	T
T	F	T	F	T	F	F
F	T	T	F	F	T	F
T	F	F	T	T	T	T
F	T	F	T	T	T	T
F	F	T	F	F	F	F
F	F	F	T	T	T	T

(*ii*) **Truth values of $\sim (p \vee \sim q) \wedge (\sim p \vee r)$**

p	q	r	$\sim p$	$\sim q$	$p \vee \sim q$	$\sim(p \vee \sim q)$	$\sim p \vee r$	$\sim (p \vee \sim q) \wedge (\sim p \vee r)$
T	T	T	F	F	F	T	T	T
T	T	F	F	F	T	F	F	F
T	F	T	F	T	T	F	T	F
F	T	T	T	F	F	T	T	T
T	F	F	F	T	T	F	F	F
F	T	F	T	F	F	T	T	T
F	F	T	T	T	T	F	T	F
F	F	F	T	T	T	F	T	F

EXERCISE 8.2

1. Write the negation of the following statements :
   (*i*) The square of 4 is 16. (*ii*) 14 divides 27.
   (*iii*) Chandigarh is in Gujarat. (*iv*) $7 > 3$.
2. Find the truth value of the statement $\sim p$ if the statement p is :
   (*i*) $5 + 7 = 12$ (*ii*) $\log_2 8 = 4$
   (*iii*) $3 \times 4 = 14$ (*iv*) $7 - 3 = 4$.
3. Find the truth values of the statement $\sim p$ if the statement p is :
   (*i*) For complex numbers z_1 and z_2, $|z_1 z_2| = |z_1| \, |z_2|$
   (*ii*) Real part of $(1+ 2i)^3$ is 4
   (*iii*) $\tan(-315°) = 1$ (*iv*) $\sec^2 45° + \text{cosec}^2 45° = 2$.
4. Write the negation of the following statements.
   (*i*) p : For every real number x, $x^2 > x$.
   (*ii*) q : There exists a rational number x such that $x^2 = 2$.
   (*iii*) r : All birds have wings.
   (*iv*) s : All students study mathematics at the elementary level.
5. Write the negation of the following statements and check whether the resulting statements are true.
   (*i*) Australia is a continent.
   (*ii*) There does not exists a quadrilateral which has all its sides equal.
   (*iii*) Every natural number is greater than 0.
   (*iv*) The sum of 3 and 4 is 9.
6. Let p and q stand for the statements : 'Lucknow is in U.P.' and '4 divides 12' respectively. Describe the following statements :
   (*i*) $p \wedge q$ (*ii*) $p \vee q$ (*iii*) $\sim p \wedge q$ (*iv*) $\sim p \vee q$
   (*v*) $p \wedge \sim q$ (*vi*) $p \vee \sim q$ (*vii*) $\sim p \wedge \sim q$ (*viii*) $\sim p \vee \sim q$.
7. Write the component statements of the following compound statements and check whether the compound statement is true or false.
   (*i*) A line is straight and extends indefinitely in both directions.
   (*ii*) 0 is less than every positive integer and every negative integer.
   (*iii*) All living things have two legs and two eyes.

8. Let p and q stand for the statements : '2 + 3 = 5' and '3 + 7 = 8' respectively. Describe the following statements :
   (i) $p \wedge q$ (ii) $\sim p \wedge q$ (iii) $p \wedge \sim q$ (iv) $\sim p \wedge \sim q$
   (v) $p \vee q$ (vi) $\sim p \vee q$ (vii) $p \vee \sim q$ (viii) $\sim p \vee \sim q$.
9. If p be any statement, then write the truth tables of the statements :
   (i) $\sim \sim p$ (ii) $\sim \sim \sim p$.
10. If p and q be any statements, then write the truth tables of the following compound statements :
   (i) $p \wedge \sim q$ (ii) $\sim p \wedge \sim q$.
11. If p and q be any statements, then write the truth tables of the following compound statements :
   (i) $p \vee \sim q$ (ii) $\sim p \vee \sim q$.
12. Find the truth values of the following compound statements :
   (i) $(2 + 4 = 6) \wedge (2 \times 3 = 6)$ (ii) (It is false that $2 + 5 = 8) \wedge (2 \times 5 = 20)$
   (iii) (It is false that $5 - 2 = 3) \wedge (4 \times 3 = 12)$ (iv) $(2 + 5 = 25) \wedge$ (It is false that $5 + 3 = 8$).
13. Find the truth values of the following compound statements :
   (i) (Real part of $(4 + i)^2 = 15) \vee$ (Roots of $x^2 - 5x + 6 = 0$ are 2, 3)
   (ii) $(\cot(-135°) = 1) \vee (\sec 450° = \frac{1}{2})$
   (iii) (It is false that $\sin^2\theta + \cos^2\theta = 1) \vee (\sec^2\theta - \tan^2\theta = 1)$
   (iv) $((2^2)^4 = 64) \vee (\log_{625} 25 = 2)$.
14. Find the truth values of the following compound statements :
   (i) $\sim (p \wedge \sim q)$ (ii) $\sim (\sim p \vee \sim q)$.
15. Find the truth values of the following compound statements :
   (i) $(p \wedge q) \vee \sim (p \vee q)$ (ii) $(\sim p \vee q) \wedge (\sim p \wedge \sim q)$.
16. Find the truth values of the following compound statements :
   (i) $p \wedge (q \wedge r)$ (ii) $(p \vee q) \wedge \sim r$ (iii) $p \wedge (q \vee r)$ (iv) $(p \wedge q) \vee r$.
   (KU 2007)
17. Find the truth values of the following compound statements :
   (i) $(p \wedge \sim q) \vee r$ (ii) $\sim p \vee (q \wedge \sim r)$ (iii) $(\sim p \wedge \sim q) \vee \sim r$ (iv) $\sim ((p \wedge q) \vee \sim r)$.

Answers

1. (i) The square of 4 is not 16. (ii) 14 does not divide 27.
   (iii) Chandigarh is not in Gujarat. (iv) $7 \le 3$.
2. (i) F (ii) T (iii) T (iv) F.
3. (i) F (ii) T (iii) F (iv) T.
4. (i) There exists a real number x for which $x^2 \le x$.
   (ii) For every rational number x, we have $x^2 \ne 2$.
   (iii) There exists a bird having no wings.
   (iv) There exists a student who does not study mathematics at the elementary level.
5. (i) Australia is not a continent. F
   (ii) There exists a quadrilateral having all its sides equal. T
   (iii) There exists a natural number not greater than 0. F
   (iv) The sum of 3 and 4 is not 9. T
6. (i) Lucknow is in U.P. and 4 divides 12. (ii) Lucknow is in U.P. or 4 divides 12.
   (iii) Lucknow is not in U.P. and 4 divides 12. (iv) Lucknow is not in U.P. or 4 divides 12.
   (v) Lucknow is in U.P. and 4 does not divide 12. (vi) Lucknow is in U.P. or 4 does not divide 12.
   (vii) Lucknow is not in U.P. and 4 does not divide 12.
   (viii) Lucknow is not in U.P. or 4 does not divide 12.

7. (*i*) A line is straight. A line extends indefinitely in both directions. True
(*ii*) 0 is less than evey positive integer. 0 is less than every negative integer. False
(*iii*) All living things have two legs. All living things have two eyes. False

8. (*i*) $2 + 3 = 5$ and $3 + 7 = 8$
(*ii*) $2 + 3 \neq 5$ and $3 + 7 = 8$
(*iii*) $2 + 3 = 5$ and $3 + 7 \neq 8$
(*iv*) $2 + 3 \neq 5$ and $3 + 7 \neq 8$
(*v*) $2 + 3 = 5$ or $3 + 7 = 8$
(*vi*) $2 + 3 \neq 5$ or $3 + 7 = 8$
(*vii*) $2 + 3 = 5$ or $3 + 7 \neq 8$
(*viii*) $2 + 3 \neq 5$ or $3 + 7 \neq 8$

9.

p	$\sim p$	(*i*)	(*ii*)
T	*F*	*T*	*F*
F	*T*	*F*	*T*

10.

p	q	(*i*)	(*ii*)
T	*T*	*F*	*F*
T	*F*	*T*	*F*
F	*T*	*F*	*F*
F	*F*	*F*	*T*

11.

p	q	(*i*)	(*ii*)
T	*T*	*T*	*F*
T	*F*	*T*	*T*
F	*T*	*F*	*T*
F	*F*	*T*	*T*

12. (*i*) *T* (*ii*) *F* (*iii*) *F* (*iv*) *F*

13. (*i*) *T* (*ii*) *T* (*iii*) *T* (*iv*) *F*

14.

p	q	(*i*)	(*ii*)
T	*T*	*T*	*T*
T	*F*	*F*	*F*
F	*T*	*T*	*F*
F	*F*	*T*	*F*

15.

p	q	(*i*)	(*ii*)
T	*T*	*T*	*F*
T	*F*	*F*	*F*
F	*T*	*F*	*F*
F	*F*	*T*	*T*

16.

p	q	r	(*i*)	(*ii*)	(*iii*)	(*iv*)
T	*T*	*T*	*T*	*F*	*T*	*T*
T	*T*	*F*	*F*	*T*	*T*	*T*
T	*F*	*T*	*F*	*F*	*T*	*T*
F	*T*	*T*	*F*	*F*	*F*	*T*
T	*F*	*F*	*F*	*T*	*F*	*F*
F	*T*	*F*	*F*	*T*	*F*	*F*
F	*F*	*T*	*F*	*F*	*F*	*T*
F	*F*	*F*	*F*	*F*	*F*	*F*

17.

p	q	r	(*i*)	(*ii*)	(*iii*)	(*iv*)
T	*T*	*T*	*T*	*F*	*F*	*F*
T	*T*	*F*	*F*	*T*	*T*	*F*
T	*F*	*T*	*T*	*F*	*F*	*T*
F	*T*	*T*	*T*	*T*	*F*	*T*
T	*F*	*F*	*T*	*F*	*T*	*F*
F	*T*	*F*	*F*	*T*	*T*	*F*
F	*F*	*T*	*T*	*T*	*T*	*T*
F	*F*	*F*	*F*	*T*	*T*	*F*

8.10. CONDITIONAL STATEMENT

If two statements are combined by using the connecting words 'if .... then', then the resulting statement is called a **conditional statement.**

The conditional statement of two statements p and q (in this order) is denoted by $p \to q$.

For example, let p : $2 + 5 = 7$ and q : 9 is an integer, then their conditional statement $p \to q$ denotes the statement : 'If $2 + 5 = 7$, then 9 is an integer'.

The conditional statement $p \to q$ is defined to be true except in case p is true and q is false. The adjoining truth table represents the truth values of the conditional statement $p \to q$.

p	q	$p \to q$
T	T	T
T	F	F
F	T	T
F	F	T

Remark. The truth values of the conditional statement $q \to p$ are not same as that of $p \to q$.

Example 1. *Let p and q stand for the statements ' Bhopal is in M.P.' and ' 3 + 4 = 7' respectively. Describe the following conditional statements :*

(*i*) $p \to q$ (*ii*) $\sim p \to q$

(*iii*) $p \to \sim q$ (*iv*) $\sim p \to \sim q$.

Sol. We have p : Bhopal is in M.P.

and q : $3 + 4 = 7$.

(*i*) $p \to q$: If Bhopal is in M.P. then $3 + 4 = 7$.

(*ii*) $\sim p \to q$: If Bhopal is not in M.P. then $3 + 4 = 7$.

(*iii*) $p \to \sim q$: If Bhopal is in M.P. then $3 + 4 \neq 7$.

(*iv*) $\sim p \to \sim q$: If Bhopal is not in M.P. then $3 + 4 \neq 7$.

Example 2. *Find the truth values of :*

(*i*) $\sim p \to q$ (*ii*) $\sim (p \to q)$.

Sol. (*i*) **Truth values of ~ p → q**

p	q	$\sim p$	$\sim p \to q$
T	T	F	T
T	F	F	T
F	T	T	T
F	F	T	F

(*ii*) **Truth values of ~ (p → q)**

p	q	$p \to q$	$\sim (p \to q)$
T	T	T	F
T	F	F	T
F	T	T	F
F	F	T	F

Example 3. *Let p and q stand for the statements '3 divides 15' and '5 – 1 = 4' respectively. Describe the following conditional statements :*

(*i*) $p \to q$ (*ii*) $q \to p$ (*iii*) $p \to \sim q$

(*iv*) $q \to \sim p$ (*v*) $\sim p \to \sim q$ (*vi*) $\sim q \to \sim p$.

Sol. We have p : 3 divides 15

and $q : 5 - 1 = 4$.

(*i*) $p \to q$: If 3 divides 15 then $5 - 1 = 4$.

(*ii*) $q \to p$: If $5 - 1 = 4$ then 3 divides 15.

(*iii*) $p \to \sim q$: If 3 divides 15 then $5 - 1 \neq 4$.

(*iv*) $q \to \sim p$: If $5 - 1 = 4$ then 3 does not divide 15.

(*v*) $\sim p \to \sim q$: If 3 does not divide 15 then $5 - 1 \neq 4$.

(*vi*) $\sim q \to \sim p$: If $5 - 1 \neq 4$ then 3 does not divide 15.

Example 4. *Let p and q stand for the statements 'God is great' and 'work is worship' respectively. Find the truth values of the following conditional statements :*

(*i*) $p \to q$ (*ii*) $p \to \sim q$ (*iii*) $q \to \sim p$

(*iv*) $\sim p \to q$ (*v*) $\sim q \to p$ (*vi*) $\sim p \to \sim q$.

Sol. We have p : God is great

and q : Work is worship.

$\therefore$ p and q are both true.

$\therefore$ $\sim p$ and $\sim q$ are both false.

(*i*) The truth value of $p \to q$ is **T**. ($\because$ p is true, q is true)

(*ii*) The truth value of $p \to \sim q$ is **F**. ($\because$ p is true, $\sim q$ is false)

(*iii*) The truth value of $q \to \sim p$ is **F**. ($\because$ q is true, $\sim p$ is false)

(*iv*) The truth value of $\sim p \to q$ is **T**. ($\because$ $\sim p$ is false, q is true)

(*v*) The truth value of $\sim q \to p$ is **T**. ($\because$ $\sim q$ is false, p is true)

(*vi*) The truth value of $\sim p \to \sim q$ is **T**. ($\because$ $\sim p$ is false, $\sim q$ is false)

Example 5. *Find the truth values of :*

(*i*) $\sim p \to (q \to p)$ (*ii*) $(p \to q) \to (p \wedge q)$.

Sol. (*i*) **Truth values of $\sim p \to (q \to p)$**

p	q	$\sim p$	$q \to p$	$\sim p \to (q \to p)$
T	T	F	T	T
T	F	F	T	T
F	T	T	F	F
F	F	T	T	T

(*ii*) **Truth values of $(p \to q) \to (p \wedge q)$**

p	q	$p \to q$	$p \wedge q$	$(p \to q) \to (p \wedge q)$
T	T	T	T	T
T	F	F	F	T
F	T	T	F	F
F	F	T	F	F

8.11. CONVERSE AND CONTRAPOSITIVE

Let p and q be two statements. The compound statement $q \to p$ is called the **converse** of the compound statement $p \to q$. The compound statement $\sim q \to \sim p$ is called the **contrapositive** of the compound statement $p \to q$. The following truth table represents the truth values of the **converse** and **contrapositive** of a conditional statement.

p	q	$p \to q$	$\sim p$	$\sim q$	$q \to p$	$\sim q \to \sim p$
T	T	T	F	F	T	T
T	F	F	F	T	T	F
F	T	T	T	F	F	T
F	F	T	T	T	T	T

Example 6. *Write the converse and the contrapositive of the following statements :*

(*i*) *If x is a prime number, then x is odd.*

(*ii*) *If the two lines are parallel, then they do not intersect in the same plane.*

(*iii*) *Something is cold implies that it has low temperature.*

(*iv*) *x is an even number implies that x is divisible by 4.*

Sol. (*i*) Given statement is : 'If x is a prime number then x is odd'.

Let $p : x$ is a prime number and $q : x$ is odd.

$\therefore$ Given statement is $p \to q$.

The converse of $p \to q$ is $q \to p$ *i.e.,*

'If x is odd then x is a prime number'.

The contrapositive of $p \to q$ is $\sim q \to \sim p$ *i.e.,*

'If x is not odd then x is not a prime number'.

(*ii*) Given statement is : 'If the two lines are parallel, then they do not intersect in the same plane'.

Let p : The two lines are parallel and

q : the two lines do not intersect in the same plane.

$\therefore$ Given statement is $p \to q$.

The converse of $p \to q$ is $q \to p$ *i.e.,*

'If the two lines do not intersect in the same plane then the lines are parallel'.

The contrapositive of $p \to q$ is $\sim q \to \sim p$ *i.e.,*

'If the two lines intersect in the same plane then the lines are not parallel'.

(*iii*) Given statement is : 'Something is cold implies that it has low temperature'.

Let p : Something is cold

and q : Something has low temperature.

$\therefore$ Given statement is $p \to q$.

The converse of $p \to q$ is $q \to p$ *i.e.,*

'If something has low temperature then it is cold'.

The contrapositive of $p \to q$ is $\sim q \to \sim p$ *i.e.,*

'If something does not have low temperature then it is not cold'.

(*iv*) Given statement is : 'x is an even number implies that x is divisible by 4'.

Let p : x is an even number and q : x is divisible by 4.

$\therefore$ Given statement is $p \to q$.

The converse of $p \to q$ is $q \to p$ *i.e.*,

'If x is divisible by 4, then x is an even number.'

The contrapositive of $p \to q$ is $\sim q \to \sim p$ *i.e.*,

'If x is not divisible by 4 then x is not an even number'.

8.12. BICONDITIONAL STATEMENT

If two statements are combined by using the connecting words 'if and only if', then the resulting statement is called a **biconditional statement.**

The biconditional statement of two statements p and q is denoted by $p \leftrightarrow q$.

For example, let p : 2 divides 4 and q : 5 divides 15, then their biconditional statement $p \leftrightarrow q$ denotes the statement : '2 divides 4 if and only if 5 divides 15'.

p	q	$p \leftrightarrow q$
T	T	T
T	F	F
F	T	F
F	F	T

The biconditional statement $p \leftrightarrow q$ is defined to be true only when p and q have same truth value. The adjoining truth table represents the truth values of the biconditional statement $p \leftrightarrow q$.

Remark. The biconditional statement $p \leftrightarrow q$ is false only when p and q have opposite truth values.

Example 7. *Let p and q stand for the statements 'Meena speaks Hindi' and 'Heena speaks English' respectively. Describe the following biconditional statements :*

(*i*) $p \leftrightarrow q$ (*ii*) $q \leftrightarrow p$ (*iii*) $p \leftrightarrow \sim q$

(*iv*) $\sim p \leftrightarrow q$ (*v*) $\sim q \leftrightarrow \sim p$.

Sol. We have p : Meena speaks Hindi

and q : Heena speaks English.

(*i*) $p \leftrightarrow q$: Meena speaks Hindi if and only if Heena speaks English.

(*ii*) $q \leftrightarrow p$: Heena speaks English if and only if Meena speaks Hindi.

(*iii*) $p \leftrightarrow \sim q$: Meena speaks Hindi if and only if Heena does not speak English.

(*iv*) $\sim p \leftrightarrow q$: Meena does not speak Hindi if and only if Heena speaks English.

(*v*) $\sim q \leftrightarrow \sim p$: Heena does not speak English if and only if Meena does not speak Hindi.

Example 8. *Find the truth values of :*

(*i*) $\sim p \leftrightarrow q$ (*ii*) $\sim (p \leftrightarrow q)$.

Sol. (*i*) **Truth values of $\sim$ p $\leftrightarrow$ q**

p	q	$\sim p$	$\sim p \leftrightarrow q$
T	T	F	F
T	F	F	T
F	T	T	T
F	F	T	F

(*ii*) **Truth values of ~ (p ↔ q)**

p	q	$p \leftrightarrow q$	$\sim (p \leftrightarrow q)$
T	T	T	F
T	F	F	T
F	T	F	T
F	F	T	F

Example 9. *Let p and q stand for the statements '2 × 4 = 8' and '4 divides 7' respectively. Find the truth values of the following biconditional statements :*

(*i*) $p \leftrightarrow q$ (*ii*) $\sim p \leftrightarrow q$

(*iii*) $\sim q \leftrightarrow p$ (*iv*) $\sim p \leftrightarrow \sim q$.

Sol. We have $p : 2 \times 4 = 8$

and q : 4 divides 7.

∴ p is true and q is false.

∴ $\sim p$ is false and $\sim q$ is true.

(*i*) The truth value of $p \leftrightarrow q$ is **F.** (∵ p is true, q is false)

(*ii*) The truth value of $\sim p \leftrightarrow q$ is **T.** (∵ $\sim p$ is false, q is false)

(*iii*) The truth value of $\sim q \leftrightarrow p$ is **T.** (∵ $\sim q$ is true, p is true)

(*iv*) The truth value of $\sim p \leftrightarrow \sim q$ is **F.** (∵ $\sim p$ is false, $\sim q$ is true)

Example 10. *Find the truth values of :*

(*i*) $(p \leftrightarrow \sim q) \leftrightarrow (q \rightarrow p)$ (*ii*) $(p \rightarrow q) \vee \sim (p \leftrightarrow \sim q)$.

Sol. (*i*) **Truth values of (p ↔ ~ q) ↔ (q → p)**

p	q	$\sim q$	$p \leftrightarrow \sim q$	$q \rightarrow p$	$(p \leftrightarrow \sim q) \leftrightarrow (q \rightarrow p)$
T	T	F	F	T	F
T	F	T	T	T	T
F	T	F	T	F	F
F	F	T	F	T	F

(*ii*) **Truth values of (p → q) ∨ ~ (p ↔ ~ q)**

p	q	$\sim q$	$p \rightarrow q$	$p \leftrightarrow \sim q$	$\sim (p \leftrightarrow \sim q)$	$(p \rightarrow q) \vee \sim (p \leftrightarrow \sim q)$
T	T	F	T	F	T	T
T	F	T	F	T	F	F
F	T	F	T	T	F	T
F	F	T	T	F	T	T

Example 11. *Find the truth values of the compound statement :*

$$(p \rightarrow q) \rightarrow [(q \rightarrow r) \rightarrow (p \rightarrow q)].$$

Sol.

p	q	r	$p \to q$	$q \to r$	$(q \to r) \to (p \to q)$	$(p \to q) \to [(q \to r) \to (p \to q)]$
T	T	T	T	T	T	T
T	T	F	T	F	T	T
T	F	T	F	T	F	T
F	T	T	T	T	T	T
T	F	F	F	T	F	T
F	T	F	T	F	T	T
F	F	T	T	T	T	T
F	F	F	T	T	T	T

EXERCISE 8.3

1. If statements p and q are respectively : '4 + 5 = 9' and '2 + 3 = 5' then write the conditional statements :
   (*i*) $p \to q$ (*ii*) $p \to \sim q$ (*iii*) $\sim p \to q$ (*iv*) $\sim q \to \sim p$.
2. If statements p and q are respectively : '3 < 4' and '7 > 5' then write the biconditional statements :
   (*i*) $p \leftrightarrow q$ (*ii*) $\sim p \leftrightarrow q$ (*iii*) $p \leftrightarrow \sim q$ (*iv*) $\sim p \leftrightarrow \sim q$.
3. If truth values of statements p and q are T and T respectively then write the truth values of :
   (*i*) $p \to q$ (*ii*) $p \to \sim q$ (*iii*) $\sim (p \leftrightarrow \sim q)$ (*iv*) $\sim p \leftrightarrow \sim q$.
4. If truth values of statements p and q are T and F respectively then write the truth values of :
   (*i*) $p \to q$ (*ii*) $p \to \sim q$ (*iii*) $\sim p \leftrightarrow q$ (*iv*) $\sim (\sim p \leftrightarrow \sim q)$.
5. If p and q stand for the statements : '0 is a natural number' and '5 divides 10' respectively then, find the truth values of the following compound statements :
   (*i*) $p \to \sim q$ (*ii*) $\sim p \to q$ (*iii*) $\sim p \to \sim q$ (*iv*) $\sim q \to \sim p$
   (*v*) $p \leftrightarrow q$ (*vi*) $\sim p \leftrightarrow q$ (*vii*) $p \leftrightarrow \sim q$ (*viii*) $\sim p \leftrightarrow \sim q$.
6. Find the truth values of the following compound statements :
   (*i*) $p \to \sim q$ (*ii*) $\sim p \to q$.
7. Find the truth values of the following compound statements :
   (*i*) $\sim p \leftrightarrow q$ (*ii*) $\sim p \leftrightarrow \sim q$.
8. Find the truth values of the following compound statements :
   (*i*) $(p \wedge q) \to \sim p$ (*ii*) $(p \wedge q) \to (p \vee q)$.
9. Find the truth values of the compound statement : $(p \to q) \leftrightarrow (\sim q \to \sim p)$.
10. Find the truth values of the compound statement : $l \wedge m$ where $l = \sim q \to \sim r$, $m = \sim r \to \sim q$.
11. Find the truth values of the compound statement : $(p \to q) \to r$.
12. Find the truth values of the compound statement : $(p \vee q) \leftrightarrow r$.
13. Prove that a conditional statement $(p \to q)$ is logically equivalent to its contrapositive $(\sim q \to \sim p)$.
    (*KU 2006 C*)

Answers

1. (*i*) If '4 + 5 = 9' then '2 + 3 = 5' (*ii*) If '4 + 5 = 9' then '2 + 3 ≠ 5'
   (*iii*) If '4 + 5 ≠ 9' then '2 + 3 = 5' (*iv*) If '2 + 3 ≠ 5' then '4 + 5 ≠ 9'.

2. (*i*) '3 < 4' if and only if '7 > 5' (*ii*) '3 ≥ 4' if and only if '7 > 5'
(*iii*) '3 < 4' if and only if '7 ≤ 5' (*iv*) '3 ≥ 4' if and only if '7 ≤ 5'.

3. (*i*) *T* (*ii*) *F* (*iii*) *T* (*iv*) *T*

4. (*i*) *F* (*ii*) *T* (*iii*) *T* (*iv*) *T*

5. (*i*) *T* (*ii*) *T* (*iii*) *F* (*iv*) *T* (*v*) *F* (*vi*) *T*
(*vii*) *T* (*viii*) *F*

6.

p	*q*	(*i*)	(*ii*)
T	*T*	*F*	*T*
T	*F*	*T*	*T*
F	*T*	*T*	*T*
F	*F*	*T*	*F*

7.

p	*q*	(*i*)	(*ii*)
T	*T*	*F*	*T*
T	*F*	*T*	*F*
F	*T*	*T*	*F*
F	*F*	*F*	*T*

8.

p	*q*	(*i*)	(*ii*)
T	*T*	*F*	*T*
T	*F*	*T*	*T*
F	*T*	*T*	*T*
F	*F*	*T*	*T*

9.

p	*q*	$(p \to q) \leftrightarrow (\sim q \to \sim p)$
T	*T*	*T*
T	*F*	*T*
F	*T*	*T*
F	*F*	*T*

10.

q	*r*	$l \wedge m$
T	*T*	*T*
T	*F*	*F*
F	*T*	*F*
F	*F*	*T*

11.

p	*q*	*r*	$(p \to q) \to r$
T	*T*	*T*	*T*
T	*T*	*F*	*F*
T	*F*	*T*	*T*
F	*T*	*T*	*T*
T	*F*	*F*	*T*
F	*T*	*F*	*F*
F	*F*	*T*	*T*
F	*F*	*F*	*F*

12.

p	*q*	*r*	$(p \vee q) \leftrightarrow r$
T	*T*	*T*	*T*
T	*T*	*F*	*F*
T	*F*	*T*	*T*
F	*T*	*T*	*T*
T	*F*	*F*	*F*
F	*T*	*F*	*F*
F	*F*	*T*	*F*
F	*F*	*F*	*T*

8.13. TAUTOLOGY

A compound statement is called a **tautology** if it is always true for all possible truth values of its component statements.

A tautology is also called a **theorem** or a **logically valid statement pattern.**

8.14. CONTRADICTION

A compound statement is called a **contradiction** if it is always false for all possible truth values of its components statements.

A contradiction is also called a **fallacy.**

Remark. (*i*) The negation of a tautology is a contradiction.

(*ii*) The negation of a contradiction is a tautology.

Example 1. *Show that :*

(*i*) $p \to (p \vee q)$ *is a tautology* (*ii*) $(p \vee q) \wedge (\sim p \wedge \sim q)$ *is a contradiction.*

Sol. (*i*) **Truth values of $p \to (p \vee q)$**

p	q	$p \vee q$	$p \to (p \vee q)$
T	T	T	T
T	F	T	T
F	T	T	T
F	F	F	T

$\therefore$ For all possible truth values of p and q, the compound statement : $p \to (p \vee q)$ is true.

$\therefore$ $p \to (p \vee q)$ is a tautology.

(*ii*) **Truth values of $(p \vee q) \wedge (\sim p \wedge \sim q)$**

p	q	$\sim p$	$\sim q$	$p \vee q$	$\sim p \wedge \sim q$	$(p \vee q) \wedge (\sim p \wedge \sim q)$
T	T	F	F	T	F	F
T	F	F	T	T	F	F
F	T	T	F	T	F	F
F	F	T	T	F	T	F

$\therefore$ For all possible truth values of p and q, the compound statement :

$$(p \vee q) \wedge (\sim p \wedge \sim q) \text{ is false.}$$

$\therefore$ $(p \vee q) \wedge (\sim p \wedge \sim q)$ is a contradiction.

Example 2. *Show that* $(p \wedge q) \vee (\sim p) \vee (p \wedge \sim q)$ *is a tautology.*

Sol. **Truth values of $(p \wedge q) \vee (\sim p) \vee (p \wedge \sim q)$**

p	q	$\sim p$	$\sim q$	$p \wedge q$	$(p \wedge q) \vee (\sim p)$	$p \wedge \sim q$	$(p \wedge q) \vee (\sim p) \vee (p \wedge \sim q)$
T	T	F	F	T	T	F	T
T	F	F	T	F	F	T	T
F	T	T	F	F	T	F	T
F	F	T	T	F	T	F	T

$\therefore$ For all possible truth values of p and q, the compound statement :

$$(p \wedge q) \vee (\sim p) \vee (p \wedge \sim q) \text{ is true.}$$

$\therefore$ $(p \wedge q) \vee (\sim p) \vee (p \wedge \sim q)$ is a tautology.

Example 3. *Show that* $[(p \vee q) \vee r] \leftrightarrow [p \vee (q \vee r)]$ *is a tautology.*

Sol. **Truth values of $[(p \vee q) \vee r] \leftrightarrow [p \vee (q \vee r)]$**

p	q	r	$p \vee q$	$(p \vee q) \vee r$	$q \vee r$	$p \vee (q \vee r)$	$[(p \vee q) \vee r] \leftrightarrow [p \vee (q \vee r)]$
T	T	T	T	T	T	T	T
T	T	F	T	T	T	T	T
T	F	T	T	T	T	T	T
F	T	T	T	T	T	T	T
T	F	F	T	T	F	T	T
F	T	F	T	T	T	T	T
F	F	T	F	T	T	T	T
F	F	F	F	F	F	F	T

∴ For all possible truth values of p, q and r, the compound statement :

$[(p \vee q) \vee r] \leftrightarrow [p \vee (q \vee r)]$ is true.

∴ $[(p \vee q) \vee r] \leftrightarrow [p \vee (q \vee r)]$ is a tautology.

Example 4. *Show that* $[(p \to q) \wedge (q \to r)] \to (p \to r)$ *is a tautology.*

Sol. **Truth values of $[(p \to q) \wedge (q \to r)] \to (p \to r)$**

p	q	r	$p \to q$	$q \to r$	$(p \to q) \wedge (q \to r)$	$p \to r$	$[(p \to q) \wedge (q \to r)] \to (p \to r)$
T	T	T	T	T	T	T	T
T	T	F	T	F	F	F	T
T	F	T	F	T	F	T	T
F	T	T	T	T	T	T	T
T	F	F	F	T	F	F	T
F	T	F	T	F	F	T	T
F	F	T	T	T	T	T	T
F	F	F	T	T	T	T	T

∴ For all possible truth values of p and q the compound statement :
$[(p \to q) \to (q \to r)] \to (p \to r)$ is true.

∴ $[(p \to q) \to (q \to r)] \to (p \to r)$ is a tautology.

EXERCISE 8.4

1. Show that :

   (*i*) $p \vee \sim p$ is a tautology (*ii*) $p \wedge \sim p$ is a contradiction.

2. Show that :

   (*i*) $(p \wedge q) \to p$ is a tautology (*ii*) $p \to (p \vee q)$ is a tautology.

3. Show that :

   (*i*) $(p \wedge q) \to (p \vee q)$ is a tautology (*ii*) $\sim [(p \wedge q) \to (p \vee q)]$ is a contradiction.

4. Show that :

   (*i*) $(p \wedge q) \wedge \sim (p \vee q)$ is a fallacy (*ii*) $(p \wedge q) \wedge \sim (p \wedge q)$ is a fallacy

   (*iii*) $(p \wedge q) \wedge (\sim p \wedge \sim q)$ is a fallacy.

5. Show that :
   (i) $(p \to q) \leftrightarrow (\sim p \vee q)$ is a tautology   (ii) $\sim (p \vee q) \leftrightarrow (\sim p \wedge \sim q)$ is a tautology.
6. Find which of the following compound statements are tautologies and which are fallacies :
   (i) $(p \wedge q) \wedge (\sim (p \wedge q))$ (ii) $((\sim q) \wedge p) \vee (p \vee \sim p)$
   (iii) $(p \wedge \sim q) \wedge ((\sim p) \vee q)$ (iv) $((\sim p) \vee q) \vee (p \wedge \sim q)$.
7. Show that :
   $((\sim p) \wedge q) \wedge (q \wedge r)) \wedge \sim q$ is a tautology.
8. Show that :
   $[(p \leftrightarrow q) \wedge ((q \to r) \wedge r)] \to r$ is a tautology.

Answers

6. (i) Fallacy   (ii) Tautology
   (iii) Fallacy   (iv) Tautology.

8.15. LOGICAL EQUIVALENCE

Two compound statements $S_1(p, q, r, ...)$ and $S_2(p, q, r, ...)$ of components statements $p, q, r, ...$ are called **logically equivalent** or simply **equivalent** or **equal** if they have identical truth values and we write $S_1(p, q, r, ...) \equiv S_2(p, q, r, ...)$.

Example 1. *Show that the compound statements $(p \vee q) \wedge \sim p$ and $\sim p \wedge q$ are logically equivalent.*

Sol. Truth values of $(p \vee q) \wedge \sim p$

p	q	$\sim p$	$p \vee q$	$(p \vee q) \wedge \sim p$
T	T	F	T	F
T	F	F	T	F
F	T	T	T	T
F	F	T	F	F

Truth values of $\sim p \wedge q$

p	q	$\sim p$	$\sim p \wedge q$
T	T	F	F
T	F	F	F
F	T	T	T
F	F	T	F

$\therefore$ $(p \vee q) \wedge \sim p$ and $\sim p \wedge q$ have identical truth values.

$\therefore$ $\mathbf{(p \vee q) \wedge \sim p \equiv \sim p \wedge q}$.

Example 2. *Show that : $\sim (p \leftrightarrow q) \equiv (p \wedge \sim q) \vee (\sim p \wedge q)$.*

Sol. **Truth values of $\sim (p \leftrightarrow q)$**

p	q	$p \leftrightarrow q$	$\sim (p \leftrightarrow q)$
T	T	T	F
T	F	F	T
F	T	F	T
F	F	T	F

Truth values of $(p \wedge \sim q) \vee (\sim p \wedge q)$

p	q	$\sim p$	$\sim q$	$p \wedge \sim q$	$\sim p \wedge q$	$(p \wedge \sim q) \vee (\sim p \wedge q)$
T	T	F	F	F	F	F
T	F	F	T	T	F	T
F	T	T	F	F	T	T
F	F	T	T	F	F	F

$\therefore$ $\sim (p \leftrightarrow q)$ and $(p \wedge \sim q) \vee (\sim p \wedge q)$ have identical truth values.

$\therefore$ $\mathbf{\sim (p \leftrightarrow q) \equiv (p \wedge \sim q) \vee (\sim p \wedge q)}$.

Example 3. *Show that the compound statements :* $\sim [\sim(\sim p \wedge q) \vee \sim r]$ *and* $((\sim p) \wedge q) \wedge r$ *are equivalent.*

Sol. **Truth values of $\sim [\sim (\sim p \wedge q) \vee \sim r]$**

p	q	r	$\sim p$	$\sim r$	$\sim p \wedge q$	$\sim (\sim p \wedge q)$	$\sim (\sim p \wedge q) \vee \sim r$	$\sim [\sim (\sim p \wedge q) \vee \sim r]$
T	T	T	F	F	F	T	T	F
T	T	F	F	T	F	T	T	F
T	F	T	F	F	F	T	T	F
F	T	T	T	F	T	F	F	T
T	F	F	F	T	F	T	T	F
F	T	F	T	T	T	F	T	F
F	F	T	T	F	F	T	T	F
F	F	F	T	T	F	T	T	F

Truth values of $(\sim p \wedge q) \wedge r$

p	q	r	$\sim p$	$\sim p \wedge q$	$(\sim p \wedge q) \wedge r$
T	T	T	F	F	F
T	T	F	F	F	F
T	F	T	F	F	F
F	T	T	T	T	T
T	F	F	F	F	F
F	T	F	T	T	F
F	F	T	T	F	F
F	F	F	T	F	F

$\therefore$ $\sim [\sim (\sim p \wedge q) \vee \sim r]$ and $(\sim p \wedge q) \wedge r$ have identical truth values.

$\therefore$ $\mathbf{\sim [\sim (\sim p \wedge q) \vee \sim r] \equiv (\sim p \wedge q) \wedge r}$.

Example 4. *Find which of the following pairs are logically equivalent :*

(i) $(p \wedge q) \vee [\sim p \vee (p \wedge \sim q)]$ *and* $(\sim p \wedge q) \vee t$ *where* t *is a tautology in terms of statements* p *and* q.

(ii) $\sim (p \vee q) \wedge (p \vee \sim r) \wedge (\sim p \vee \sim q)$ *and* $p \vee r$.

Sol. (*i*) **Truth table of $(p \wedge q) \vee [\sim p \vee (p \wedge \sim q)]$**

p	q	$\sim p$	$\sim q$	$p \wedge q$	$p \wedge \sim q$	$\sim p \vee (p \wedge \sim q)$	$(p \wedge q) \vee [\sim p \vee (p \wedge \sim q)]$
T	T	F	F	T	F	F	T
T	F	F	T	F	T	T	T
F	T	T	F	F	F	T	T
F	F	T	T	F	F	T	T

Truth table of $(\sim p \wedge q) \vee t$

p	q	$\sim p$	$\sim p \wedge q$	t	$(\sim p \wedge q) \vee t$
T	T	F	F	T	T
T	F	F	F	T	T
F	T	T	T	T	T
F	F	T	F	T	T

$\therefore$ $(p \wedge q) \vee [\sim p \vee (p \wedge \sim q)]$ and $(\sim p \wedge q) \vee t$ have identical truth values.

$\therefore$ Given compound statements are logically equivalent.

(ii) **Truth table of $\sim (p \vee q) \wedge (p \vee \sim r) \wedge (\sim p \vee \sim q)$**

p	q	r	$\sim p$	$\sim q$	$\sim r$	$p \vee q$	$\sim (p \vee q)$	$p \vee \sim r$	$\sim p \vee \sim q$	$\sim (p \vee q) \wedge (p \vee \sim r)$	$\sim (p \vee q) \wedge (p \vee \sim r) \wedge (\sim p \vee \sim q)$
T	T	T	F	F	T	T	F	T	F	F	F
T	T	F	F	F	T	T	F	T	F	F	F
T	F	T	F	T	T	T	F	T	T	F	F
F	T	T	T	F	F	T	F	F	T	F	F
T	F	F	F	T	T	T	F	T	T	F	F
F	T	F	T	F	F	T	F	F	T	F	F
F	F	T	T	T	F	F	T	F	T	F	F
F	F	F	T	T	F	F	T	F	T	F	F

Truth table of $p \vee r$

p	q	r	$p \vee r$
T	T	T	T
T	T	F	T
T	F	T	T
F	T	T	T
T	F	F	T
F	T	F	F
F	F	T	T
F	F	F	F

$\therefore$ $\sim (p \vee q) \wedge (p \vee \sim r) \wedge (\sim p \vee \sim q)$ and $p \vee r$ do not have identical truth values.

$\therefore$ Given compound statements are not logically equivalent.

EXERCISE 8.5

1. Show that : $p \wedge (p \vee q) \equiv p$.
2. Show that :
   (i) $\sim (\sim p \wedge q) \equiv p \vee \sim q$ (ii) $\sim (p \wedge \sim q) \equiv \sim p \vee q$ (iii) $\sim (\sim p \vee \sim q) \equiv p \wedge q$.
3. Show that :
   (i) $p \to q \equiv (\sim p) \vee q$ (KU 2006) (ii) $p \to q \equiv q \to p$.
4. Show that : $p \leftrightarrow q \equiv (p \to q) \wedge (q \to p)$.
5. Show that :
   (i) $p \to q \equiv \sim q \to \sim p$ (ii) $p \to \sim q \equiv q \to \sim p$ (iii) $\sim (p \to q) \equiv p \wedge \sim q$.
6. Show that : $\sim (p \leftrightarrow q) \equiv (\sim p) \leftrightarrow q \equiv p \leftrightarrow \sim q$.
7. Find which of the following pairs of compound statements are logically equivalent :
   (i) $\sim p \vee q$ and $\sim (p \vee q)$.
   (ii) $p \vee (p \wedge q)$ and p.
   (iii) $(\sim p \vee q) \vee (p \wedge \sim q)$ and $(p \wedge q) \vee t$ where t is a tautology in terms of statements p and q.
   (iv) $(\sim q \wedge p) \wedge (p \wedge \sim p)$ and $(p \wedge q) \wedge f$ where f is a fallacy in terms of statements p and q.
8. Show that : $p \to (q \wedge r) \equiv (p \to q) \wedge (p \to r)$.
9. Show that : $\sim (p \to (q \wedge \sim r)) \equiv p \wedge \sim (q \wedge \sim r) \equiv p \wedge (\sim q \vee r)$.

10. Find which of the following pairs of compound statements are logically equivalent :

(*i*) $\sim [(p \vee q) \vee r]$ and $\sim [p \vee (q \vee r)]$

(*ii*) $(\sim p \vee q) \wedge (p \vee \sim r) \wedge (\sim p \vee \sim q)$ and $\sim (p \vee r)$.

Answers

7. (*ii*), (*iii*) **10.** (*i*), (*ii*)

8.16. ALGEBRA OF STATEMENTS

I. Idempotent laws. If p is any statement then

(*i*) $\mathbf{p \vee p \equiv p}$ (*ii*) $\mathbf{p \wedge p \equiv p}$.

Proof. **Truth values of p ∨ p and p ∧ p**

p	$p \vee p$	$p \wedge p$
T	T	T
F	F	F

$\therefore \quad p \vee p \equiv p$ and $p \wedge p \equiv p$.

II. Complement laws. If p is any statement, then

(*i*) $\mathbf{p \vee \sim p \equiv t}$ (*ii*) $\mathbf{p \wedge \sim p = f}$ (*iii*) $\mathbf{\sim\sim p \equiv p}$ (*iv*) $\mathbf{\sim t \equiv f, \sim f \equiv t}$,

where t and f are respectively some tautology and fallacy in terms of the statement p.

Proof. **Truth values of p ∨ ~ p, p ∧ ~ p, ~ ~ p, ~ t, ~ f**

p	$\sim p$	$p \vee \sim p$	t	$p \wedge \sim p$	f	$\sim\sim p$	$\sim t$	$\sim f$
T	F	T	T	F	F	T	F	T
F	T	T	T	F	F	F	F	T

$\therefore \quad p \vee \sim p \equiv t, \quad p \wedge \sim p \equiv f, \quad \sim\sim p \equiv p, \quad \sim t \equiv f, \ \sim f \equiv t.$

III. Identity laws. If p is any statement, then

(*i*) $\mathbf{p \wedge t = p}$ (*ii*) $\mathbf{p \wedge f = f}$ (*iii*) $\mathbf{p \vee t = t}$ (*iv*) $\mathbf{p \vee f = p}$,

where t and f are respectively some tautology and fallacy in terms of the statement p.

Proof. **Truth values of p ∧ t, p ∧ f, p ∨ t, p ∨ f**

p	t	f	$p \wedge t$	$p \wedge f$	$p \vee t$	$p \vee f$
T	T	F	T	F	T	T
F	T	F	F	F	T	F

$\therefore \quad p \wedge t \equiv p, \quad p \wedge f \equiv f, \quad p \vee t \equiv t, \quad p \vee f \equiv p.$

IV. Commutative laws. If p and q be any two statements then

(*i*) $\mathbf{p \wedge q = q \wedge p}$ (*ii*) $\mathbf{p \vee q = q \vee p}$.

Proof. **Truth values of p ∧ q, q ∧ p, q ∨ p, p ∨ q**

p	q	$p \wedge q$	$q \wedge p$	$p \vee q$	$q \vee p$
T	T	T	T	T	T
T	F	F	F	T	T
F	T	F	F	T	T
F	F	F	F	F	F

$\therefore \quad p \wedge q \equiv q \wedge p, \quad p \vee q \equiv q \vee p.$

V. De Morgan's laws. If p and q be any two statements, then

(*i*) $\sim (p \wedge q) \equiv \sim p \vee \sim q$ (*ii*) $\sim (p \vee q) \equiv \sim p \wedge \sim q$.

Proof. (*i*)

Truth values of $\sim (p \wedge q)$

p	q	$p \wedge q$	$\sim (p \wedge q)$
T	T	T	F
T	F	F	T
F	T	F	T
F	F	F	T

Truth values of $\sim p \vee \sim q$

p	q	$\sim p$	$\sim q$	$\sim p \vee \sim q$
T	T	F	F	F
T	F	F	T	T
F	T	T	F	T
F	F	T	T	T

$\therefore \quad \sim (p \wedge q) \equiv \sim p \vee \sim q.$

(*ii*) **Truth values of $\sim (p \vee q)$**

p	q	$p \vee q$	$\sim (p \vee q)$
T	T	T	F
T	F	T	F
F	T	T	F
F	F	F	T

Truth values of $\sim p \wedge \sim q$

p	q	$\sim p$	$\sim q$	$\sim p \wedge \sim q$
T	T	F	F	F
T	F	F	T	F
F	T	T	F	F
F	F	T	T	T

$\therefore \quad \sim (p \vee q) \equiv \sim p \wedge \sim q.$

VI. Associative laws. If p, q, r be any three statements then

(*i*) $(p \wedge q) \wedge r \equiv p \wedge (q \wedge r)$ (*ii*) $(p \vee q) \vee r \equiv p \vee (q \vee r)$.

Proof. (*i*)

Truth values of $(p \wedge q) \wedge r$

p	q	r	$p \wedge q$	$(p \wedge q) \wedge r$
T	T	T	T	T
T	T	F	T	F
T	F	T	F	F
F	T	T	F	F
T	F	F	F	F
F	T	F	F	F
F	F	T	F	F
F	F	F	F	F

Truth values of $p \wedge (q \wedge r)$

p	q	r	$q \wedge r$	$p \wedge (q \wedge r)$
T	T	T	T	T
T	T	F	F	F
T	F	T	F	F
F	T	T	T	F
T	F	F	F	F
F	T	F	F	F
F	F	T	F	F
F	F	F	F	F

$\therefore \quad (p \wedge q) \wedge r \equiv p \wedge (q \wedge r).$

(*ii*) Proof is left for the reader.

VII. Distributive laws. If p, q, r be any three statements, then

(*i*) $p \wedge (q \vee r) \equiv (p \wedge q) \vee (p \wedge r)$ (*ii*) $p \vee (q \wedge r) \equiv (p \vee q) \wedge (p \vee r)$.

Proof. (*i*)

Truth values of $p \wedge (q \vee r)$

p	q	r	$q \vee r$	$p \wedge (q \vee r)$
T	T	T	T	T
T	T	F	T	T
T	F	T	T	T
F	T	T	T	F
T	F	F	F	F
F	T	F	T	F
F	F	T	T	F
F	F	F	F	F

Truth values of $(p \wedge q) \vee (p \wedge r)$

p	q	r	$p \wedge q$	$p \wedge r$	$(p \wedge q) \vee (p \wedge r)$
T	T	T	T	T	T
T	T	F	T	F	T
T	F	T	F	T	T
F	T	T	F	F	F
T	F	F	F	F	F
F	T	F	F	F	F
F	F	T	F	F	F
F	F	F	F	F	F

$\therefore \quad p \wedge (q \vee r) \equiv (p \wedge q) \vee (p \wedge r).$

(*ii*) Proof is left for the reader.

Example 1. *By using laws of algebra of statements, shows that :*

$$(p \vee q) \wedge \sim p \equiv \sim p \wedge q.$$

Sol. $(p \vee q) \wedge \sim p \equiv (\sim p) \wedge (p \vee q)$ (Using Commutative law)

$\equiv (\sim p \wedge p) \vee (\sim p \wedge q)$ (Using Distributive law)

$\equiv f \vee (\sim p \wedge q)$ (Using Complement law)

$\equiv \sim p \wedge q$ (Using Identity law)

$\therefore \quad \mathbf{(p \vee q) \wedge \sim p \equiv \sim p \wedge q.}$

Example 2. *By using laws of algebra of statements, show that :*

$$\sim (p \vee q) \vee (\sim p \wedge q) \equiv \sim p.$$

Sol. $\sim (p \vee q) \vee (\sim p \wedge q)$

$\equiv (\sim p \wedge \sim q) \vee (\sim p \wedge q)$ (Using De Morgan's law)

$\equiv \sim p \wedge (\sim q \vee q)$ (Using Distributive law)

$\equiv \sim p \wedge t$ $(\because \quad \sim q \vee q = t)$

$\equiv \sim p$ (Using Identity law)

$\therefore \quad \mathbf{\sim (p \vee q) \vee (\sim p \wedge q) \equiv \sim p.}$

EXERCISE 8.6

By using laws of algebra of statements, prove the following logical equivalences :

1. $\sim \sim \sim p \equiv \sim p$
2. $\sim (\sim p \vee \sim q) \equiv p \wedge q$
3. $\sim (\sim p \wedge \sim q) \equiv p \vee q$
4. $\sim (p \vee \sim q) \equiv \sim p \wedge q$
5. $\sim (\sim p \wedge q) \equiv p \vee \sim q$
6. $p \wedge (p \vee q) \equiv p \vee (p \wedge q)$
7. $(p \wedge q) \vee \sim p \equiv \sim p \vee q$
8. $p \wedge (\sim p \vee q) \equiv p \wedge q$
9. $p \vee (p \wedge q) \equiv p.$

Hint

9. $p \vee (p \wedge q) \equiv (p \wedge t) \vee (p \wedge q) \equiv p \wedge (t \vee q) \equiv p \wedge t \equiv p.$

8.17. DUALITY

(*i*) **Duality of connectives.** The connectives $\wedge$ and $\vee$ are called duals of each other.

(*ii*) **Duality of compound statements.** Two compound statements are called duals of each other if one can be obtained from the other by replacing $\wedge$ by $\vee$, $\vee$ by $\wedge$, tautology t by fallacy f and fallacy f by tautology t.

Illustrations. (*a*) The compound statements $(p \vee q) \wedge r$ and $(p \wedge q) \vee r$ are duals of each other.

(*b*) The compound statements $(p \wedge q) \vee (r \vee t)$ and $(p \vee q) \wedge (r \wedge f)$ are duals of each other.

(*iii*) **Duality of logical equivalences.** Two logical equivalences are called duals of each other if one can be obtained from the other by replacing $\wedge$ by $\vee$ and $\vee$ by $\wedge$.

Illustrations. (*a*) The logical equivalences $\sim (p \wedge q) \equiv \sim p \vee \sim q$ and $\sim (p \vee q) \equiv \sim p \wedge \sim q$ are duals of each other.

(*b*) The logical equivalences $p \wedge (q \vee r) = (p \wedge q) \vee (p \wedge r)$ and $p \vee (q \wedge r) = (p \vee q) \wedge (p \vee r)$ are duals of each other.

An important result. Let $S(p, q, r, \ldots\ldots)$ be a compound statement in terms of finitely many statements $p, q, r, \ldots\ldots$. If $S^* (p, q, r, \ldots\ldots)$ be the dual compound statement of $S(p, q, r, \ldots\ldots)$, then

$$\sim \mathbf{S^*(p, q, r, \ldots\ldots) \equiv S(\sim p, \sim q, \sim r, \ldots\ldots)}.$$

Example 1. *Write the dual statements of the following compound statements :*

(*i*) *Ram is honest and Shyam is intelligent.*

(*ii*) *Kamla is beautiful or Bimla is rich.*

Sol. (*i*) Let p : Ram is honest

and q : Shyam is intelligent.

$\therefore$ Given compound statement is $p \wedge q$.

$\therefore$ The dual statement of $p \wedge q$ is $p \vee q$ *i.e.*, **Ram is honest or Shyam is intelligent**.

(*ii*) Let p : Kamla is beautiful

and q : Bimla is rich.

$\therefore$ Given compound statement is $p \vee q$.

$\therefore$ The dual statement of $p \vee q$ is $p \wedge q$ *i.e.*, **Kamla is beautiful and Bimla is rich**.

Example 2. *Write the duals of the following compound statements :*

(*i*) $(p \vee q) \wedge \sim p$ (*ii*) $[(p \vee r) \wedge (p \wedge q)] \vee f$

(*iii*) $[(p \wedge \sim q) \vee (\sim p \wedge q)] \vee [p \wedge q \wedge t]$.

Sol. (*i*) The dual of $(p \vee q) \wedge \sim p$ is $\mathbf{(p \wedge q) \vee \sim p}$.

(*ii*) The dual of $[(p \vee r) \wedge (p \wedge q)] \vee f$ is $\mathbf{[(p \wedge r) \vee (p \vee q)] \wedge t}$.

(*iii*) The dual of $[(p \wedge \sim q) \vee (\sim p \wedge q)] \vee [p \wedge q \wedge t]$ is $\mathbf{[(p \vee \sim q) \wedge (\sim p \vee q) \wedge [p \vee q \vee f]}$.

Example 3. *If* $S(p, q) = (p \vee \sim q) \wedge p$ *and* $S^* (p, q)$ *be the dual of* $S (p, q)$ *then verify that*

$$\sim S^*(p, q) \equiv S(\sim p, \sim q).$$

Sol. We have $S(p, q) = (p \vee \sim q) \wedge p$.

$\therefore$ $S^*(p, q) =$ dual of $S (p, q) = (p \wedge \sim q) \vee p$.

$\therefore$ $\sim S^*(p, q) = \sim [(p \wedge \sim q) \vee p]$

$\equiv \sim (p \wedge \sim q) \wedge \sim p \equiv (\sim p \vee q) \wedge \sim p.$ $(\because \sim \sim q \equiv q)$

Also, $S (\sim p, \sim q) = (\sim p \vee \sim \sim q) \wedge \sim p \equiv (\sim p \vee q) \wedge \sim p$.

$\therefore$ $\mathbf{\sim S^* (p, q) \equiv S (\sim p, \sim q)}$.

Example 4. *If $S(p, q, r) = p \vee (q \vee r)$ and $S^*(p, q, r)$ be the dual of $S(p, q, r)$ then verify that*

$$\sim S^*(p, q, r) \equiv S(\sim p, \sim q, \sim r).$$

Sol. We have $S(p, q, r) = p \vee (q \vee r)$.

$\therefore \quad S^*(p, q, r) = \text{dual of } S(p, q, r) = p \wedge (q \wedge r)$

$\therefore \quad \sim S^*(p, q, r) = \sim [p \wedge (q \wedge r)]$

$\equiv \sim p \vee \sim (q \wedge r)$

$\equiv \sim p \vee (\sim q \vee \sim r)$.

Also $S(\sim p, \sim q, \sim r) = \sim p \vee (\sim q \vee \sim r)$.

$\therefore \quad \mathbf{\sim S^*(p, q, r) \equiv S(\sim p, \sim q, \sim r)}$.

EXERCISE 8.7

1. Write the duals of the following compound statements :

   (i) $(p \wedge q) \wedge \sim p$ (ii) $(p \vee q) \wedge (p \wedge r)$

   (iii) $[(p \vee r) \vee (\sim p \vee \sim q)] \wedge r$ (iv) $\sim (p \vee q) \wedge [p \vee \sim (q \wedge \sim s)]$

   (v) $[(p \vee q) \wedge \sim r] \vee (p \wedge t)$ (vi) $(\sim p \vee f) \wedge [\sim q \wedge (p \vee q) \wedge \sim r]$.

2. Verify that $\sim S^*(p, q) \equiv S(\sim p, \sim q)$ if $S^*(p, q)$ is the dual of the compound statement $S(p, q)$ and $S(p, q)$ is equal to :

   (i) $p \wedge q$ (ii) $p \vee q$.

3. Verify that $\sim S^*(p, q, r) \equiv S(\sim p, \sim q, \sim r)$ if $S^*(p, q, r)$ is the dual of the compound statement $S(p, q, r)$ and $S(p, q, r)$ is equal to :

   (i) $p \wedge (q \vee r)$ (ii) $\sim p \wedge \sim (q \vee r)$.

Answers

1. (i) $(p \vee q) \vee \sim p$ (ii) $(p \wedge q) \vee (p \vee r)$

   (iii) $[(p \wedge r) \wedge (\sim p \wedge \sim q)] \vee r$ (iv) $\sim (p \wedge q) \vee [p \wedge \sim (q \vee \sim s)]$

   (v) $[(p \wedge q) \vee \sim r] \wedge (p \vee f)$ (vi) $(\sim p \wedge t) \vee [\sim q \vee (p \wedge q) \vee \sim r]$.

8.18. NEGATION OF COMPOUND STATEMENTS

(i) **Negation of conjunction.** Let p and q be any statements. The negation $\sim (p \wedge q)$ of the conjunction $p \wedge q$ is given by De Morgan's law and we have

$$\mathbf{\sim (p \wedge q) \equiv \sim p \vee \sim q.}$$

(ii) **Negation of disjunction.** Let p and q be any statements. The negation $\sim (p \vee q)$ of the disjunction $p \vee q$ is given by De Morgan's law and we have

$$\mathbf{\sim (p \vee q) = \sim p \wedge \sim q.}$$

Remark. The compound statement $\sim p \wedge \sim q$ represent 'neither p nor q'. The compound statement $\sim p \wedge \sim q$ is also called the **joint denial** of statements p and q and is denoted by $p \downarrow q$.

$\therefore \quad \mathbf{p \downarrow q = \sim p \wedge \sim q.}$

(iii) **Negation of conditional statement.** Let p and q be any statements. The compound statements $p \rightarrow q$ and $\sim p \vee q$ are logically equivalent. The negation $\sim (p \rightarrow q)$ of the conditional statement $p \rightarrow q$ is given by

$$\sim (p \rightarrow q) \equiv \sim (\sim p \vee q) \equiv \sim \sim p \wedge \sim q \equiv p \wedge \sim q.$$

$\therefore \quad \mathbf{\sim (p \rightarrow q) \equiv p \wedge \sim q.}$

(*iv*) **Negation of biconditional statement.** Let p and q be any statements. The compound statements $p \leftrightarrow q$ and $(p \rightarrow q) \wedge (q \rightarrow p)$ are logically equivalent. The negation $\sim (p \leftrightarrow q)$ of the biconditional statement $p \leftrightarrow q$ is given by

$$\sim (p \leftrightarrow q) \equiv \sim [(p \rightarrow q) \wedge (q \rightarrow p)] \equiv \sim (p \rightarrow q) \vee \sim (q \rightarrow p)$$

$$\equiv \sim (\sim p \vee q) \vee \sim (\sim q \vee p) \equiv (\sim \sim p \wedge \sim q) \vee (\sim \sim q \wedge \sim p) \equiv (p \wedge \sim q) \vee (\sim p \wedge q)$$

$\therefore$ $\mathbf{\sim (p \leftrightarrow q) \equiv (p \wedge \sim q) \vee (\sim p \wedge q).}$

Example. *Find the negation of the following compound statements :*

(*i*) $p \wedge \sim q$ (*ii*) $\sim p \rightarrow q$ (*iii*) $(p \rightarrow q) \rightarrow (q \rightarrow p)$.

Sol. (*i*) Negation of $(p \wedge \sim q) \equiv \sim (p \wedge \sim q)$

$$\equiv \sim p \vee \sim \sim q \equiv \sim p \vee q.$$

(*ii*) Negation of $(\sim p \rightarrow q) \equiv \sim (\sim p \rightarrow q)$

$$\equiv \sim (\sim \sim p \vee q) \qquad (\because \; p \rightarrow q \equiv \sim p \vee q)$$

$$\equiv \sim (p \vee q) \equiv \sim p \wedge \sim q.$$

(*iii*) Negation of $[(p \rightarrow q) \rightarrow (q \rightarrow p)] \equiv \sim [(p \rightarrow q) \rightarrow (q \rightarrow p)]$

$$\equiv \sim [\sim (p \rightarrow q) \vee (q \rightarrow p)] \equiv \sim \sim (p \rightarrow q) \wedge \sim (q \rightarrow p)$$

$$\equiv (p \rightarrow q) \wedge \sim (q \rightarrow p) \equiv (\sim p \vee q) \wedge \sim (\sim q \vee p)$$

$$\equiv (\sim p \vee q) \wedge (\sim \sim q \wedge \sim p) \equiv (\sim p \vee q) \wedge (q \wedge \sim p).$$

EXERCISE 8.8

Find the negation of the following compound statements :

1. $\sim p \wedge q$ 2. $\sim p \wedge \sim q$ 3. $\sim p \vee q$ 4. $p \vee \sim q$
5. $\sim p \vee \sim q$ 6. $p \rightarrow \sim q$ 7. $\sim p \rightarrow \sim q$ 8. $p \leftrightarrow \sim q$
9. $\sim p \leftrightarrow q$ 10. $\sim p \leftrightarrow \sim q$.

Answers

1. $p \vee \sim q$ 2. $p \vee q$ 3. $p \wedge \sim q$ 4. $\sim p \wedge q$
5. $p \wedge q$ 6. $p \wedge q$ 7. $\sim p \wedge q$ 8. $(p \wedge q) \vee (\sim p \wedge \sim q)$
9. $(p \wedge q) \vee (\sim p \wedge \sim q)$ 10. $(\sim p \wedge q) \vee (p \wedge \sim q)$.

8.19. VALIDITY OF ARGUMENTS

Argument. An **argument** is a statement which asserts that a given set of n statements $S_1, S_2, \ldots\ldots, S_n$ yield another statement S. This argument is denoted as :

$$S_1, S_2, \ldots\ldots, S_n \vdash S$$

The statements $S_1, S_2, \ldots\ldots, S_n$ are called **hypotheses** or **premises** or **assumptions.** The statement S is called **conclusion.**

The symbol '$\vdash$' is called **turnsile.** The argument $S_1, S_2, \ldots\ldots, S_n \vdash S$ is defined to be true if S is true whenever $S_1, S_2, \ldots\ldots, S_n$ are all true, otherwise the argument is defined to be false.

A true argument is also called a **valid argument.**

We have seen that the argument $S_1, S_2, \ldots\ldots, S_n \vdash S$ is valid if S is true whenever S_1, $S_2, \ldots\ldots, S_n$ are all true.

∴ The argument $S_1, S_2, \ldots\ldots, S_n \vdash S$ is valid if S is true whenever $S_1 \wedge S_2 \wedge \ldots\ldots \wedge S_n$ is true.

∴ **The argument $S_1, S_2, \ldots\ldots, S_n \vdash S$ is valid if $(S_1 \wedge S_2 \wedge \ldots\ldots \wedge S_n) \to S$ is a tautology.** (∵ $p \to q$ can be false only when q is false whenever p is true.)

This gives an alternative method to check the validity of an argument.

Thus, we have the following two methods to check the validity of an argument :

Method I. The argument $S_1, S_2, \ldots\ldots, S_n \vdash S$ is valid if S is true whenever $S_1, S_2, \ldots\ldots, S_n$ are all true.

Method II. The argument $S_1, S_2, \ldots\ldots, S_n \vdash S$ is valid if the compound statement $(S_1 \wedge S_2 \wedge \ldots\ldots \wedge S_n) \to S$ is a tautology.

Example 1. *Test the validity of the following argument :*

"If it is a good watch, then it is a Titan watch. It is a Titan watch therefore it is a good watch".

Sol. Let p = It is a good watch

and q = It is a Titan watch.

∴ The assumptions are $p \to q$, q and the conclusion is p.

Let $S_1 = p \to q$, $S_2 = q$ and $S = p$

∴ Given argument is $S_1, S_2 \vdash S$.

Truth values of S_1, S_2, S

p	q	S_1 $(p \to q)$	S_2 (q)	S (p)
T	T	T	T	T
T	F	F	F	T
F	T	T	T	F
F	F	T	F	F

In the 1st row, S_1, S_2 are true and S is true. In the 3rd row, S_1, S_2 are true but S is not true.

∴ Given argument is not valid.

Example 2. *Test the validity of the following argument :*

"If it is cloudy tonight, it will rain tomorrow, and if it rains tomorrow, I shall be on leave tomorrow ; and the conclusion is if it is cloudy tonight, I shall be on leave tomorrow".

Sol. Let p = It is cloudy tonight,

q = It will rain tomorrow

and r = I shall be on leave tomorrow.

∴ The assumptions are $p \to q$, $q \to r$ and the conclusion is $p \to r$.

Let $S_1 = p \to q$, $S_2 = q \to r$ and $S = p \to r$.

∴ Given argument is $S_1, S_2 \vdash S$.

Truth values of S_1, S_2, S

p	q	r	S_1 $(p \to q)$	S_2 $(q \to r)$	S $(p \to r)$
T	T	T	T	T	T
T	T	F	T	F	F
T	F	T	F	T	T
F	T	T	T	T	T
T	F	F	F	T	F
F	T	F	T	F	T
F	F	T	T	T	T
F	F	F	T	T	T

In the 1st, 4th, 7th, 8th rows, S_1, S_2 are both true and in each of these rows, S is also true.

∴ Given argument is valid.

Remark. In the following table, we show that $(S_1 \wedge S_2) \to S$ is a tautology.

S_1	S_2	S	$S_1 \wedge S_2$	$(S_1 \wedge S_2) \to S$
T	T	T	T	T
T	F	F	F	T
F	T	T	F	T
T	T	T	T	T
F	T	F	F	T
T	F	T	F	T
T	T	T	T	T
T	T	T	T	T

Example 3. *Test the validity of the following argument :*

"If my son stands first in his class, I give him a gift. Either he stood first or I was out of station. I did not give son a gift this time. Therefore I, was out of station."

Sol. Let p = My son stands first in his class,

q = I give him a gift

and r = I was out of station.

∴ The assumptions are $p \to q$, $p \vee r$, $\sim q$ and the conclusion is r.

Let $S_1 = p \to q$, $S_2 = p \vee r$, $S_3 = \sim q$ and $S = r$.

∴ Given argument is $S_1, S_2, S_3 \vdash S$.

Truth values of S_1, S_2, S_3, S

p	q	r	S_1 $(p \to q)$	S_2 $(p \vee r)$	S_3 $(\sim q)$	S (r)
T	T	T	T	T	F	T
T	T	F	T	T	F	F
T	F	T	F	T	T	T
F	T	T	T	T	F	T
T	F	F	F	T	T	F
F	T	F	T	F	F	F
F	F	T	T	T	T	T
F	F	F	T	F	T	F

In the 7th row S_1, S_2, S_3 are all true and S is also true.

$\therefore$ Given argument is valid.

EXERCISE 8.9

1. Show that the following argument is not valid :
"If it rains, crops will be good. It did not rain. Therefore the crops were not good".
2. Show that the following argument is valid :
"If he works hard, he will be successful. He was not successful. Therefore he did not work hard".
3. Test the validity of the following argument :
"If today is sunday, then yesterday was saturday. Yesterday was not saturday. Therefore, today is not sunday".
4. Test the validity of the following argument :
"If it rains tomorrow, I shall carry my umbrella if its cloth is mended. It will rain tomorrow and the cloth will not be mended. Therefore, I shall not carry my umbrella".
5. Test the validity of the following argument :
"If Nidhi works hard then she will be successful. If she is successful then she will be happy. Therefore, hard work leads to happiness".
6. Test the validity of the following argument :
"Wages will increase if and only if there is an inflation. If there is an inflation then the cost of living will increase. Wages increased. Therefore, the cost of living will increase.

Answers

3. Valid **4.** Not valid **5.** Valid **6.** Valid.

Hints

1. Let p = It rains, q = Crops are good.
$\therefore$ $S_1 = p \to q, S_2 = \sim p, S = \sim q.$
2. Let p = he works hard, q = He is successful.
$\therefore$ $S_1 = p \to q, S_2 = \sim q, S = \sim p.$
3. Let p = Today is sunday, q = Yesterday is saturday.
$\therefore$ $S_1 = p \to q, S_2 = \sim q, S = \sim p.$
4. Let p = It rains tomorrow, q = I shall carry my umbrella, r = Cloth of umbrella is mended.
$\therefore$ $S_1 = p \to (r \to q), S_2 = p \wedge \sim r, S = \sim q.$
5. Let p = Nidhi works hard,
q = She is successful
and r = She is happy.
$\therefore$ $S_1 = p \to q, S_2 = q \to r, S = p \to r.$
6. Let p = Wages will increase, q = There is an inflation, r = Cost of living increases.
$\therefore$ $S_1 = p \leftrightarrow q, S_2 = q \to r, S_3 = p, S = r.$

SECTION III

9

Lattices

9.1. INTRODUCTION

We know that a partially ordered set (poset) is a set equipped with a relation which is reflexive, antisymmetric and transitive. Also, the supremum and infimum of a subset of a partially ordered set may or may not exist. If for every subset containing two elements, the supremum and infimum exist then a partially ordered set becomes of special interest and we call it a *lattice*. We shall also define a lattice as an algebraic structure equipped with two binary operations and prove the equivalence of both definitions.

9.2. LATTICE (AS A PARTIALLY ORDERED SET)

A partially ordered set L is called a **lattice** if every pair of elements in L has the supremum and the infimum.

In other words, a partially ordered set L is a lattice if sup (a, b), inf (a, b) exist for all $a, b \in L$.

Remark. In general, the partial order on a lattice is denoted by the symbol '$\leq$' until the contrary is specified explicitly.

Example 1. *Prove that* (**N**, $\leq$) *is a lattice, where '$\leq$' is the relation of divisibility.*

(KU 2006)

Sol. The relation '$\leq$' is the relation of divisibility. Let $a, b, c \in \mathbf{N}$.

(*i*) $a/a \Rightarrow a \leq a \quad \therefore \leq$ is reflexive.

(*ii*) $a \leq b, b \leq a \Rightarrow a/b, b/a \Rightarrow b = a\lambda, a = b\mu$ for some $\lambda, \mu \in \mathbf{N}$

$\Rightarrow b = (b\mu)\lambda = b(\mu\lambda) \Rightarrow \mu\lambda = 1 \Rightarrow \lambda = 1, \mu = 1 \Rightarrow a = b(1) = b$

$\therefore \leq$ is antisymmetric.

(*iii*) $a \leq b, b \leq c \Rightarrow a/b, b/c \Rightarrow b = am, c = bn$ for some $m, n \in \mathbf{N}$

$\Rightarrow c = (am)n = a(mn) \Rightarrow a/c$ i.e., $a \leq c$

$\therefore \leq$ is transitive.

$\therefore \leq$ is a partial order an **N** .

$\therefore$ (**N**, $\leq$) is a partially ordered set.

Let $a, b \in \mathbf{N}$ and $l = \text{lcm}(a, b)$ and $g = \text{gcd}(a, b) \quad \therefore \quad l, g \in \mathbf{N}$

$\Rightarrow \quad a/l, b/l$ i.e., $a \leq l, b \leq l \quad \therefore \quad l$ is an upper bound of $\{a, b\}$

Let l' be any upper bound of $\{a, b\}$.

$\Rightarrow \quad a \leq l', b \leq l'$ i.e., $a/l', b/l'$

$\Rightarrow \quad l/l'$ i.e., $l \leq l' \qquad (\because l = \text{lcm}(a, b))$

$\therefore \quad l = \sup(a, b)$.

Also $g/a, g/b$ *i.e.*, $g \le a, g \le b$ $\therefore$ g is a lower bound of $\{a, b\}$

Let g' be any lower bound of $\{a, b\}$.

$\Rightarrow$ $g' \le a, g' \le b$ *i.e.*, $g'/a, g'/b$

$\Rightarrow$ g'/g *i.e.*, $g' \le g$ $(\because \; g = \gcd(ab))$

$\therefore$ $g = \inf(a, b)$

$\therefore$ sup (a, b) and inf (a, b) exist for every $a, b \in \mathbf{N}$.

$\therefore$ **N** is a lattice, where '$\le$' is the relation of divisibility.

Example 2. *Let X be a non-empty set and P(X), the power set of X. Show that (P(X), ≤) is a lattice, where '≤' is the relation of inclusion.* (KU 2006 C)

Sol. The relation '$\le$' is the relation of inclusion. Let $A, B, C \in P(X)$.

(*i*) $A \subseteq A \Rightarrow A \le A$ $\therefore$ $\le$ is reflexive.

(*ii*) $A \le B, B \le A \Rightarrow A \subseteq B, B \subseteq A \Rightarrow A = B$ $\therefore$ $\le$ is antisymmetric.

(*iii*) $A \le B, B \le C \Rightarrow A \subseteq B, B \subseteq C \Rightarrow A \subseteq C \Rightarrow A \le C$ $\therefore$ $\le$ is transitive.

$\therefore$ $\le$ is a partial order on $P(X)$.

$\therefore$ $(P(X), \le)$ is a partially ordered set.

Let $A, B \in P(X)$ $\therefore$ $A \cup B, A \cap B \in P(X)$

$A \subseteq A \cup B, B \subseteq A \cup B \Rightarrow A \le A \cup B, \; B \le A \cup B$

$\therefore$ $A \cup B$ is an upper bound of A and B.

Let U be any upper bound of A and B.

$\Rightarrow$ $A \le U, B \le U \Rightarrow A \subseteq U, B \subseteq U \Rightarrow A \cup B \subseteq U \Rightarrow A \cup B \le U$

$\therefore$ $\sup(A, B) = A \cup B$

Also, $A \cap B \subseteq A, A \cap B \subseteq B \Rightarrow A \cap B \le A, \; A \cap B \le B$

$\therefore$ $A \cap B$ is a lower bound of A and B.

Let V be any lower bound of A and B.

$\therefore$ $V \le A, V \le B \Rightarrow V \subseteq A, V \subseteq B \Rightarrow V \subseteq A \cap B \Rightarrow V \le A \cap B$

$\therefore$ $\inf(A, B) = A \cap B$

$\therefore$ sup (A, B), inf (A, B) exist for every $A, B \in P(X)$.

$\therefore$ $(P(X), \le)$ is a lattice, where '$\le$' is the relation of inclusion.

Example 3. *Show that a lattice with three or fewer elements is a chain.*

Sol. Let $(L, \le)$ be a lattice. If $n(L) = 1$, then L is trivially a chain.

Let $n(L) = 2$ and $L = \{a, b\}$.

$\therefore$ sup (a, b) is either a or b.

Let $\sup(a, b) = a$. $\therefore$ $b \le a$

$\therefore$ $(L, \le)$ is a chain.

Let $n(L) = 3$ and $L = \{a, b, c\}$. If possible let L be not a chain.

$\therefore$ There exist non-comparable elements, say a and b.

$\therefore$ Neither $a \le b$ nor $b \le a$

$\therefore$ $\sup(a, b) = c$ and $\inf(a, b) = c$

$\Rightarrow$ $a \le c, b \le c$ and $c \le a, c \le b$

$\Rightarrow$ $a = c, b = c \Rightarrow a = b = c$

This is impossible.

$\therefore$ L is a chain.

$\therefore$ The result holds.

Example 4. (*a*) *Show that the set D_{20} of all positive divisors of 20 form a lattice with the relation of divisibility.*

(*b*) *Represent the lattice D_{20} by a Hasse diagram.* (KU 2006)

Sol. (*a*) We have $D_{20} = \{1, 2, 4, 5, 10, 20\}$.

D_{20} is a partially ordered set with the relation of divisibility.

For $a, b \in D_{20}$, we have sup $(a, b) = \text{lcm}(a, b)$ and inf $(a, b) = \text{gcd}(a, b)$.

The following tables give the supremum and infimum of pairs of elements of D_{20}.

a \ *b*

sup (a, b)	1	2	4	5	10	20
1	1	2	4	5	10	20
2	2	2	4	10	10	20
4	4	4	4	20	20	20
5	5	10	20	5	10	20
10	10	10	20	10	10	20
20	20	20	20	20	20	20

a \ *b*

inf (a, b)	1	2	4	5	10	20
1	1	1	1	1	1	1
2	1	2	2	1	2	2
4	1	2	4	1	2	4
5	1	1	1	5	5	5
10	1	2	2	5	10	10
20	1	2	4	5	10	20

From the above tables, we see that sup (*a*, *b*) and inf (*a*, *b*) exist for every $a, b \in D_{20}$.

$\therefore$ D_{20} is a lattice with the relation of divisibility.

(*b*) We know that

$$D_{20} = \{1, 2, 4, 5, 10, 20\}$$

is a lattice with the relation of divisibility.

The Hasse diagram of the lattice D_{20} is shown in the adjoining figure.

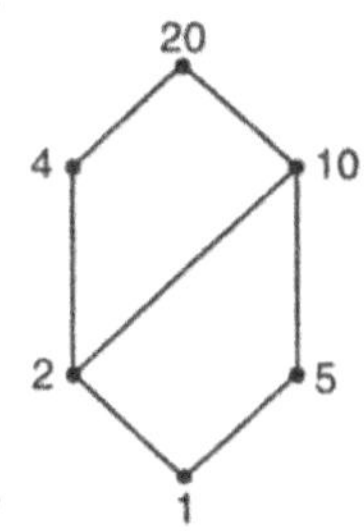

Remark. If *n* is any positive integer then the set D_n of all positive divisors of *n* forms a lattice under the relation of divisibility.

Example 5. *Determine whether or not each of the following partially ordered set under the relation of divisibility is a lattice :*

(*i*) $A_1 = \{1, 2, 4, 5, 20\}$ (*ii*) $A_2 = \{2, 4, 8, 10, 20, 40\}$.

Sol. (*i*) A_1 is not a lattice because sup $(2, 5) = \text{lcm}(2, 5) = 10 \notin A_1$.

(*ii*) A_2 is a lattice because the supremum and infimum of every pair of elements in A_2 are in A_2. This is shown in the following tables :

a \ *b*

sup (a, b)	2	4	8	10	20	40
2	2	4	8	10	20	40
4	4	4	8	20	20	40
8	8	8	8	40	40	40
10	10	20	40	10	20	40
20	20	20	40	20	20	40
40	40	40	40	40	40	40

a \ *b*

inf (a, b)	2	4	8	10	20	40
2	2	2	2	2	2	2
4	2	4	4	2	4	4
8	2	4	8	2	4	8
10	2	2	2	10	10	10
20	2	4	4	10	20	20
40	2	4	8	10	20	40

Example 6. *Which of the following Hasse diagrams represent lattices ?*

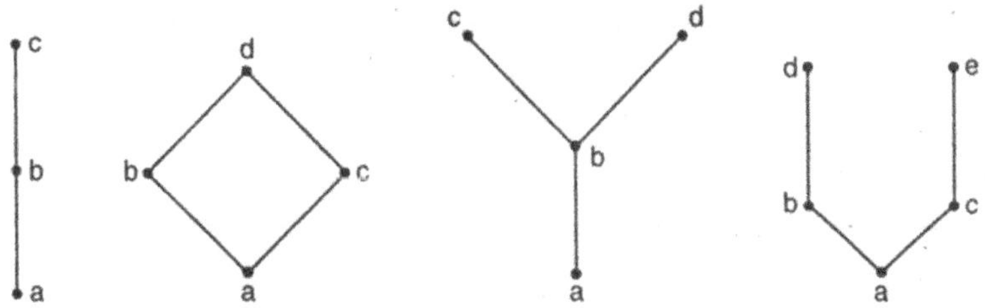

(*i*) (*ii*) (*iii*) (*iv*)

Sol. (*i*) Let $A = \{a, b, c\}$

Here sup (x, y) and inf (x, y) exist for every $x, y \in A$. This is shown in the following tables :

sup (x, y) (rows x, columns y)	a	b	c
a	a	b	c
b	b	b	c
c	c	c	c

inf (x, y) (rows x, columns y)	a	b	c
a	a	a	a
b	a	b	b
c	a	b	c

$\therefore$ The partially ordered set A with Hasse diagram (*i*) is a lattice.

(*ii*) Let $A = \{a, b, c, d\}$

Here sup (x, y) and inf (x, y) exist for every $x, y \in A$. This is shown in the following tables :

sup (x, y) (rows x, columns y)	a	b	c	d
a	a	b	c	d
b	b	b	d	d
c	c	d	c	d
d	d	d	d	d

inf (x, y) (rows x, columns y)	a	b	c	d
a	a	a	a	a
b	a	b	a	b
c	a	a	c	c
d	a	b	c	d

$\therefore$ The partially ordered set A with Hasse diagram (*ii*) is a lattice.

(*iii*) Let $A = \{a, b, c, d\}$

Here sup (x, y) and inf (x, y) do not exist for every $x, y \in A$. This is shown in the following tables :

sup (x, y) (rows x, columns y)	a	b	c	d
a	a	b	c	d
b	b	b	c	d
c	c	c	c	
d	d	d		d

inf (x, y) (rows x, columns y)	a	b	c	d
a	a	a	a	a
b	a	b	b	b
c	a	b	c	b
d	a	b	b	d

$\therefore$ The partially ordered set A with Hasse diagram (*iii*) is not a lattice.

(*iv*) Let $A = \{a, b, c, d, e\}$

Here sup (x, y) and inf (x, y) do not exist for every $x, y \in A$. This is shown in the following tables :

	y					
	sup (x, y)	a	b	c	d	e
	a	a	b	c	d	e
	b	b	b		d	
x	c	c		c		e
	d	d	d		d	
	e	e		e		e

	y					
	inf (x, y)	a	b	c	d	e
	a	a	a	a	a	a
	b	a	b	a	b	a
x	c	a	a	c	a	c
	d	a	b	a	d	a
	e	a	a	c	a	e

$\therefore$ The partially ordered set A with Hasse diagram (*iv*) is not a lattice.

Example 7. *Let A be any set and L be a lattice. Let T be the set of all functions from A to L. Define a relation '≤' on T by*

$$f \leq g \quad \textit{iff} \quad f(x) \leq g(x) \quad \forall\, x \in A, \quad f, g \in T.$$

Show that $(T, \leq)$ *is a lattice.*

Sol. Let $f, g, h \in T$

(*i*) $f(x) \leq f(x)\ \forall\, x \in A \Rightarrow f \leq f$

$\therefore$ $\leq$ is reflexive.

(*ii*) $f \leq g,\ g \leq f \Rightarrow f(x) \leq g(x),\ g(x) \leq f(x)\ \forall\, x \in A \Rightarrow f(x) = g(x)\ \forall\, x \in A \Rightarrow f = g$

$\therefore$ $\leq$ is antisymmetric.

(*iii*) $f \leq g,\ g \leq h \Rightarrow f(x) \leq g(x),\ g(x) \leq h(x)\ \forall\, x \in A \Rightarrow f(x) \leq h(x)\ \forall\, x \in A \Rightarrow f \leq h$

$\therefore$ $\leq$ is transitive.

$\therefore$ T is a partially ordered set with relation $\leq$.

Let $f, g \in T$

Define $\phi : A \to L$ by $\phi(x) = \sup\,(f(x), g(x)),\ x \in A$

$\Rightarrow$ $f(x) \leq \phi(x),\ g(x) \leq \phi(x),\ x \in A$

$\Rightarrow$ $f \leq \phi,\ g \leq \phi$

$\therefore$ ϕ is an upper bound of $\{f, g\}$

Let ψ be any upper bound of $\{f, g\}$.

$\Rightarrow$ $f \leq \psi,\ g \leq \psi \Rightarrow f(x) \leq \psi(x),\ g(x) \leq \psi(x),\ x \in A \Rightarrow \phi(x) \leq \psi(x),\ x \in A$

$(\because\ \phi(x) = \sup\,(f(x), g(x)))$

$\Rightarrow$ $\phi \leq \psi \quad \therefore \quad \sup\,(f, g) = \phi$

$\therefore$ $\sup\,(f, g) \in T$.

Similarly, we can show that the function $\theta : A \to L$ defined by $\theta(x) = \inf\,(f(x), g(x)),\ x \in A$ in equal to $\inf\,(f, g)$.

$\therefore$ For any $f, g \in T$, $\sup\,(f, g)$ and $\inf\,(f, g)$ exist $\therefore$ $(T, \leq)$ is a lattice.

Remark. The symbol '≤' in $f \leq g$ is representing the relation on T and the symbol '≤' in $f(x) \leq g(x)$ is representing the relation on the set L.

Theorem 1. A partially ordered set L with partial order ≤ is a lattice if and only if every non-empty finite subset of L has supremum and infimum.

Proof. Let $(L, \leq)$ be a lattice. Let A be any non-empty finite subset of L.

Case I. n(A) = 1. Let $A = \{a\}$

$\therefore \quad \sup A = \inf A = a$

Case II. n(A) = 2. Let $A = \{a, b\}$

$\therefore$ By definition, $\sup A$ and $\inf A$ exist.

Case III. n(A) = n(>2). Let $a, b, c \in A$.

Let $\sup(a, b) = d$ and $\sup(c, d) = e$ ...(1)

We shall show that $\sup\{a, b, c\} = e$

(1) $\Rightarrow \quad a \leq d, b \leq d, c \leq e, d \leq e$

$\Rightarrow \quad a \leq e, b \leq e, c \leq e$ (By transitivity)

$\Rightarrow$ e is an upper bound of $\{a, b, c\}$.

Let f be any upper bound of $\{a, b, c\}$.

$\Rightarrow \quad a \leq f, b \leq f, c \leq f$ ...(2)

(2) $\Rightarrow \quad a \leq f, b \leq f \Rightarrow d \leq f$ $(\because \ d = \sup(a, b))$

Also, $\quad c \leq f, d \leq f \Rightarrow e \leq f$ $(\because \ e = \sup(c, d))$

$\therefore \quad \sup\{a, b, c\} = e$

Similarly, $\inf\{a, b, c\}$ exists.

$\therefore$ If $n = 3$, the result is true

This result can be extended to any finitely many elements in A.

Conversely, let every finite subset of L has supremum and infimum.

$\therefore$ In particular, $\sup(a, b)$, $\inf(a, b)$ exist for every $a, b \in L$.

$\therefore$ $(L, \leq)$ is a lattice.

Theorem 2. Let $(L_1, \leq)$ and $(L_2, \leq)$ be lattices. Prove that $L_1 \times L_2$ is also a lattice under the product partial order.

Proof. For $(a, b), (c, d) \in L_1 \times L_2$,

$$(a, b) \leq (c, d) \text{ iff } a \leq c \text{ and } b \leq d.$$

Under the above product partial order, $L_1 \times L_2$ is a partially ordered set.

Let $\quad (a, b), (c, d) \in L_1 \times L_2$

$\Rightarrow \quad a, c \in L_1 \Rightarrow \sup(a, c), \inf(a, c) \in L_1$

Also $\quad b, d \in L_2 \Rightarrow \sup(b, d), \inf(b, d) \in L_2$

Let $\sup(a, c) = s_1$, $\inf(a, c) = i_1$, $\sup(b, d) = s_2$, $\inf(b, d) = i_2$

$\Rightarrow \quad a \leq s_1, c \leq s_1, b \leq s_2, d \leq s_2$

$\Rightarrow \quad (a, b), (c, d) \leq (s_1, s_2)$

$\therefore$ (s_1, s_2) is an upper bound of (a, b) and (c, d).

Let (u_1, u_2) be any upper bound of (a, b) and (c, d).

$\Rightarrow \quad (a, b), (c, d) \leq (u_1, u_2)$

$\Rightarrow \quad a, c \leq u_1, \ b, d \leq u_2$ ...(1)

(1) $\Rightarrow$ u_1 is an upper bound of a and c

$\therefore \quad s_1 \leq u_1$ $(\because \ s_1 = \sup(a, c))$

(1) $\Rightarrow$ u_2 is an upper bound of b and d

$\therefore \quad s_2 \leq u_2$ $(\because \ s_2 = \sup(b, d))$

$\Rightarrow \qquad (s_1, s_2) \leq (u_1, u_2)$

$\therefore \quad \sup((a, b), (c, d)) = (s_1, s_2) \in L_1 \times L_2$

Similarly,

$\inf((a, b), (c, d)) = (i_1, i_2) \in L_1 \times L_2$

$\therefore \qquad (L_1 \times L_2, \leq)$ is a lattice.

Example 8. *If L_1 and L_2 are the lattices shown in the figure, draw the Hasse diagram of the lattice $L_1 \times L_2$ under the product partial order.*

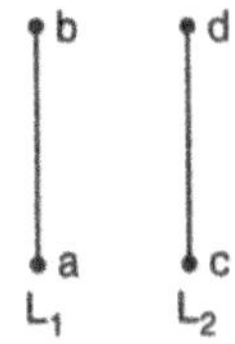

Sol. We have $L_1 = \{a, b\}$, $L_2 = \{c, d\}$

Also, $a \leq b$ and $c \leq d$

$\therefore \quad L_1 \times L_2 = \{(a, c), (a, d), (b, c), (b, d)\}$

$a \leq b$, $c \leq d$ implies

$(a, c) \leq (a, c)$, $(a, c) \leq (a, d)$, $(a, c) \leq (b, c)$,
$(a, c) \leq (b, d)$, $(a, d) \leq (a, d)$, $(a, d) \leq (b, d)$,
$(b, c) \leq (b, c)$, $(b, c) \leq (b, d)$, $(b, d) \leq (b, d)$

$\Rightarrow \quad (a, c) << (a, d)$, $(a, c) << (b, c)$, $(a, d) << (b, d)$

and $\quad (b, c) << (b, d)$

The Hasse diagram of the lattice $L_1 \times L_2$ is shown in the adjoining figure.

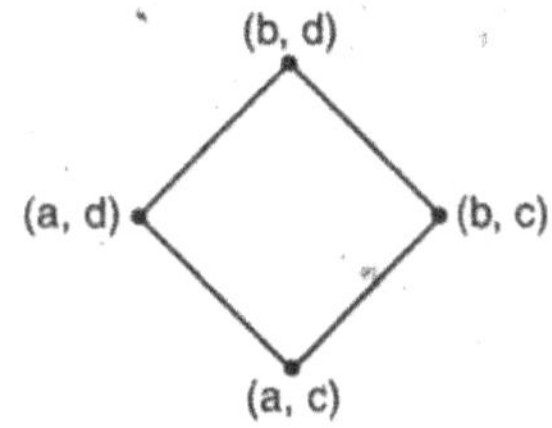

Example 9. *We know that D_4 and D_9 are lattices under the relation of divisibility. Draw the Hasse diagram of the lattice $D_4 \times D_9$ under the product partial order.*

Sol. We have $D_4 = \{1, 2, 4\}$ and $D_9 = \{1, 3, 9\}$.

The Hasse diagrams of the lattices D_4 and D_9 under the relation of divisibility are shown in the figure.

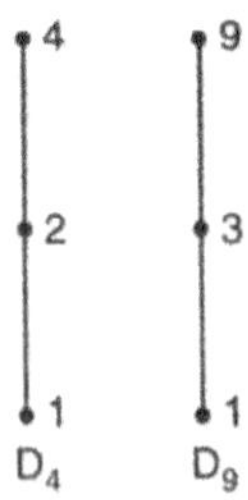

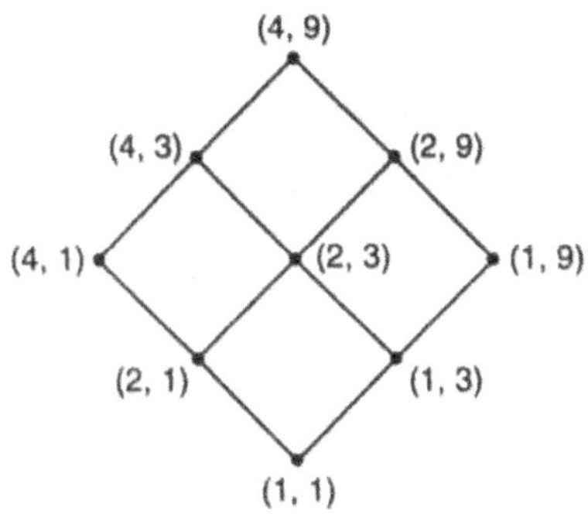

$D_4 \times D_9 = \{(1, 1), (1, 3), (1, 9), (2, 1), (2, 3) (2, 9), (4, 1), (4, 3), (4, 9)\}$

For $(a, b), (c, d) \in D_4 \times D_9$, we have $(a, b) \leq (c, d)$ iff a/c and b/d.

Under product partial order, $D_4 \times D_9$ is a lattice and its Hasse diagram is shown in the figure.

EXERCISE 9.1

1. Show that the set of integers is not a lattice with the relation of divisibility.
2. Show that the set {2, 3, 4, 6} is not a lattice with the relation of divisibility.
3. Show that every chain is a lattice.

4. (a) Show that the set D_8 of all positive divisors of 8 forms a lattice with the relation of divisibility.
(b) Represent the lattice D_8 by a Hasse diagram.

5. (a) Show that the set D_{30} of all positive divisors of 30 forms a lattice with the relation of divisibility.
(b) Represent the lattice D_{30} by a Hasse diagram. *(KU 2006 C)*

6. Let $A = [0, 1]$ and '$\leq$' be the operation of "less than or equal to" on A. Show that $(A, \leq)$ is a lattice.

7. Let $L = \{1, 2, 3, 5, 30\}$. Show that L is a lattice under divisibility. Also draw the Hasse diagram of this lattice.

8. Show that the set of all subsets of the set $\{a, b, c\}$ is a lattice under 'set inclusion'. Also draw the Hasse diagram of this lattice.

9. Determine whether the following partially ordered sets shown by Hasse diagrams are lattices or not :

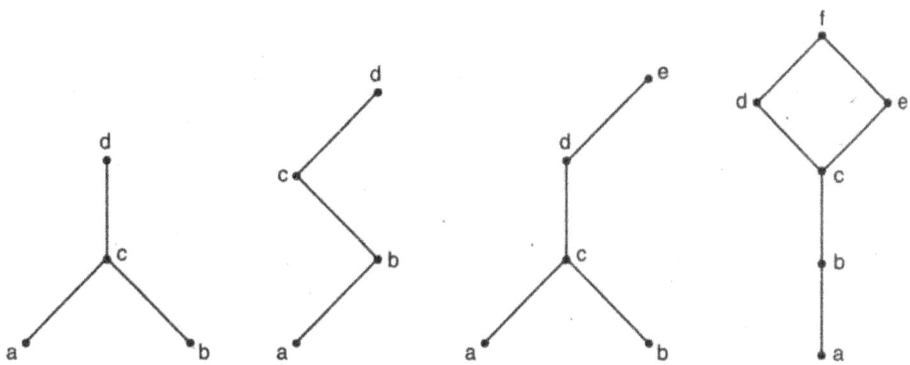

(i) (ii) (iii) (iv)

10. Which of the following Hasse diagrams represent lattices ?

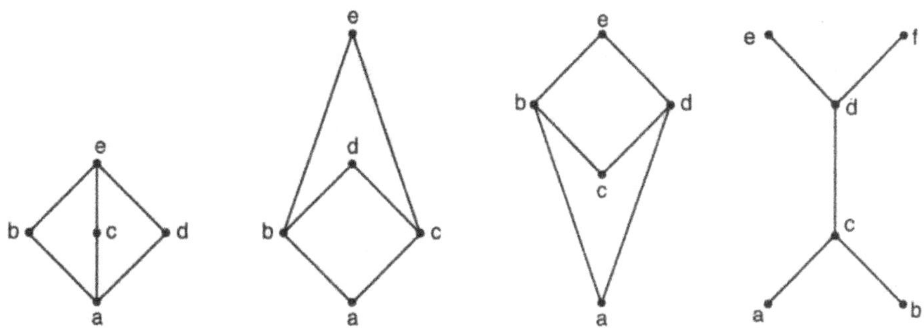

(i) (ii) (iii) (iv)

11. Let L_1, L_2 be lattices show in the figure. Represent the lattice $L_1 \times L_2$ under the product partial order by a Hasse diagram.

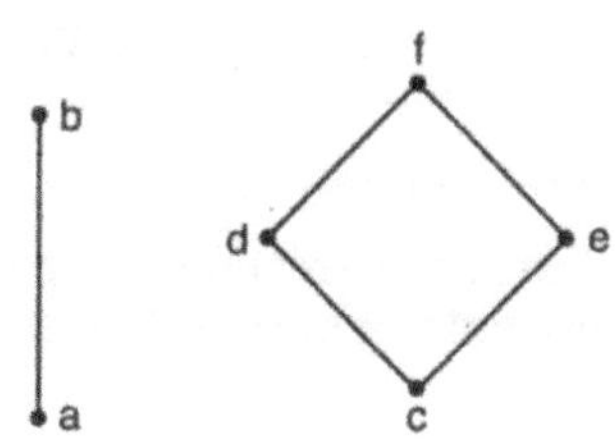

L_1 L_2

Answers

4. (*b*)

5. (*b*)

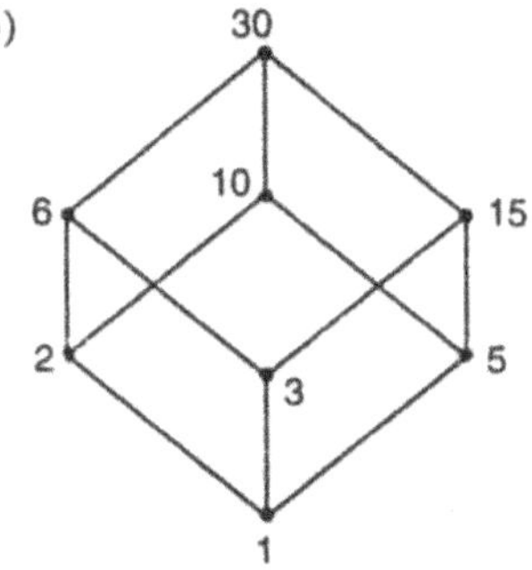

7.

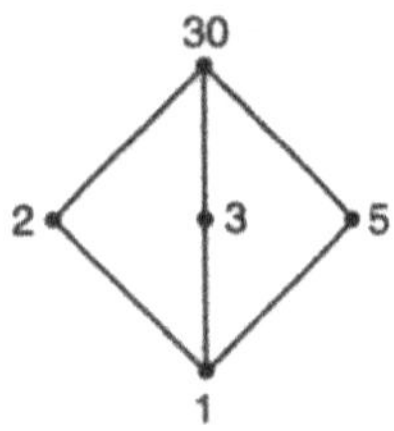

8.

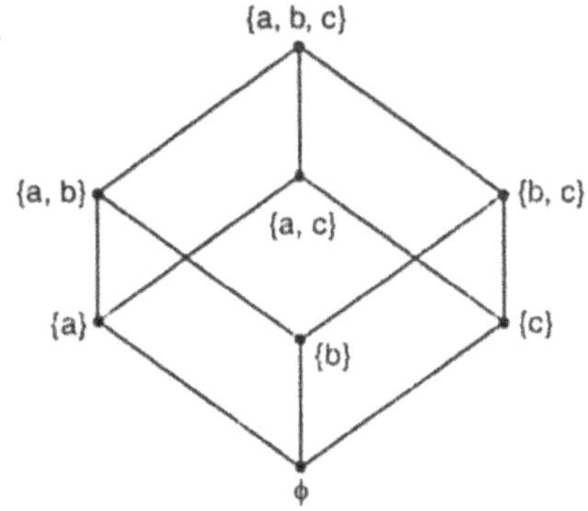

9. (*ii*) (*iv*)

10. (*i*)

11.

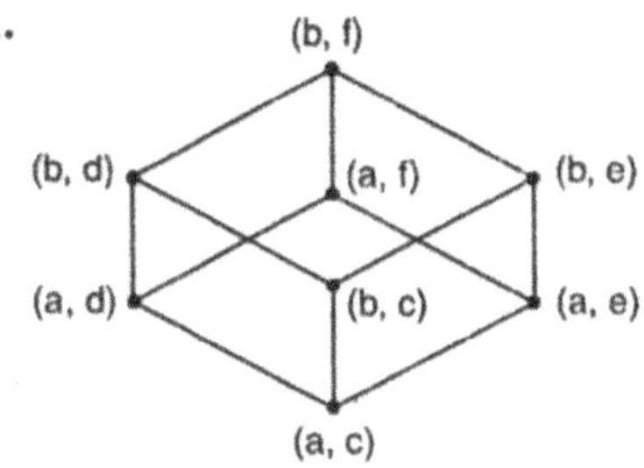

Hints

1. The relation is not antisymmetric.

2. sup (3, 4) = 12 $\notin$ {2, 3, 4, 6}

3. For any a, b in the chain, either $a \le b$ or $b \le a$.

$$a \le b \Rightarrow \sup(a, b) = b \quad \text{and} \quad \inf(a, b) = a.$$

9. (*i*) inf (a, b) does not exist.

10. (*ii*) sup (d, e) does not exist. (*iii*) inf (a, c) does not exist.

9.3. LATTICE (AS AN ALGEBRAIC STRUCTURE)

A non-empty set L with two binary operations '$\vee$' (called *join*) and '$\wedge$' (called *meet*) is called a **lattice** if the following axioms hold, where a, b, c are any elements in L :

L_1 Commutative laws

(*i*) $a \vee b = b \vee a$ (*ii*) $a \wedge b = b \wedge a$

L_2 Associative laws

(*i*) $a \vee (b \vee c) = (a \vee b) \vee c$ (*ii*) $a \wedge (b \wedge c) = (a \wedge b) \wedge c$

L_3 Absorption laws

(*i*) $a \vee (a \wedge b) = a$ (*ii*) $a \wedge (a \vee b) = a$.

The lattice L is also written as $(L, \vee, \wedge)$ when we want to specify the binary operations involved in the lattice.

Theorem 1. If a partially ordered set (L, ≤) is a lattice then prove that it is also a lattice as an algebraic structure.

Proof. The partially ordered set $(L, \leq)$ is a lattice.

$\therefore$ sup (a, b) and inf (a, b) exist for any $a, b \in L$

We define binary operations $\vee$ and $\wedge$ in L as follows :

For $a, b \in$ L, $a \vee b = \sup (a, b)$ and $a \wedge b = \inf (a, b)$

Now we shall prove the axioms required for $(L, \vee, \wedge)$ to be a lattice.

1. Commutative laws. For $a, b \in L$,

$a \vee b = \sup (a, b) = \sup (b, a) = b \vee a$ and $a \wedge b = \inf (a, b) = \inf (b, a) = b \wedge a$

$\therefore$ $a \vee b = b \vee a$ and $a \wedge b = b \wedge a$ for $a, b \in L$.

2. Associative laws. Let $a, b, c \in L$.

$$a \leq \sup (a, b \vee c) \Rightarrow a \leq a \vee (b \vee c)$$
$$b \vee c \leq \sup (a, b \vee c) \Rightarrow b \vee c \leq a \vee (b \vee c)$$

Also, $b \leq b \vee c$ and $c \leq b \vee c$

$\therefore$ By transitivity, $b \leq a \vee (b \vee c)$ and $c \leq a \vee (b \vee c)$

$a \leq a \vee (b \vee c)$ and $b \leq a \vee (b \vee c)$ implies that $a \vee (b \vee c)$ is an upper bound of a and b.

$\therefore$ By definition, $a \vee b \leq a \vee (b \vee c)$ ...(1)

(1) and $c \leq a \vee (b \vee c)$ implies that $a \vee (b \vee c)$ is an upper bound of $a \vee b$ and c.

$\therefore$ By definition, $(a \vee b) \vee c \leq a \vee (b \vee c)$.

Similarly, we can show that

$$a \vee (b \vee c) \leq (a \vee b) \vee c$$

$\therefore$ By antisymmetry, we have

$$a \vee (b \vee c) = (a \vee b) \vee c.$$

Similarly, we can show that

$$a \wedge (b \wedge c) = (a \wedge b) \wedge c.$$

3. Absorption laws

(*i*) $a \leq \sup (a, a \wedge b) \Rightarrow a \leq a \vee (a \wedge b)$ ...(1)

$\inf (a, b) \leq a \Rightarrow a \wedge b \leq a$. Also $a \leq a$

$\therefore$ a is an upper bound of a and $a \wedge b$.

$\Rightarrow$ $\sup (a, a \wedge b) \leq a \Rightarrow a \vee (a \wedge b) \leq a$ ...(2)

By antisymmetry, (1) and (2) implies $a \vee (a \wedge b) = a$.

(*ii*) $\inf (a, a \vee b) \leq a \Rightarrow a \wedge (a \vee b) \leq a$ ...(1)

$a \leq \sup (a, b) \Rightarrow a \leq a \vee b$. Also $a \leq a$

$\therefore$ a is a lower bound of a and $a \vee b$.

$\Rightarrow \quad a \le \inf(a, a \vee b) \;\Rightarrow\; a \le a \wedge (a \vee b)$ ...(2)

By antisymmetry, (1) and (2) implies $a \wedge (a \vee b) = a$.

$\therefore$ $(L, \vee, \wedge)$ is a lattice.

$\therefore$ The partially ordered set $(L, \le)$ is a lattice as an algebraic structure.

Remark. If a partially ordered set $(L, \le)$ is a lattice then for the sake of convenience we write sup (a, b) and inf (a, b) as $a \vee b$ and $a \wedge b$ respectively. For example, we write sup $(a$, sup $(b, c))$ as $a \vee (b \vee c)$.

Theorem 2. If (L, ∨, ∧) is a lattice then prove that it is also a lattice as a partially ordered set.

Proof. $(L, \vee, \wedge)$ is a lattice.

$\therefore$ L satisfies commutative laws, associative laws and absorption laws w.r.t. binary operations '$\vee$' and '$\wedge$'.

We define a relation '$\le$' on L as follows :

For $a, b \in L$, we say $a \le b$ if $a \vee b = b$

We note that

$$a \vee b = b \;\Rightarrow\; a \wedge b = a \wedge (a \vee b) = a$$

and
$$a \wedge b = a \;\Rightarrow\; a \vee b = (a \wedge b) \vee b = b \vee (b \wedge a) = b$$

$\therefore \quad a \vee b = b$ iff $a \wedge b = a$

$\therefore$ $a \le b$ if $a \vee b = b$ (or equivalently $a \wedge b = a$)

Now we shall prove that $(L, \le)$ is a partially ordered set.

(*i*) Let $a \in L$. For any $b \in L$, we have

$$a \vee a = a \vee (a \wedge (a \vee b)) = a \quad \text{(By absorption law)}$$

$\therefore \quad a \le a \quad \therefore \le$ is reflexive.

(*ii*) Let $a \le b$ and $b \le a$ for $a, b \in L$.

$\Rightarrow \quad a \wedge b = a$ and $b \wedge a = b$

$\Rightarrow \quad a = b \quad (\because \; a \wedge b = b \wedge a)$

$\therefore$ $\le$ is antisymmetric.

(*iii*) Let $a \le b$ and $b \le c$ for $a, b, c \in L$.

$\Rightarrow \quad a \wedge b = a$ and $b \wedge c = b$

Now $\quad a \wedge c = (a \wedge b) \wedge c = a \wedge (b \wedge c) = a \wedge b = a$

$\therefore \quad a \le c \quad \therefore \le$ is transitive.

$\therefore$ $(L, \le)$ is a partially ordered set.

Let $a, b \in L$.

By absorption laws

$$a \wedge (a \vee b) = a \quad \text{and} \quad b \wedge (a \vee b) = b$$

$\Rightarrow \quad a \le a \vee b$ and $b \le a \vee b$

$\therefore$ $a \vee b$ is an upper bound of a and b.

Let u be any upper bound of a and b.

$\Rightarrow \quad a \le u$ and $b \le u$

$\Rightarrow \quad a \vee u = u$ and $b \vee u = u$

$\therefore \quad (a \vee b) \vee u = a \vee (b \vee u) = a \vee u = u$

$\Rightarrow \qquad a \vee b \leq u$

$\therefore \qquad \sup(a, b) = a \vee b.$

Similarly, we can show that $\inf(a, b) = a \wedge b$.

$\therefore$ For any $a, b \in L$, $\sup(a, b)$ and $\inf(a, b)$ exist.

$\therefore$ The partially ordered set $(L, \leq)$ is a lattice.

$\therefore$ The lattice $(L, \vee, \wedge)$ is also a lattice as a partially ordered set.

Remarks. In the above theorems we have proved that :

(*i*) if $(L, \leq)$ is a lattice as a partially ordered set then L is a lattice as an algebraic structure with two binary operations '$\vee$' and '$\wedge$' defined as $a \vee b = \sup(a, b)$ and $a \wedge b = \inf(a, b)$, where $a, b \in L$.

(*ii*) if $(L, \vee, \wedge)$ is a lattice as an algebraic structure then L is a lattice as a partially ordered set with the relation $\leq$ on L defined as $a \leq b$ if $a \vee b = b$ (or $a \wedge b = a$), where $a, b \in L$.

Also, in this partially ordered set :

$$\sup(a, b) = a \vee b \quad \text{and} \quad \inf(a, b) = a \wedge b, \text{ where } a, b \in L.$$

9.4. PRINCIPLE OF DUALITY

Let $(L, \vee, \wedge)$ be a lattice. The **dual** of any statement in this lattice is defined to be the statement that is obtained by interchanging the binary operations '$\vee$' and '$\wedge$' in the given statement.

For example, the dual of $(a \wedge b) \vee a = a \wedge (b \vee a)$ is $(a \vee b) \wedge a = a \vee (b \wedge a)$.

In the definition of the lattice $(L, \vee, \wedge)$, the dual of each axiom is also an axiom.

$\therefore$ The dual of any theorem in a lattice is also a theorem because the dual theorem can be proved by using the dual of each step of the proof of the original theorem.

Thus, we see that the dual of any theorem in a lattice is also a theorem. This is known as the **principle of duality.**

9.5. PROPERTIES OF LATTICES

Let $(L, \leq)$ be a lattice and $a, b \in L$. From the definitions of supremum ($\vee$) and infimum ($\wedge$), we have

(*i*) $a \leq a \vee b$ and $b \leq a \vee b$ because $a \vee b$ is an upper bound of a and b.

(*ii*) if $a \leq c$ and $b \leq c$ then $a \vee b \leq c$ because $a \vee b = \sup(a, b)$.

(*iii*) $a \wedge b \leq a$ and $a \wedge b \leq b$ because $a \wedge b$ is a lower bound of a and b.

(*iv*) if $c \leq a$ and $c \leq b$ then $c \leq a \wedge b$ because $a \wedge b = \inf(a, b)$.

Theorem 3. Let $(L, \leq)$ be a lattice and $a, b \in L$. Prove the following :

(i) $a \vee b = b$ if and only if $a \leq b$

(ii) $a \wedge b = a$ if and only if $a \leq b$

(iii) $a \vee b = b$ if and only if $a \wedge b = a$.

Proof. (*i*) Let $a \vee b = b$.

By definition, $a \vee b$ *i.e.*, $\sup(a, b)$ is also an upper bound of a and b, so we have $a \leq a \vee b$.

$\therefore \qquad a \leq b$ because $a \vee b = b$.

Conversely, let $a \leq b$.

$\Rightarrow \qquad a \leq b, b \leq b \Rightarrow b$ is an upper bound of a and b.

$\Rightarrow \qquad a \vee b \leq b. \qquad (\because a \vee b = \sup(a, b))$

By definition, $a \vee b$ *i.e.*, $\sup(a, b)$ is also an upper bound of a and b, so we have $b \leq a \vee b$.

$\therefore$ By antisymmetry, $a \vee b = b$.

$\therefore \qquad$ **$a \vee b = b$ iff $a \leq b$.**

(*ii*) Let $a \wedge b = a.$

By definition, $a \wedge b$ *i.e.*, inf (a, b) is also a lower bound of a and b, so we have $a \wedge b \leq b$.

$\therefore$ $a \leq b$ because $a \wedge b = a$

Conversely, let $a \leq b$.

$\Rightarrow$ $a \leq a, a \leq b \Rightarrow a$ is a lower bound of a and b

$\therefore$ $a \leq a \wedge b$ $(\because \ a \wedge b = \inf(a, b))$

By definition, $a \wedge b$ *i.e.*, inf (a, b) is also a lower bound of a and b, so we have $a \wedge b \leq a$.

$\therefore$ By antisymmetry, $a \wedge b = a$.

$\therefore$ $a \wedge b = a$ iff $a \leq b$.

(*iii*) By using parts (*i*) and (*ii*), we have

$a \vee b = b$ iff $a \leq b$ iff $a \wedge b = a$

$\therefore$ $\mathbf{a \vee b = b}$ **iff** $\mathbf{a \wedge b = a.}$

Remark. We have already shown that $\mathbf{a \vee b = b}$ **iff** $\mathbf{a \wedge b = a}$ when a and b are elements of the lattice L as an algebraic structure with binary operations $\vee$ and $\wedge$.

Theorem 4. Let (L, ≤) be a lattice and a, b, c ∈ L. Prove the following :

(i) a ≤ c and b ≤ c iff a ∨ b ≤ c

(ii) c ≤ a and c ≤ b iff c ≤ a ∧ b.

Proof. (*i*) $a \leq c$ and $b \leq c \Rightarrow c$ is an upper bound of a and b

$\therefore$ $\sup(a, b) \leq c$ *i.e.*, $a \vee b \leq c$.

Conversely, let $a \vee b \leq c$.

By definition, $a \leq a \vee b$ and $b \leq a \vee b$

$\therefore$ By transitivity, $a \leq c$ and $b \leq c$

$\therefore$ $\mathbf{a \leq c}$ **and** $\mathbf{b \leq c}$ **iff** $\mathbf{a \vee b \leq c.}$

(*ii*) $c \leq a$ and $c \leq b \Rightarrow c$ is a lower bound of a and b.

$\therefore$ $c \leq \inf(a, b)$ *i.e.*, $c \leq a \wedge b$

Conversely, let $c \leq a \wedge b$.

By definition, $a \wedge b \leq a$ and $a \wedge b \leq b$

$\therefore$ By transitivity, $c \leq a$ and $c \leq b$

$\therefore$ $\mathbf{c \leq a}$ **and** $\mathbf{c \leq b}$ **iff** $\mathbf{c \leq a \wedge b.}$

Theorem 5. Let (L, ≤) be a lattice and a, b, c, d ∈ L. Prove the following :

(i) a ∨ a = a and a ∧ a = a (Idempotent laws)

(ii) a ≤ b ⇒ a ∨ c ≤ b ∨ c and a ∧ c ≤ b ∧ c (Isotonicity laws)

(iii) a ≤ b and c ≤ d ⇒ a ∨ c ≤ b ∨ d and a ∧ c ≤ b ∧ d.

Proof. (*i*) $a \vee a = \sup(a, a) = a$ and $a \wedge a = \inf(a, a) = a$

$\therefore$ $\mathbf{a \vee a = a}$ **and** $\mathbf{a \wedge a = a.}$

(*ii*)
$$\begin{aligned}(a \vee c) \vee (b \vee c) &= [(a \vee c) \vee b] \vee c = [a \vee (c \vee b)] \vee c \\ &= [a \vee (b \vee c)] \vee c = (a \vee b) \vee (c \vee c) \\ &= b \vee c \qquad (\because \ a \leq b \Rightarrow a \vee b = b \text{ and } c \vee c = c)\end{aligned}$$

$\therefore$ $(a \vee c) \vee (b \vee c) = b \vee c$

$\Rightarrow$ $\mathbf{a \vee c \leq b \vee c.}$ $(\because \ x \vee y = y \Rightarrow x \leq y)$

$$\begin{aligned}(a \wedge c) \wedge (b \wedge c) &= [(a \wedge c) \wedge b] \wedge c = [a \wedge (c \wedge b)] \wedge c \\ &= [a \wedge (b \wedge c)] \wedge c = (a \wedge b) \wedge (c \wedge c) \\ &= a \wedge c \qquad (\because \ a \leq b \Rightarrow a \wedge b = a \text{ and } c \wedge c = c)\end{aligned}$$

$\therefore$ $(a \wedge c) \wedge (b \wedge c) = a \wedge c$

$\Rightarrow$ $\mathbf{a \wedge c \leq b \wedge c.}$ $(\because \ x \wedge y = x \Rightarrow x \leq y)$

(*iii*) $a \le b \Rightarrow a \vee c \le b \vee c$ (By part (*ii*))

$c \le d \Rightarrow c \vee b \le d \vee b \Rightarrow b \vee c \le b \vee d$

$\therefore$ By transitivity, $\mathbf{a \vee c \le b \vee d}$.

$a \le b \Rightarrow a \wedge c \le b \wedge c$ (By part (*ii*))

$c \le d \Rightarrow c \wedge b \le d \wedge b \Rightarrow b \wedge c \le b \wedge d$

$\therefore$ By transitivity, $\mathbf{a \wedge c \le b \wedge d}$.

Theorem 6. (Distributive inequalities). Let (L, ≤) be a lattice and a, b, c ∈ L. Prove that

(i) $\mathbf{a \vee (b \wedge c) \le (a \vee b) \wedge (a \vee c)}$

(ii) $\mathbf{a \wedge (b \vee c) \ge (a \wedge b) \vee (a \wedge c)}$.

Proof. (*i*) We have $a \le a \vee b$ and $a \le a \vee c$

$\Rightarrow$ $a \le \inf(a \vee b, a \vee c)$ *i.e.,* $a \le (a \vee b) \wedge (a \vee c)$ ...(1)

Also $b \wedge c \le b \le a \vee b$ and $b \wedge c \le c \le a \vee c$

$\Rightarrow$ $b \wedge c \le (a \vee b) \wedge (a \vee c)$ ...(2)

(1) and (2) $\Rightarrow$ $\mathbf{a \vee (b \wedge c) \le (a \vee b) \wedge (a \vee c)}$.

(*ii*) We have $a \wedge b \le a$ and $a \wedge c \le a$

$\Rightarrow$ $\sup(a \wedge b, a \wedge c) \le a$ *i.e.,* $(a \wedge b) \vee (a \wedge c) \le a$ ...(1)

Also $a \wedge b \le b \le b \vee c$ and $a \wedge c \le c \le b \vee c$

$\Rightarrow$ $(a \wedge b) \vee (a \wedge c) \le b \vee c$ ...(2)

(1) and (2) $\Rightarrow$ $(a \wedge b) \vee (a \wedge c) \le a \wedge (b \vee c)$ *i.e.,* $\mathbf{a \wedge (b \vee c) \ge (a \wedge b) \vee (a \wedge c)}$.

Remark. Later on we shall see that there exists a class of lattices for which we have

$$a \vee (b \wedge c) = (a \vee b) \wedge (a \vee c) \quad \text{and} \quad a \wedge (b \vee c) = (a \wedge b) \vee (a \wedge c).$$

Example 1. *If L is a lattice then show that* $a \vee (b \wedge c) = (a \vee b) \wedge (a \vee c)$, *when a, b, c ∈ L and any two of a, b, c are equal.*

Sol. We have to show that

$$a \vee (b \wedge c) = (a \vee b) \wedge (a \vee c) \quad ...(1)$$

Case I. a = b

L.H.S. of (1) $= a \vee (a \wedge c) = a$ (By absorption law)

R.H.S. of (1) $= (a \vee a) \wedge (a \vee c) = a \wedge (a \vee c) = a$ (By absorption law)

$\therefore$ (1) holds.

Case II. b = c

L.H.S. of (1) $= a \vee (b \wedge b) = a \vee b$

R.H.S. of (1) $= (a \vee b) \wedge (a \vee b) = a \vee b$

$\therefore$ (1) holds.

Case III. a = c

L.H.S. of (1) $= a \vee (b \wedge a) = a \vee (a \wedge b) = a$

R.H.S. of (1) $= (a \vee b) \wedge (a \vee a) = (a \vee b) \wedge a = a \wedge (a \vee b) = a$

$\therefore$ (1) holds.

EXERCISE 9.2

1. Let $(L, \leq)$ be a lattice. If for $a, b \in L$, sup (a, b) and inf (a, b) are denoted by $a \vee b$ and $a \wedge b$ respectively then prove the commutative laws :
   (i) $a \vee b = b \vee a$ (ii) $a \wedge b = b \wedge a$, where $a, b \in L$.
2. Let $(L, \leq)$ be a lattice. If for $a, b \in L$, sup (a, b) and inf (a, b) are denoted by $a \vee b$ and $a \wedge b$ respectively then prove the associative laws:
   (i) $a \vee (b \vee c) = (a \vee b) \vee c$ (ii) $a \vee (b \vee c) = (a \vee b) \vee c$, where $a, b, c \in L$.
3. Let $(L, \leq)$ be a lattice. If for $a, b \in L$, sup (a, b) and inf (a, b) are denoted by $a \vee b$ and $a \wedge b$ respectively then prove the absorption laws :
   (i) $a \vee (a \wedge b) = a$ (ii) $a \wedge (a \vee b) = a$, where $a, b \in L$.
4. If L is a lattice then show that $a \wedge (b \vee c) = (a \wedge b) \vee (a \wedge c)$, when $a, b, c \in L$ and any two of a, b, c are equal.
5. Let $(L, \leq)$ be a lattice and $a, b, c \in L$.
   If $a \leq b \leq c$ show that :
   (i) $a \vee b = b \wedge c$. (ii) $(a \wedge b) \vee (b \wedge c) = (a \vee b) \wedge (a \vee c)$.

9.6. SUBLATTICE

Let $(L, \leq)$ be a lattice. A non-empty subset M of L is called a **sublattice** of L if sup $(a, b) \in M$ and inf $(a, b) \in M$ whenever $a, b \in M$.

If M is a sublattice of L then $(M, \leq)$ is also a partially ordered set because all the elements of M are elements of L and $(L, \leq)$ is a partially ordered set.

∴ The sublattice M of the lattice $(L, \leq)$ is also a lattice under the partial order relation $\leq$.

Equivalently, we can also define a sublattice as follows : If L is a lattice as an algebraic structure with binary operations $\vee$ and $\wedge$ then a non-empty set M of L is called a **sublattice** of L if $a \vee b \in M$ and $a \wedge b \in M$ whenever $a, b \in M$.

If M is a sublattice of L then commutative laws, associative laws and absorption laws are satisfied in M because all the elements of M are elements of L and L is a lattice.

∴ The sublattice M of the lattice $(L, \vee, \wedge)$ is also a lattice with binary operations $\vee$ and $\wedge$.

Illustration. The set **N** of positive integers is a lattice under the relation of divisibility. For any positive integer m, the set D_m of all positive divisors of m is a sublattice of **N** because sup (a, b) and inf (a, b) are in D_m whenever $a, b, \in D_m$.

Example 1. *Let $L = \{a, b, c, d, e, f, g\}$, be a lattice with the Hasse diagram shown in the figure. Which of the following subsets of L are sublattices of L ?*

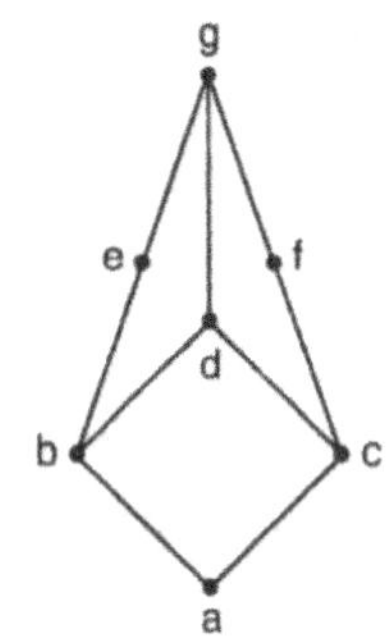

(i) $A = \{a, b, c\}$
(ii) $B = \{a, b, c, d\}$
(iii) $C = \{c, d, f, g\}$
(iv) $D = \{b, c, d, e, f, g\}$.

Sol. We have $L = \{a, b, c, d, e, f, g\}$

(i) $A = \{a, b, c\}$

A is a partially ordered set with Hasse diagram shown in the figure.

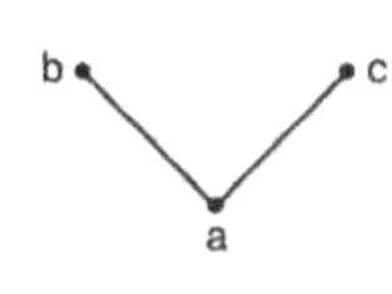

Here sup $(b, c) \notin A$.

∴ A is not a sublattice of L.

(*ii*) $B = \{a, b, c, d\}$

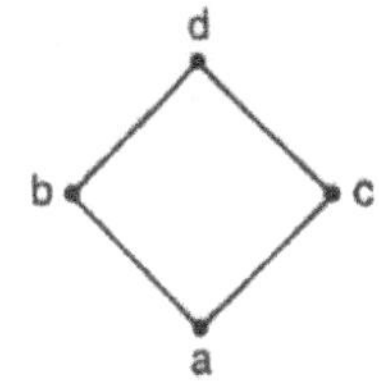

B is a partially ordered set with Hasse diagram shown in the figure.

Here sup (x, y), inf $(x, y) \in B$ for every $x, y \in B$.

$\therefore$ B is a sublattice of L.

(*iii*) $C = \{c, d, f, g\}$

C is a partially ordered set with Hasse diagram shown in the figure.

Here sup (x, y), inf $(x, y) \in C$ for every $x, y \in C$.

$\therefore$ C is a sublattice of L.

(*iv*) $D = \{b. c, d, e, f, g\}$.

D is a partially ordered set with Hasse diagram shown in the figure.

Here inf $(b, c) \notin D$.

$\therefore$ D is not a sublattice of L.

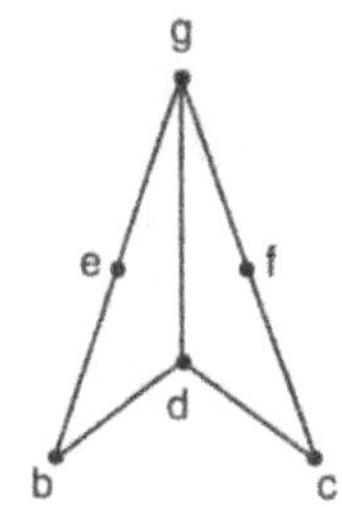

Example 2. *Consider the lattice* $L = \{1, 2, 3, 4, 5\}$ *as shown in the figure. Find all the sublattices of L with at least three elements.*

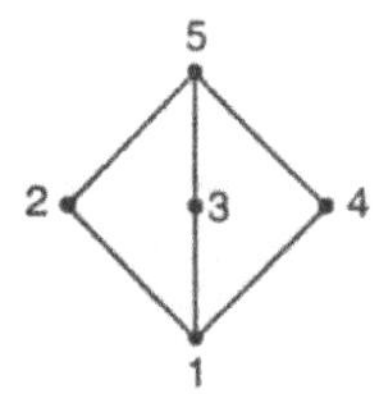

Sol. We have $L = \{1, 2, 3, 4, 5\}$.

A subset of L containing at least three elements is a sublattice of L provided it contains the supremum and infimum of every pair of elements in that subset. The following are sublattices of L containing at least three elements.

(*i*) $\{1, 2, 5\}$ (*ii*) $\{1, 3, 5\}$ (*iii*) $\{1, 4, 5\}$

(*iv*) $\{1, 2, 3, 5\}$ (*v*) $\{1, 3, 4, 5\}$ (*vi*) $\{1, 2, 4, 5\}$

(*vii*) $\{1, 2, 3, 4, 5\}$.

Example 3. *Consider the lattice L as shown in the figure. Determine whether or not each of the following is sublattice of L ?*

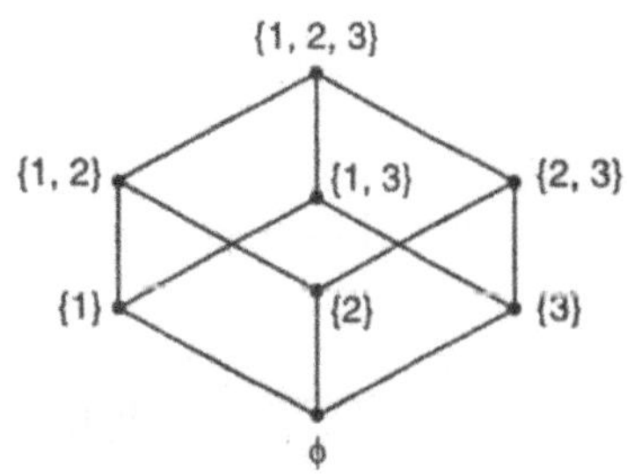

$A = \{\phi, \{1, 2\}, \{2, 3\}, \{1, 2, 3\}\}$

$B = \{\phi, \{1\}, \{1, 2\}, \{1, 2, 3\}\}$

$C = \{\phi, \{3\}, \{1, 3\}, \{1, 2, 3\}\}$

$D = \{\{1\}, \{3\}, \{1, 3\}, \{1, 2, 3\}\}$

$E = \{\phi, \{3\}, \{1, 2\}, \{1, 2, 3\}\}$.

Sol. A is not a sublattice of L because $\{1, 2\} \wedge \{2, 3\} = \{2\}$ and $\{2\} \notin A$.

B is a sublattice of L because supremum and infimum of every pair of elements in B are in B.

C is a sublattice of L because supremum and infimum of every pair of elements in C are in C.

D is not a sublattice of L because $\{1\} \wedge \{3\} = \phi \notin D$.

E is a sublattice of L because supremum and infimum of every pair of elements in E are in E.

Example 4. *Show that the union of two sublattices may not be a sublattice.*

Sol. Let D_{12} be the set of all positive divisors of 12.

$\therefore \quad D_{12} = \{1, 2, 3, 4, 6, 12\}$

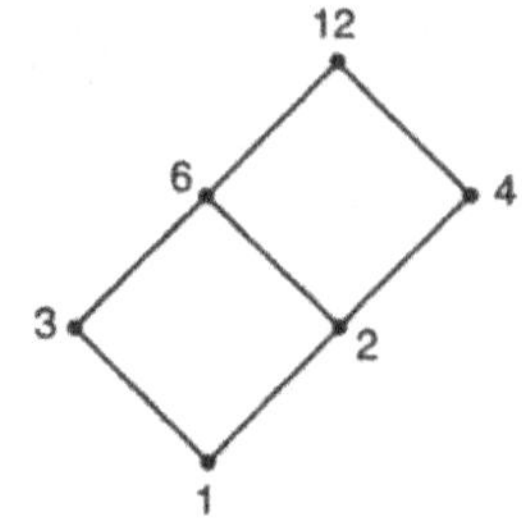

D_{12} is a lattice with Hasse diagram shown in the figure.

Let $A = \{1, 2\}$ and $B = \{1, 3\}$ be subsets of D_{12}.

Here sup (a, b) and inf (a, b) are in A for any $a, b \in A$.

$\therefore$ A is a sublattice of D_{12}.

Similarly, B is a sublattice of D_{12}.

$A \cup B = \{1, 2, 3\}$ and this is a subset of D_{12}.

$\sup(2, 3) = 6 \notin A \cup B$

$\therefore \quad 2, 3 \in A \cup B$ and $\sup(2, 3) = 6 \notin A \cup B$

$\therefore$ $A \cup B$ is not a sublattice of D_{12}.

$\therefore$ Union of two sublattices may not be a sublattice.

Example 5. *Show that a lattice L is a chain if and only if every non-empty subset of L is a sublattice of L.*

Sol. Let the lattice L be a chain.

Let A be a non-empty subset of L. let $x, y \in A$

$$\Rightarrow \quad x, y \in L \Rightarrow \text{Either } x \le y \text{ or } y \le x$$
$$x \le y \Rightarrow \sup(x, y) = y \in A, \inf(x, y) = x \in A$$
$$y \le x \Rightarrow \sup(x, y) = x \in A, \inf(x, y) = y \in A$$

$\therefore$ $\sup(x, y), \inf(x, y) \in A$ for any $x, y \in A$.

$\therefore$ A is a sublattice of L.

Conversely, let every non-empty subset of L is sublattice of L.

Let $x, y \in L$.

$\therefore$ The subset $\{x, y\}$ of L is a sublattice of L.

$\therefore$ $\sup(x, y), \inf(x, y) \in \{x, y\}$

$$\Rightarrow \quad \sup(x, y) = x \text{ or } y$$
$$\sup(x, y) = x \Rightarrow y \le x \quad \text{and} \quad \sup(x, y) = y \Rightarrow x \le y$$

$\therefore$ For $x, y \in L$, either $x \le y$ or $y \le x$

$\therefore$ The lattice L is a chain.

Example 6. *Let L be a lattice and let a and b be elements of L such that $a \le b$. The* **interval** *$[a, b]$ is defined as the set of all $x \in L$ such that $a \le x \le b$. Prove that $[a, b]$ is a sublattice of L.*

Sol. Let $\alpha, \beta \in [a, b]$

$$\therefore \quad \alpha, \beta \in L \quad \text{and} \quad a \le \alpha \le b, a \le \beta \le b$$
$$\Rightarrow \quad \alpha \vee \beta, \alpha \wedge \beta \in L \qquad (\because L \text{ is a lattice})$$
$$\alpha \le b, \beta \le b \Rightarrow b \text{ is an upper bound of } \alpha \text{ and } \beta$$
$$\Rightarrow \alpha \vee \beta \le b \qquad \text{...(1)}$$

Also $\quad \alpha \le \alpha \vee \beta, \beta \le \alpha \vee \beta$

$$\Rightarrow \quad a \le \alpha \le \alpha \vee \beta \Rightarrow a \le \alpha \vee \beta \qquad \text{...(2)}$$

Combining (1) and (2), we get $a \le \alpha \vee \beta \le b$.

$\Rightarrow \qquad \alpha \vee \beta \in [a, b] \qquad (\because \ \alpha \vee \beta \in L)$

Similarly, $\alpha \wedge \beta \in [a, b]$.

$\therefore$ For all $\alpha, \beta \in [a, b]$, $\alpha \vee \beta, \alpha \wedge \beta \in [a, b]$.

$\therefore$ $[a, b]$ is a sublattice of L.

9.7. ISOMORPHIC LATTICES

Two lattices $(L, \leq)$ and $(L', \leq')$ are called **isomorphic lattices** if there exists a one-one onto mapping $f : L \to L'$ such that $a \leq b \iff f(a) \leq' f(b)$ for any $a, b \in L$.

Equivalently, let L and L' be two lattices as algebraic structures. These lattices are called **isomorphic lattices** if there exists a one-one onto mapping $f : L \to L'$ such that

$$f(a \vee b) = f(a) \vee f(a) \quad \text{and} \quad f(a \wedge b) = f(a) \wedge f(b) \text{ for any } a, b \in L.$$

Example 7. *If $f : (L, \leq) \to (L', \leq')$ be a lattice isomorphism then show that*

$$f(a \vee b) = f(a) \vee f(b) \text{ for } a, b \in L.$$

Sol. Let $a, b \in L$ and sup $(a, b) = s$, sup $(f(a), f(b)) = f(k)$.

$\Rightarrow \quad a, b \leq s \Rightarrow f(a), f(b) \leq f(s) \Rightarrow f(s)$ is an upper bound of $f(a)$ and $f(b)$

$\Rightarrow \quad f(k) \leq f(s)$ ...(1)

Also, $f(a), f(b) \leq f(k) \Rightarrow a, b \leq k \Rightarrow k$ is an upper bound of a and b

$\Rightarrow \quad s \leq k \Rightarrow f(s) \leq f(k)$ ...(2)

$\therefore$ By antisymmetry, (1) and (2) implies

$$f(s) = f(k)$$

$\Rightarrow \quad f(a \vee b) = f(a) \vee f(b)$.

Example 8. *Determine whether the lattices shown in the figure are isomorphic or not.*

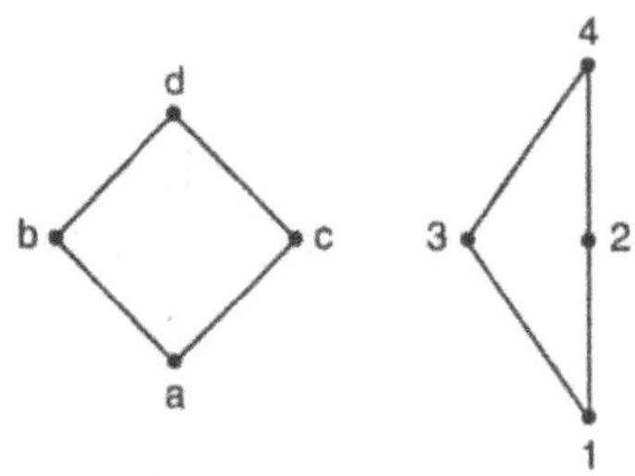

Sol. Let $L_1 = \{a, b, c, d\}$

and $L_2 = \{1, 2, 3, 4\}$.

Define $f : L_1 \to L_2$ by $f(a) = 1, f(b) = 3, f(c) = 2, f(d) = 4$.

The mapping f is one-one and onto.

Also $x \leq y$ in $L_1 \iff f(x) \leq f(y)$ in L_2

For example, $a \leq b$ in $L_1 \iff 1 \leq 3$ in L_2

$\therefore$ The lattices L_1 and L_2 are isomorphic.

EXERCISE 9.3

1. If L is a lattice then show that every singleton subset of L is a sublattice of L.
2. If L is a lattice and A and B be sublattices of L such that $A \cap B \neq \phi$, then show that $A \cap B$ is also a sublattice of L.
3. Let $L = \{a, b, c, d\}$ be a lattice with the Hasse diagram shown in the figure. Which of the following subsets of L are sublattices of L ?

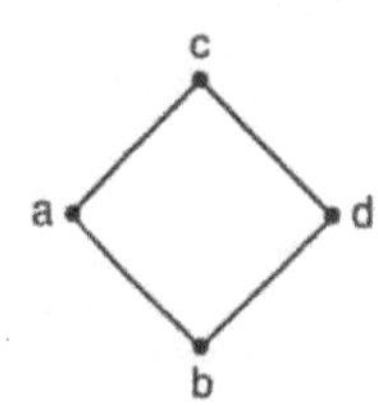

(*i*) $\{a, b\}$ (*ii*) $\{a, d\}$ (*iii*) $\{a, b, c\}$

(*iv*) $\{a, b, d\}$ (*v*) $\{a, c, d\}$.

4. Let $L = P(S)$ be the lattice of all subsets of a non-empty set S under the relation of containment. Let T be a non-empty subset of S. Show that $P(T)$ is a sublattice of L.
5. If $f : (L, \leq) \to (L', \leq')$ be a lattice isomorphism then show that
$$f(a \wedge b) = f(a) \wedge f(b) \text{ for } a, b \in L.$$
6. Let L and L' be lattices as algebraic structures and $f : L \to L'$ be a lattice isomorphism. If $\leq$ and $\leq'$ be the partial orders in L and L' respectively as partially ordered sets then show that
$$a \leq b \text{ iff } f(a) \leq' f(b) \text{ for any } a, b \in L.$$
7. Determine whether the lattices shown in the figures are isomorphic or not

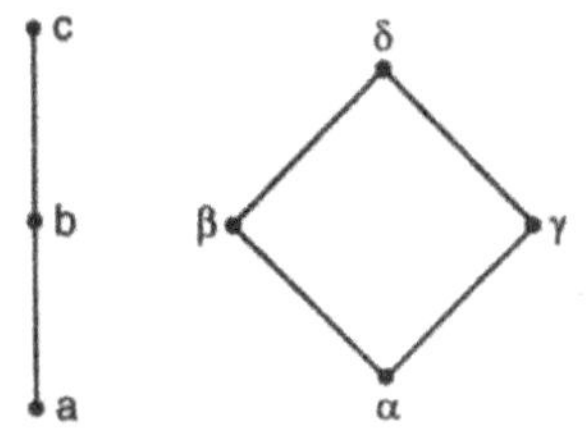

Answers

3. (*i*), (*iii*) 7. No.

Hint

6. $a \leq b$ in L $\Rightarrow$ $a \vee b = b$ $\Rightarrow$ $f(a \vee b) = f(b)$ $\Rightarrow$ $f(a) \vee f(b) = f(b)$ $\Rightarrow$ $f(a) \leq f(b)$ in L'.
Also, $f(a) \leq f(b)$ in L' $\Rightarrow$ $f(a) \vee f(b) = f(b)$ $\Rightarrow$ $f(a \vee b) = f(b)$ $\Rightarrow$ $a \vee b = b$ $\Rightarrow$ $a \leq b$.
($\because$ f is 1 – 1)

9.8. DISTRIBUTIVE LATTICE

A lattice $(L, \leq)$ is called a **distributive lattice** if for any a, b, c in L, we have

(*i*) $a \vee (b \wedge c) = (a \vee b) \wedge (a \vee c)$

(*ii*) $a \wedge (b \vee c) = (a \wedge b) \vee (a \wedge c)$.

A lattice which is not distributive is called a **nondistributive lattice.**

In order to show a lattice to be distributive, we need show only one of the distributive laws because the other is bound to hold by the *principle of duality.*

Remark. A sublattice of a distributive lattice is also distributive.

Example 1. *Let X be a non-empty set and P(X), the power set of X. Show that the lattice P(X) under the relation of set inclusion is distributive.* (KU 2006 C)

Sol. For $A, B \in P(X)$, we have
$$A \vee B = \sup(A, B) = A \cup B$$
and
$$A \wedge B = \inf(A, B) = A \cap B.$$
Let $A, B, C \in P(X)$.
$$\therefore \quad A \vee (B \wedge C) = A \cup (B \cap C) = (A \cup B) \cap (A \cup C) = (A \vee B) \wedge (A \vee C)$$
and
$$A \wedge (B \vee C) = A \cap (B \cup C) = (A \cap B) \cup (A \cap C) = (A \wedge B) \vee (A \vee C)$$
$\therefore$ The lattice $P(X)$ is distributive.

Example 2. *Give an example of a lattice which is not distributive.* (KU 2006, 2007)

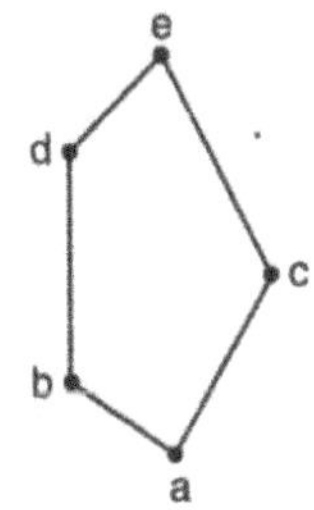

Sol. Let $L = \{a, b, c, d, e\}$ be a lattice with the Hasse diagram shown in the figure.

In this lattice, $b \vee (c \wedge d) = b \vee a = b$ and $(b \vee c) \wedge (b \vee d) = e \wedge d = d$

$\therefore \quad b \vee (c \wedge d) \neq (b \vee c) \wedge (b \vee d)$

$\therefore$ The lattice L is not distributive.

Example 3. *Let L_1 and L_2 be two lattices and $L_1 \times L_2$ be the lattice under the product partial order. Show that the lattice $L_1 \times L_2$ is distributive if and only if the lattices L_1 and L_2 are distributive.*

Sol. For $(a, b), (c, d) \in L_1 \times L_2$, $(a, b) \leq (c, d)$ iff $a \leq c$ and $b \leq d$

Also $\quad (a, b) \vee (c, d) = (a \vee c, b \vee d)$ and $(a, b) \wedge (c, d) = (a \wedge c, b \wedge d)$

Let the lattices L_1 and L_2 be distributive.

Let $\quad (a, b), (c, d), (e, f) \in L_1 \times L_2$.

$$\begin{aligned}\therefore \quad (a, b) \vee [(c, d) \wedge (e, f)] &= (a, b) \vee (c \wedge e, d \wedge f)\\ &= (a \vee (c \wedge e), b \vee (d \wedge f))\\ &= ((a \vee c) \wedge (a \vee e), (b \vee d) \wedge (b \vee f)) \quad (\because L_1 \text{ and } L_2 \text{ are distributive})\\ &= (a \vee c, b \vee d) \wedge (a \vee e, b \vee f)\\ &= [(a, b) \vee (c, d)] \wedge [(a, b) \vee (e, f)]\end{aligned}$$

$\therefore \quad (a, b) \vee [(c, d) \wedge (e, f)] = [(a, b) \vee (c, d)] \wedge [(a, b) \vee (e, f)]$

By the *principle of duality*, we have

$$(a, b) \wedge [(c, d) \vee (e, f)] = [(a, b) \wedge (c, d)] \vee [(a, b) \wedge (e, f)]$$

$\therefore$ The lattice $L_1 \times L_2$ is distributive.

Conversely, let the lattice $L_1 \times L_2$ be distributive.

Let $a, b, c \in L_1$ and $d, e, f \in L_2$

$\therefore \quad (a, d), (b, e), (c, f) \in L_1 \times L_2$

Since $L_1 \times L_2$ is distributive, we have

$$(a, d) \vee [(b, e) \wedge (c, f)] = [(a, d) \vee (b, e)] \wedge [(a, d) \vee (c, f)]$$

$\Rightarrow \quad (a, d) \vee (b \wedge c, e \wedge f) = (a \vee b, d \vee e) \wedge (a \vee c, d \vee f)$

$\Rightarrow \quad (a \vee (b \wedge c), d \vee (e \wedge f)) = ((a \vee b) \wedge (a \vee c), (d \vee e) \wedge (d \vee f))$

$\Rightarrow \quad a \vee (b \wedge c) = (a \vee b) \wedge (a \vee c)$ and $d \vee (e \wedge f) = (d \vee e) \wedge (d \vee f)$

By the *principle of duality*, we have

$$a \wedge (b \vee c) = (a \wedge b) \vee (a \wedge c) \quad \text{and} \quad d \wedge (e \vee f) = (d \wedge e) \vee (d \wedge f).$$

$\therefore$ The lattices L_1 and L_2 are distributive.

Example 4. *Let L be a distributive lattice. Show that if there exists an a with $a \vee x = a \vee y$ and $a \wedge x = a \wedge y$, then $x = y$.*

Sol.
$$\begin{aligned} x &= x \vee (x \wedge a) = x \vee (a \wedge x) = x \vee (a \wedge y) = (x \vee a) \wedge (x \vee y)\\ &= (a \vee x) \wedge (x \vee y) = (a \vee y) \wedge (x \vee y) = (y \vee a) \wedge (y \vee x)\\ &= y \vee (a \wedge x) = y \vee (a \wedge y) = y \vee (y \wedge a) = y\end{aligned}$$

$\therefore \quad x = y.$

Example 5. *Show that a totally ordered set i.e., a chain is always a distributive lattice.* (KU 2007)

Sol. Let $(T, \leq)$ be a totally ordered set. Let $a, b \in T$

$\therefore$ Either $a \le b$ or $b \le a$

$$a \le b \Rightarrow \sup(a, b) = b,\ \inf(a, b) = a$$
$$b \le a \Rightarrow \sup(a, b) = a,\ \inf(a, b) = b$$

$\therefore$ In either case, sup (a, b) and inf (a, b) are in T.

$\therefore$ T is a lattice.

Let $a, b, c \in T$.

Case I. a ≤ b ≤ c

$$a \vee (b \wedge c) = a \vee b = b \quad \text{and} \quad (a \vee b) \wedge (a \vee c) = b \wedge c = b$$

$$\therefore \quad a \vee (b \wedge c) = (a \vee b) \wedge (a \vee c).$$

Case II. a ≤ c ≤ b

$$a \vee (b \wedge c) = a \vee c = c \quad \text{and} \quad (a \vee b) \wedge (a \vee c) = b \wedge c = c$$

$$\therefore \quad a \vee (b \wedge c) = (a \vee b) \wedge (a \vee c).$$

Similarly, $a \vee (b \wedge c) = (a \vee b) \wedge (a \vee c)$ holds in the cases :

$$b \le a \le c,\ b \le c \le a,\quad c \le a \le b \text{ and } \quad c \le b \le a.$$

$$\therefore \quad a \vee (b \wedge c) = (a \vee b) \wedge (a \vee c) \text{ for } a, b, c \in T \quad ...(1)$$

By the *principle of duality*, (1) implies

$$a \wedge (b \vee c) = (a \wedge b) \vee (a \wedge c) \text{ for } a, b, c \in T.$$

$\therefore$ The lattice T is distributive.

9.9. MODULAR LATTICE

A lattice $(L, \le)$ is said to be a **modular lattice** if for any $a, b, c \in L$ and $a \le c$, we have

$$a \vee (b \wedge c) = (a \vee b) \wedge c.$$

For example, the lattice shown in the figure is a modular lattice.

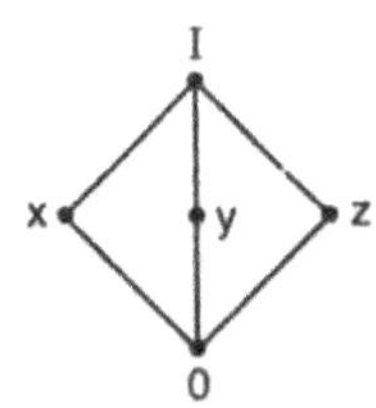

Let $\quad a = 0, b = y, c = z. \quad \therefore \quad a \le c$

Here $\quad a \vee (b \wedge c) = 0 \vee (y \wedge z) = 0 \vee 0 = 0$

and $\quad (a \vee b) \wedge c = (0 \vee y) \wedge z = y \wedge z = 0$

$$\therefore \quad a \vee (b \wedge c) = (a \vee b) \wedge c.$$

Remark. A sublattice of a modular lattice is also modular.

Example 6. *Show that a distributive lattice is always a modular lattice.*

Sol. Let $(L, \le)$ be a distributive lattice. Let $a, b, c \in L$ and $a \le c$.

$$\therefore \quad a \vee (b \wedge c) = (a \vee b) \wedge (a \vee c)$$
$$= (a \vee b) \wedge c \qquad (\because \ a \le c)$$

$\therefore$ The lattice L is modular.

Example 7. *Show that a totally ordered set is always a modular lattice.*

Sol. Let $(T, \le)$ be a totally ordered set. Let $a, b \in T$

$\therefore$ Either $a \le b$ or $b \le a$

$$a \le b \Rightarrow \sup(a, b) = b,\quad \inf(a, b) = a$$
$$b \le a \Rightarrow \sup(a, b) = a,\quad \inf(a, b) = b$$

$\therefore$ In either case, sup(a, b) and inf (a, b) are in T.

$\therefore$ T is a lattice.

Let $a, b, c \in T$ and $a \le c$.

$\therefore$ The elements a, b, c are comparable in pairs. We have four possibilities :

Case I. $a \le b, b \le c$

Here $a \vee (b \wedge c) = a \vee b = b$ and $(a \vee b) \wedge c = b \wedge c = b$

$\therefore$ $a \vee (b \wedge c) = (a \vee b) \wedge c$

Case II. $a \le b, c \le b$

Here $a \vee (b \wedge c) = a \vee c = c$ and $(a \vee b) \wedge c = b \wedge c = c$ (Using $a \le c$)

$\therefore$ $a \vee (b \wedge c) = (a \vee b) \wedge c$

Case III. $b \le a, b \le c$

Here $a \vee (b \wedge c) = a \vee b = a$ and $(a \vee b) \wedge c = a \wedge c = a$ (Using $a \le c$)

$\therefore$ $a \vee (b \wedge c) = (a \vee b) \wedge c$

Case IV. $b \le a, c \le b$

Here $c \le b, b \le a \Rightarrow c \le a$ Also $a \le c$. $\therefore$ $a = c$

$\therefore$ $a \vee (b \wedge c) = a \vee (b \wedge a) = a \vee b = a$

and $(a \vee b) \wedge c = a \wedge a = a$

$\therefore$ $a \vee (b \wedge c) = (a \vee b) \wedge c$

$\therefore$ For $a, b, c \in T$ with $a \le c$, we have $a \vee (b \wedge c) = (a \vee b) \wedge c$.

$\therefore$ The lattice T is modular.

Example 8. *Let L_1 and L_2 be two lattices and $L_1 \times L_2$ be the lattice under the product partial order. Show that the lattice $L_1 \times L_2$ is modular if and only if lattices L_1 and L_2 are modular.*

Sol. For $(a, b), (c, d) \in L_1 \times L_2$, $(a, b) \le (c, d)$ iff $a \le c$ and $b \le d$

Also, $(a, b) \vee (c, d) = (a \vee c, b \vee d)$ and $(a, b) \wedge (c, d) = (a \wedge c, b \wedge d)$

Let the lattices L_1 and L_2 be modular.

Let $(a, b), (c, d), (e, f) \in L_1 \times L_2$ and $(a, b) \le (e, f)$.

$\Rightarrow$ $a, c, e \in L_1, a \le e$ and $b, d, f \in L_2, b \le f$

Since L_1 and L_2 are modular lattices, we have

$$a \vee (c \wedge e) = (a \vee c) \wedge e \text{ and } b \vee (d \wedge f) = (b \vee d) \wedge f.$$

Now
$$\begin{aligned}(a, b) \vee [(c, d) \wedge (e, f)] &= (a, b) \vee (c \wedge e, d \wedge f)\\ &= (a \vee (c \wedge e), b \vee (d \wedge f)) = ((a \vee c) \wedge e, (b \vee d) \wedge f)\\ &= (a \vee c, b \vee d) \wedge (e, f) = [(a, b) \vee (c, d)] \wedge (e, f)\end{aligned}$$

$\therefore$ $(a, b) \vee [(c, d) \wedge (e, f)] = [(a, b) \vee (c, d)] \wedge (e, f)$

$\therefore$ The lattice $L_1 \times L_2$ is modular.

Conversely, let the lattice $L_1 \times L_2$ be modular.

Let $a, b, c \in L_1$, $a \le c$ and $d, e, f \in L_2, d \le f$

$\therefore$ $(a, d), (b, e), (c, f) \in L_1 \times L_2$ and $(a, d) \le (c, f)$

Since $L_1 \times L_2$ is modular, we have

$$(a, d) \vee [(b, e) \wedge (c, f)] = [(a, d) \vee (b, e)] \wedge (c, f)$$

$\Rightarrow$ $(a, d) \vee (b \wedge c, e \wedge f) = (a \vee b, d \vee e) \wedge (c, f)$

$\Rightarrow$ $(a \vee (b \wedge c), d \vee (e \wedge f)) = ((a \vee b) \wedge c, (d \vee e) \wedge f)$

$\Rightarrow$ $a \vee (b \wedge c) = (a \vee b) \wedge c$ and $d \vee (e \wedge f) = (d \vee e) \wedge f$

$\therefore$ The lattices L_1 and L_2 are modular.

9.10. BOUNDED LATTICE

A lattice $(L, \leq)$ is called a **bounded lattice** if it has the first element and the last element.

The first element and the last element of a bounded lattice are denoted by 0 and I respectively.

$\therefore$ If $(L, \leq)$ is a bounded lattice then $0 \leq x$ and $x \leq I$ for every $x \in L$.

The first element and the last element of a lattice are also called its **least element** and **greatest element** respectively.

If 0 and I are the first element and the last element of a bounded lattice then 0 and I are also a lower bound and an upper bound of the lattice respectively.

If $(L, \leq)$ is a bounded lattice then for any $x \in L$, we have

(*i*) $0 \leq x \leq I$

(*ii*) $x \vee 0 = \sup(x, 0) = x$ $(\because\ 0 \leq x)$

(*iii*) $x \wedge 0 = \inf(x, 0) = 0$ $(\because\ 0 \leq x)$

(*iv*) $x \vee I = \sup(x, I) = I$ $(\because\ x \leq I)$

(*v*) $x \wedge I = \inf(x, I) = x$ $(\because\ x \in I)$

$\therefore$ For any $x \in L$, we have $0 \leq x \leq I$ and $x \vee 0 = x$, $x \wedge 0 = 0$, $x \vee I = I$, $x \wedge I = x$.

Illustrations :

1. Let X be any non-empty set and $P(X)$ be its power set. $P(X)$ is a lattice under the relation of set inclusion.

We have $\phi, X \in P(X)$ and $\phi \subseteq A$, $A \subseteq X$ for any $A \in P(X)$.

$\therefore$ ϕ and X are respectively the least element and the greatest element of the lattice $(P(X), \subseteq)$.

$\therefore$ The lattice $P(X)$ is bounded.

2. **N** is a lattice under the relation 'usual $\leq$'.

Since $1 \leq n$ for any $n \in$ **N**, the element 1 is the first element.

The lattice **N** has no last element.

Example 9. *Show that if a bounded lattice has two or more elements then $0 \neq I$.*

Sol. Let L be a lattice having two or more elements. If possible, let $0 = I$.

There exist at least one x in L such that $x \neq 0$.

By definition, $0 \leq x$ and $x \leq I$

$\Rightarrow$ $0 \leq x$ and $x \leq 0$

$\therefore$ By antisymmetry, $x = 0$, which is impossible.

$\therefore$ Our supposition is wrong.

$\therefore$ $0 \neq I$.

Theorem. Prove that a finite lattice is always bounded.

Proof. Let $L = \{a_1, a_2, ..., a_n\}$ be a finite lattice. Let the partial order of L be denoted by $\leq$.

The elements $a_1 \vee a_2 \vee ..... \vee a_n$ and $a_1 \wedge a_2 \wedge ..... \wedge a_n$ are in L and for any $a_i \in L$, we have $a_i \leq a_1 \vee a_2 \vee ..... \vee a_n$ and $a_1 \wedge a_2 \wedge ..... \wedge a_n \leq a_i$.

$\therefore$ $a_1 \vee a_2 \vee .... \vee a_n$ and $a_1 \wedge a_2 \wedge ...... \wedge a_n$ are respectively upper bound and lower bound of the lattice L.

$\therefore$ The lattice L is bounded.

9.11. JOIN-IRREDUCIBLE ELEMENT

Let L be a lattice. An element a in L is said to be a **join-irreducible element** if $a = x \vee y$ for distinct elements in L then either $a = x$ or $a = y$.

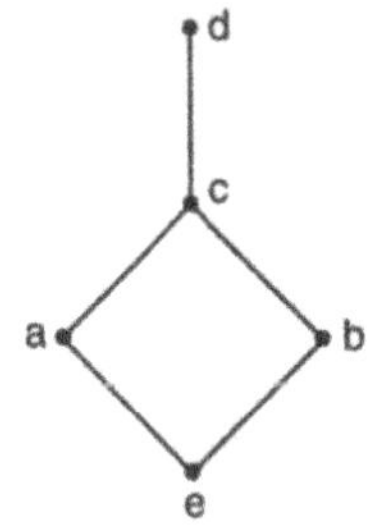

For example, in the lattice shown in the figure, the element a is join-irreducible because we cannot write a as $x \vee y$, where $a \neq x$, $a \neq y$. c is not join-irreducible because $c = a \vee b$ and $c \neq a$, $c \neq b$.

Remark. If 0 is the lower bound of a lattice L then it is trivially a join-irreducible element of L.

Theorem I. Let L be a lattice with lower bound 0. Prove that a($\neq$ 0) $\in$ L is a join-irreducible element if and only if a has a unique immediate predecessor.

Proof. Let $a(\neq 0)$ be join-irreducible.

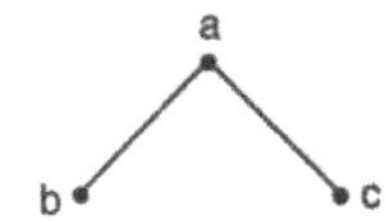

Since $a \neq 0$, we have $0 \leq a$ and thus a has at least one immediate predecessor.

If possible, let b and c be two immediate predecessors of a.

$\therefore \quad a = \sup(b, c) = b \vee c$ and $a \neq b, a \neq c$.

$\therefore$ a is not join-irreducible.

$\therefore$ Our supposition is wrong.

$\therefore$ a has a unique immediate predecessor.

Conversely, let $a(\neq 0)$ has a unique immediate predecessor. If possible, let a be not join-irreducible.

$\therefore$ There exist distinct elements x, y in L such that $a = x \vee y$ and $a \neq x, a \neq y$.

$$a = x \vee y \Rightarrow x \leq a, y \leq a \Rightarrow x < a, y < a$$

Let x', y' be such that $x \leq x' << a$ and $y \leq y' << a$. (If $x << a$, we take $x' = x$)

$\therefore$ a has two immediate predecessors x', y'. This is impossible.

$\therefore$ a must be a join-irreducible element.

Theorem II. Let L be a finite distributive lattice. Prove that every element in L can be written uniquely (except for order) as the join of irredundant join-irreducible elements* of L.

Proof. Let a be any element of L. If a is join-irreducible, we have nothing to prove. So, let a be not join-irreducible.

$\therefore$ By definition, there exist elements b_1, b_2 in L such that $a = b_1 \vee b_2$ and $a \neq b_1$, $a \neq b_2$.

If both b_1 and b_2 are join-irreducible, we stop, otherwise we can write b_1 and b_2 as the join of other elements of L. Since the lattice L is finite, this process is to stop after a finite number of steps.

$\therefore$ We can express a as the join of join-irreducible elements.

Let $$a = c_1 \vee c_2 \vee .... \vee c_k,$$

where $c_1, c_2, ..., c_k$ are join-irreducible elements. In this expression if any c_i precedes c_j then we have by commutativity, $c_i \vee c_j = c_j$.

$\therefore$ c_i can be deleted in the above expression of a.

Thus, we can express a as the join of join-irreducible elements where no c precedes any other c.

Let $a = d_1 \vee d_2 \vee ...... \vee d_m$ be an expression of a as the join of join-irreducible elements and no d precedes any other d. Thus, the elements $d_1, d_2, ..., d_m$ in the expression of a are irredundant join-irreducible elements.

We shall show that the expression

$$a = d_1 \vee d_2 \vee ...... \vee d_m \qquad ...(1)$$

is unique, except for the order.

Let $$a = e_1 \vee e_2 \vee .... \vee e_n \qquad ...(2)$$

be any other representation of a as the join of irredundant join-irreducible elements.

---

*Join-irreducible elements are called **irredundant join-irreducible elements** if no join-irreducible element precedes any other join-irreducible element.

For any i, $1 \le i \le m$, we have $d_i \le d_1 \vee d_2 \vee ..... \vee d_m$

$\Rightarrow \quad d_i \le e_1 \vee e_2 \vee ..... \vee e_n$

$\Rightarrow \quad d_i = d_i \wedge (e_1 \vee e_2 \vee ..... \vee e_n)$

$\Rightarrow \quad d_i = (d_i \wedge e_1) \vee (d_i \vee e_2) \vee ..... \vee (d_i \wedge e_n)$ ($\because$ L is distributive)

Since d_i is join-irreducible, so by definition, there exists j, $1 \le j \le n$ such that $d_i = d_i \wedge e_j$

$\Rightarrow \quad d_i \le e_j$ ...(3)

By using similar arguments, we can show that

$e_j \le d_r$ ...(4)

for some r such that $1 \le r \le m$.

(3) and (4) $\Rightarrow \quad d_i \le e_j \le d_r$

Since d's are irredundant, we must have $d_i = e_j = d_r$. $\therefore \quad d_i = e_j$

So by choosing different values of i, the d's and e's may be paired off.

$\therefore$ The representation (1) of a is unique except for the order.

Remark. The result of the above theorem may not be true if the lattice is non-distributive.

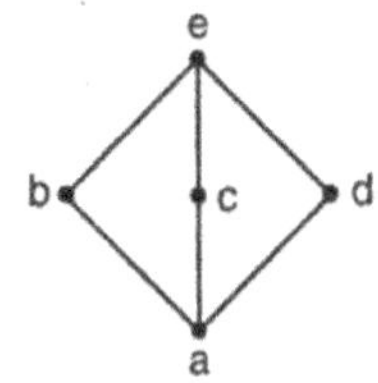

Consider the finite lattice given in the figure.

Here $e = b \vee c$ and $e = c \vee d$ are two distinct representations of e as the join of irredundant join-irreducible elements. This has become possible because the lattice under consideration is not distributive.

9.12. ATOM

Let L be a lattice with lower bound 0. An element a ($\neq 0$) in L is called an **atom** if a is an immediate successor of 0.

If a is an atom, then $a \neq 0$ and $a >> 0$ implies that a has an immediate predecessor, 0.

If possible, let $b(\neq 0)$ be another immediate predecessor of a.

$\Rightarrow \quad 0 < b, b << a \quad \Rightarrow \quad 0 < b < a \quad \Rightarrow \quad 0 << a$ is false.

$\therefore \quad a \neq 0$ and a has unique immediate predecessor.

$\therefore \quad a$ must be join-irreducible. $\therefore$ Every atom must be join-irreducible.

Remark. A join-irreducible element may or may not be an atom.

Example 10. *Find the join-irreducible elements and atoms of the lattice shown in the figure.*

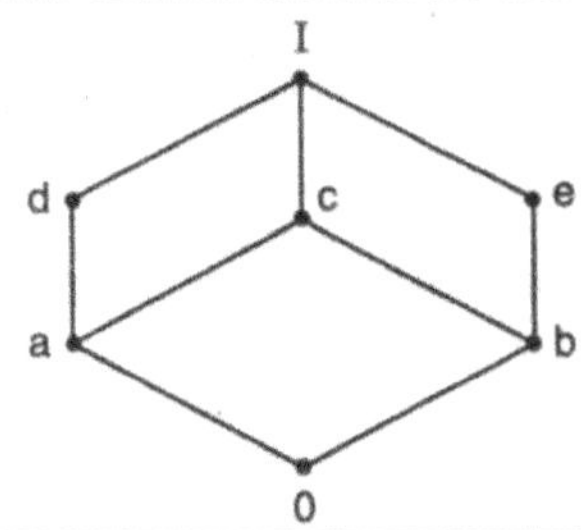

Sol. Let $L = \{0, a, b, c, d, e, I\}$. 0 is always join-irreducible.

Also $x(\neq 0) \in L$ is join-irreducible if and only if x has a unique immediate predecessor. Elements of L having unique immediate predecessor are a, b, d and e.

$\therefore$ The join-irreducible elements are **0, a, b, d, e.**

Also, immediate successors of 0 are atoms.

$\therefore$ Atoms are **a, b.**

EXERCISE 9.4

1. Let P be the power set of the set $\{1, 2, 3\}$. Show that the lattice P under the relation of set inclusion is distributive.
2. Show that the lattice **N** under the relation of divisibility is not a bounded lattice.
3. Show that the lattice **Z** under the relation 'usual $\le$' is not a bounded lattice.
4. If L is a bounded lattice then show that

$$a \vee (b \wedge c) = (a \vee b) \wedge (a \vee c)$$

when $a, b, c \in L$ and any one of a, b, c is either 0 or I.

5. If L is a lattice with lower bound 0 then show that 0 is join- irreducible.

6. Find the join-irreducible elements and atoms of the lattice shown in the figure.

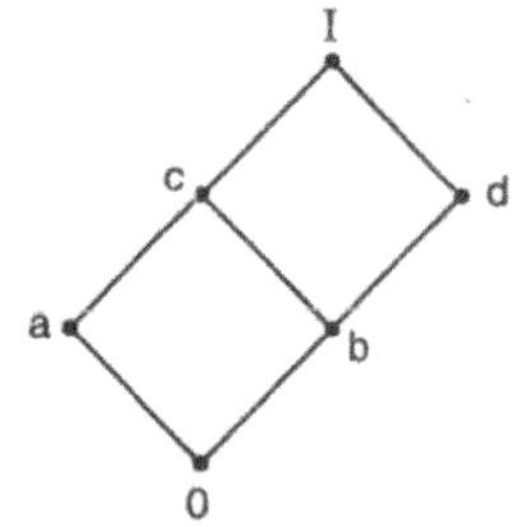

7. Let $L = \{0, a, b, c, I\}$ be a bounded lattice with the Hasse diagram given in the figure. Show that the lattice L is not distributive.

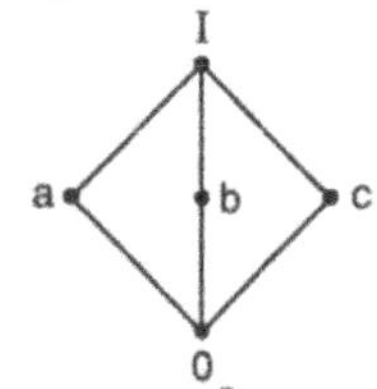

8. Let $L = \{0, a, b, c, d, I\}$ be a bounded lattice with the Hasse diagram given in the figure. Show that for any x, y, z from the subset $\{a, b, d\}$ of L, we have

$$x \vee (y \wedge z) = (x \vee y) \wedge (x \vee z)$$

and $$x \wedge (y \vee z) = (x \wedge y) \vee (x \wedge z).$$

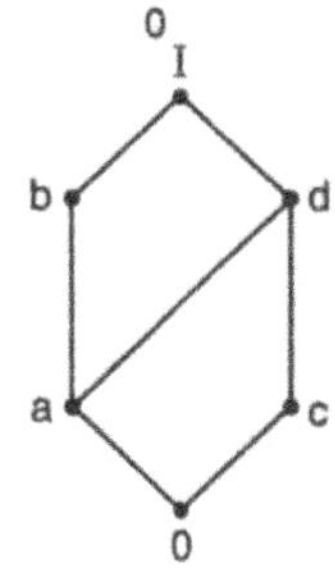

Answer

6. Join-irreducible elements are 0, a, b, d. Atoms are a, b.

Hint

5. $0 = x \vee y \Rightarrow x \leq 0, y \leq 0$

Also $0 \leq x, 0 \leq y$. ($\because$ 0 is a lower bound)

9.13. COMPLEMENT OF AN ELEMENT

Let L be a bounded lattice with the least element 0 and the greatest element I. Let a be an element of L. An element $a' \in L$ is called a **complement** of a if $a \vee a' = I$ and $a \wedge a' = 0$.

In general, complements need not exist and need not be unique.

We observe that $0' = I$ and $I' = 0$ because $0 \vee I = I$, $0 \wedge I = 0$ and $I \vee 0 = I$, $I \wedge 0 = 0$.

Illustration. If S is a non-empty set then the power set $P(S)$ of S is a bounded lattice under the relation of set inclusion.

Here $\quad 0 = \phi$ and $I = S$

Let $\quad A \in P(S) \quad \therefore \quad A' = S - A \in P(S)$

We have $\quad A \vee A' = A \cup A' = S = I$ and $A \wedge A' = A \cap A' = \phi = 0$

$\therefore$ Complement of A in the lattice $P(S)$ = set complement of A *i.e.*, A'.

Example 1. *Let L be the bounded lattice shown in the figure. Show that each element of L has at least one complement.*

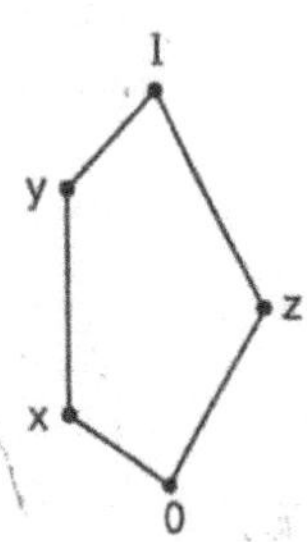

Sol. We have $L = \{0, x, y, z, I\}$.

(*i*) $0 \vee I = I, \quad 0 \wedge I = 0 \quad \Rightarrow \quad 0' = I$

(*ii*) $I \vee 0 = I, \quad I \wedge 0 = 0 \quad \Rightarrow \quad I' = 0$

(*iii*) $x \vee z = I, \quad x \wedge z = 0 \quad \Rightarrow \quad x' = z$

(*iv*) $y \vee z = I, \quad y \wedge z = 0 \quad \Rightarrow \quad y' = z$

(*v*) $z \vee x = I, \quad z \wedge x = 0 \quad \Rightarrow \quad z' = x$

Also, $z \vee y = I, z \wedge y = 0 \quad \Rightarrow \quad z' = y$

$\therefore$ Every element of L has at least one complement.

Example 2. *Show that every element in the bounded lattice D_{30} has a unique complement.*

Sol. $D_{30} = \{1, 2, 3, 5, 6, 10, 15, 30\}$

The Hasse diagram of the bounded lattice D_{30} is shown in the figure.

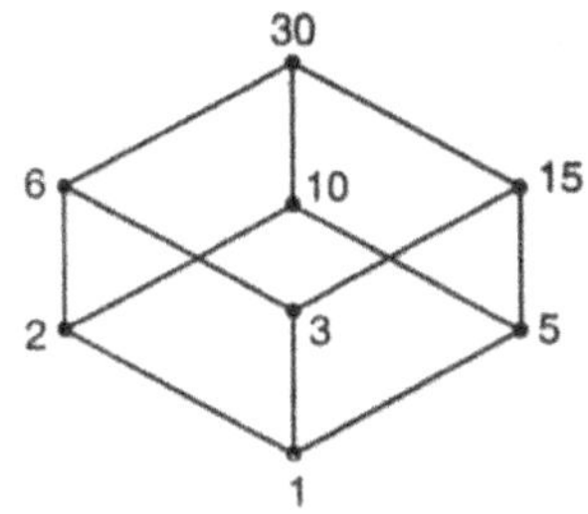

1 and 30 are respectively the least element and the greatest element of the lattice D_{30}.

(*i*) $1 \vee 30 = 30, \quad 1 \wedge 30 = 1 \quad \Rightarrow \quad 1' = 30$ and $30' = 1$

(*ii*) $2 \vee 15 = 30, \quad 2 \wedge 15 = 1 \quad \Rightarrow \quad 2' = 15$ and $15' = 2$

(*iii*) $3 \vee 10 = 30, \quad 3 \wedge 10 = 1 \quad \Rightarrow \quad 3' = 10$ and $10' = 3$

(*iv*) $5 \vee 6 = 30, \quad 5 \wedge 6 = 1 \quad \Rightarrow \quad 5' = 6$ and $6' = 5$.

$\therefore$ Every element in the lattice D_{30} has a unique complement.

Example 3. *Show that the element 2 in the bounded lattice D_{20} has no complement.*

Sol. $D_{20} = \{1, 2, 4, 5, 10, 20\}$

The Hasse diagram of the bounded lattice D_{20} is shown in the figure.

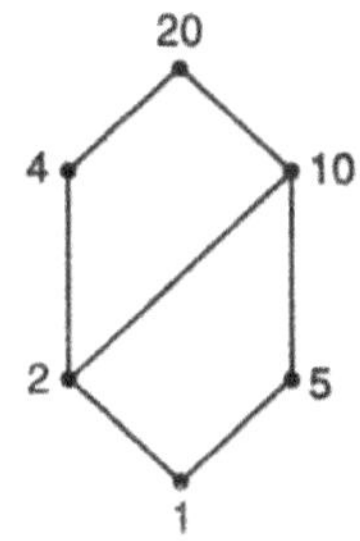

1 and 20 are respectively the least element and the greatest element of the lattice D_{20}.

$2' \neq 1$ because $2 \vee 1 = 2 \neq 20$

$2' \neq 2$ because $2 \vee 2 = 2 \neq 20$

$2' \neq 4$ because $2 \vee 4 = 4 \neq 20$

$2' \neq 5$ because $2 \vee 5 = 10 \neq 20$

$2' \neq 10$ because $2 \vee 10 = 10 \neq 20$

$2' \neq 20$ because $2 \vee 20 = 20$ but $2 \wedge 20 = 2 \neq 1$

$\therefore$ 2 has no complement in D_{20}.

Example 4. *Let L be a bounded lattice with at least two elements. Show that no element of L is its own-complement.*

Sol. Since L has at least two elements, we have $0 \neq I$.

Let $x \in L$ and $x' = x$.

$\therefore \quad x \vee x' = I \quad \Rightarrow \quad x \vee x = I \quad \Rightarrow \quad x = I$

Also $\quad x \wedge x' = 0 \quad \Rightarrow \quad x \wedge x = 0 \quad \Rightarrow \quad x = 0$

$\therefore \quad 0 = I$, which is impossible.

$\therefore$ No element of L is its own-complement.

Example 5. *If L is a bounded distributive lattice then show that the set of complemented elements of L is a sublattice of L.*

Sol. Let $M = \{x \in L : x \text{ is complemented}\}$

Let 0 and I be respectively the least element and the greatest element of the lattice L.

$$0 \vee I = I \text{ and } 0 \wedge I = 0 \quad \therefore \quad 0' = I \text{ and } I' = 0 \quad \therefore \quad 0, I \in M$$

$\therefore$ M is a non-empty subset of L.

Let x and y be any elements of M. $\therefore$ x and y are complemented elements of L.

Let x' and y' be complements of x and y respectively.

$$\therefore \quad x \vee x' = I, \quad x \wedge x' = 0, \quad y \vee y' = I, \quad y \wedge y' = 0$$

Now $\quad (x \vee y) \vee (x' \wedge y') = [(x \vee y) \vee x'] \wedge [(x \vee y) \vee y'] = [(x \vee x') \vee y] \wedge [x \vee (y \vee y')]$

$$= (I \vee y) \wedge (x \vee I) = I \wedge I = I$$

and $\quad (x \vee y) \wedge (x' \wedge y') = (x' \wedge y') \wedge (x \vee y) = [(x' \wedge y') \wedge x] \vee [(x' \wedge y') \wedge y]$

$$= [(x \wedge x') \wedge y'] \vee [x' \wedge (y \wedge y')]$$

$$= (0 \wedge y') \vee (x' \wedge 0) = 0 \vee 0 = 0$$

$\therefore \quad (x \vee y) \vee (x' \wedge y') = I$ and $(x \vee y) \wedge (x' \wedge y') = 0$

$\therefore \quad x' \wedge y'$ is a complement of $x \vee y$. Similarly, $x' \vee y'$ is a complement of $x \wedge y$.

$\therefore \quad x \vee y, x \wedge y \in M \quad \forall\, x, y \in M \quad \therefore \quad M$ is a sublattice of L.

Theorem. Prove that in a bounded distributive lattice, complement (if exists) of an element is unique. (*KU 2006*)

Proof. Let L be a bounded distributive lattice.

Let 0 and I be respectively the least element and the greatest element of the lattice L.

Let x and y be two complements of an element a in L.

$\therefore \quad a \vee x = I, a \wedge x = 0, a \vee y = I, a \wedge y = 0$

Now $\quad x = x \vee 0 = x \vee (a \wedge y) = (x \vee a) \wedge (x \vee y) = (a \vee x) \wedge (x \vee y) = I \wedge (x \vee y) = x \vee y$

and $\quad y = y \vee 0 = y \vee (a \wedge x) = (y \vee a) \wedge (y \vee x) = (a \vee y) \wedge (x \vee y) = I \wedge (x \vee y) = x \vee y$

$\therefore \quad x = y$ (each $= x \vee y$) $\quad \therefore$ The theorem holds.

9.14. COMPLEMENTED LATTICE

A lattice is said to be a **complemented lattice** if it is bounded and every element of the lattice has a complement.

For example, the lattice D_{30} is a complemented lattice because the lattice D_{30} is bounded by 1 and 30 and $1' = 30$, $2' = 15$, $3' = 10$, $5' = 6$, $6' = 5$, $10' = 3$, $15' = 2$ and $30' = 1$.

Theorem I. Let L be a complemented distributive lattice. Prove that every join-irreducible element of L, other than 0, is an atom of L. (*KU 2007*)

Proof. Since the complemented lattice L is distributive, each element of L has a unique complement.

Let $a (\neq 0)$ be a join-irreducible element of L. If possible, let a be not an atom. Then a has a unique immediate predecessor $b(\neq 0)$, say. Let b' be the complement (unique) of b. Since $b \neq 0$ and $0' = I$, we have $b' \neq I$.

If $a \leq b'$ then we have $b \leq a \leq b'$ and thus $b \leq b'$ and so $b \vee b' = b'$, which is impossible because $b \vee b' = I$ and $b' \neq I$.

$\Rightarrow \quad a \nleq b' \quad \Rightarrow \quad a \wedge b' \neq a \quad \Rightarrow \quad a \wedge b' < a \; (\because \; a \wedge b' \leq a) \quad \Rightarrow \quad a \wedge b' \leq b$, because b is the unique immediate predecessor of a.

$\Rightarrow \quad a \wedge b' \leq b \wedge b' \; (\because \; a \wedge b' \leq b') \quad \Rightarrow \quad a \wedge b' \leq 0 \; (\because \; b \wedge b' = 0) \quad \Rightarrow \quad a \wedge b' = 0$

Also $\quad b \leq a \quad \Rightarrow \quad a \vee b = a \quad \Rightarrow \quad a \vee b' = (a \vee b) \vee b' = a \vee (b \wedge b') = a \vee I = I$

$\Rightarrow \quad b'$ is complement of $a \quad \Rightarrow \quad b = a$ (unqiueness of complement)

This is impossible because b is an immediate predecessor of a.

$\therefore \quad$ Every join-irreducible element of L, other than 0, is an atom of L.

Corollary. If L is a complemented lattice with unique complements then every join-irreducible element of L, other than 0, is an atom of L.

Theorem II. Let L be a finite complemented distributive lattice. Prove that every element in L can be written uniquely (except for order) as the join of irredundant atoms of L.

Proof. Let a be any element of L. Since the lattice L is finite and distributive, we can write

$$a = b_1 \vee b_2 \vee \ldots\ldots \vee b_n \qquad \ldots(1)$$

where b's are irredundant join-irreducible elements of L.

Since, L is a complemented distributive lattice, every join-irreducible element of L, other than 0, is an atom of L. (**By Theorem I**)

The b's in the representation (1) are irredundant, so no b can be equal to 0.

$\therefore \quad b_1, b_2, \ldots\ldots, b_n$ are all atoms of L.

$\therefore$ (1) implies that a is the join of irredundant atoms of L. $\therefore$ The theorem holds.

Example 5. *Find whether the given bounded lattice is complemented or not.*

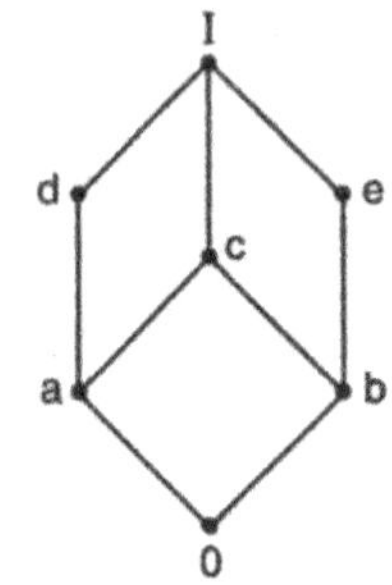

Sol. Let $L = \{0, a, b, c, d, e, I\}$.

We have $0' = I, I' = 0$

$$a \vee e = I, \quad a \wedge e = 0 \Rightarrow a' = e, e' = a$$

$$b \vee d = I, \quad b \wedge d = 0 \Rightarrow b' = d, d' = b$$

Also $c \vee 0 = c, c \vee a = c, c \vee b = c, c \vee d = I$ but $c \wedge d = a \neq 0$, $c \vee e = I$ but $c \wedge e = b \neq 0$, $c \vee I = I$ but $c \wedge I = c \neq 0$.

$\therefore$ c has no complement. $\therefore$ Given lattice is not complemented.

EXERCISE 9.5

1. Let L be the bounded lattice shown in the figure. Show that each element of L has at least one complement.

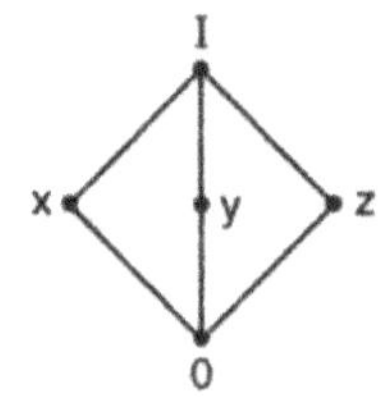

2. Show that the element 10 in the bounded lattice D_{20} has no complement.
3. If a and b are elements in a bounded distributive lattice and a has a complement a', then show that : (*i*) $a \vee (a' \wedge b) = a \vee b$ (*ii*) $a \wedge (a' \vee b) = a \wedge b$.
4. Find the complement of each element in the lattice D_{42}.
5. Let X be a non-empty set. Show that the lattice $P(X)$, the power set of X, under the relation of set inclusion is distributive and complemented. (*KU 2006 C*)
6. Find whether the given bounded lattice is complemented or not.

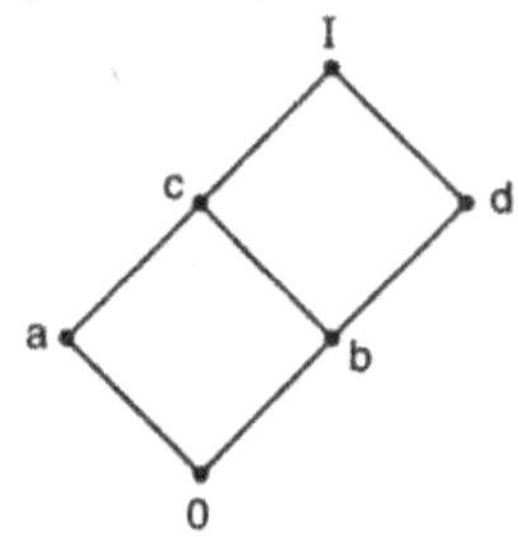

Answers

1. $0' = I, I' = 0, \quad x' = y, z, y' = x, z, \quad z' = x, y$
4. $1' = 42, 2' = 21, 3' = 14, 6' = 7, 7' = 6, 14' = 3, 21' = 2, 42' = 1$
6. No.

Hint

6. b has no complement.

10

Boolean Algebra

10.1. INTRODUCTION

In the mid-1800's, the English mathematician, *George Boole* (1815–1864) developed an algebra which was found very useful in designing logic circuits used by the processors of computer systems. This algebra became known as *Boolean algebra* after its developer's name. In the year 1938, *Claude E Shannon* used Boolean algebra in the design of relay switching circuits. Boolean algebra provides an economical and straight forward approach to the design and analysis of switching circuits. Presently, Boolean algebra is used very extensively in designing the circuitry of computers.

In the present chapter, we shall make a humble attempt to learn Boolean algebra and its application to switching circuits.

10.2. BOOLEAN ALGEBRA (As an algebraic structure)

A non-empty set B with the binary operations '+' (called *addition*) and '.' (called *multiplication*) and one unary operation* ''' (called *complementation*) is called a **Boolean algebra** if the following axioms hold, where a, b, c are any elements of B :

B_1 Commutative laws

(*i*) $a + b = b + a$ (*ii*) $a.b = b.a$

B_2 Distributive laws

(*i*) $a + (b.c) = (a + b).(a + c)$ (*ii*) $a.(b + c) = (a.b) + (a.c)$

B_3 Identity laws

There exist two distinct elements 0 and 1 in B such that

(*i*) $a + 0 = a$ (*ii*) $a.1 = a$

0 is called the **zero element** of B and 1 is called the **unit element** of B.

---

* Let S be a non-empty set. A function from S into S is called a **unary operation** on S.
For example, let S be the set of all subsets of the set {1, 3, 4}.

$\therefore$ $S = \{\phi, \{1\}, \{3\}, \{4\}, \{1, 3\}, \{3, 4\}, \{1, 4\}, \{1, 3, 4\}\}$.

Let $*$ be a function from S into S defined by $*(A) = \{1, 3, 4\} - A$, for $A \in S$.

$\therefore$ $*$ is a unary operation on S

Here $*(\phi) = \{1, 3, 4\}$, $*(\{3, 4\}) = \{1\}$ etc.

B_4 Complement laws

For $a \in B$

(i) $a + a' = 1$ (ii) $a.a' = 0$.

a' is called the **complement** of a.

Remark 1. The binary operations '+' and '.' used above have got nothing to do with ordinary addition and multiplication of numbers. In facts, $*$, $+$, $.$, are just symbols representing binary operations. Similarly 0 and 1 are two specific elements of B satisfying certain properties. These symbols have no bearing with the numbers 0 and 1 of the set of integers.

Remark 2. Since '+' and '.' are commutative, we have

$$0 + a = a + 0 = a, \quad 1.a = a.1 = a$$

and

$$a' + a = a + a' = 1, \quad a'.a = a.a' = 0.$$

Remark 3. The distributive law $a + (b.c) = (a + b).(a + c)$ appears to be slightly unusual. The reader is advised not to verify this for usual numbers. This property would be satisfied only for the elements of a Boolean algebra.

Remark 4. A Boolean algebra is generally denoted by tuples (B, +, ., ', 0, 1) or (B +, ., ') or simply by B.

Remark 5. In writing $a.b$, we generally drop the symbol '.' and use juxtaposition instead. Thus ab stands for $a.b$.

Remark 6. A Boolean algebra (B, +, ., ') always satisfy *associative laws i.e.*, if $a, b, c \in$ B then $a + (b + c) = (a + b) + c$ and $a.(b.c) = (a.b).c$. We shall prove it later.

Remark 7. The symbols '$\vee$' and '$\wedge$' are also used to denote '+' and '.' respectively in a Boolean algebra.

10.3. EXAMPLES OF BOOLEAN ALGEBRA

Example 1. *Let S be any non-empty set. Let B be the set of all **subsets** of the set S. Show that $\{B, \cup, \cap, '\}$ is a Boolean algebra.*

Sol. For $X, Y \in B$, the sets $X \cup Y$ and $X \cap Y$ are also subsets of S.

$\therefore \quad X \cup Y, X \cap Y \in B$

$\therefore \quad \cup$ and $\cap$ are binary operations on B.

For $X \in B$, we define X', the complement of X as $S - X$, which is an element of B.

B_1 For $X, Y \in B$, we have

$$X \cup Y = Y \cup X \text{ and } X \cap Y = Y \cap X$$

$\therefore$ Commutative laws hold.

B_2 For $X, Y, Z \in B$, we have

$$X \cup (Y \cap Z) = (X \cup Y) \cap (X \cup Z) \quad \text{and} \quad X \cap (Y \cup Z) = (X \cap Y) \cup (X \cap Z)$$

$\therefore$ Distributive laws hold.

B_3 For $X \in B$, we have

$$X \cup \phi = X \quad \text{and} \quad X \cap S = X$$

$\therefore$ Identity laws hold.

Here, ϕ is the *zero element* and S is the *unit element* of B.

B_4 For $X \in B$, we have

$$X \cup X' = X \cup (S - X) = S \quad \text{and} \quad X \cap X' = X \cap (S - X) = \phi$$

$\therefore$ Complement laws hold. $\therefore$ $(B, \cup, \cap, ')$ is a Boolean algebra.

Example 2. *Let $B = \{0, 1\}$. The binary operations '+' and '.' on B are defined as follows :*

+	0	1
0	0	1
1	1	1

.	0	1
0	0	0
1	0	1

A unary operation ' ′ ' is defined as $0' = 1$ and $1' = 0$. Show that $(B, +, ., ')$ is a Boolean algebra.

Sol. By definition, for $a, b \in B$, the elements $a + b$ and $a.b$ are unique elements of B.

∴ '+' and '.' are binary operations on B.

$\mathbf{B_1}$

a	b	$a + b$	$b + a$
0	0	0	0
1	0	1	1
0	1	1	1
1	1	1	1

a	b	$a.b$	$b.a$
0	0	0	0
1	0	0	0
0	1	0	0
1	1	1	1

∴ $a + b = b + a \quad \forall\, a, b \in B$ ∴ $a.b = b.a \quad \forall\, a, b \in B$

∴ Commutative laws hold.

$\mathbf{B_2}$

a	b	c	$b.c$	$a + (b.c)$	$a + b$	$a + c$	$(a + b).(a + c)$
0	0	0	0	0	0	0	0
1	0	0	0	1	1	1	1
0	1	0	0	0	1	0	0
0	0	1	0	0	0	1	0
1	1	0	0	1	1	1	1
1	0	1	0	1	1	1	1
0	1	1	1	1	1	1	1
1	1	1	1	1	1	1	1

∴ $a + (b.c) = (a + b).(a + c) \quad \forall\, a, b, c \in B$

a	b	c	$b + c$	$a.(b + c)$	$a.b$	$a.c$	$(a.b) + (a.c)$
0	0	0	0	0	0	0	0
1	0	0	0	0	0	0	0
0	1	0	1	0	0	0	0
0	0	1	1	0	0	0	0
1	1	0	1	1	1	0	1
1	0	1	1	1	0	1	1
0	1	1	1	0	0	0	0
1	1	1	1	1	1	1	1

∴ $a.(b + c) = (a.b) + (a.c) \quad \forall\, a, b, c \in B$

∴ Distributive laws hold.

$\mathbf{B_3}$

a	0	$a + 0$
0	0	0
1	0	1

a	1	$a.1$
0	1	0
1	1	1

$\therefore \quad a + 0 = a \quad \forall\, a \in B \qquad\qquad a.1 = a \quad \forall\, a \in B$

$\therefore$ Identity laws hold.

Here '0' is the zero element and '1' is the unit element.

$\mathbf{B_4}$

a	a'	$a + a'$
0	1	1
1	0	1

a	a'	$a.a'$
0	1	0
1	0	0

$\therefore \quad a + a' = 1 \quad \forall\, a \in B \qquad\qquad \therefore \quad a.a' = 0 \quad \forall\, a \in B$

$\therefore$ Complement laws hold.

$\therefore$ (B +, ., ′) is a Boolean algebra.

Example 3. *Let $B = \{0, a, b, 1\}$. Two binary operations '+' and '.' on B are defined as follows :*

+	0	a	b	1
0	0	0	0	0
a	0	a	0	a
b	0	0	b	b
1	0	a	b	1

.	0	a	b	1
0	0	a	b	1
a	a	a	1	1
b	b	1	b	1
1	1	1	1	1

The unary operation ' ′ ' is defined as $0' = 1$, $a' = b$, $b' = a$, $1' = 0$. Show that $(B, +, ., ')$ is a Boolean algebra.

Sol. By definition, for $x, y \in B$, the elements $x + y$ and $x.y$ are unique elements of B.

$\therefore$ '+' and '.' are binary operations on B.

$\mathbf{B_1}$ For $x, y \in B$, we have

$$x + y = y + x \quad \text{and} \quad x.y = y.x$$

$\therefore$ Commutative laws hold.

For example, let $x = a, y = 1$.

$\therefore \qquad x + y = a + 1 = a, \qquad y + x = 1 + a = a$

and $\qquad x.y = a.1 = 1, \qquad y.x = 1.a = 1$

$\therefore \qquad x + y = y + x, \qquad x.y = y.x$

$\mathbf{B_2}$ For $x, y, z \in B$, we have

$$x + (y.z) = (x + y).(x + z) \quad \text{and} \quad x.(y + z) = (x.y) + (x.z)$$

$\therefore$ Distributive laws hold.

For example, let $x = a, y = 0, z = b$.

$\therefore \qquad x + (y.z) = a + (0.b) = a + b = 0, \quad (x + y).(x + z) = (a + 0).(a + b) = 0.0 = 0$

and $\qquad x.(y + z) = a.(0 + b) = a.0 = a, \quad (x.y) + (x.z) = (a.0) + (a.b) = a + 1 = a$

$\therefore \qquad x + (y.z) = (x + y).(x + z) \quad \text{and} \quad x.(y + z) = (x.y) + (x.z).$

$\mathbf{B_3}$

x	1	$x+1$
0	1	0
a	1	a
b	1	b
1	1	1

x	0	$x.0$
0	0	0
a	0	a
b	0	b
1	0	1

$\therefore \quad x + 1 = x \quad \forall\, x \in B \qquad\qquad x.0 = x \quad \forall\, x \in B$

$\therefore$ Identity laws hold.

Here '1' is the zero element and '0' is the unit element.

$\mathbf{B_4}$

x	x'	$x+x'$
0	1	0
a	b	0
b	a	0
1	0	0

x	x'	$x.x'$
0	1	1
a	b	1
b	a	1
1	0	1

$\therefore \quad x + x' = 0 \quad \forall\, x \in B \qquad\qquad x.x' = 1 \quad \forall\, x \in B$

$\therefore$ Complement laws hold.

$\therefore$ $(B, +, ., ')$ is a Boolean algebra.

Example 4. *Let D_6 be the set of positive divisors of 6. Two binary operations '+' and '.' on B are defined as follows :*

$a + b = lcm(a, b), \quad a.b = gcd\,(a, b) \quad \forall\, a, b \in B.$

A unary operation ' ' ' on D_6 is defined as $a' = \dfrac{6}{a} \quad \forall\, a \in D_6.$

Show that $(D_6, +, ., ')$ is a Boolean algebra.

Sol. We have $D_6 = \{1, 2, 3, 6\}$. The composition tables of the binary operations are given below :

+	1	2	3	6
1	1	2	3	6
2	2	2	6	6
3	3	6	3	6
6	6	6	6	6

.	1	2	3	6
1	1	1	1	1
2	1	2	1	2
3	1	1	3	3
6	1	2	3	6

$\therefore$ '+' and '.' are binary operations on D_6.

$\mathbf{B_1}$ From the composition tables, we have

$$a + b = b + a \quad \text{and} \quad a.b = b.a \quad \forall\, a, b \in D_6$$

Commutative laws hold.

$\mathbf{B_2}$ For $a, b, c \in D_6$, we have $a + (b.c) = (a + b).(a + c)$ and $a.(b + c) = (a.b) + (a.c)$

$\therefore$ Distributive laws hold.

For example, let $a = 3, b = 6, c = 2$.

$\therefore \quad a + (b.c) = 3 + (6.2) = 3 + 2 = 6, \qquad (a + b).(a + c) = (3 + 6).(3 + 2) = 6.6 = 6$

and $\quad a.(b + c) = 3.(6 + 2) = 3.6 = 3, \qquad (a.b) + (a.c) = (3.6) + (3.2) = 3 + 1 = 3$

$\therefore \quad a + (b.c) = (a + b).(a + c)$ and $a.(b + c) = (a.\,b) + (a.c).$

$\mathbf{B_3}$

a	1	$a+1$
1	1	1
2	1	2
3	1	3
6	1	6

a	6	$a.6$
1	6	1
2	6	2
3	6	3
6	6	6

$\therefore$ $a + 1 = a \quad \forall\, a \in D_6$ $\qquad a.6 = a \quad \forall\, a \in D_6$

$\therefore$ Identity laws hold.

Here '1' is the zero element and 6 is the unit element.

$\mathbf{B_4}$

a	a'	$a+a'$
1	6	6
2	3	6
3	2	6
6	1	6

a	a'	$a.a'$
1	6	1
2	3	1
3	2	1
6	1	1

$\therefore$ $a + a' = 6 \quad \forall\, a \in D_6$ $\qquad a.a' = 1 \quad \forall\, a \in D_6$

$\therefore$ Complement laws hold.

$\therefore$ $(D_6, +, ., ')$ is a Boolean algebra.

Example 5. *Let D_{12} be the set of positive divisors of 12. Two binary operations '+' and '.' are defined as follows :*

$$a + b = lcm\ (a, b),\ a.b = gcd\ (a, b) \quad \forall\, a, b \in D_{12}.$$

A unary operation ' ' ' is defined on D_{12} as $a' = \dfrac{12}{a} \quad \forall\, a \in D_{12}$.

Show that $(D_{12}, +, ., ')$ is not a Boolean algebra.

Sol. We have $D_{12} = \{1, 2, 3, 4, 6, 12\}$. The composition tables of the binary operations are given below :

$a+b$	1	2	3	4	6	12
1	1	2	3	4	6	12
2	2	2	6	4	6	12
3	3	6	3	12	6	12
4	4	4	12	4	12	12
6	6	6	6	12	6	12
12	12	12	12	12	12	12

$a\,.\,b$	1	2	3	4	6	12
1	1	1	1	1	1	1
2	1	2	1	2	2	2
3	1	1	3	1	3	3
4	1	2	1	4	2	4
6	1	2	3	2	6	6
12	1	2	3	4	6	12

$\therefore$ '+' and '.' are binary operations on D_{12}.

From the tables, we have

$$a + 1 = a \quad \forall\, a \in D_{12}, \quad a.12 = a \quad \forall\, a \in D_{12}$$

$\therefore$ 1 is the zero element and 12 is the unit element.

a	$a' = \frac{12}{a}$	$a + a' = lcm\ (a, a')$
1	12	12
2	6	6
3	4	12
4	3	12
6	2	6
12	1	12

$\therefore$ $a + a' = 12 \quad \forall\ a \in D_{12}$ is not true.

$\therefore$ Complement laws does not hold.

$\therefore$ $(D_{12}, +, ., ')$ is not a Boolean algebra.

Remark. The set D_m of all positive divisors of a positive integer m is a Boolean algebra, w.r.t. operations defined in **example 5,** only if m can be expressed as the product of distinct prime numbers. In the aboe example, the number 12 cannot be expressed as the product of distinct primes.

Example 6. *Let* $S_n = \{1, 2, 3, ..., n\}$. *Two binary operations '+' and '.' are defined as follows :* $a + b = max\ \{a, b\}$ *and* $a.b = min\ \{a, b\} \quad \forall, a, b \in S_n$. *Show that* S_n *is not a Boolean algebra w.r.t. binary operations '+' and '.' and any unary operation defined on* S_n.

Sol. B$_1$ Let $a, b, \in S_n$. Let $a \geq b$.

$\therefore \quad a + b = a, \quad b + a = a, \quad a.b = b, \quad b.a = b$

$\therefore \quad a + b = b + a \quad$ and $\quad a.b = b.a.$

Similarly if $b \geq a$, then

$a + b = b + a$ (each $= b$) and $a.b = b.a$ (each $= a$)

$\therefore$ Commutative laws hold.

B$_2$ Let $a, b, c \in S_n$. Let $a \geq b \geq c$.

$\therefore \quad a + b.c = a + c = a \quad$ and $\quad (a + b).(a + c) = a.a = a$

$\therefore \quad a + b.c = (a + b).(a + c)$

Also $\quad a.(b + c) = a.b = b \quad$ and $\quad a.b + a.c = b + c = b$

$\therefore \quad a.(b + c) = (a + b).(a + c)$

Now let $c \geq a \geq b$.

$\therefore \quad a + b.c = a + b = a \quad$ and $\quad (a + b).(a + c) = a.c = a$

and $\quad a.(b + c) = a.c = a \quad$ and $\quad a.b + a.c = b + a = a$

$\therefore \quad a + b.c = (a + b).(a + c) \quad$ and $\quad a.(b + c) = a.b + a.c$

Also for other possibilities, the distributive laws hold.

B$_3$ Let $\quad a \in S_n \quad \therefore \quad a + 1 = a$

$\therefore$ 1 is zero element of S_n.

Also $a.\ n = a \quad \therefore \quad n$ is identity element of S_n.

B$_4$ Let '′' be any unary operation on S_n. Let $a \in S_n$ and $a + a' = n, \quad a.a' = 1.$

Let $\quad n = 10. \quad \therefore \quad S_n = \{1, 2, 3, ..., 10\}.$

Now $2 \in S_n$

$\therefore \quad 2 + 2' = 10 \quad$ and $\quad 2.\ 2' = 1$

($\because$ Identity element = 10 and zero element = 1)

$\Rightarrow \quad$ **max** $\{2, 2'\} = 10 \quad$ and $\quad$ min $\{2, 2'\} = 1 \quad \Rightarrow \quad 2' = 10 \quad$ and $\quad 2' = 1$

This is impossible.

∴ There cannot be defined an unary operation so that $(S_n, +, ., ')$ may be a Boolean algebra.

∴ S_n is not a Boolean algebra.

Theorem. If B_1 and B_2 are Boolean algebras then prove that $B_1 \times B_2$ is also a Boolean algebra.

Proof. Let $(a, b), (c, d) \in B_1 \times B_2$. We define binary operations '+' and '.' and unary operation ''' in $B_1 \times B_2$ as follows :

$$(a, b) + (c, d) = (a + c, b + d)$$
$$(a, b).(c, d) = (a.c, b.d)$$
$$(a, b)' = (a', b')$$

1. Commutative laws.

Let $(a, b), (c, d) \in B_1 \times B_2$.

$$(a, b) + (c, d) = (a + c, b + d) = (c + a, d + b) = (c, d) + (a, b)$$

and $(a, b).(c, d) = (a.c, b.d) = (c.a, d.b) = (c, d).(a, b)$

∴ Commutative laws holds.

2. Distributive laws.

Let $(a, b), (c, d), (e, f) \in B_1 \times B_2$.

$$(a, b).[(c, d) + (e, f)] = (a, b).(c + e, d + f) = (a.(c + e), b.(d + f))$$
$$= (a.c + a.e, b.d + b.f) = (a.c, b.d) + (a.e, b.f)$$
$$= (a, b).(c, d) + (a, b).(e, f)$$

Also $(a, b) + [(c, d) .(e, f)] = (a, b) + (c.e, d.f) = (a + c.e, b + d.f)$

$$= ((a + c).(a + e), (b + d).(b + f)) = (a + c, b + d).(a + e, b + f)$$
$$= [(a, b) + (c, d)].[(a, b) + (e, f)]$$

∴ Distributive laws hold.

3. Identity laws.

Let $0_1, 0_2$ be the zero elements of B_1 and B_2 respectively.

∴ $(0_1, 0_2) \in B_1 \times B_2$

Let $(a, b) \in B_1 \times B_2$

$$(a, b) + (0_1, 0_2) = (a + 0_1, b + 0_2) = (a, b)$$

Let 1_1 and 1_2 be the unit elements of B_1 and B_2 respectively.

∴ $(1_1, 1_2) \in B_1 \times B_2$ Let $(a, b) \in B_1 \times B_2$

$$(a, b).(1_1, 1_2) = (a.1_1, b.1_2) = (a, b)$$

∴ Identity laws hold.

4. Complement laws.

Let $(a, b) \in B_1 \times B_2$

$$a \in B_1 \Rightarrow a' \in B_1 \text{ and } b \in B_2 \Rightarrow b'_2 \in B_2$$

∴ $(a', b') \in B_1 \times B_2$

and $(a, b) + (a', b') = (a + a', b + b') = (1_1, 1_2)$

and $(a, b).(a', b') = (a.a', b.b') = (0_1, 0_2)$

∴ Complement laws hold.

∴ $B_1 \times B_2$ is a Boolean algebra.

Example 7. *Let $B = \{0, 1\}$, the set of binary digits with operations '+' '.' and ''' defined as follows :*

+	0	1
0	0	1
1	1	1

.	0	1
0	0	0
1	0	1

a	a'
0	1
1	0

$\{B, +, ., ', 0, 1\}$ is a Boolean algebra. Show that $B^n = B \times B \times ...... \times B$ (n factors) is a Boolean algebra with component wise addition, multiplication and complement.

Sol. Let $(a_1, ..., a_n), (b_1, ..., b_n) \in B^n$

$\therefore \quad (a_1, ..., a_n) + (b_1, ..., b_n) = (a_1 + b_1, ..., a_n + b_n),$

$(a_1, ..., a_n).(b_1, ..., b_n) = (a_1.b_1, ..., a_n.b_n)$

and $\quad (a_1, ..., a_n)' = (a_1', ..., a_n').$

1. Commutative laws

Let $(a_1, ..., a_n), (b_1, ..., b_n) \in B^n$

$$(a_1, ..., a_n) + (b_1, ..., b_n) = (a_1 + b_1, ..., a_n + b_n)$$
$$= (b_1 + a_1, ..., b_n + a_n) = (b_1, ..., b_n) + (a_1, ..., a_n)$$

Also $\quad (a_1, ..., a_n).(b_1, ..., b_n) = (a_1.b_1, ..., a_n.b_n)$

$$= (b_1.a_1, ..., b_n.a_n) = (b_1, ..., b_n).(a_1, ..., a_n)$$

$\therefore$ Commutative laws hold.

2. Distributive laws

Let $(a_1, ..., a_n), (b_1, ..., b_n), (c_1, ..., c_n) \in B^n$

$$(a_1, ..., a_n) . [(b_1, ..., b_n) + (c_1, ..., c_n)]$$
$$= (a_1, ..., a_n) . (b_1 + c_1, ..., b_n + c_n)$$
$$= (a_1.(b_1 + c_1), ..., a_n(b_n + c_n))$$
$$= (a_1.b_1 + a_1.c_1, ..., a_n.b_n + a_n.c_n)$$
$$= (a_1.b_1, ..., a_n.b_n) + (a_1.c_1, ..., a_n.c_n)$$
$$= [(a_1, ..., a_n).(b_1, ..., b_n)] + [(a_1, ..., a_n).(c_1, ..., c_n)]$$

Similarly, $\quad (a_1, ..., a_n) + [(b_1, ..., b_n) . (c_1, ..., c_n)]$

$$= [(a_1, ..., a_n) + (b_1, ..., b_n)] . [(a_1, ..., a_n) + (c_1, ..., c_n)]$$

$\therefore$ Distributive laws hold.

3. Identity laws

$$(0, ..., 0), (1, ..., 1) \in B^n$$

For $\quad (a_1, ..., a_n) \in B^n,$

$(a_1, ..., a_n) + (0, ..., 0) = (a_1 + 0, ..., a_n + 0) = (a_1, ..., a_n)$

and $\quad (a_1, ..., a_n).(1, ..., 1) = (a_1 . 1 ......, a_n . 1) = (a_1, ..., a_n)$

$\therefore$ Identity laws hold.

4. Complement laws

Let $(a_1, ..., a_n) \in B^n$

Let $(b_1, ..., b_n) \in B^n$ be such that $b_i = 1$ if $a_i = 0$ and $b_i = 0$ if $a_i = 1$.

$\therefore \quad (a_1, ..., a_n) + (b_1, ..., b_n) = (a_1 + b_1, ..., a_n + b_n) = (1, ..., 1)$

and $\quad (a_1, ..., a_n) . (b_1, ..., b_n) = (a_1.b_1, ..., a_n.b_n) = (0, ..., 0).$

$\therefore$ Complement laws hold.

$\therefore$ B^n is a Boolean algebra.

EXERCISE 10.1

1. Let B be the set of all logical statements. Show that $(B, \vee, \wedge, \sim)$ is a Boolean algebra.
2. (*i*) Let D_{35} be the set of all positive divisiors of 35. Two binary operations '+' and '.' are defined as follows :

$$a + b = \text{lcm}(a, b), \quad a.b = \gcd(a, b) \quad \forall\, a, b \in D_{35}.$$

A unary operation ' ' ' on D_{35} is defined as $a' = \frac{35}{a}$ $\forall\, a \in D_{35}$. Show that $(D_{35}, +, ., ')$ is a Boolean algebra.

(*ii*) Let D_{105} be the set of all positive divisiors of 105. Two binary operations '+' and '.' are defined as follows :

$$a + b = \text{lcm}(a, b), \quad a.b = \gcd(a, b) \;\forall\, a, b \in D_{105}.$$

A unary operation ' ' ' on D_{105} is defined as $a' = \frac{105}{a}$ $\forall\, a \in D_{105}$. Show that $(D_{105}, +, ., ')$ is a Boolean algebra.

3. (*i*) Let D_4 be the set of all positive divisiors of 4. Two binary operations '+' and '.' are defined as follows :

$a + b = \text{lcm}(a, b)$, $a.b = \gcd(a, b)$ $\forall\, a, b \in D_4$. A unary operation ' ' ' on D_4 is defined as

$$a' = \frac{4}{a} \quad \forall\, a \in D_4.$$

Show that $(D_4, +, ., ')$ is not a Boolean algebra.

(*ii*) Let D_8 be the set of positive divisiors of 8. Two binary operations '+' and '.' are defined as follows :

$$a + b = \text{lcm}(a, b), \quad a.b = \gcd(a, b) \quad \forall\, a, b \in D_8.$$

A unary operation ' ' ' on D_8 is defined as $a' = \frac{8}{a}$ $\forall\, a \in D_8$. Show that $(D_8, +, ., ')$ is not a Boolean algebra.

4. Let B = $\{\phi, \{1\}, \{2\}, \{3\}, \{1, 2\}\, \{1, 3\}, \{2, 3\}, \{1, 2, 3\}\}$. Show that $(B, \cup, \cap, ')$ is a Boolean algebra.

Hint

1. Let f and t represent a fallacy and a tautology respectively. Verify that f works as zero element *i.e.*, identity for '$\vee$' and t works as unit element *i.e.*, identity for '$\wedge$'.

10.4. BASIC PROPERTIES OF A BOOLEAN ALGEBRA

Theorem 1. (uniqueness of zero element and unit element). If (B, +, .,′) is a Boolean algebra, then prove that

(*i*) the zero element is unique

(*ii*) the unit element is unique.

Proof. (*i*) Let 0_1 and 0_2 be two zero elements of the Boolean algebra B.

$\therefore \quad 0_2 + 0_1 = 0_2 = 0_1 + 0_2 \qquad (\because\ 0_1 \text{ is zero element})$

and $\quad 0_1 + 0_2 = 0_1 = 0_2 + 0_1 \qquad (\because\ 0_2 \text{ is zero element})$

$\therefore \quad 0_1 = 0_2. \quad \therefore$ zero element is unique.

(*ii*) Let 1_1 and 1_2 be two unit elements of the Boolean algebra B.

$\therefore \quad 1_2.1_1 = 1_2 = 1_1.1_2$ $\quad$ ($\because$ 1_1 is unit element)

and $\quad 1_1.1_2 = 1_1 = 1_2.1_1$ $\quad$ ($\because$ 1_2 is unit element)

$\therefore \quad 1_1 = 1_2 \quad \therefore$ unit element is unique.

Theorem 2. Let (B, +, ., ′) be a Boolean algebra. For a ∈ B, let x ∈ B be such that a + x = 1 and a.x = 0. Prove that x is equal to the complement of a.

Proof. We have $a + x = 1$ and $a.x = 0$. Let complement of a be a'.

$\therefore \quad a + a' = 1, \quad a.a' = 0.$

We have
$$x = x.1 = x.(a + a') = x.a + x.a' = a.x + x.a' = 0 + x.a' = a.a' + x.a'$$
$$= a'.a + a'.x = a'.(a + x) = a'.1 = a'$$

(By using $\mathbf{B_3, B_4, B_2, B_1}$, **a.x = 0**, $\mathbf{B_4, B_1, B_2}$, **a + x = 1**, $\mathbf{B_3}$ in this order)

$\therefore \quad x = a'.$

Theorem 3. In a Boolean algebra (B, +, ., ′), prove that the zero element and unit element are complement of each other.

Proof. Let 0 and 1 be the zero element and unit element of the Boolean algebra respectively.

We have $\quad 1'.1 = 1' = 1.1'$ $\quad$ ($\because$ 1 is unit element)

Also, by complement law, $1.1' = 0$.

$\therefore \quad \mathbf{1' = 0.}$

We have $\quad 0' + 0 = 0' = 0 + 0'$ $\quad$ ($\because$ 0 is zero element)

Also, by complement law, $0 + 0' = 1$.

$\therefore \quad \mathbf{0' = 1.}$

Theorem 4. (Idempotent laws). If (B, +, ., ′) is a Boolean algebra, then prove that

(*i*) a + a = a ∀ a ∈ B $\qquad$ **(*ii*) a.a = a ∀ a ∈ B.**

Proof (*i*) $a = a + 0 = a + (a.a') = (a + a).(a + a') = (a + a).1 = a + a$

(By using $\mathbf{B_3, B_4, B_2, B_4, B_3}$ in this order)

$\therefore \quad \mathbf{a + a = a \quad \forall\ a \in B.}$

(*ii*) $a = a.1 = a.(a + a') = (a.a) + (a.a') = a.a + 0 = a.a$

(By using $\mathbf{B_3, B_4, B_2, B_4, B_3}$ in this order)

$\therefore \quad \mathbf{a.a = a \quad \forall\ a \in B.}$

Remarks. (*i*) $\underbrace{a + a + \ldots\ldots + a}_{n \text{ times}} = (a + a) + \underbrace{a + \ldots\ldots + a}_{n - 2 \text{ times}}$

$$= \underbrace{a + a + \ldots\ldots + a}_{n - 2 \text{ times}} = (a + a) + \underbrace{a + \ldots\ldots + a}_{n - 3 \text{ times}}$$

....

....

$$= a + a = a.$$

$\therefore$ **a + a + ... n times = a.**

(*ii*) $\underbrace{a.a\ldots\ldots.a}_{n \text{ times}} = (a.a).\underbrace{a.\ \ldots\ldots.a}_{n - 2 \text{ times}}$

$$= \underbrace{a.a. \ldots\ldots .a}_{n-2 \text{ times}} = (a.a).\underbrace{a. \ldots\ldots .a}_{n-3 \text{ times}}$$

....

....

$$= a.a = a.$$

$\therefore$ **a.a. ...... n times = a.**

Theorem 5. (Boundedness laws). If (B, +, ., ′) is a Boolean algebra, then prove that

(*i*) $\mathbf{a + 1 = 1 \quad \forall\ a \in B}$ (*ii*) $\mathbf{a.0 = 0 \quad \forall\ a \in B.}$

Proof. (*i*) $a + 1 = (a + 1).1 = (a + 1).(a + a') = a + (1.a') = a + (a'.1) = a + a' = 1.$

(By using $\mathbf{B_3}$, $\mathbf{B_4}$, $\mathbf{B_2}$, $\mathbf{B_1}$, $\mathbf{B_3}$, $\mathbf{B_4}$ in this order)

$\therefore \quad \mathbf{a + 1 = 1 \quad \forall\ a \in B.}$

(*ii*) $a.0 = (a.0) + 0 = (a.0) + (a.a') = a.(0 + a') = a.(a' + 0) = a.a' = 0.$

(By using $\mathbf{B_3}$, $\mathbf{B_4}$, $\mathbf{B_2}$, $\mathbf{B_1}$, $\mathbf{B_3}$, $\mathbf{B_4}$ in this order)

$\therefore \quad \mathbf{a.0 = 0 \quad \forall\ a \in B.}$

Theorem 6. (Absorption laws). If (B, +, ., ′) is a Boolean algebra, then prove that

(*i*) $\mathbf{a + (a.b) = a \quad \forall\ a, b \in B}$ (*ii*) $\mathbf{a.(a + b) = a \quad \forall\ a, b \in B.}$

Proof. (*i*) $a + (a.b) = (a.1) + (a.b) = a.(1 + b) = a.(b + 1) = a.1 = a$

(By using $\mathbf{B_3}$, $\mathbf{B_2}$, $\mathbf{B_1}$, **boundedness law**, $\mathbf{B_3}$ in this order)

$\therefore \quad \mathbf{a + (a.b) = a \quad \forall\ a, b \in B.}$

(*ii*) $a.(a + b) = (a + 0).(a + b) = a + (0.b) = a + (b.0) = a + 0 = a$

(By using $\mathbf{B_3}$, $\mathbf{B_2}$, $\mathbf{B_1}$, **boundedness law**, $\mathbf{B_3}$ in this order)

$\therefore \quad \mathbf{a.(a + b) = a \quad \forall\ a, b \in B.}$

Theorem 7. (Involution law). If (B, +, ., ′) is a Boolean algebra, then prove that

$\mathbf{a'' = a \quad \forall\ a \in B, \text{ where } \quad a'' = (a')'.}$

Proof. $a'' = a'' + 0 = a'' + a.a' = (a'' + a).(a'' + a') = (a'' + a).(a' + a'') = (a'' + a).1$

$= (a + a'').(a + a') = a + (a''.a') = a + (a'.a'') = a + 0 = a$

(By using $\mathbf{B_3}$, $\mathbf{B_4}$, $\mathbf{B_2}$, $\mathbf{B_1}$, $\mathbf{B_4}$, $\mathbf{B_4}$, $\mathbf{B_2}$, $\mathbf{B_1}$, $\mathbf{B_4}$, $\mathbf{B_3}$ in this order)

$\therefore \quad \mathbf{a'' = a \quad \forall\ a \in B.}$

Theorem 8. (Associative laws). If (B, +, .,′) is a Boolean algebra, then prove that

(*i*) $\mathbf{a + (b + c) = (a + b) + c \quad \forall\ a, b, c \in B}$

(*ii*) $\mathbf{a.(b.c) = (a.b).c \quad \forall\ a, b, c \in B.}$

Proof. (*i*) $a + (b + c) = (a + (b + c)) . (c + c')$ $(\because \quad c + c' = 1)$

$= [(a + (b + c)).c] + [(a + (b + c)).c']$ (By using $\mathbf{B_2}$)

$= [c.(a + (b + c))] + [c'.(a + (b + c))]$ (By using $\mathbf{B_1}$)

$= [(c.a) + (c.(b + c))] + [(c'.a) + (c'.(b + c))]$ (By using $\mathbf{B_2}$)

$= [(c.a) + c] + [(c'.a) + ((c'.b) + (c'.c))]$ (By using absorption law and $\mathbf{B_2}$]

$= [(c.a) + (c.1)] + [(c'.a) + ((c'.b) + 0)]$ (By using $\mathbf{B_3}$, $\mathbf{B_4}$]

$= [c.(a + 1)] + c'.(a + b)$ (By using $\mathbf{B_2}$)

$= (c.1) + c'.(a + b)$ (By using boundedness law)

$= c + c'.(a + b)$ (By using $\mathbf{B_3}$)

$= (c + c').(c + (a + b))$ (By using $\mathbf{B_2}$)

$= 1.((a + b) + c)$ (By using $\mathbf{B_4}$, $\mathbf{B_1}$)

$= (a + b) + c$ (By using $\mathbf{B_3}$)

$\therefore$ $\mathbf{a + (b + c) = (a + b) + c \quad \forall\ a, b, c \in B.}$

(*ii*) $a.(b.c) = (a.(b.c)) + (c.c')$ ($\because$ $c.c' = 0$)

$= [(a.(b.c)) + c].[(a.(b.c)) + c']$ (By using $\mathbf{B_2}$)

$= [c + (a.(b.c))].[c' + (a.(b.c))]$ (By using $\mathbf{B_1}$)

$= [(c + a).(c + (b.c))].[(c' + a).(c' + (b.c))]$ (By using $\mathbf{B_2}$)

$= [(c + a).c].[(c' + a).((c' + b).(c' + c))]$ (By using absorption law and $\mathbf{B_2}$)

$= [(c + a).(c + 0)].[(c' + a).((c' + b).1)]$ (By using $\mathbf{B_3}$, $\mathbf{B_4}$)

$= [c + (a.0)].[c' + (a.b)]$ (By using $\mathbf{B_2}$)

$= (c + 0).[c' + (a.b)]$ (By using boundedness law)

$= c.(c' + (a.b))$ (By using $\mathbf{B_3}$)

$= (c.c') + (c.(a.b))$ (By using $\mathbf{B_2}$)

$= 0 + ((a.b).c)$ (By using $\mathbf{B_4}$, $\mathbf{B_1}$)

$= (a.b).c$ (By using $\mathbf{B_3}$)

$\therefore$ $\mathbf{a.(b.c) = (a.b).c \quad \forall\ a, b, c \in B.}$

Theorem 9. (De Morgan's Laws). If (B, +, ., ') is a Boolean algebra, then prove that

(*i*) $\mathbf{(a + b)' = a'.b' \quad \forall\ a, b \in B}$ (*ii*) $\mathbf{(a.b)' = a' + b' \quad \forall\ a, b \in B.}$ (KU 2007)

Proof. (*i*) $(a + b) + (a'.b') = ((a + b) + a').((a + b) + b')$ (By using $\mathbf{B_2}$)

$= (a + (b + a')).(a + (b + b'))$ (By using associative law)

$= (a + (a' + b)).(a + 1)$ (By using $\mathbf{B_1}$, $\mathbf{B_4}$)

$= ((a + a') + b).1$ (By using associative law and boundedness law)

$= 1 + b$ (By using $\mathbf{B_4}$ and $\mathbf{B_3}$)

$= 1$ (By using boundedness law)

$\therefore$ $(a + b) + (a'.b') = 1$ ...(1)

Also, $(a + b).(a'.b') = (a'.b').(a + b)$ (By using $\mathbf{B_1}$)

$= ((a'.b').a) + ((a'.b').b)$ (By using $\mathbf{B_2}$)

$= ((b'.a').a) + (a'.(b'.b))$ (By using $\mathbf{B_1}$, associative law)

$= (b'.(a'.a)) + (a'.(b'.b))$ (By using associative law)

$= (b'.0) + (a'.0)$ (By using $\mathbf{B_4}$)

$= 0 + 0$ (By using boundedness law)

$= 0$ (By using $\mathbf{B_3}$)

$\therefore$ $(a + b).(a'.b') = 0$ ...(2)

(1) and (2) implies $\mathbf{(a + b)' = a'.b'}$

(*ii*) Proof is left for the reader as an exercise.

Theorem 10. (Cancellation laws). Let (B, +, ., ') be a Boolean algebra.

(*i*) **If** $\mathbf{a + b = a + c}$ **and** $\mathbf{a' + b = a' + c}$, **then prove that** $\mathbf{b = c}$

(*ii*) **If** $\mathbf{a.b = a.c}$ **and** $\mathbf{a'.b = a'.c}$, **then prove that** $\mathbf{b = c.}$

Proof. (*i*) $b = b + 0$ (By using $\mathbf{B_3}$)

$= b + (a.a')$ (By using $\mathbf{B_4}$)

$= (b + a).(b + a')$ (By using $\mathbf{B_2}$)

$= (a + b).(a' + b)$ (By using $\mathbf{B_1}$)

$= (a + c).(a' + c)$ (Given)
$= (c + a).(c + a')$ (By using $\mathbf{B_1}$)
$= c + (a.a')$ (By using $\mathbf{B_2}$)
$= c + 0$ (By using $\mathbf{B_4}$)
$= c$ (By using $\mathbf{B_3}$)

$\therefore$ **b = c.**

(*ii*) $b = b.1$ (By using $\mathbf{B_3}$)
$= b.(a + a')$ (By using $\mathbf{B_4}$)
$= (b.a) + (b.a')$ (By using $\mathbf{B_2}$)
$= (a.b) + (a'.b)$ (By using $\mathbf{B_1}$)
$= (a.c) + (a'.c)$ (Given)
$= (c.a) + (c.a')$ (By using $\mathbf{B_1}$)
$= c.(a + a')$ (By using $\mathbf{B_2}$)
$= c.1$ (By using $\mathbf{B_4}$)
$= c$ (By using $\mathbf{B_3}$)

$\therefore$ **b = c.**

Theorem 11. In a Boolean algebra (B, +, ., ′, 0, 1), prove that the following are equivalent :

(*i*) **a + b = b** (*ii*) **a.b = a**
(*iii*) **a′ + b = 1** (*iv*) **a.b′ = 0.**

Proof. (*i*) $\Rightarrow$ (*ii*) Let $a + b = b$

$a \,.\, b = a.(a + b) = a$ (By absorption law)

$\therefore$ $a \,.\, b = a.$

(*ii*) $\Rightarrow$ (*iii*) Let $a.b = a$

$a' + b = (a.b)' + b = (a' + b') + b$
$= a' + (b' + b) = a' + 1$ (By using B_4)
$= 1$ (By boundedness law)

$\therefore$ $a' + b = 1.$

(*iii*) $\Rightarrow$ (*iv*) Let $a' + b = 1$

$\Rightarrow$ $(a' + b)' = 1' \Rightarrow a''.b' = 0 \Rightarrow a.b' = 0.$

$\therefore$ $a.b' = 0.$

(*iv*) $\Rightarrow$ (*i*) Let $a.b' = 0$

$\Rightarrow$ $a + b = a.1 + b = a.(b + b') + b = a.b + a.b' + b = a.b + 0 + b$
$= b + b.a = b$

$\therefore$ $a + b = b$

$\therefore$ (*i*) $\Rightarrow$ (*ii*) $\Rightarrow$ (*iii*) $\Rightarrow$ (*iv*) $\Rightarrow$ (*i*)

$\therefore$ (*i*), (*ii*), (*iii*), (*iv*) are equivalent.

10.5. BOOLEAN EXPRESSION

Let $(B, +, ., ')$ be a Boolean algebra.

Elements of *B* are called **constants** in *B*. *A* symbol representing an arbitrary element of *B* is called a **variable** in *B*. *A* variable in a Boolean algebra is denoted by letters *a*, *b*, *c*, ..., *p*, *q*, *r*, ..., *x*, *y*, *z* etc.

In the Boolean algebra $(B, +, ., ')$, a constant or a variable or a well formed expression using finite number of variables and Boolean operations '+', '.' and ''' is called a **Boolean expression** over the Boolean algebra $(B, +, ., ')$. For example, $(xy' + z) + xyz + x'$, $(x + y + z)' - xy' + xz'$ are Boolean expressions in the variables x, y and z. A Boolean expression involving variables $x_1, x_2, ..., x_n$ is generally written $E(x_1, x_2, ..., x_n)$.

It may be observed that the sum and product of two Boolean expressions is also a Boolean expression over the same Boolean algebra.

Let $E(x_1, x_2, ..., x_n)$ be a Boolean expression of n variables over a Boolean algebra $(B, +, ., ')$. By an assignment of values of the variables $x_1, x_2, ..., x_n$, we mean an assignment of elements of B to be the values of the variables. For an assignment of values to the variables, we can evaluate the Boolean expression $E(x_1, x_2, ..., x_n)$ by substituting the variables in the expression by their values.

For example, let $E(x, y) = (x + y)x + x'y$ be a Boolean expression over the Boolean algebra $(\{0, 1\}, +, ., ')$ of binary digits 0 and 1. For $x = 0, y = 1$, we have

$$E(0, 1) = (0 + 1).0 + 0'.1 = 1.0 + 1.1 = 0 + 1 = 1.$$

Two Boolean expressions of n variables over a Boolean algebra are said to be **equivalent** if they assume the same value for any assignment of values to the n variables. Thus, when we say to simplify a Boolean expression we always mean to simplify it into an equivalent form. Since the elements of the Boolean algebra will be assigned as values of the variables in a Boolean expression, all laws of Boolean algebra can be applied to simplify Boolean expressions. The following rule is also followed while simplifying a Boolean expression.

An expression within a bracket is simplified by observing precedence of ''' over '.' and precedence of '.' over '+'.

For example (*i*) $a + b.c + d.e$ shall mean $a + (b.c) + (d.e)$.

(*ii*) $a.b' + c'$ shall mean $(a.(b')) + c'$ and not $(a.b' + c)'$ or $(a.b)' + c'$.

These rules help a lot in simplifying expressions. But at the same time, greater care is expected while using these rules. This is why, we did not introduce these rules in the beginning. For an illustration, let us prove the associative law, $a.(b.c) = (a.b).c$ in the Boolean algebra $(B, +, .,')$, by using above rules :

$$\begin{aligned}
a.(b.c) &= a.(b.c) + c.c' && (\because\ c.c' = 0)\\
&= (a.(b.c) + c).(a.(b.c) + c') && (\text{By using } \mathbf{B_2})\\
&= (c + a.(b.c)).(c' + (a.(b.c)) && (\text{By using } \mathbf{B_1})\\
&= (c + a).(c + b.c).(c' + a).(c' + b.c) && (\text{By using } \mathbf{B_2})\\
&= (c + a).c.(c' + a).(c' + b).(c' + c) && (\text{By using absorption law and } \mathbf{B_2})\\
&= (c + a).(c + 0).(c' + a).(c' + b).1 && (\text{By using } \mathbf{B_3}, \mathbf{B_4})\\
&= (c + a.0).(c' + a.b) && (\text{By using } \mathbf{B_2})\\
&= (c + 0).(c' + a.b) && (\text{By using boundedness law})\\
&= c.(c' + a.b) && (\text{By using } \mathbf{B_3})\\
&= c.c' + c.(a.b) && (\text{By using } \mathbf{B_2})\\
&= 0 + (a.b).c && (\text{By using } \mathbf{B_4}, \mathbf{B_1})\\
&= (a.b).c && (\text{By using } \mathbf{B_3})
\end{aligned}$$

$\therefore$ $\mathbf{a.(b.c) = (a.b).c \quad \forall\ a, b, c \in B.}$

SUMMARY OF BASIC PROPERTIES OF A BOOLEAN ALGEBRA

Let (B, +, ., ′) be a Boolean algebra.

I. (i) *Zero element, unit element are unique*

(ii) *Complement of an element is unique.*

II. $1' = 0$ and $0' = 1$.

III. Idempotent laws

(i) $a + a = a \quad \forall\, a \in B$ (ii) $a.a = a \quad \forall\, a \in B$.

IV. Boundedness laws

(i) $a + 1 = 1 \quad \forall\, a \in B$ (ii) $a.0 = 0 \quad \forall\, a \in B$.

V. Absorption laws

(i) $a + a.b = a \quad \forall\, a, b \in B$ (ii) $a.(a + b) = a \quad \forall\, a, b \in B$.

VI. Involution law $a'' = a \quad \forall\, a \in B$.

VII. Associative laws

(i) $a + (b + c) = (a + b) + c \quad \forall\, a, b, c \in B$

(ii) $a.(b.c) = (a.b).c \quad \forall\, a, b, c \in B$.

VIII. De Morgan's laws

(i) $(a + b)' = a'.b' \quad \forall\, a, b \in B$ (ii) $(a.b)' = a' + b' \quad \forall\, a, b \in B$.

IX. Cancellation laws

(i) If $a + b = a + c$ and $a' + b = a' + c$, then $b = c$

(ii) If $a.b = a.c$ and $a'.b = a'.c$, then $b = c$.

X. *An expression within bracket is simplified by observing precedence of ′ over '.' and precedence of '.' over '+'.*

Example 1. *In the Boolean algebra (B, +, ., ′), show that*

$$a.b + a.b' + a'.b + a'.b' = 1 \quad \forall\, a, b \in B.$$

Sol. $a.b + a.b' + a'.b + a'.b'$

$= a.(b + b') + a'.(b + b')$ (By using $\mathbf{B_2}$)

$= a.1 + a'.1$ (By using $\mathbf{B_4}$)

$= a + a' = 1$ (By using $\mathbf{B_3}$, $\mathbf{B_4}$)

$\therefore$ $\mathbf{a.b + a.b' + a'.b + a'.b' = 1.}$

Example 2. *If a, b, c are elements of the Boolean algebra (B, +, ., ′), find the complement of (a + b).(a+ c′).*

Sol. Complement of $(a + b).(a + c') = [(a + b).(a + c')]'$

$= (a + b.c')'$ (By using $\mathbf{B_2}$)

$= a'.(b.c')'$ (By using De Morgan's law)

$= a'.(b' + c'')$ (By using De Morgan's law)

$= \mathbf{a'.(b' + c)}$. (By using involution law)

Example 3. *Prove that a Boolean algebra cannot have three elements.*

Sol. If possible, let (B, +, ., ′, 0, 1) be a Boolean algebra have exactly three elements.

Let a be the third element of B other than 0 and 1.

Since $0' = 1$, $1' = 0$, we must have $a' = a$.

($a' = 0 \Rightarrow a'' = 0' \Rightarrow a = 1$, which is not true. Similarly, $a' \neq 1$)

We have $a.a' = 0$ (By using $\mathbf{B}_4$)

$\therefore$ $a.a = 0$ ($\because$ $a' = a$)

$a = 0$ (By using idempotent law)

This is impossible, because $a \neq 0$.

$\therefore$ Our supposition is wrong.

$\therefore$ There cannot be a Boolean algebra containing three elements.

Example 4. *Show that in the Boolean algebra $(B, \vee, \wedge, ')$:*

(i) $a \vee (a' \wedge b) = a \vee b$

(ii) $a \wedge (a' \vee b) = a \wedge b$, *where* $a, a', b \in B$.

Sol. *(i)* $a \vee (a' \wedge b)$

$= (a \vee a') \wedge (a \vee b)$ (By using $\mathbf{B}_2$)

$= 1 \wedge (a \vee b)$ (By using $\mathbf{B}_4$)

$= a \vee b.$ (By using $\mathbf{B}_3$)

(ii) $a \wedge (a' \vee b)$

$= (a \wedge a') \vee (a \wedge b)$ (By using $\mathbf{B}_2$)

$= 0 \vee (a \wedge b)$ (By using $\mathbf{B}_4$)

$= a \wedge b.$ (By using $\mathbf{B}_3$)

Example 5. *In the Boolean algebra $(B, +, ., ')$, show that :*

(i) $(a + b).a'.b' = 0$ *(ii)* $(a + b)' + (a + b')' = a'$

(iii) $a.b + [(a + b').b]' = 1$ *(iv)* $[a + a'.b'].[a' + a.b] = a'.b' + a.b.$

Sol. *(i)* $(a + b).a'.b' = (a + b).(a + b)'$ (By using De Morgan's law)

$= 0$ (By using $\mathbf{B}_4$)

$\therefore$ $\mathbf{(a + b).a'.b' = 0.}$

(ii) $(a + b)' + (a + b')'$

$= a'.b' + a'.b''$ (By using De Morgan's law)

$= a'.(b' + b'')$ (By using $\mathbf{B}_2$)

$= a'.(b' + b)$ (Using involution law)

$= a'.1$ (By using $\mathbf{B}_4$)

$= a'$ (By using $\mathbf{B}_3$)

$\therefore$ $\mathbf{(a + b)' + (a + b')' = a'.}$

(iii) $a.b + [(a + b').b]'$

$= a.b + [b.(a + b')]'$ (By using $\mathbf{B}_1$)

$= a.b + (b.a + b.b')'$ (By using $\mathbf{B}_2$)

$= a.b + (a.b + 0)'$ (By using $\mathbf{B}_1$, $\mathbf{B}_4$)

$= a.b + (a.b)'$ (By using $\mathbf{B}_3$)

$= 1$ (By using $\mathbf{B}_4$)

$\therefore$ $\mathbf{a.b + [(a + b').b]' = 1.}$

(iv) $[a + a'.b'].[a' + a.b] = (a + a').(a + b').(a' + a).(a' + b)$ (By using $\mathbf{B}_2$)

$= 1.(a + b').1.(a' + b)$ (By using $\mathbf{B}_4$)

$= (a + b').(a' + b)$ (By using $\mathbf{B}_3$)

$= (a + b').a' + (a + b').b$ (By using $\mathbf{B}_2$)

$= a'.(a + b') + b.(a + b')$ (By using $\mathbf{B_1}$)

$= a'.a + a'.b' + b.a + b.b'$ (By using $\mathbf{B_2}$)

$= 0 + a'.b' + a.b + 0$ (By using $\mathbf{B_4}, \mathbf{B_1}$)

$= a'.b' + a.b$ (By using $\mathbf{B_3}$)

$\therefore$ $\mathbf{[a + a'.b'].[a' + a.b] = a'.b' + a.b.}$

Example 6. *Let (B, +, ., ') be a Boolean algebra. If for a, b, c ∈ B,*

$$a + b = a + c \quad \text{and} \quad a.b = a.c, \quad \text{then} \quad b = c.$$

Sol. $b = b.(b + a)$ (By using absorption law)

$= b.(a + b)$ (By using $\mathbf{B_1}$)

$= b.(a + c)$ (Given)

$= b.a + b.c$ (By using $\mathbf{B_2}$)

$= a.b + b.c$ (By using $\mathbf{B_1}$)

$= a.c + b.c$ (Given)

$= c.a + c.b$ (By using $\mathbf{B_1}$)

$= c.(a + b)$ (By using $\mathbf{B_2}$)

$= c.(a + c)$ (Given)

$= c.(c + a)$ (By using $\mathbf{B_1}$)

$= c$ (By using absorption law)

$\therefore$ $\mathbf{b = c.}$

EXERCISE 10.2

1. In a Boolean algebra prove that
   (*i*) zero element is unique. (*ii*) unit element is unique.
2. In the Boolean algebra (*B*, +, ., '), show that
   (*i*) $(a + b).c = (a.c) + (b.c)\ \forall\ a, b, c \in B$ (*ii*) $(a.b) + c = (a + c).(b + c)\ \forall\ a, b, c \in B$.
3. In the Boolean algebra (*B*, +, .,') show that
   (*i*) $(a' + b')' = a.b\ \forall\ a, b \in B$ (*ii*) $(a'.b')' = a + b\ \forall\ a, b \in B$.
4. In the Boolean algebra (*B*, +, .,'), show that
   (*i*) $a.(a' + b) = a.b\ \forall\ a, b \in B$ (*ii*) $a + (a'.b) = a + b\ \forall\ a, b \in B$.
5. In the Boolean algebra (*B*, +, .,'), show that
   (*i*) $a + (a + b) = a + b\ \ \forall\ a, b \in B$ (*ii*) $a.(a.b) = a.b\ \ \forall\ a, b \in B$.
6. In the Boolean algebra (*B*, +, .,'), show that
   (*i*) $a + b + c.a' = a + b + c\ \ \forall\ a, b, c \in B$ (*ii*) $a.b + c.(a' + b') = a.b + c\ \ \forall\ a, b, c \in B$.
7. In the Boolean algebra (*B*, +, .,'), simplify : $x.(x + y) + ((y' + x).y))'$.
8. If *a*, *b*, *c* are elements of the Boolean algebra (*B*, +, .,'), then find the complements of the following expressions :
   (*i*) $a + b'$ (*ii*) $a''.b'$
   (*iii*) $a.b.c'$ (*iv*) $b + c.a'$.
9. State and prove the following laws in a Boolean algebra :
   (*i*) Idempotent laws (*ii*) Boundedness laws
   (*iii*) Absorption laws (*iv*) Involution law.

10. State and prove the following laws in a Boolean algebra :
   (*i*) Associative law of addition   (*ii*) Associative law of multiplication.
11. In the Boolean algebra $(B, +, ., ')$, show that
   $(a + b).(c + d) = a.c + a.d + b.c + b.d \quad \forall\, a, b, c, d \in B.$
12. In the Boolean algebra $(B, +, .,')$, show that
   $(a + b).(a' + c) = a.c + a'.b \quad \forall\, a, b, c \in B.$
13. In the Boolean algebra $(B, +, .,')$, show that for $a, b \in B$
   (*i*) $a + b = 0$ if and only if $a = 0, b = 0$ (*ii*) $a.b' = 0$ if and only if $a' + b = 1$.
14. In the Boolean algebra $(B, +, .,')$, show that $(a + b).[a.b' + b]' = 0 \quad \forall\, a, b \in B.$
15. In the Boolean algebra $(B, +, .,')$, show that
   (*i*) $(a + b + c)' = a'.b'.c' \quad \forall\, a, b, c \in B$ (*ii*) $(a.b.c)' = a' + b' + c' \quad \forall\, a, b, c \in B$
   (*iii*) $a + a.(b + 1) = a \quad \forall\, a, b \in B$ (*iv*) $(a + b).(a + 1) = a + b \quad \forall\, a, b \in B.$
16. Let $(B, + , ., ')$ be a Boolean algebra. Prove that for $a, b \in B$:
   $a.b' = 0$ if and only if $a.b = a$.
17. In the Boolean algebra $(B, +, . , ')$, show that
   $(x + y)(x + z)(y + z) = xz + xy + yz, \forall\, x, y, z \in B.$ (*KU 2006 C*)

Answers

7. 1
8. (*i*) $a'.b$ (*ii*) $a' + b$ (*iii*) $a' + b' + c$ (*iv*) $b'.(c' + a)$.

Hints

5. Use idempotent laws.
6. (*i*) L.H.S. $= (a + c.a') + b = (a + c).(a + a') + b$
   (*ii*) L.H.S. $= a.b + c.(a.b)' = (a.b + c).(a.b + (a.b)')$
7. $x.(x + y) + ((y' + x).y)' = x + (y'.y + x.y)'$
   $= x + (0 + x.y)' = x + (x.y)' = x + (x' + y') = (x + x') + y' = 1 + y' = 1.$
12. L.H.S. $= (a + b).a' + (a + b).c = a'.a + a'.b + c.(a + b).(a + a')$
   $= a'.b + c.(a + b.a') = a'.b + c.a + c.b.a'$
   $= a'.b.1 + a'.b.c + a.c = a'.b.(1 + c) + a.c$
13. (*i*) Let $a + b = 0$.
   $a = a + 0 = a + b.b' = (a + b).(a + b').$

10.6. PRINCIPLE OF DUALITY

Let $(B, +, ., ', 0, 1)$ be a Boolean algebra. The **dual** of any statement in this Boolean algebra is defined to be the statement that is obtained by interchanging the binary operations '+' and '.' and interchanging identity elements 0 and 1 in the given statement.

For example, the dual of $x + [x.(y + 1)] = x$ is $x.[x + (y.0)] = x$.

In the definition of the Boolean algebra $(B, +, ., ', 0, 1)$, the dual of each axiom is also an axiom.

$\therefore$ The dual of any theorem in a Boolean algebra is also a theorem because the dual theorem can be proved by using the dual of each step of the proof of the original theorem.

Thus, we see that the dual of any theorem in a Boolean algebra is also a theorem. This is known as the **principle of duality.**

Example 1. *Find the duals of the followings axioms of the Boolean algebra (B, +, ., ′, 0, 1) :*

(*i*) $a + b = b + a \ \forall\, a, b \in B$ (*ii*) $a + (b.c) = (a + b).(a + c) \ \forall\, a, b, c \in B$

(*iii*) $a + 0 = a \ \forall\, a \in B$ (*iv*) $a + a' = 1 \ \forall\, a \in B$.

Sol. (*i*) We have $a + b = b + a \ \forall\, a, b \in B$.

Replacing '+' by '.', we get

$$\mathbf{a.b = b.a \ \forall\, a, b \in B}$$

This is the dual statement of the given statement.

(*ii*) We have $a + (b.c) = (a + b).(a + c) \ \forall\, a, b, c \in B$.

Replacing '+ ' by '.' and '.' by '+', we get

$$\mathbf{a.(b + c) = (a.b) + (a.c) \quad \forall\, a, b, c \in B}$$

This is the dual statement of the given statement.

(*iii*) We have $a + 0 = a \ \forall\, a \in B$.

Replacing '+' by '.' and 0 by 1, we get

$$\mathbf{a.1 = a \ \forall\, a \in B}$$

This is the dual statement of the given statement.

(*iv*) We have $a + a' = 1 \quad \forall\, a \in B$

Replacing '+' by '.' and 1 by 0, we get

$$\mathbf{a.a' = 0 \ \forall\, a \in B}$$

This is the dual statement of the given statement.

Example 2. *Write the absorption laws for the Boolean algebra (B, +, .,′) and verify that each is dual of the other.*

Sol. The *absorption laws* for the Boolean algebra (*B*, +, .,′) are :

(*i*) $a.(a + b) = a \ \forall\, a, b \in B$ (*ii*) $a + a.b = a \quad \forall\, a, b \in B$.

Interchanging '+' and '.' in (*i*), we get $a + a.b = a \quad \forall\, a, b \in B$.

This is the statement (*ii*).

Interchanging '+ ' and '.' in (*ii*), we get $a.(a + b) = a \quad \forall\, a, b \in B$.

This is the statement (*i*).

∴ Statement (*i*) and (*ii*) are dual of each other.

Example 3. *In the Boolean algebra (B, +, .,′), find the truth value of the dual statement of the statement*

$$(a + b).(c + d) = a.c + a.d + b.c + b.d \ \forall\, a, b, c, d \in B.$$

Sol. Given statement is $(a + b).(c + d) = a.c + a.d + b.c + b.d \quad \forall\, a, b, c, d \in B$.

The dual of this statement is $a.b + c.d = (a + c).(a + d).(b + c).(b + d) \quad \forall\, a, b, c, d \in B$ ...(1)

L.H.S. of (1) $= a.b + c.d$

$= (a.b + c).(a.b + d)$ (By using $\mathbf{B_2}$)

$= (c + a.b).(d + a.b)$ (By using $\mathbf{B_1}$)

$= (c + a).(c + b).(d + a).(d + b)$ (By using $\mathbf{B_2}$)

$= (a + c).(a + d).(b + c).(b + d)$

$=$ R.H.S. of (1) (By using $\mathbf{B_1}$)

∴ The truth value of the dual statement (1) is '**T**'.

Example 4. *In the Boolean algebra (B, +, ., ′, 0, 1), show that the dual statement of the statement*

$$a.b.c + a' + b' + c' = 1 \quad \forall\ a, b, c \in B \text{ is true.}$$

Sol. Given statement is $a.b.c + a' + b' + c' = 1 \quad \forall\ a, b, c \in B.$

The dual of this statement is $(a + b + c).\ a'.b'.c' = 0 \quad \forall\ a, b, c \in B$ ...(1)

L.H.S. of (1) $= (a + b + c).a'.b'.c'$

$= (a + b + c).((a + b)'.c')$ (By using De Morgan's law)

$= (a + b + c).((a + b) + c)'$ (By using De Morgan's law)

$= 0$ (By using $\mathbf{B_4}$)

$=$ R.H.S. of (1)

$\therefore$ The dual statement (1) is true.

EXERCISE 10.3

1. Write the dual of the following statements in the Boolean algebra (B, + , ., ′) :
   (i) $(x' + y')' = x.y \quad \forall\ x, y \in B$ (ii) $(x.y)' = x' + y' \quad \forall\ x, y \in B$
   (iii) $x.y' + y = x + y \quad \forall\ x, y \in B$ (iv) $x + [(y' + x).y]' = 1 \quad \forall\ x, y \in B.$
2. In the Boolean algebra (B, +, .,′), verify that the *idempotent laws* are dual of each other.
3. In the Boolean algebra (B, +, .,′), verify that the *boundedness laws* are dual of each other.
4. In the Boolean algebra (B, +, ., ′, 0, 1), show that the dual statement of the statement $a.(a'.b) = 0 \quad \forall \quad a, b \in B$ is true.
5. In the Boolean algebra (B, +, ., ′, 0, 1), find the truth value of the dual statement of the statement $x.y + x.y' + x'.y + x'.y' = 1\ \forall\ x, y \in B.$

Answers

1. (i) $(x'.y')' = x + y \quad \forall\ x, y \in B$ (ii) $(x + y)' = x'.y' \quad \forall\ x, y \in B$
   (iii) $(x + y').y = x.y \quad \forall\ x, y \in B$ (iv) $x.[y'.x + y]' = 0 \quad \forall\ x, y \in B$
5. *T.*

10.7. BOOLEAN ALGEBRA (As a lattice)

A lattice is called a **Boolean algebra** if it is distributive and complemented.

Theorem 1. If (B, +, ., ′, 0, 1) is a Boolean algebra then prove that it is also a Boolean algebra as a lattice.

Proof. (B, +, ., ′, 0, 1) is a Boolean algebra.

$\therefore$ B satisfies commutative laws, associative laws and absorption laws w.r.t. binary operations '+' and '.'

We define a relation '≤' on B as follows :

For $a, b \in B$, we say $a \le b$ if $a + b = b$.

We note that

$$a + b = b \Rightarrow a.b = a.(a + b) = a$$

and

$$a.b = a \Rightarrow a + b = (a.b) + b = b + (b.a) = b$$

$\therefore$ $a + b = b$ iff $a.b = a$

$\therefore$ $a \leq b$ if $a + b = b$ (or equivalently $a.b = a$)

The relation '$\leq$' is a partial order on B. (Check it !)

Also for any $a, b \in B$, we can show that $\sup(a, b) = a + b$ and $\inf(a, b) = a \cdot b$

$\therefore$ For any $a, b \in B$, $\sup(a, b)$ and $\inf(a, b)$ exist.

$\therefore$ The partially ordered set $(B, \leq)$ is a lattice.

By definition of a Boolean algebra, $(B, +, ., ')$ satisfies the distributive laws :

$$a + (b.c) = (a + b).(a + c) \quad \text{and} \quad a.(b + c) = a \cdot b + a \cdot c$$

$\therefore$ The lattice $(B, \leq)$ is distributive.

The Boolean algebra B satisfies the *boundedness laws :*

$$a + 1 = 1 \quad \text{and} \quad a.0 = 0 \text{ for } a \in B$$

$\therefore$ By definition of '$\leq$', we have $a \leq 1$ and $0 \leq a$ for $a \in B$.

$\therefore$ 0 and 1 are respectively the least element and the greatest element of B.

$\therefore$ The lattice $(B, \leq)$ is bounded.

Let $a \in B$. $\therefore$ There exists an element a' in B such that

$$a + a' = 1 \quad \text{and} \quad a \cdot a' = 0$$

$\therefore$ a' is complement of a in the lattice $(B, \leq)$.

$\therefore$ The lattice $(B, \leq)$ is also complemented.

$\therefore$ The lattice $(B, \leq)$ is distributive and complemented.

$\therefore$ The lattice $(B, \leq)$ is a Boolean algebra.

$\therefore$ The Boolean algebra $(B, +, ., ', 0, 1)$ is also a Boolean algebra as a lattice.

Theorem 2. If the lattice (B, ≤) is a Boolean algebra then prove that it is also a Boolean algebra as an algebraic structure.

Proof. $(B, \leq)$ is a lattice.

$\therefore$ $\sup(a, b)$ and $\inf(a, b)$ exist for any $a, b \in B$.

We define binary operations + and . in B as follows :

For $a, b \in B$, $\quad a + b = \sup(a, b)$ and $a.b = \inf(a, b)$

For $a, b \in B$, $\quad a + b = \sup(a, b) = \sup(b, a) = b + a$

and $\quad a.b = \inf(a, b) = \inf(b, a) = b.a$

$\therefore$ $\quad a + b = b + a$ and $a.b = b.a$

$\therefore$ **Commutative laws** hold.

Since the lattice $(B, \leq)$ is a Boolean algebra, it is distributive and complemented.

$\therefore$ The lattice $(B, \leq)$ is distributive.

$\therefore$ For $a, b, c \in B$,

$$a + (b.c) = (a + b).(a + c) \quad \text{and} \quad a.(b + c) = a.b + a.c$$

$\therefore$ **Distributive laws** hold.

The lattice $(B, \leq)$ being complemented is bounded. Let 0 and 1 be the least element and the greatest element respectively of the lattice $(B, \leq)$.

$\therefore$ $0 \leq a$ and $a \leq 1$ for every $a \in B$.

$\therefore$ By definition of '+' and '.', we have

$$a + 0 = \sup(a, 0) = a \quad \text{and} \quad a.1 = \inf(a, 1) = a$$

$\therefore$ $$a + 0 = a \quad \text{and} \quad a.1 = a$$

$\therefore$ **Identity laws** hold. Here 0 and 1 are respectively the *zero element* and the *unit element.*

The lattice $(B, \leq)$ is complemented.

$\therefore$ Every element of the lattice has a complement.

Since the lattice $(B, \leq)$ is bounded and distributive, each element of B has a *unique* complement.

$\therefore$ For each $a \in B$, there exists a unique complement, say a'.

This defines a unary operation on B.

Also, by definition of complement, we have

$$a + a' = 1 \quad \text{and} \quad a.\ a' = 0.$$

$\therefore$ **Complement laws** are satisfied.

$\therefore$ $(B, +, ., ', 0, 1)$ is a Boolean algebra.

$\therefore$ The lattice $(B, \leq)$ is a Boolean algebra as an algebraic structure.

Remarks. In the above theorems we have proved that :

(*i*) if $(B, +, ., ', 0, 1)$ is a Boolean algebra as an algebraic structure then B is a Boolean algebra as a lattice with the partial order $\leq$ on B defined as $a \leq b$ if $a + b = b$ (or $a.b = a$), where $a, b \in B$.

The lattice B is distributive. The lattice B is bounded by the least element 0 and the greatest element 1. The lattice B is also complemented. The complement of a in B is a'.

(*ii*) if $(B, \leq)$ is a Boolean algebra as a lattice then B is a Boolean algebra as an algebraic structure with two binary operations '+' and '.' defined as $a + b = \sup(a, b)$ and $a.b = \inf.(a, b)$, where $a, b \in B$. The unary operation ''' on B is defined as a' = complement of a in B.

Example 1. *Let S be a non-empty set. Let B be the set of all subsets of the set S. Show that the set B is a Boolean algebra under the partial order of set inclusion.*

Sol. For $X, Y \in B$, we define $X \leq Y$ if $X \subseteq Y$.

'$\leq$' is a partial order on S.

Also, $\sup(X, Y) = X \cup Y \in B$ and $\inf(X, Y) = X \cap Y \in B$ for every $X, Y \in B$.

$\therefore$ $(B, \leq)$ is a lattice.

For $X, Y, Z \in B$,

$$X \vee (Y \wedge Z) = X \cup (Y \cap Z) = (X \cup Y) \cap (X \cup Z) = (X \vee Y) \wedge (X \vee Z)$$

and $$X \wedge (Y \vee Z) = X \cap (Y \cup Z) = (X \cap Y) \cup (X \cap Z) = (X \wedge Y) \vee (X \wedge Z)$$

$\therefore$ The lattice $(B, \leq)$ is distributive.

Also, $\phi \leq X$, $X \leq S$ for $X \in B$.

$\therefore$ The lattice $(B, \leq)$ is bounded with the least element ϕ and the greatest element S.

For $X \in B$, we have $X' = S - X \in B$

and $$X \vee X' = X \cup X' = S \quad \text{and} \quad X \wedge X' = X \cap X' = \phi$$

$\therefore$ X' is the complement of X.

$\therefore$ The lattice $(B, \leq)$ is complemented.

$\therefore$ The lattice $(B, \leq)$ is a Boolean algebra.

Example 2. *For m $(\geq 2) \in$* **N**, *let D_m be the set of all positive divisors of m. Show that D_m is a Boolean algebra if m is the product of distinct primes.*

Sol. We know that the set D_m is a lattice under the relation of divisibility.

1 and m are respectively the least element and the greatest element of D_m. This lattice is also distributive. Let m be the product of distinct primes.

Let $x \in D_m$. Let $x' = \dfrac{m}{x}$

Since m is the product of distinct primes, we have $x + x' = \sup(x, x') = m$ and $x.x' = \inf(x, x') = 1$.

$\therefore$ x' is the complement of x.

$\therefore$ The lattice D_m is distributive and complemented.

$\therefore$ D_m is a Boolean algebra.

Remark. We can also show that if D_m is a Boolean algebra then m is square free.

For example let $m = 12$.

$\therefore$ $m = 2^2 . 3$ is not square free.

If $x = 2$, then there is no element y of D_{12} for which $x + y = 12$ and $x . y = 1$.

$\therefore$ D_{12} is not a Boolean algebra.

Example 3. *If D_m is a Boolean algebra, find the atoms of D_m.*

Sol. Since D_m is a Boolean algebra, m must be a product of distinct primes.

Let $m = p_1 p_2 \ldots\ldots p_k$, where $p_1, p_2, \ldots\ldots, p_k$ are distinct prime numbers. We know that B is a lattice with lower bound 1. The immediate successors of 1 are the atoms of B.

Since $1 << p_1, 1 << p_2, \ldots\ldots, 1 << p_k$, the atoms of the Boolean algebra B are the prime numbers $p_1, p_2, \ldots\ldots, p_k$.

10. 8. ISOMORPHIC BOOLEAN ALGEBRA

Two Boolean algebras B and B' are called **isomorphic Boolean algebras** if there exists a one-one onto mapping $f : B \to B'$ such that

$$f(a + b) = f(a) + f(b), \ f(a.b) = f(a).f(b)$$

and $$f(a') = (f(a))' \text{ for } a, b \in B.$$

For the sake of convenience, we have denoted the binary operations and the unary operation on B and B' by identical symbols.

10.9. REPRESENTATION THEOREM

The representation theorem is of fundamental importance. This theorem gives the result regarding the number of elements in a finite Boolean algebra.

Statement. Let (B, +, ., ′, 0, 1) be a finite Boolean algebra. Let A be the set of all atoms of the Boolean algebra B. If A contains n elements then prove that B has 2^n elements.

Proof. A is the set of all atoms of B. As a lattice, B is distributive and complemented.

$\therefore$ B is a finite distributive and complemented lattice.

$\therefore$ Every element in B can be written uniquely (except for order) as the sum of irredundant atoms of B.

Let $P(A)$ be the Boolean algebra of the set of all subsets of A under the operations '$\cup$', '$\cap$' and '′'.

If $x \in B$, then there exist unique atoms $a_1, \ldots\ldots, a_n$ of B such that $x = a_1 + \ldots\ldots + a_n$.

Define $f : B \to P(A)$ by

$$f(x) = \{a_1, \ldots\ldots, a_n\} \in P(A)$$

Let $x = a_1 + \ldots\ldots + a_r + b_1 + \ldots\ldots + b_s$

and $y = b_1 + \ldots\ldots + b_s + c_1 + \ldots\ldots + c_t$ be any elements of B and

$$A = \{a_1, \ldots\ldots, a_r, b_1, \ldots\ldots, b_s\ c_1, \ldots\ldots, c_t, d_1, \ldots\ldots, d_k\}$$

In x and y, the common atoms are $b_1, \ldots, b_s$.

$d_1, \ldots, d_k$ are the atoms of A which are not involved in x and y.

If $e_i \in A$ then $e_i + e_i = e_i$ and $e_i.e_i = e_i$.

If $e_i, e_j \in A$ then $e_i.e_j = 0$, whenever $i \neq j$ because e_i and e_j are both immediate successors of 0.

$\therefore$ $x + y = a_1 + \ldots\ldots + a_r + b_1 + \ldots\ldots + b_s + c_1 + \ldots\ldots + c_t$

and $xy = b_1 b_1 + \ldots\ldots + b_s b_s = b_1 + \ldots\ldots + b_s$

$\therefore$
$$\begin{aligned} f(x+y) &= \{a_1, \ldots\ldots, a_r, b_1, \ldots\ldots, b_s, c_1, \ldots\ldots, c_t\} \\ &= \{a_1, \ldots\ldots, a_r, b_1, \ldots\ldots, b_s\} \cup \{b_1, \ldots\ldots, b_s, c_1, \ldots\ldots, c_t\} \\ &= f(x) \cup f(y) \end{aligned}$$

and
$$\begin{aligned} f(xy) &= \{b_1, \ldots\ldots, b_s\} \\ &= \{a_1, \ldots\ldots, a_r, b_1, \ldots\ldots, b_s\} \cap \{b_1, \ldots\ldots, b_s, c_1, \ldots\ldots, c_t\} \\ &= f(x) \cap f(y) \end{aligned}$$

Let $x' = c_1 + \ldots\ldots + c_t + d_1 + \ldots\ldots + d_k$

$\therefore$ $x + x' = a_1 + \ldots\ldots + a_r + b_1 + \ldots\ldots + b_s + c_1 + \ldots\ldots + c_t + d_1 + \ldots\ldots + d_k$

The element '1' of B is also expressible as the sum of irredundant atoms of B.

Let $1 = e_1 + \ldots\ldots + e_m$ and $A = \{e_1, \ldots\ldots, e_m, e_{m+1}, \ldots\ldots, e_p\}$

$\therefore$ $e_1 + \ldots\ldots + e_m \leq (e_1 + \ldots\ldots + e_m) + (e_{m+1} + \ldots\ldots + e_p)$

$\Rightarrow$ $1 \leq e_1 + \ldots\ldots + e_p$

Also $e_1 + \ldots\ldots + e_p \leq 1$ ($\because$ Each $e_i \leq 1$)

$\therefore$ $e_1 + \ldots\ldots + e_p = 1$

$\therefore$ Sum of all atoms of B is 1.

$\therefore$ $a_1 + \ldots\ldots + a_r + b_1 + \ldots\ldots + b_s + c_1 + \ldots\ldots + c_t + d_1 + \ldots\ldots + d_k = 1$

$\Rightarrow$ $x + x' = 1$

Also $xx' = 0$ ($\because$ $e_i.e_j = 0$ for $i \neq j$)

$\therefore$ x' is the complement of x in B.

Also
$$\begin{aligned} f(x') &= f(c_1 + \ldots\ldots + c_t + d_1 + \ldots\ldots + d_k) \\ &= \{c_1, \ldots\ldots, c_t, d_1, \ldots\ldots, d_k\} \\ &= \{a_1, \ldots\ldots, a_r, b_1, \ldots\ldots, b_s\}' = (f(x))' \end{aligned}$$

The mapping f is also one-one and onto.

$\therefore$ Mapping $f : B \to P(A)$ is a Boolean algebra isomorphism.

$\therefore$ No. of elements in B

$$= \text{no. of elements in } P(A) = 2^n.$$

Remarks 1. By using the above theorem, a finite Boolean algebra can have only 2, 4, 8, 16, 32, ..., elements.

2. Let D_m be a Boolean algebra $\therefore$ m is the product of distinct prime number.

Let $m = p_1 p_2 \ldots\ldots p_k$

$\therefore$ The atoms of D_m are prime numbers $p_1, p_2, \ldots\ldots, p_k$. These are k in number.

If $A = \{p_1, p_2, \ldots, p_k\}$, then B is isomorphic to $P(A)$.

$$\therefore \qquad n(B) = n(P(A)) = 2^k.$$

For example, $\quad 20 = 2 \times 3 \times 5$.

$\therefore$ D_{20} has 8 (= 2^3) elements.

Example 4. *Which of the lattices : $D_6, D_{12}, D_{20}, D_{55}, D_{99}, D_{210}$ are Boolean algebras and what are their atoms ?*

Sol. D_6 $\quad 6 = 2 \times 3$ is a product of distinct primes.

$\therefore$ D_6 is a Boolean algebra and its atoms are **2** and **3**.

D_{12} $\quad 12 = 2^2 \times 3$ is not a product of distinct primes.

$\therefore$ D_{12} is not a Boolean algebra.

D_{20} $\quad 20 = 2^2 \times 5$ is not a product of distinct primes.

$\therefore$ D_{20} is not a Boolean algebra

D_{55} $\quad 55 = 5 \times 11$ is a product of distinct primes.

$\therefore$ D_{55} is a Boolean algebra and its atoms are **5** and **11**.

D_{99} $\quad 99 = 3^3 \times 11$ is not a product of distinct primes.

$\therefore$ D_{99} is not a Boolean algebra.

D_{210} $\quad 210 = 2 \times 3 \times 5 \times 7$ is a product of distinct primes.

$\therefore$ D_{210} is a Boolean algebra and its atoms are **2, 3, 5** and **7**.

10.10. SUBALGEBRA

Let $(B, +, ., ', 0, 1)$ be a Boolean algebra. A non-empty set C of B is called a **subalgebra** of B if C itself is a Boolean algebra with respect to the operations of B. It is easy to check that a non-empty subset C of B is a subalgebra of B if and only if C is closed under the operations '+', '.' and '''.

Example 5. *Show that D_{105} is a Boolean algebra. Draw the Hasse diagram of D_{105}. Find the complements, atoms and subalgebras of D_{105}.*

Sol. We have $105 = 3 \times 5 \times 7$

$\therefore$ 105 is the product of distinct prime numbers.

$\therefore$ D_{105} is a Boolean algebra.

$$D_{105} = \{1, 3, 5, 7, 15, 21, 35, 105\}$$

The Hasse diagram of D_{105} is shown in the figure.

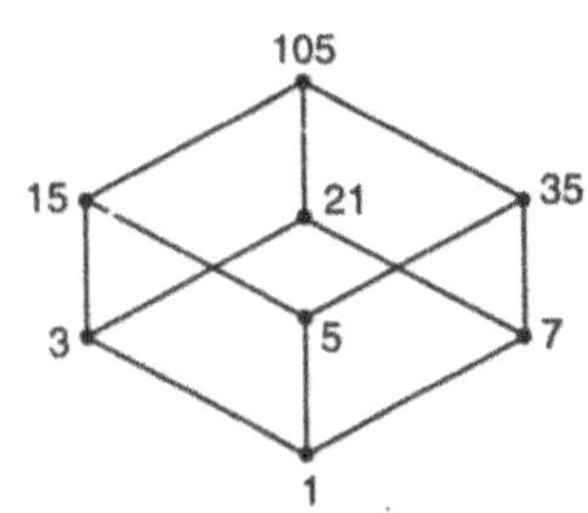

We have $\quad 1' = 105, 3' = 35,$

$$5' = 21, 7' = 15, 15' = 7, 21' = 5, 35' = 3, 105' = 1.$$

The immediate successors of the lower bound '1' are 3, 5, 7.

$\therefore$ The atoms of D_{105} are **3, 5, 7.**

$$0(D_{105}) = 2^3 = 8$$

$\therefore$ A subalgebra of D_{105} can have 2 or 4 or 8 elements.

$\{1, 105\}$ and D_{105} are subalgebras of order 2 and 8 respectively.

Let $\{1, x, x', 105\}$ be a subalgebra of B_{105} of order 4.

$\therefore$ x can be either 3 or 5 or 7.

$\therefore$ The subalgebras of D_{105} of order 4 are {1, 3, 35, 105}, {1, 5, 21, 105}, and {1, 7, 15, 105}.

$\therefore$ The subalgebras of D_{105} are **{1, 105}, {1, 3, 35, 105}, {1, 5, 21, 105}, {1, 7, 15, 105}** and D_{105}.

EXERCISE 10.4

1. Is D_{110} a Boolean algebra ? Draw its Hasse diagram. If D_{110} is a Boolean algebra find :
(*i*) the atoms and (*ii*) complements of all elements in D_{110}. Also find all subalgebras of order 4.
2. If B is a Boolean algebra, show that :
(*i*) $0 \le x \le 1$ for $x \in B$ (*ii*) $a < b \Leftrightarrow b' < a'$ for $a, b \in B$.
3. Show that none of the lattices shown in the following diagrams is a Boolean algebra :

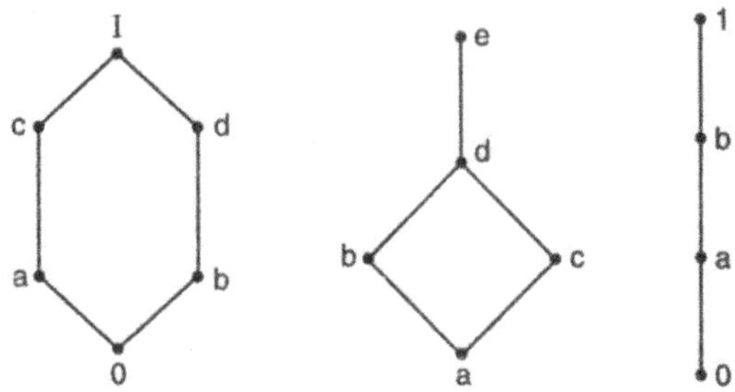

(i) (ii) (iii)

4. Show that the subsets {1, 2, 35, 70} and {1, 7, 10, 70} are subalgebras of the Boolean algebra D_{70}.
5. Find all atoms and subalgebras of order 4 of the Boolean algebra D_{30}.
6. Show that the subset {1, 2, 3, 6} of D_{30} is not a subalgebra of Boolean algebra D_{30}, but it is a Boolean algebra in itself.

Answers

1. Yes, (*i*) 2, 5, 11
(*ii*) $1' = 110,\ 2' = 55,\ 5' = 22,\ 10' = 11,$
$11' = 10,\ 22' = 5,\ 55' = 2,\ 110' = 1,$
Subalgebras of order 4 are {1, 2, 55, 110}, {1, 5, 22, 110}, {1, 10, 11, 110}.

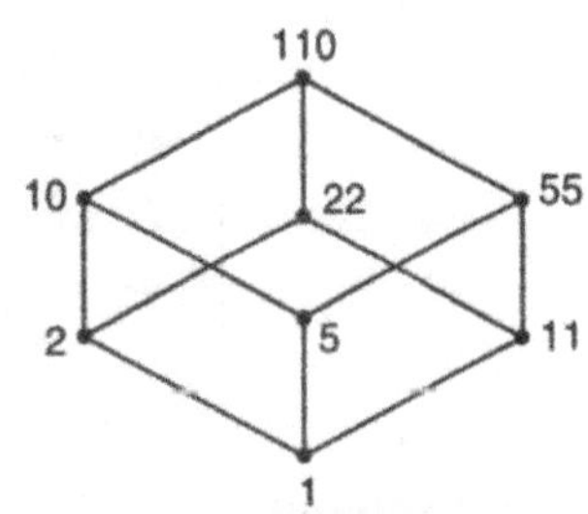

5. 2, 3, 5, subalgebras of order 4 are {1, 2, 15, 30}, {1, 3, 10, 30} and {1, 5, 6, 30}.

Hints

2. (*i*) $x + 0 = x \Rightarrow \sup(x, 0) = x \Rightarrow 0 \le x$
$x \cdot 1 = x \Rightarrow \inf(x, 1) = x \Rightarrow x \le 1$
(*ii*) $a < b \Leftrightarrow \sup(a, b) = b,\ a \ne b \Leftrightarrow a + b = b,\ a \ne b \Leftrightarrow (a + b)' = b',\ a' \ne b'$
$\Leftrightarrow a'.b' = b', a' \ne b' \Leftrightarrow \inf(a', b') = b',\ a' \ne b' \Leftrightarrow b' \le a', a' \ne b' \Leftrightarrow b' < a'.$
3. (*i*) $6 \ne 2^n$ for any positive integer.
(*iii*) a', b' do not exist.
6. $6' = 5 \notin \{1, 2, 3, 6\}$.
{1, 2, 3, 6} being D_6 is a Boolean algebra because $6 = 2 \times 3$ is a product of distinct primes.

10.11. FUNDAMENTAL PRODUCT

We recall that a Boolean expression over a Boolean algebra $(B, +, ., ')$ is a constant *i.e.*, an element of B or a variable or a well formed expression using finite number of variables and Boolean operations '+', '.' and '''.

Let x be any variable. The variable x and its complemented variable x' are called **literals** corresponding to the variable x.

$\therefore$ If x, y, z are variables then x, x', y, y', z, z' are literals.

A literal or a product of two or more literals is called a **fundamental product** if no two literals involve the same variable.

For example : $y, xyz', x'z, x'y'z$ are fundamental products.

The product $xyzz'$ is not a fundamental product because the literals z and z' involve the same variable z.

By using the idempotent law $xx = x$ and the complement law $xx' = 0$, any product of literals can be reduced to either 0 or to a fundamental product.

Illustrations :

(*i*) $xyz'x' = (xx')yz' = 0yz' = 0$

(*ii*) $xxyzy = (xx)(yy)z = xyz.$

Example 1. *In the Boolean algebra (B, +, . , '), reduce the following products to either 0 or a fundamental product :*

(*i*) $yxyz$ (*ii*) $xyxx'$

(*iii*) $xxy'z'$ (*iv*) $xyyzz'x.$

Sol. (*i*) $yxyz = x(yy)z = \mathbf{xyz}$

(*ii*) $xyxx' = (xy)(xx') = (xy)0 = \mathbf{0}$

(*iii*) $xxy'z' = (xx)y'z' = \mathbf{xy'z'}$

(*iv*) $xyyzz'x = x(yy)(zz')\, x = (xy)0(x) = (xyx)(0) = \mathbf{0}.$

10.12. 'CONTAINED IN' FUNDAMENTAL PRODUCT

A fundamental product P_1 is said to be **contained in** another fundamental product P_2 if the literals of P_1 are also literals of P_2.

For example the fundamental product xy' is contained in the fundamental product $xy'z$. Also, xy' is not contained in xyz' because y' is a not a literal of xyz'.

Let the fundamental product P_1 be contained in the fundamental product P_2.

$\therefore$ The literals of P_1 are also literals of P_2.

$\therefore$ We can write $P_2 = P_1Q$, where Q is the product of literals of P_2 which are not literals of P_1.

$$\therefore \qquad P_1 + P_2 = P_1 + P_1Q = P_1 \qquad \text{(By absorption law)}$$

For example, $x'y + x'yz' = x'y + (x'y)z' = x'y.$

10.13. SUM-OF-PRODUCTS EXPRESSION

A Boolean expression is called a **sum-of-products expression** if it is a fundamental product or the sum of two or more fundamental products none of which is contained in another.

Illustration. In the Boolean algebra $(B, +, ., ')$:

(*i*) $xz + xy + xyz'$ is not a sum-of-products expression because xy is contained in xyz'.

(*ii*) $xy + xz + x'y'z + xy'z'$ is a sum-of-products expression because none of the terms is contained in another.

10.14. SUM-OF-PRODUCTS FORM

Let E be a Boolean expression. A sum-of-products expression is said to be a **sum-of-products form** of E if E is equivalent to this sum-of-products expression.

Illustration. In the Boolean algebra $(B, +, ., ')$, let $E(x, y) = x + xy + x'y$.

Since x is contained in xy, by absorption law, we have $x + xy + x'y = (x + xy) + x'y = x + x'y$.

$\therefore$ $x + x'y$ is a sum-of-products expression.

$\therefore$ $x + x'y$ is the sum-of-products form of the Boolean expression $x + xy + x'y$.

Remark. The sum-of-products form of a Boolean expression is also called the **disjunctive normal form** (DNF) of the given expression.

WORKING RULES FOR FINDING THE SUM-OF-PRODUCTS FORM OF A BOOLEAN EXPRESSION

Step I. *Wherever complement operation occur outside a bracket, move it inside the bracket by using De Morgan's laws.*

Step II. *Use 'involution law' wherever needed. After these operations, the given Boolean expression would have only sums and products of literals.*

Step III. *Use 'distributive laws' to reduce the resultant of* **step II** *into a sum of products of literals.*

Step IV. *Use commutative laws, complement laws and idempotent laws to transform each product obtained after* **step III** *to '0' or a fundamental product.*

Step V. *Use absorption laws and identity laws to finally transform the given Boolean expression in to a sum-of-products expression.*

Example 2. *In the Boolean algebra $(B, +, ., ')$, express the Boolean expression $x(y'z)'$ in its sum-of-products form.*

Sol. $x(y'z)' = x(y'' + z')$ (By using **step I**)

$= x(y + z')$ (By using **step II**)

$= \mathbf{xy + xz'}$. (By using **step III**)

This is the required sum-of-products form of the given expression.

Example 3. *In the Boolean algebra $(B, +, ., ')$, find the sum-of-products form of the Boolean expression $(x + y)'(x' + y)(x' + y')$.*

Sol. $(x + y)'(x' + y)(x' + y')$

$= (x'y')(x' + y)(x' + y')$ (By using De Morgan law)

$= (x'y'x' + x'y'y)(x' + y')$ (By using Distributive law)

$= (x'x'y' + x'0)(x' + y')$ (By using commutative and complement laws)

$= (x'y' + 0)(x' + y')$ (By using Idempotent and boundedness laws)

$= (x'y')(x' + y')$ (By using identity law)

$= x'y'x' + x'y'y'$ (By using distributive law)

$= x'x'y' + x'y'$ (By using commutative and idempotent law)

$= x'y' + x'y'$ (By using idempotent law)

$= \mathbf{x'y'}$. (By using idempotent law)

This is the required sum-of-products form of the given expression.

Example 4. *In the Boolean algebra $(B, +, ., ')$, find the sum-of-products form of the Boolean expression $(x + y)(x + z') + y + z'$.*

Sol. $(x + y)(x + z') + y + z'$

$$= (x + y)x + (x + y)z' + y + z'$$
$$= x(x + y) + z'(x + y) + y + z'$$
$$= xx + xy + z'x + z'y + y + z'$$
$$= x + xy + xz' + yz' + y + z'$$
$$= (x + xy) + (z' + xz') + (y + yz')$$
$$= \mathbf{x + z' + y}.$$

This is the required sum-of-products form of the given expression.

10.15. MINTERM

A fundamental product is called a **minterm** if it involves all the variables under consideration.

Let $(B, +, ., ')$ be a Boolean algebra.

If the variables under consideration are x and y then the minterms are $xy, xy', x'y, x'y'$.

If the variables under consideration are x, y and z then the minterms are $xyz, xyz', xy'z, x'yz, xy'z', x'yz', x'y'z, x'y'z'$.

If n variables are under consideration then the number of minterms is 2^n.

10.16. COMPLETE SUM-OF-PRODUCTS FORM

Let E be a Boolean expression. A sum-of-products expression is said to be a **complete sum-of-products** form of E if E is equivalent to this sum-of-products expression and each product in this sum involves all the variables under consideration *i.e.*, each products is a minterm.

Remark. The complete sum-of-products form of a Boolean expression can have at most 2^n terms.

Remark. The complete sum-of-products form of a Boolean expression is also called the **full disjunctive normal form** or the **disjunctive canonical form** or the **minterm canonical form** of the given expression.

WORKING RULES FOR FINDING THE COMPLETE SUM-OF-PRODUCTS FORM OF A BOOLEAN EXPRESSION

Step I *Find the sum-of-products form of the given Boolean expression* $E(x_1, x_2, ..., x_n)$. *Every product in the required form should involve all the variables* $x_1, x_2, ..., x_n$.

Step II *If a product obtained in* **step I** *is not a minterm and does not involve* x_i *(or* x_i'*), then multiple it by* $x_i + x_i'$ *and then use distributive law* $a.(b + c) = a.b + a.c$. *Repeat this process wherever needed. Ultimately, we shall get the complete sum-of-products form of the given Boolean expression.*

Example 5. *In the Boolean algebra* $(B, +, ., ')$, *express the Boolean expression* $E(x, y, z)$ *given by* $x'y + (x' + y)'$ *in its complete sum-of-products form.*

Sol. $E(x, y, z) = x'y + (x' + y)' = x'y + (x''y')$

$$= x'y + xy' = x'y(1) + xy'(1)$$
$$= x'y(z + z') + xy'(z + z')$$
$$= \mathbf{x'yz + x'yz' + xy'z + xy'z'}.$$

This is the required complete sum-of-products form of the given expression in variables x, y and z.

Example 6. *In the Boolean algebra (B, +, ., '), express the Boolean expression* $E(x, y, z) = ((xy)'z)'((x'+z)(y'+z'))'$ *in its complete sum-of-products form.*

Sol.
$$\begin{aligned}E(x, y, z) &= ((xy)'z)'((x'+z)(y'+z'))' = ((xy)''+z')((x'+z)'+(y'+z')')\\ &= (xy+z')(x''z'+y''z'') = (xy+z')(xz'+yz)\\ &= (xy+z')(xz')+(xy+z')(yz) = (xz')(xy+z')+(yz)(xy+z')\\ &= xz'xy+xz'z'+yzxy+yzz' = (xx)yz'+xz'+x(yy)z+y(zz')\\ &= xyz'+xz'+xyz+y(0) = xyz'+xz'(1)+xyz+0\\ &= xyz'+xz'(y+y')+xyz = xyz'+xz'y+xz'y'+xyz\\ &= (xyz'+xyz')+xy'z'+xyz = \mathbf{xyz'+xy'z'+xyz}.\end{aligned}$$

This is the required complete sum-of-products form of the given expression in variables x, y and z.

Example 7. *In the Boolean algebra (B, +, ., ') express the Boolean expression* $E(x, y, z) = (x'+y)z+y'$ *in its complete DNF form.*

Sol.
$$\begin{aligned}E(x, y, z) &= (x'+y)z+y' = z(x'+y)+y'\\ &= zx'+zy+y' = zx'(1)+zy(1)+y'(1)\\ &= zx'(y+y')+zy(x+x')+y'(x+x')\\ &= zx'y+zx'y'+zyx+zyx'+y'x+y'x'\\ &= x'yz+x'y'z+xyz+x'yz+xy'(z+z')+x'y'(z+z')\\ &= (x'yz+x'yz)+x'y'z+xyz+(xy'z+xy'z')+(x'y'z+x'y'z')\\ &= x'yz+(x'y'z+x'y'z)+xyz+xy'z+xy'z'+x'y'z'\\ &= \mathbf{x'yz+x'y'z+xyz+xy'z+xy'z'+x'y'z'}.\end{aligned}$$

This is the required complete DNF of the given Boolean expression in variables x, y and z.

EXERCISE 10.5

1. In the Boolean algebra (*B*, +, ., ') reduce the following products to either 0 or to a fundamental product :
   (*i*) xyx' (*ii*) $xyxyxy$
   (*iii*) xzx' (*iv*) $xyz'z'xyyy$.
2. Which of the following Boolean expressions in the Boolean algebra (B, +, ., ') are sum-of-products expressions ?
   (*i*) $x'+xyz$ (*ii*) $xy+xyz'+x'y'z$
   (*iii*) $xy+x'y'z'+xy'z$ (*iv*) $x+y+z+xy+x'yz$.
3. In the Boolean algebra (*B*, +, ., '), express the following Boolean expressions in their sum-of-products form :
   (*i*) $x'(yz)'$ (*ii*) $x'y(y+z)'+(xy)'$
   (*iii*) $x'y+(x'+y)'$ (*iv*) $(x'y'+z)(x+y')'$
   (*v*) $(x+y)(x+z')+y+z'$.
4. In the Boolean algebra (*B*, +, ., '), express the following Boolean expressions in the variables *x*, *y*, *z* in their complete sum-of-products form :
   (*i*) $x(y'z')'$ (*KU 2007*) (*ii*) $(x+y)'(x'+y)(x'+y')$
   (*iii*) $(x'+y)'+x'y$ (*iv*) $(x'y)'(x'+xyz')$
   (*v*) $(x+y)(x+z')+y+z'$ (*vi*) $x+y$. (*KU 2006 C*)

Answers

1. (*i*) 0 (*ii*) xy
   (*iii*) 0 (*iv*) xyz'
2. (*i*), (*iii*)
3. (*i*) $x'y' + x'z'$ (*ii*) $x' + y'$
   (*iii*) $x'y + xy'$ (*iv*) $x'yz$
   (*v*) $x + y + z'$
4. (*i*) $xyz + xyz' + xy'z$ (*ii*) $x'y'z + x'y'z'$
   (*iii*) $xy'z + xy'z' + x'yz + x'yz'$ (*iv*) $xyz' + x'y'z + x'y'z'$
   (*v*) $xyz + xy'z + xyz' + x'yz + x'yz' + xy'z' + x'y'z'$
   (*vi*) $xyz + xy'z + x'yz + xy'z' + xyz' + x'yz'$.

10.17. BOOLEAN FUNCTION

Let $(B, +, ., ')$ be a Boolean algebra. Let $B \times B \times \ldots\ldots \times B$ (n factors) be denoted by B^n.

A function $f : B^n \to B$ is called **a Boolean function** if it can be specified by a Boolean expression of n variables.

Thus if $E(x_1, x_2, \ldots\ldots, x_n)$ be a Boolean expression of n variables $x_1, x_2, \ldots\ldots, x_n$, over a Boolean algebra $(B, +, ., ')$ then a function $f : B^n \to B$ of the form $f(x_1, x_2, \ldots\ldots, x_n) = E(x_1, x_2, \ldots\ldots, x_n)$ is a Boolean function. Here each x_i takes, value in the set B.

Illustration 1. Let $(B, +, ., ')$ be a Boolean algebra.

(*i*) $f(x, y) = (x + y)'x$ is a Boolean function from B^2 to B.

(*ii*) $g(x, y, z) = (xy)'z + (x + y' + z)'y$ is a Boolean function from B^3 to B.

Illustration 2. Let $(B, +, ., ')$ be the Boolean algebra of binary digits 0 and 1.

$\therefore \quad B^2 = \{(1, 1), (1, 0), (0, 1), (0, 0)\}$

Let $\phi : B^2 \to B$ be the function given by the adjoining table. Technically speaking, ϕ is a function from B^2 to B because each element of B^2 is having unique image under ϕ.

(x, y)	$\phi(x, y)$
(1, 1)	0
(1, 0)	0
(0, 1)	1
(0, 0)	1

Here $\phi(x, y)$ cannot be expressed as a Boolean expression of two variables x and y.

$\therefore$ The function $\phi : B^2 \to B$ is not a Boolean function.

Remark. If $E(x, y) = (x + y')'$, then $E(1, 1) = (1 + 1')' = (1 + 0)' = 1' = 0$,

$$E(1, 0) = (1 + 0')' = (1 + 1)' = 1' = 0,\ E(0, 1) = (0 + 1')' = (0 + 0)' = 0' = 1,$$

$$E(0, 0) = (0 + 0')' = (0 + 1)' = 1' = 0 \neq 1.$$

Here, we will not be able to find any Boolean expression which may correspond the function ϕ.

Illustration 3. The function $f : (\{0, 1\})^2 \to \{0, 1\}$ defined by $f(x, y) = (xy') + x$ is a Boolean function with domain given by

$$(\{0, 1\})^2 = \{0, 1\} \times \{0, 1\} = \{(0, 0), (0, 1), (1, 0), (1, 1)\}$$

and codomain $\{0, 1\}$. Here the Boolean algebra under consideration is the Boolean algebra of binary digits 0 and 1.

Here
$$f(0, 0) = 0.0' + 0 = 0.1 + 0 = 0 + 0 = 0$$
$$f(0, 1) = 0.1' + 0 = 0.0 + 0 = 0 + 0 = 0$$
$$f(1, 0) = 1.0' + 1 = 1.1 + 1 = 1 + 1 = 1$$
$$f(1, 1) = 1.1' + 1 = 1.0 + 1 = 0 + 1 = 1.$$

10.18. BOOLE'S EXPANSION THEOREM

Let $(B, +, ., ', 0, 1)$ be a Boolean algebra. We know that if f is a Boolean function from B^n to B, then it is can be specified by a Boolean expression. Also, a Boolean expression is expressible in its complete sum-of-product form.

The following theorem gives a method of finding a Boolean function $f : B^n \to B$ if the values of $f(x_1, x_2, \ldots\ldots, x_n)$ are given whenever $x_i = 0$ or 1, $1 \le i \le n$.

Boole's expansion theorem. If $(B, +, ., ', 0, 1)$ be a Boolean algebra and $f(x_1, x_2, \ldots\ldots, x_n)$ be a Boolean function in n variables $x_1, x_2, \ldots\ldots, x_n$, then prove that

$$\begin{aligned} f(x_1, x_2, \ldots\ldots, x_n) &= f(1, 1, \ldots\ldots 1)x_1 . x_2 \ldots\ldots x_n + f(0, 1, \ldots\ldots, 1). x_1'.x_2. \ldots\ldots x_n \\ &\quad + f(1, 0, \ldots\ldots, 1) . x_1 . x_2' \ldots\ldots x_n + f(0, 0, \ldots\ldots, 1).x_1'.x_2'. \ldots\ldots x_n \\ &\quad + \ldots + f(0, 0, \ldots\ldots, 0) .x_1'. x_2' \ldots\ldots x_n'. \end{aligned}$$

Proof. If the Boolean function is identically 0, then the result holds trivially.

So, let the given Boolean function be not identically '0'.

$\therefore$ We can express this function as the sum of some minterms in the variables $x_1, x_2, \ldots\ldots, x_n$.

Let
$$\begin{aligned} f(x_1, x_2, \ldots\ldots, x_n) &= A_1.x_1.x_2 \ldots\ldots x_n + A_2.x_1'.x_2 \ldots\ldots x_n + A_3.x_1.x_2'. \ldots\ldots x_n \\ &\quad + A_4.x_1'.x_2' \ldots\ldots x_n + \ldots\ldots + A_{2^n} . x_1'. x_2' \ldots\ldots x_n' \qquad \ldots(1) \end{aligned}$$

Here $A_1, A_2, \ldots\ldots, A_{2^n}$ are constants each equal to either 0 or 1 and these constants are not all zero.

Putting $x_1 = 1$, $x_2 = 1$, ..., $x_n = 1$ in (1), we get

$$\begin{aligned} f(1, 1, \ldots\ldots, 1) &= A_1.1.1 \ldots\ldots . 1 + A_2. 0.1. \ldots\ldots .1 + A_3 .1.0 . \ldots\ldots 1 + A_4. 0.0. \ldots\ldots . \\ &\quad 2^n + \ldots\ldots + A_{2^n} . 0. 0. \ldots\ldots . 0 \\ &= A_1 + 0 + 0 + 0 + \ldots\ldots + 0 = A_1 \end{aligned}$$

$\therefore$ $A_1 = f(1, 1, \ldots\ldots, 1)$

Putting $x_1 = 0$, $x_2 = 1$, ..., $x_n = 1$ in (1), we get

$$\begin{aligned} f(0, 1, \ldots\ldots, 1) &= A_1.0.1. \ldots\ldots .1 + A_2.1.1. \ldots\ldots . 1 + A_3.0.0. \ldots\ldots .1 + \ldots\ldots \\ &\quad + A_{2^n} .1.0. \ldots\ldots .0 \\ &= 0 + A_2 + 0 + 0 + \ldots\ldots + 0 = A_2 \end{aligned}$$

$\therefore$ $A_2 = f(0, 1, \ldots\ldots, 1)$

Similarly, $A_3 = f(1, 0, \ldots\ldots, 1)$

$A_4 = f(0, 0, \ldots\ldots, 1)$

....

$A_{2^n} = f(0, 0, \ldots\ldots, 0)$

$\therefore$ (1) implies

$$\begin{aligned} f(x_1, x_2, \ldots\ldots, x_n) &= f(1, 1, \ldots\ldots, 1)\, x_1 . x_2 . \ldots\ldots .x_n + f(0, 1, \ldots\ldots, 1). x_1'. x_2 \ldots\ldots x_n \\ &\quad + f(1, 0, \ldots\ldots, 1) . x_1 . x_2' \ldots\ldots . x_n + f(0, 0, \ldots\ldots, 1).x_1'. x_2' \ldots\ldots . x_n \\ &\quad + \ldots\ldots + f(0, 0, \ldots\ldots, 0)\, x_1'. x_2' \ldots\ldots x_n' . \end{aligned}$$

Hence the result holds.

Example 1. *Let (B, +, ., ′ ,0, 1) be a Boolean algebra and let f(x, y) be a Boolean function of the variables x and y. By using the given table, find the expression for f(x, y).*

x	y	$f(x, y)$
1	1	0
1	0	1
0	1	0
0	0	1

Sol. From the given table, we have

$$f(1, 1) = 0, f(1, 0) = 1, f(0, 1) = 0, f(0, 0) = 1.$$

By **Boole's expansion theorem**, we have

$$f(x, y) = f(1, 1).x.y + f(1, 0).x.y' + f(0, 1).x'.y + f(0, 0).x'.y'$$
$$= 0.x.y + 1.x.y' + 0.x'.y + 1.x'.y'$$
$$= 0 + x.y' + 0 + x'.y' \qquad \text{(By using } a.0 = 0, a.1 = 1)$$
$$= x.y' + x'.y'$$
$$= y'.x + y'.x' = y'.(x + x') = y'.1 = y'$$

∴ **f(x, y) = y′.**

Example 2. *Let (B,+, ., ′ ,0, 1) be a Boolean algebra and let f(x, y, z) be a Boolean function of the variable x, y and z. By using the given table, find the expression for f(x, y, z).*

x	y	z	$f(x, y, z)$
1	1	1	1
1	1	0	0
1	0	1	1
0	1	1	0
1	0	0	0
0	1	0	0
0	0	1	1
0	0	0	1

Sol. From the given table, we have

$$f(1, 1, 1) = 1, f(1, 1, 0) = 0, f(1, 0, 1) = 1, f(0, 1, 1) = 0,$$
$$f(1, 0, 0) = 0, f(0, 1, 0) = 0, f(0, 0, 1) = 1, f(0, 0, 0) = 1.$$

By **Boole's expansion theorem,** we have

$$f(x, y, z) = f(1, 1, 1).x.y.z + f(1, 1, 0).x.y.z' + f(1, 0, 1).x.y'.z$$
$$+ f(0, 1, 1).x'.y.z + f(1, 0, 0).x.y'.z'$$
$$+ f(0, 1, 0).x'.y.z' + f(0, 0, 1).x'.y'.z + f(0, 0, 0).x'.y'.z'$$
$$= 1.x.y.z + 0.x.y.z' + 1.x.y'.z + 0.x'.y.z$$
$$+ 0.x.y'.z' + 0.x'.y.z' + 1.x'.y'.z + 1.x'.y'.z'$$
$$= x.y.z + 0 + x.y'.z + 0 + 0 + 0 + x'.y'.z + x'.y'.z' \qquad \text{(By using } a.1 = a, a.0 = 0)$$

∴ **f(x, y, z) = x.y.z + x.y′.z + x′.y′.z + x′.y′.z′**

This function may be simplified as follows :

$$f(x.y.z) = (x.z.y + x.z.y') + (x'.y'.z + x'.y'.z')$$
$$= x.z\,(y + y') + x'.y'.(z + z')$$
$$= x.z.1 + x'.y'.1 = x.z + x'.y'$$

∴ **f(x, y, z) = x.z. + x′.y′.**

10.19. TRUTH TABLE

Let $(B, +, ., ')$ be the Boolean algebra of binary digits 0 and 1. Let $f : B^n \to B$ be a Boolean function. A table showing the values of the Boolean function f for all values in its domain is called the **truth table** of the Boolean function f. The truth table of f will have 2^n rows. The truth table of a Boolean function is also called its **input/output table**.

Example 3. *Construct the truth table of the Boolean function $f : B^2 \to B$ given by $f(x_1, x_2) = x_1x_2'$, where $B = \{0, 1\}$.*

Sol. We have $B^2 = \{(1, 1), (1, 0)\ (0, 1)\ (0, 0)\}$ and $f(x_1, x_2) = x_1.x_2'$

$$f(1, 1) = 1.1' = 1.0 = 0$$
$$f(1, 0) = 1.\ 0' = 1.1 = 1$$
$$f(0, 1) = 0.\ 1' = 0.0 = 0$$
$$f(0, 0) = 0.\ 0' = 0.1 = 0$$

The truth table of f is shown in the figure.

x_1	x_2	$f(x_1, x_2)$
1	1	0
1	0	1
0	1	0
0	0	0

Example 4. *Construct the truth table for the Boolean function* $f : B^3 \to B (= \{0, 1\})$, *where* $f(x, y, z) = (x \wedge y) \vee (x \vee (y' \wedge z))$. (KU 2006, 2007)

Sol. Using symbols '+' and '.' for '∨' and '∧' respectively, we have

$$f(x, y, z) = (x.y) + (x + (y'.z)) = xy + x + y'z = (x + xy) + y'z = x + y'z$$

$\therefore$
$$f(1, 1, 1) = 1 + 1'.1 = 1 + 0.1 = 1 + 0 = 1$$
$$f(1, 1, 0) = 1 + 1'.0 = 1 + 0.0 = 1 + 0 = 1$$
$$f(1, 0, 1) = 1 + 0'.1 = 1 + 1.1 = 1 + 1 = 1$$
$$f(0, 1, 1) = 0 + 1'.1 = 0 + 0.1 = 0 + 0 = 0$$
$$f(1, 0, 0) = 1 + 0'.0 = 1 + 1.0 = 1 + 0 = 1$$
$$f(0, 1, 0) = 0 + 1'.0 = 0 + 0.0 = 0 + 0 = 0$$
$$f(0, 0, 1) = 0 + 0'.1 = 0 + 1.1 = 0 + 1 = 1$$
$$f(0, 0, 0) = 0 + 0'.0 = 0 + 1.0 = 0 + 0 = 0$$

The truth table of the Boolean function is given in the figure.

x	y	z	$f(x, y, z)$
1	1	1	1
1	1	0	1
1	0	1	1
0	1	1	0
1	0	0	1
0	1	0	0
0	0	1	1
0	0	0	0

Example 5. *Construct the truth table for the Boolean function* $f : B^3 \to B$, *where* $f(x, y, z) = xyz + xy'z' + x'yz$ *and* $B = \{0, 1\}$.

Sol. We have

$$f(x, y, z) = xyz + xy'z' + x'yz$$
$$f(1, 1, 1) = 1.1.1 + 1.1'.1' + 1'.1.1$$
$$= 1 + 1.0.0 + 0.1.1 = 1 + 0 + 0 = 1$$
$$f(1, 1, 0) = 1.1.0 + 1.1'.0' + 1'.1.0$$
$$= 0 + 1.0.1 + 0.1.0 = 0 + 0 + 0 = 0$$
$$f(1, 0, 1) = 1.0.1 + 1.0'.1' + 1'.0.1$$
$$= 0 + 1.1.0 + 0.0.1 = 0 + 0 + 0 = 0$$
$$f(0, 1, 1) = 0.1.1 + 0.1'.1' + 0'.1.1$$
$$= 0 + 0.0.0 + 1.1.1 = 0 + 0 + 1 = 1$$
$$f(1, 0, 0) = 1.0.0 + 1.0'.0' + 1'.0.0$$
$$= 0 + 1.1.1 + 0.0.0 = 0 + 1 + 0 = 1$$
$$f(0, 1, 0) = 0.1.0 + 0.1'.0' + 0'.1.0$$
$$= 0 + 0.0.1 + 1.1.0 = 0 + 0 + 0 = 0$$
$$f(0, 0, 1) = 0.0.1 + 0.0'.1' + 0'.0.1$$
$$= 0 + 0.1.0 + 1.0.1 = 0 + 0 + 0 = 0$$
$$f(0, 0, 0) = 0.0.0 + 0.0'.0' + 0'.0.0$$
$$= 0 + 0.1.1 + 1.0.0 = 0 + 0 + 0 = 0.$$

x	y	z	$f(x, y, z)$
1	1	1	1
1	1	0	0
1	0	1	0
0	1	1	1
1	0	0	1
0	1	0	0
0	0	1	0
0	0	0	0

$\therefore$ The truth table of the Boolean function is given in the figure.

EXERCISE 10.6

1. Let $(B, +, ., ', 0, 1)$ be a Boolean algebra and let $f(x, y)$ be a Boolean function of the variables x and y. By using the given table, find the expression for $f(x, y)$.

x	y	$f(x, y)$
1	1	0
1	0	0
0	1	1
0	0	1

2. Let $(B, +, ., ', 0, 1)$ be a Boolean algebra and let $f(x, y)$ be a Boolean function of the variables x and y. By using the given table, find the expression for $f(x, y)$.

x	y	$f(x, y)$
1	1	0
1	0	1
0	1	1
0	0	0

3. Let $(B, +, ., ', 0, 1)$ be a Boolean algebra and let $\phi(x, y, z)$ be a Boolean function of the variables x, y and z. By using the following table, find the expression for $\phi(x, y, z)$.

x	y	z	$\phi(x, y, z)$
1	1	1	1
1	1	0	0
1	0	1	0
0	1	1	1
1	0	0	0
0	1	0	1
0	0	1	0
0	0	0	0

4. Let $(B, +, *, ', 0, 1)$ be a Boolean algebra and $g(p, q, r)$ be a Boolean function of the variables p, q and r. By using the following table, find the expression for $g(p, q, r)$.

p	q	r	$g(p, q, r)$
1	1	1	1
1	1	0	0
1	0	1	0
0	1	1	1
1	0	0	0
0	1	0	1
0	0	1	1
0	0	0	1

5. Construct the truth table of the Boolean function $f : B^2 \to B$ given by $f(x, y) = xy + x'y'$, where $B = \{0, 1\}$.
6. Construct the truth table of the Boolean function $f : B^3 \to B$ where $f(x, y, z) = xyz + xy'z' + x'yz'$ and $B = \{0, 1\}$.
7. Construct the truth table of the Boolean function $f : B^3 \to B$ where $f(x, y, z) = x(y' + z) + y$ and $B = \{0, 1\}$.
8. Construct the truth table of the Boolean function $\phi : B^3 \to B$ where $\phi(x, y, z) = xz + x'y'$ and $B = \{0, 1\}$.

Answers

1. x'
2. $xy' + x'y$
3. $(x' + z)y$
4. $p' + qr$

5.

x	y	$f(x, y)$
1	1	1
1	0	0
0	1	0
0	0	1

6.

x	y	z	$f(x, y, z)$
1	1	1	1
1	1	0	0
1	0	1	0
0	1	1	0
1	0	0	1
0	1	0	1
0	0	1	0
0	0	0	0

7.

x	y	z	$f(x, y, z)$
1	1	1	1
1	1	0	1
1	0	1	1
0	1	1	1
1	0	0	1
0	1	0	1
0	0	1	0
0	0	0	0

8.

x	y	z	$\phi(x, y, z)$
1	1	1	1
1	1	0	0
1	0	1	1
0	1	1	0
1	0	0	0
0	1	0	0
0	0	1	1
0	0	0	1

10.20. SWITCHING CIRCUITS

We know that a *switching circuit* is an arrangement of wires and switches connected together to the terminals of a battery. A *switch* is a two state device used for allowing current to pass through it or not to pass through it.

If current is allowed to pass through a switch then it is said to be **'closed'** or 'on'. If current is not allowed to pass through a switch then it is said to be **'open'** or **'off'** .

The switches in a circuit are denoted by letters $a, b, c, \ldots\ldots, p, q, r, \ldots\ldots, x, y, z$ etc.

There are two methods of connecting two switches. Two switches can be connected either *in parallel* or *in series.*

(*i*) Connecting switches in parallel

Two switches x and y are connected in parallel as shown in the diagram. The lamp is 'on' if and only if at least one of the switches x and y is closed.

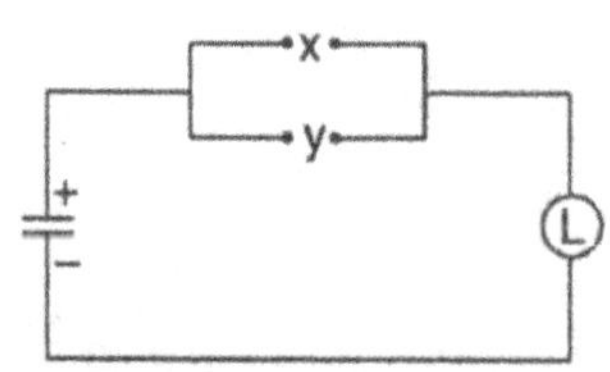

(*ii*) Connecting switches in series

Two switches x and y are connected in series as shown in the diagram. The lamp is 'on' if and only if x and y are both closed.

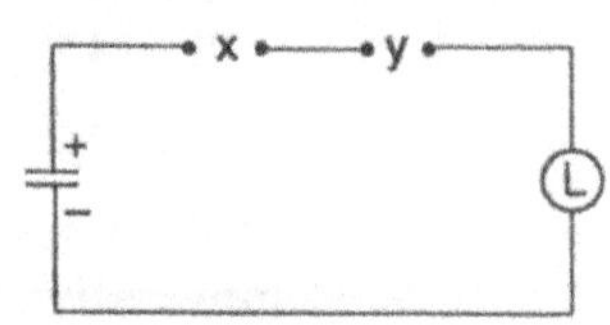

10.21. SWITCHING ALGEBRA

In the present section, we shall learn the technique of simplying switching circuits by using the theory of Boolean algebra.

We know that a switch or a switching circuit is either 'on' or 'off'. If a switch (resp. switching circuit) is 'on' then its value is denoted by '1' and if a switch (resp. switching circuit) is 'off' then its value is denoted by '0'. Here the symbols '1' and '0' have nothing to do with 'one' and 'zero' of arithmetic.

Two switches can be connected either in parallel or in series.

(*i*) Let switches x and y be connected in parallel.

$\therefore$ Current flows through the circuit of x and y only when at least one switch is 'on'.

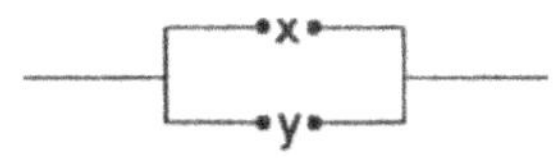

In other words, the value of the circuit of x and y is 1 only when at least one switch has value 1. The circuit having switches x and y in parallel is denoted by $x + y$.

$\therefore$ The values of x, y and $x + y$ are given in the adjoining table.

x	y	$x + y$
0	0	0
1	0	1
0	1	1
1	1	1

$\therefore$ '+' is a binary operation on the set $B = \{0, 1\}$.

(*ii*) Let switches x and y be connected in series.

$\therefore$ Current flows through the circuit of x and y only when both switches are 'on'.

In other words, the value of the circuit of x and y is 1 only when both switches have value 1. The circuit having switches x and y in series is denoted by $x.y$.

$\therefore$ The values of x, y and $x.y$ are given in the adjoining table.

x	y	$x.y$
0	0	0
1	0	0
0	1	0
1	1	1

$\therefore$ '.' is a binary operation on the set $B = \{0, 1\}$.

A switch which is 'off' when the switch x is 'on' and is 'on' when the switch x is 'off' is called the **complement** of the switch x and is denoted by x'.

The values of x and x' are given in the adjoining table.

x	x'
0	1
1	0

$\therefore$ '′' is a unary operation on the set $B = \{0, 1\}$.

$\therefore$ We have two binary operations '+' and '.' and a unary operation '′' on the set $B = \{0, 1\}$.

Now, we shall show that $(B, +, ., ')$ is a Boolean algebra.

B_1.

x	y	$x + y$	$y + x$
0	0	0	0
1	0	1	1
0	1	1	1
1	1	1	1

$\therefore \quad x + y = y + x \quad \forall\, x, y \in B$

x	y	$x.y$	$y.x$
0	0	0	0
1	0	0	0
0	1	0	0
1	1	1	1

$\therefore \quad x.y = y.x \quad \forall\, x, y \in B$

$\therefore$ Commutative laws hold.

B_2.

x	y	z	$y.z$	$x + y.z$	$x + y$	$x + z$	$(x + y).(x + z)$
0	0	0	0	0	0	0	0
1	0	0	0	1	1	1	1
0	1	0	0	0	1	0	0
0	0	1	0	0	0	1	0
1	1	0	0	1	1	1	1
1	0	1	0	1	1	1	1
0	1	1	1	1	1	1	1
1	1	1	1	1	1	1	1

$\therefore \quad x + y.z = (x + y).(x + z) \quad \forall\, x, y, z \in B$

x	y	z	$y + z$	$x.(y + z)$	$x.y$	$x.z$	$x.y + x.z$
0	0	0	0	0	0	0	0
1	0	0	0	0	0	0	0
0	1	0	1	0	0	0	0
0	0	1	1	0	0	0	0
1	1	0	1	1	1	0	1
1	0	1	1	1	0	1	1
0	1	1	1	0	0	0	0
1	1	1	1	1	1	1	1

$\therefore \quad x.(y + z) = x.y + x.z \quad \forall\, x, y, z \in B$

$\therefore$ Distributive laws hold.

B_3.

a	0	$a + 0$
0	0	0
1	0	1

a	1	$a.1$
0	1	0
1	1	1

$\therefore \quad a + 0 = a \quad \forall\, a \in B \qquad\qquad a.1 = a \quad \forall\, a \in B$

$\therefore$ Identity laws hold.

Here '0' is the zero element and '1' is the unit element.

B_4.

a	a'	$a + a'$
0	1	1
1	0	1

a	a'	$a.a'$
0	1	0
1	0	0

$\therefore \quad a + a' = 1 \quad \forall\, a \in B \qquad\qquad a.a' = 0 \quad \forall\, a \in B$

$\therefore$ Complement laws hold.

$\therefore$ $(B, +, ., ')$ is a Boolean algebra.

This Boolean algebra is called **Switching algebra** or **Boolean algebra of switching circuits.**

In the switching algebra $(B, +, ., ')$, the Boolean functions are functions of switches x_1, x_2, ... Here each variable *i.e.*, switch can have value either 0 or 1. Thus, switching circuits can be expressed as Boolean functions in the switching algebra $(B, +, ., ')$, where $B = \{0, 1\}$.

In a switching circuit, switches need not act independently of each other. The following rules are observed in this connection :

(*i*) If two or more switches open or close simultaneously, then these switches are denoted by the same variable.

(*ii*) If x and y are two switches such that y is closed when x is open and y is open when x is closed, then y is written as x'.

WORKING RULES FOR SOLVING PROBLEMS

Rule I. *If x and y represent two switches, then the circuit having x and y in parallel is denoted by $x + y$.*

Rule II. *If x and y represent two switches, then the circuit having x and y in series is denoted by $x.y$.*

Rule III. *If two or more switches open or close simultaneously, then these switches are represented by the same variable.*

Rule IV. *If x and y are switches such that y is closed when x is open and y is open when x is closed, then y is written as x'.*

Example 1. *Express the following circuit as a Boolean function in the switching algebra.*

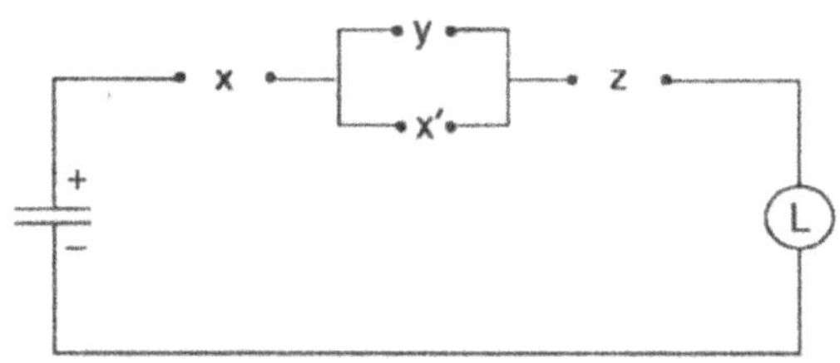

Sol. In the circuit, y and x' are connected in parallel.

∴ This portion of the circuit is represented by $y + x'$.

∴ Boolean function of the given circuit is $\mathbf{x.(y + x').z.}$

Example 2. *Express the following circuit as a Boolean function in the switching algebra.*

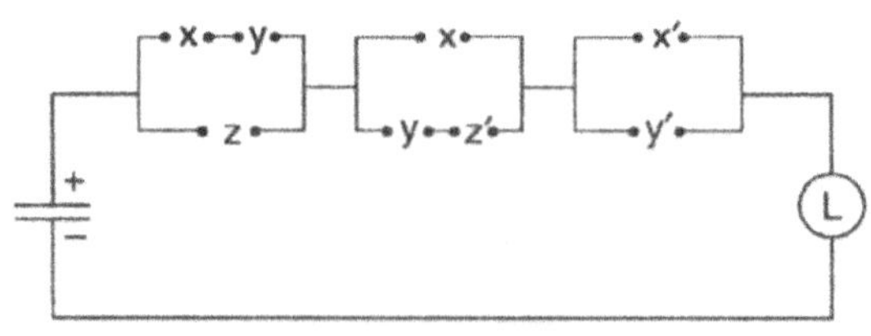

Sol. First portion of the circuit is represented by $x.y + z$.

Second portion of the circuit is represented by $x + y.z'$.

Third portion of the circuit is represented by $x' + y'$.

These portions are connected in series.

∴ The Boolean function of the given circuit is

$$\mathbf{(x.y + z).(x + y.z').(x' + y').}$$

Example 3. *Construct the circuit corresponding to the Boolean function* $(x.y.z') + x'.(y + z')$ *in the switching algebra.*

Sol. Boolean function is $(x.y.z') + x.'(y + z')$...(1)

(1) implies that the circuits corresponding $x.y.z'$ and $x'.(y + z')$ are connected in parallel. $x.y.z'$ implies that the switches x, y, z' are connected in series. $x.'(y + z')$ implies that switch x' and the circuit corresponding to $y + z'$ are connected in series. The circuit of the given Boolean function is given in the diagram.

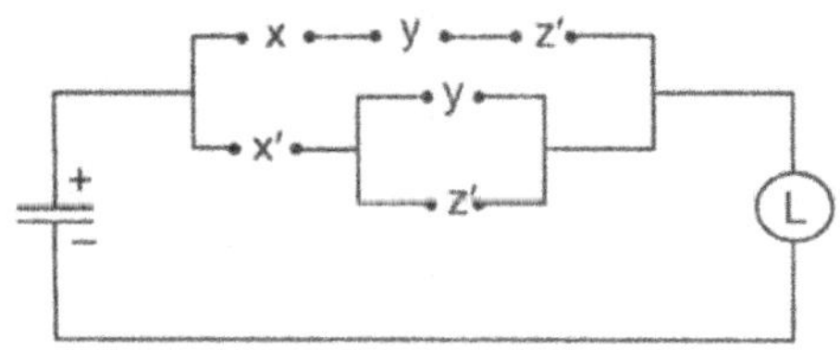

Example 4. *Construct the circuit corresponding to the Boolean function* $(x.y + z).(x'+ y.z').(x' + z)$ *in the switching algebra.*

Sol. Boolean function is $(x.y + z).(x'+ y.z').(x' + z)$ ...(1)

(1) implies that the circuits corresponding to $x.y + z$, $x' + y.z'$ and $x' + z$ are connected in series. $x.y + z$ implies that the circuit corresponding to $x.y$ and the switch z are connected in parallel. $x' + y.z'$ implies that the switch x' and the circuit corresponding to $y.z'$ are connected in parallel. $x' + z$ implies that the switches x' and z are connected in parallel.

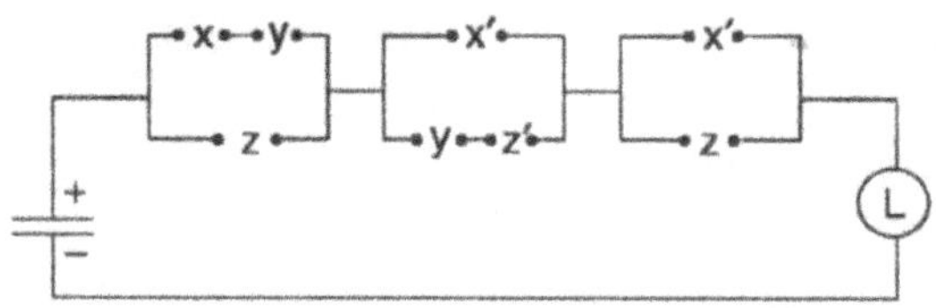

The circuit of the given Boolean function is given in the diagram.

Example 5. *Draw the circuit corresponding to the Boolean function* $f(x, y) = x + x.y$ *in the switching algebra. Also draw an equivalent simpler circuit.*

Sol. Given Boolean function is $f(x, y) = x + x.y$.

∴ The corresponding circuit is shown below [Fig. (*i*)] :

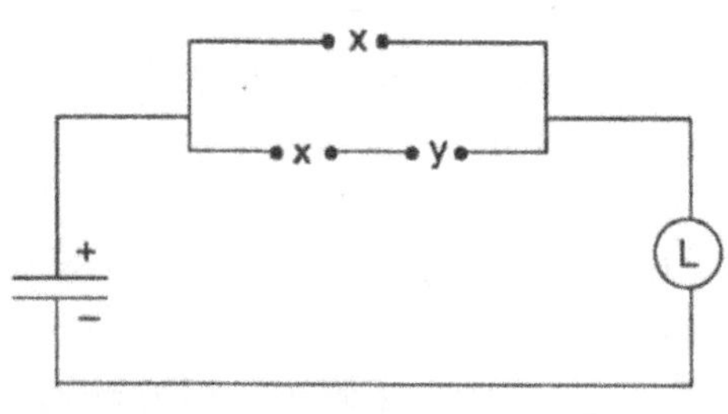

Fig. (*i*)

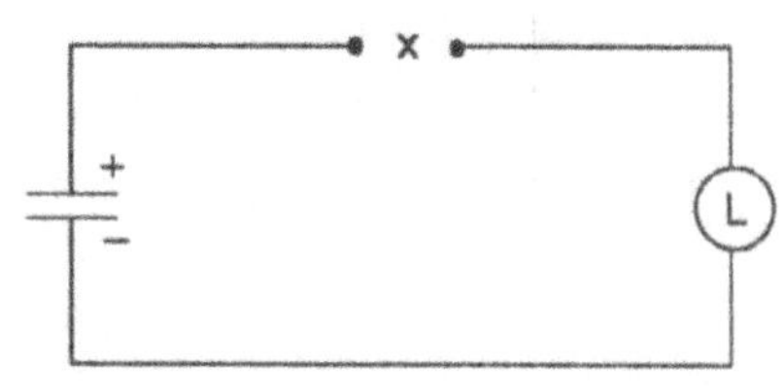

Fig. (*ii*)

We simplify the given Boolean function.

We have $f(x, y) = x + x.y = \mathbf{x}$. (By using absorption law)

∴ The given circuit is equivalent to the circuit shown in the diagram [Fig. (*ii*)].

Example 6. *Simplify the circuit given below :*

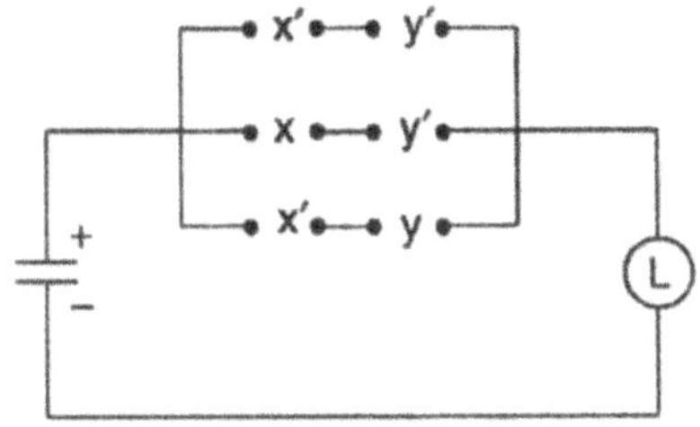

Sol. Let $f(x.y)$ be the Boolean function of the given circuit in the switching algebra.

$\therefore \qquad f(x, y) = x'.y' + x.y' + x'.y.$

We simplify this Boolean function.

We have $f(x, y) = (y'.x' + y'.x) + x'.y$ (By using $\mathbf{B_1}$)

$= y'.(x' + x) + x'.y$ (By using $\mathbf{B_2}$)

$= y'.1 + x'.y$ (By using $\mathbf{B_4}$)

$= y' + x'.y$ (By using $\mathbf{B_3}$)

$= (y' + x').(y' + y)$ (By using $\mathbf{B_2}$)

$= (x' + y').1$ (By using $\mathbf{B_1}$, $\mathbf{B_4}$)

$= \mathbf{x' + y'}.$ (By using $\mathbf{B_3}$)

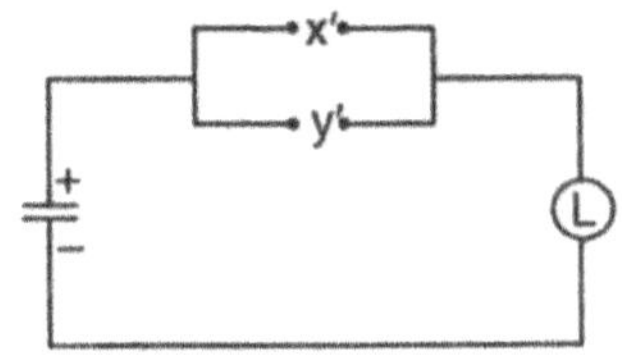

$\therefore$ The given circuit is equivalent to the circuit in which switches x' and y' are connected in parallel.

Example 7. *Simplify the circuit given below :*

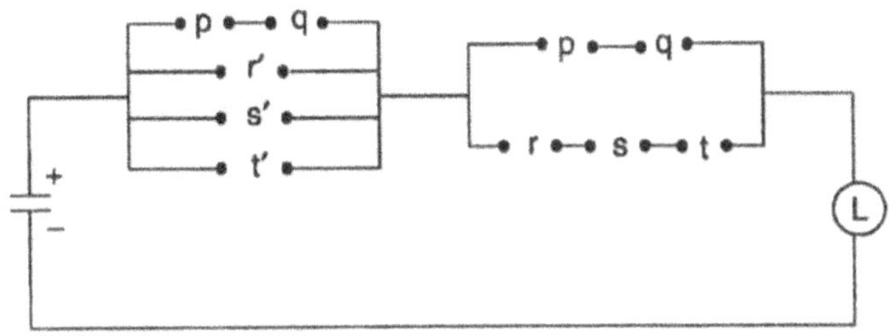

Sol. Let $f(p, q, r, s, t)$ be the Boolean function of the given circuit in the switching algebra.

$\therefore \qquad f(p, q, r, s, t) = (p.q + r' + s' + t').(p.q + r.s.t).$

We simplify this Boolean function.

We have $f(p, q, r, s, t) = p.q + (r' + s' + t').(r.s.t)$ (By using $\mathbf{B_2}$)

$= p.q + (r.s.t).(r' + s' + t')$ (By using $\mathbf{B_1}$)

$= p.q + (r.s.t).((r.s)' + t')$ (By using De Morgan's law)

$= p.q + (r.s.t).((r.s).t)'$

$= p.q + (r.s.t).(r.s.t)'$

$= p.q + 0 = \mathbf{p.q}$ (By using $\mathbf{B_4}$)

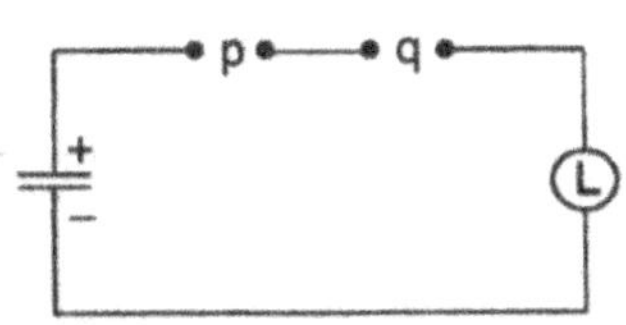

$\therefore$ The given circuit is equivalent to a circuit in which switches p and q are connected in series.

Example 8. *Simplify the circuit given below :*

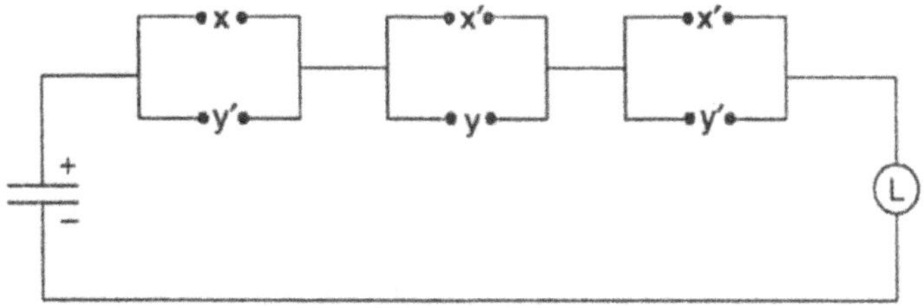

Sol. Let $f(x, y)$ be the Boolean function of the given circuit in the switching algebra.

$\therefore \quad f(x, y) = (x + y').(x' + y).(x' + y')$

We simplify this Boolean function.

We have $\quad f(x, y) = (x + y').(x' + y.y')$ (By using $\mathbf{B_2}$)

$= (x + y').(x' + 0) = x'.(x + y')$

$= x'.x + x'.y' = 0 + x'.y' = \mathbf{x'.y'}$

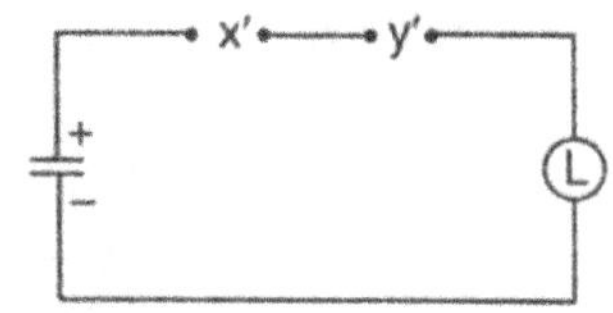

$\therefore$ The given circuit is equivalent to a circuit in which switches x' and y' are connected in series.

Example 9. *Simplify the circuit given below :*

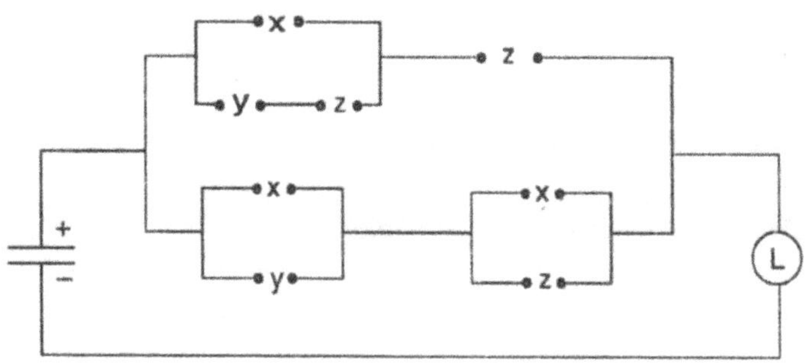

Sol. Let $f(x, y, z)$ be the Boolean function of the given circuit in the switching algebra.

$\therefore \quad f(x, y, z) = (x + y.z).z + (x + y).(x + z)$

We simplify this Boolean function.

We have $\quad f(x, y, z) = (x + y.z).z + (x + y).(x + z)$

$= z.(x + y.z) + x + y.z$

$= z.x + z.y.z + x + y.z$

$= (x + x.z) + y.z.z + y.z$

$= x + (y.z + yz)$ (By using absorption law)

$= \mathbf{x + y.z}$

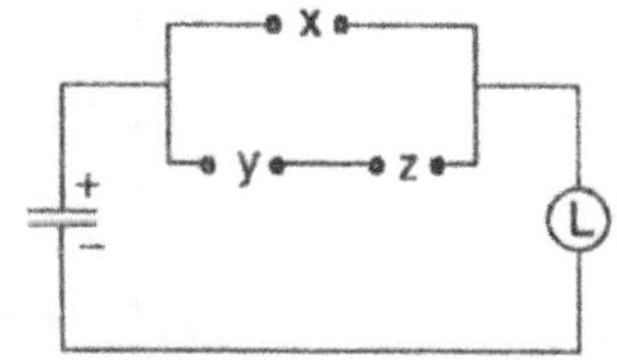

$\therefore$ The given circuit is equivalent to a circuit shown in the diagram.

Example 10. *Write the Boolean expression for the following circuit :*

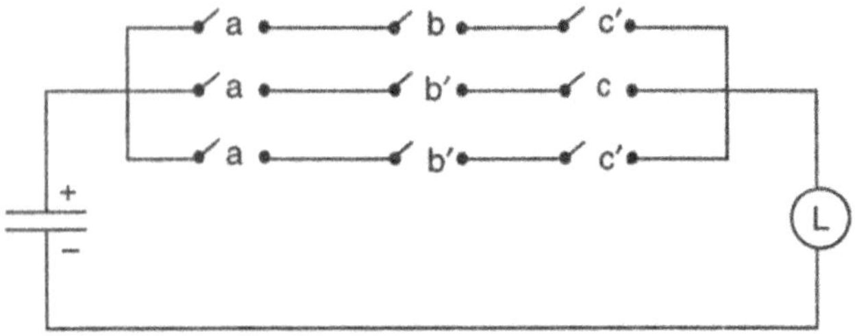

Simplify the expression and construct the switching circuit for the simplified expression.

Sol. Let $f(a, b, c)$ be the Boolean function of the given circuit in the switching algebra.

$\therefore \quad f(a, b, c) = a.b.c' + a.b'.c + a.b'.c'$

We simplify this Boolean expression.

We have

$$\begin{aligned} f(a, b, c) &= a.b.c' + a.b'.c + a.b'.c' \\ &= a.b.c' + a.b'.(c + c') && \text{(By using } \mathbf{B_2}) \\ &= a.b.c' + a.b'.1 && \text{(By using } \mathbf{B_4}) \\ &= a.b.c' + a.b' && \text{(By using } \mathbf{B_3}) \\ &= a.(b.c' + b') && \text{(By using } \mathbf{B_2}) \\ &= a.(b + b').(c' + b') && \text{(By using } \mathbf{B_2}) \\ &= a.1.(c' + b') && \text{(By using } \mathbf{B_4}) \\ &= \mathbf{a.(b' + c')} \end{aligned}$$

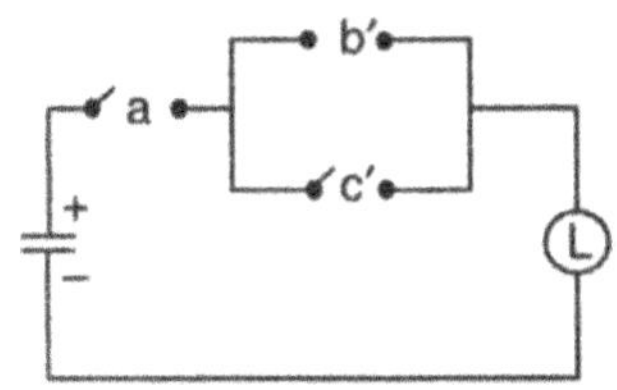

$\therefore$ The given circuit is equivalent to the circuit shown in the figure.

Example 11. *Simplify the circuit given below :*

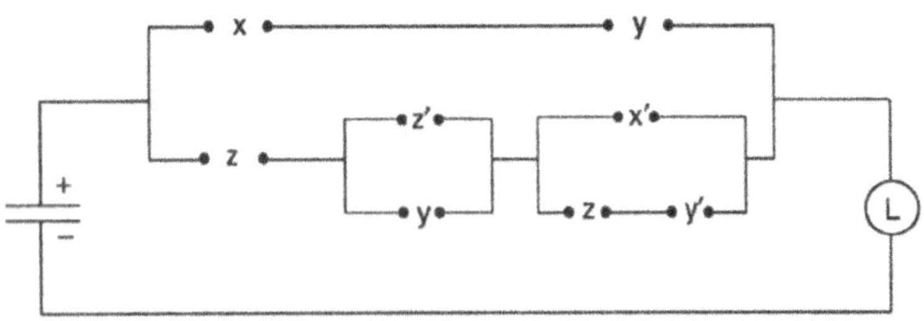

Sol. Let $f(x, y, z)$ be the Boolean function of the given circuit in the switching algebra.

$\therefore \quad f(x, y, z) = x.y + z.(z' + y).(x' + z.y')$.

We simplify this Boolean function.

We have

$$\begin{aligned} f(x, y, z) &= x.y + z.(z' + y).(x' + z.y') \\ &= x.y + (z.z' + z.y).(x' + z.y') && \text{(By using } \mathbf{B_2}) \\ &= x.y + (0 + z.y).(x' + z.y') && \text{(By using } \mathbf{B_4}) \\ &= x.y + z.y.(x' + z.y') = x.y + z.y.x' + z.y.z.y' \end{aligned}$$

$$= x.y + z.y.x' + z.z.y.y' = x.y + z.y.x' + z.0$$
$$(\because \quad z.z = z \text{ and } y.y' = 0)$$
$$= y.x + y.z.x' + 0$$
$$= y.(x + z.x') = y.((x + z).(x + x'))$$
$$= y.(x + z).1 = \mathbf{y.(x + z)}$$

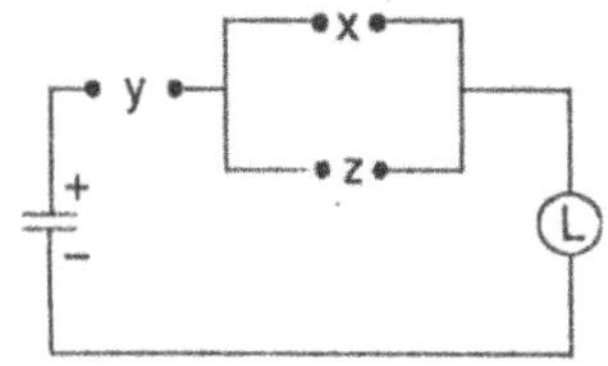

$\therefore$ The given circuit is equivalent to the circuit shown in the figure.

EXERCISE 10.7

1. Express the following circuit as a Boolean function in the switching algebra.

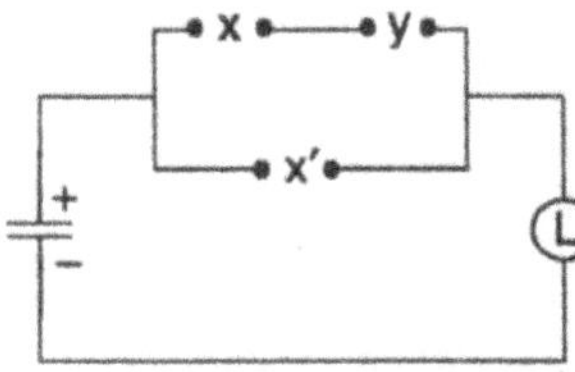

Q. 1

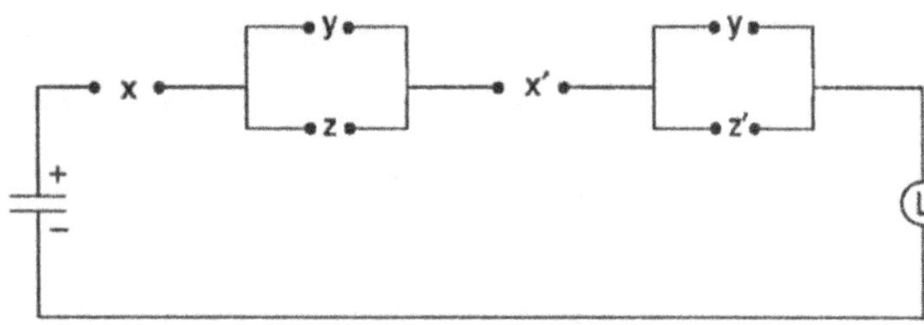

Q. 2

2. Express the above circuit as a Boolean function in the switching algebra.
3. Express the following circuit as a Boolean function in the switching algebra.

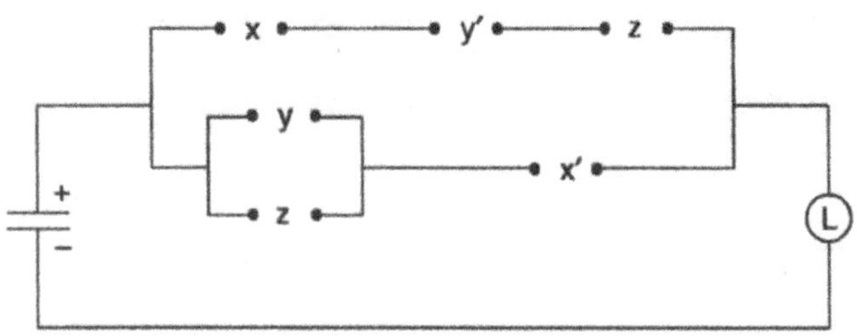

4. Express the following circuit as a Boolean function in the switching algebra.

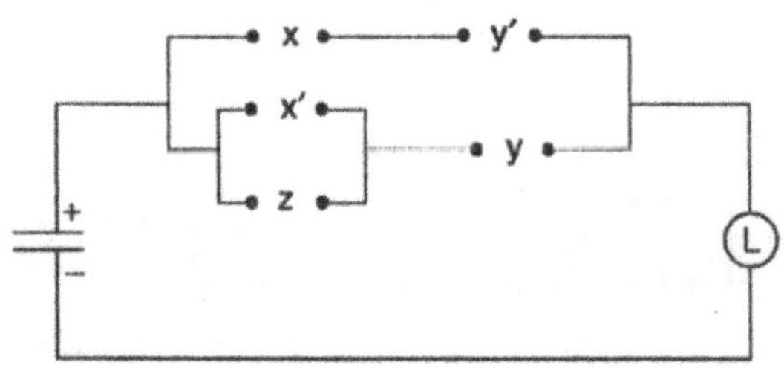

5. Construct the circuit corresponding to the Boolean function $x.(y' + z) + y.z'$ in the switching algebra.
6. Construct the circuit corresponding to the Boolean function $x.y + x'.(x + y + y')$ in the switching algebra.

7. Construct the circuit corresponding to the Boolean function $x.y + x.y' + x'.y$ in the switching algebra.
8. Construct the circuit corresponding to the Boolean function $x.y'.z + (y + z).x'$ in the switching algebra.
9. Simplify the circuit given below :

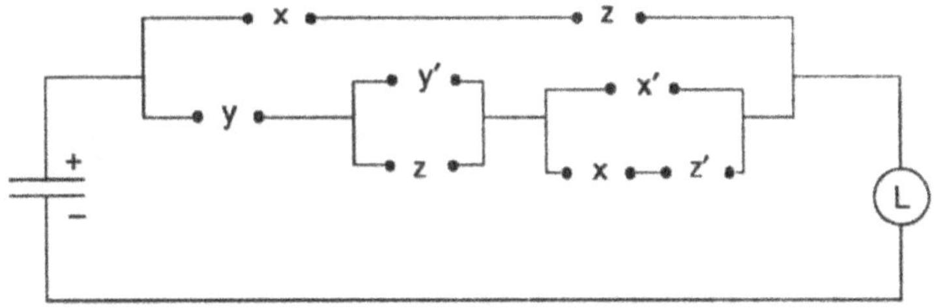

10. Simplify the circuit given below :

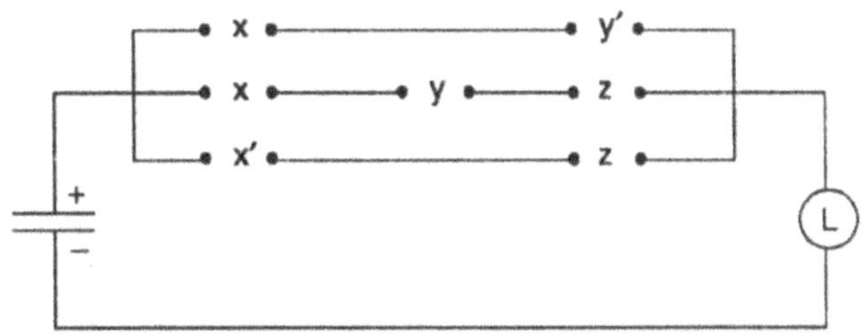

11. Simplify the circuit given below :

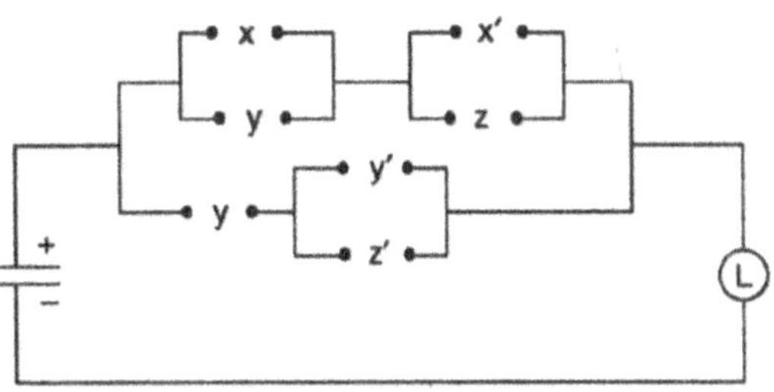

Q. 11

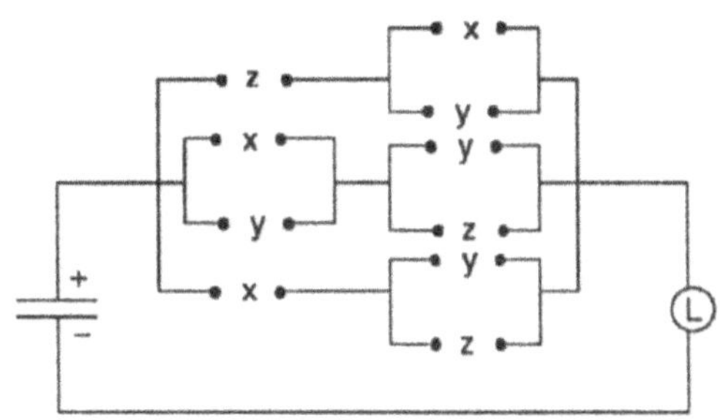

Q. 12

12. Simplify the circuit given above :

Answers

1. $x.y + x'$
2. $x.(y + z).\,x'.(y + z')$
3. $x.y'.z + (y + z).x'$
4. $x.y' + (x' + z).y$
5.

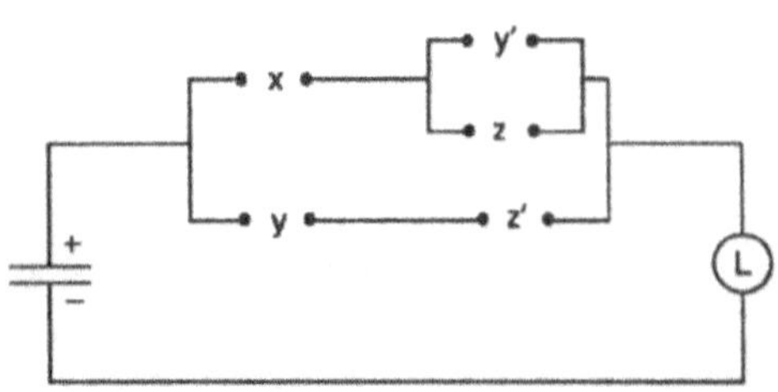

6.

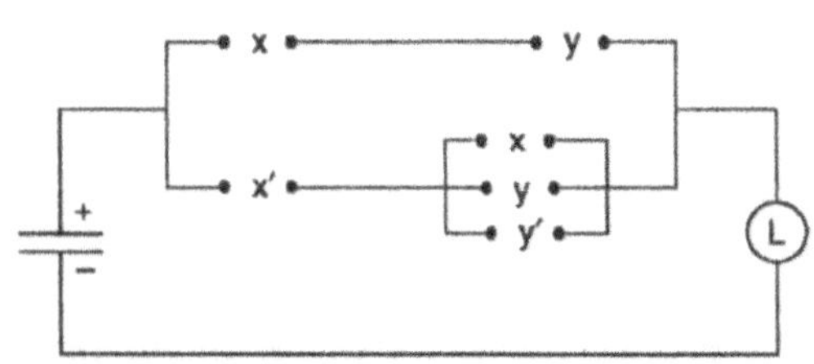

7.

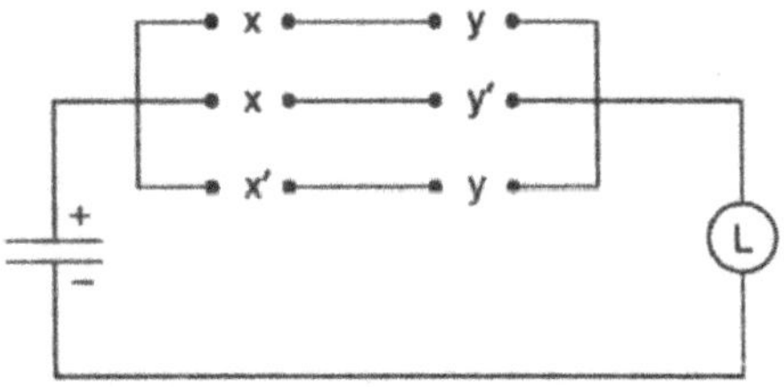

8.

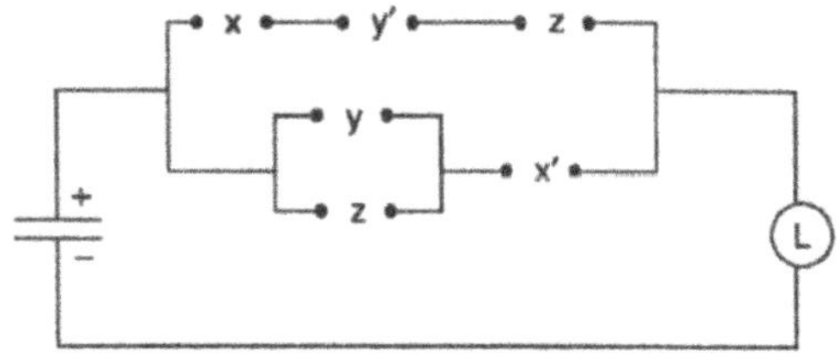

9.

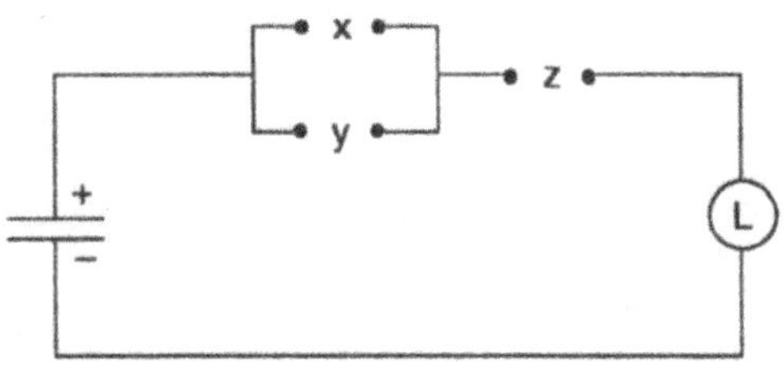

10.

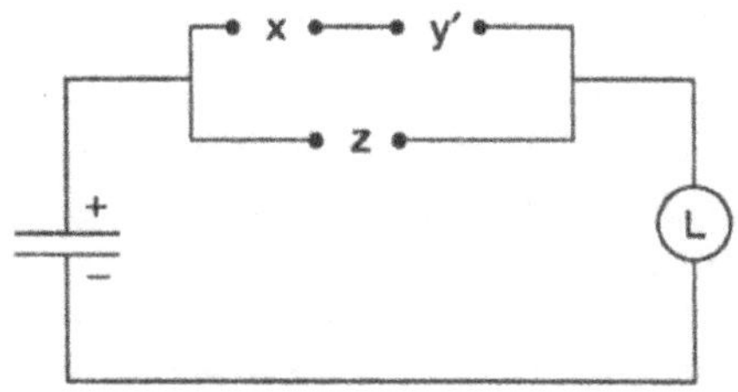

11.

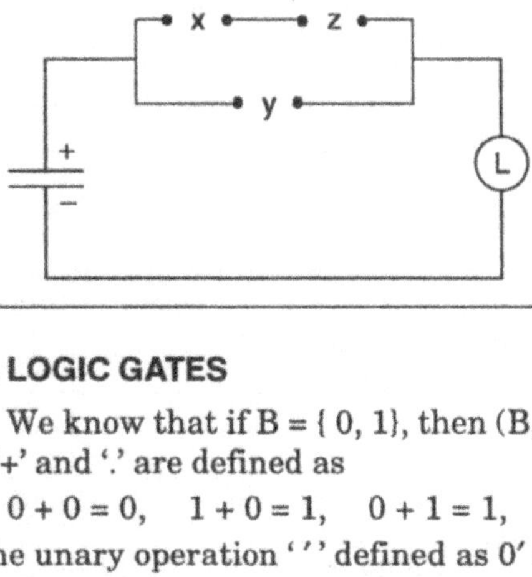

12.

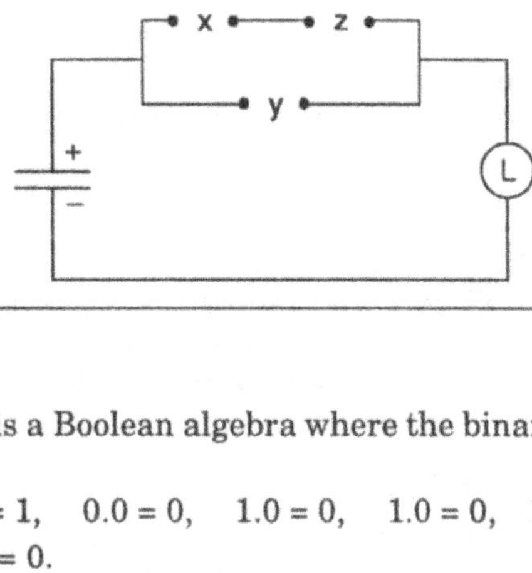

10.22. LOGIC GATES

We know that if B = { 0, 1}, then (B, +, ., ′) is a Boolean algebra where the binary operations '+' and '.' are defined as

$$0 + 0 = 0, \quad 1 + 0 = 1, \quad 0 + 1 = 1, \quad 1 + 1 = 1, \quad 0.0 = 0, \quad 1.0 = 0, \quad 1.0 = 0, \quad 1.1 = 1$$

and the unary operation ' ′ ' defined as $0' = 1$, $1' = 0$.

This Boolean algebra is called *switching algebra* because if two switches are connected in parallel then the value of the circuit is given by $x + y$, where x and y are the values (0 or 1) of the switches in the circuit and if two switches are connected in series then the value of the circuit is given by $x.y$, where x and y are the values (0 or 1) of the switches in the circuit.

In a computer, the operations are carried out by means of combinations of signals passing through standard blocks of built-in circuits that are known as **logic gates.**

In other words, a logic gate is simply an electronic circuit which operate on one or more input signals to produce standard output signals. These logic gates are the building blocks of all the circuits in a computer.

There are three basic logic gates called OR gate, AND gate and NOT gate. We shall use the convention that the lines enterning the gate symbol from the left are input lines and the single line on the right is the output line.

10.23. OR GATE

An **OR gate** is an electronic circuit that generates the output signal of 1 if any of the input signals is 1. Two or more switches connected in parallel behave as an OR gate.

Truth table for an OR gate

Inputs		*Output*
x	y	$x + y$
0	0	0
1	0	1
0	1	1
1	1	1

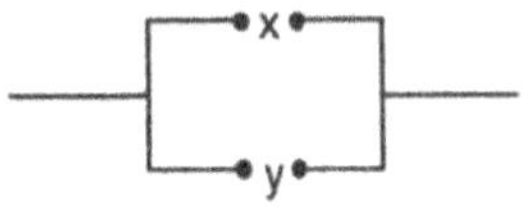

(*i*)
Switches in parallel

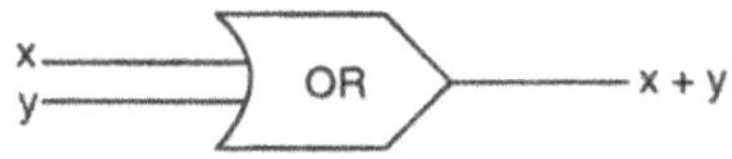

(*ii*)
OR gate

In the Fig. (*i*), switches x and y are connected in parallel and the input current will reach the output point when any one of the two switches are in the on(1) state. There will be no output only when both the switches x and y are in the off (0) state. The OR gate for two input signals x and y is shown in the Fig. (*ii*). An output of signal 1 is obtained when any of the input signals is 1. Output is 0 only when both the inputs are 0.

Truth table for an OR gate

Inputs			*Output*
x	y	z	$x + y + z$
0	0	0	0
0	0	1	1
0	1	0	1
1	0	0	1
0	1	1	1
1	0	1	1
1	1	0	1
1	1	1	1

Since associative laws hold in a Boolean algebra, an OR gate may have more than two inputs. The figure given belows shows an OR gate having three inputs.

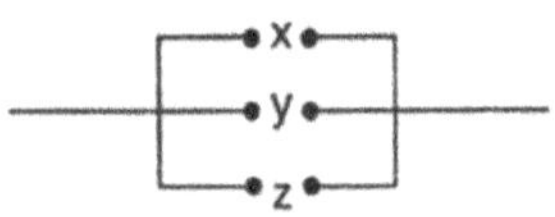

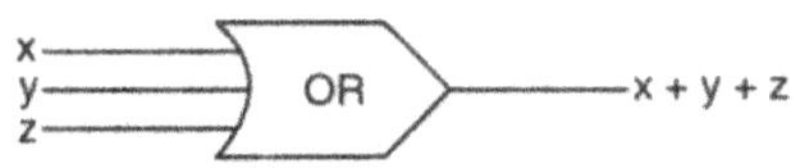

The OR gate with inputs x, y, z has output 1 if x or y or z is with value 1. Similarly, we can have OR gate having more than three signals and the output has signal 1 only when at least one input signal is of value 1.

Truth table for an AND gate

Inputs		*Output*
x	y	x, y
0	0	0
1	0	0
0	1	0
1	1	1

10.24. AND GATE

An **AND gate** is an electronic circuit that generates the output signal of 1 only if all input signals are 1. Two or more switches connected in series behave as an AND gate.

(*i*)
Switches in series

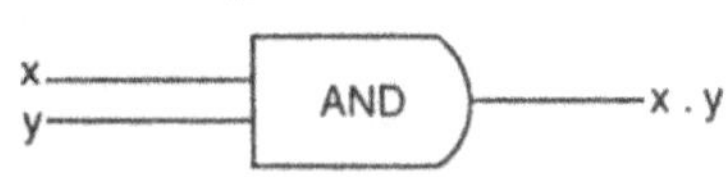

(*ii*)
AND gate

In the Fig. (*i*), switches x and y are connected in series and the input current will reach the output point only when both switches are in the on(1) state. There will be no output when at least one of the switches x and y is in the off(0) state. The AND gate for two input signals x and y is shown in the Fig. (*ii*). An output of signal 1 is obtained when both input signals are 1. Output is 0 when at least one of the inputs is 0.

Truth table for an AND gate

Input			*Output*
x	y	z	$x.y.z$
0	0	0	0
0	0	1	0
0	1	0	0
1	0	0	0
0	1	1	0
1	0	1	0
1	1	0	0
1	1	1	1

Since associative law hold in a Binary algebra, an AND gate may have more than two inputs. The figure given below shows an AND gate having three inputs.

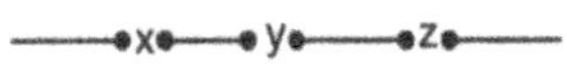

The AND gate with inputs x, y, z has output 1 only if x, y and z have value 1. Similarly, we can have AND gate having more than three signals and the output has signal 1 only if all input signals are of value 1.

10.25. NOT GATE

An **NOT gate** is an electronic circuit that generates an output signal which is reverse of the input signal.

NOT gate

Truth table for a NOT gate

Input	*Output*
x	x'
0	1
1	0

A NOT gate is also called an **inverter.**

10.26. LOGIC CIRCUITS

We know that logic gates are electronic circuits which operate on one or more input signals to produce standard output signals. The logic gates are seldom used by themselves but are used in combinations. A combination of logic gates is called a **combinatorial logic circuit** or simply a **logic circuit.** For a given logic circuit, a Boolean expression can be derived by systematically progressing from input to output on the gates.

Example 1. *Find the value of the Boolean expression $(x.y' + z).x'$ in the switching algebra if:*

(*i*) $x = 1, y = 0, z = 0$ (*ii*) $x = 0, y = 1, z = 1$ (*iii*) $x = 0, y = 0, z = 1$.

Sol. Boolean expression $= (x.y' + z).x'$

(*i*) We have $x = 1, y = 0, z = 0$.

$\therefore \quad (x.y' + z).x' = (1.0' + 0).1' = (1.1 + 0).0 = (1 + 0).0 = 1.0 = \mathbf{0}.$

(*ii*) We have $x = 0, y = 1, z = 1$.

$\therefore \quad (x.y' + z).x' = (0.1' + 1).0' = (0.0 + 1).1 = (0 + 1).1 = 1.1 = \mathbf{1}.$

(*iii*) We have $x = 0, y = 0, z = 1$.

$\therefore \quad (x.y' + z).x' = (0.0' + 1).0' = (0.1 + 1).1 = (0 + 1).1 = 1.1 = \mathbf{1}.$

Example 2. *Construct an input/output table for the Boolean function $f(x, y) = (x. y)' + y$ in the switching algebra.*

Sol. **Input/output table**

Inputs				*Output*
x	y	$x.y$	$(x.y)'$	$f(x, y) = (x.y)' + y$
0	0	0	1	1
0	1	0	1	1
1	0	0	1	1
1	1	1	0	1

Example 3. *Construct an input/output table for the Boolean function $f(x, y, z) = (x.y') + z$ of the switching algebra.*

Sol. **Input/output table**

Inputs					*Output*
x	y	z	y'	$x.y'$	$(x.\ y') + z$
0	0	0	1	0	0
0	0	1	1	0	1
0	1	0	0	0	0
1	0	0	1	1	1
0	1	1	0	0	1
1	0	1	1	1	1
1	1	0	0	0	0
1	1	1	0	0	1

Example 4. *Write the Boolean expression and the Boolean function in the switching algebra by using its input/output table.*

Inputs			*Output*
x	y	z	
1	1	1	1
1	1	0	1
1	0	1	1
1	0	0	0
0	1	1	0
0	1	0	0
0	0	1	0
0	0	0	1

Sol. Let the Boolean function be $f(x, y, z)$.

$\therefore$ From the given table, we have

$f(1, 1, 1) = 1, f(1, 1, 0) = 1, f(1, 0, 1) = 1,$
$f(1, 0, 0) = 0, f(0, 1, 1) = 0, f(0, 1, 0) = 0,$
$f(0, 0, 1) = 0, f(0, 0, 0) = 1.$

By **Boole's expension theorem**, we have

$$f(x, y, z) = f(1, 1, 1).x.y.z + f(1, 1, 0).\ x.y.z' + f(1, 0, 1).\ x.y'.z + f(1, 0, 0).\ x.y'.z'$$
$$+ f(0, 1, 1).x'.y.z + f(0, 1, 0).x'.y.z' + f(0, 0, 1).x'.y'.z + f(0, 0, 0).x'.y'.z'$$
$$= 1.x.y.z + 1.x.y.z' + 1.x.y'.z + 0.x.y'.z' + 0.x'.y.z + 0.x'.y.z' + 0.x'.y'.z + 1.x'.y'.z'$$
$$= x.y.z + x.y.z' + x.y'.z + 0 + 0 + 0 + 0 + x'.y'.z'$$
$$= x.y.(z + z') + y'.(x.z + x'.z') = x.y.1 + y'.(x.z + x'.z')$$
$$= x.y + y'.(x.z + x'.z').$$

$\therefore$ The Boolean expression is $\mathbf{x.y + y'.(x.z + x'.z')}$

The Boolean function is $\mathbf{f(x, y, z) = x.y + y'.(x.z + x'.z')}$.

Example 5. *Find the output of the given inputs for the following circuit :*

(*i*) $x = 1, y = 0, z = 1$

(*ii*) $x = 0, y = 1, z = 1$

(*iii*) $x = 1, y = 1, z = 1$

(*iv*) $x = 0, y = 0, z = 1.$

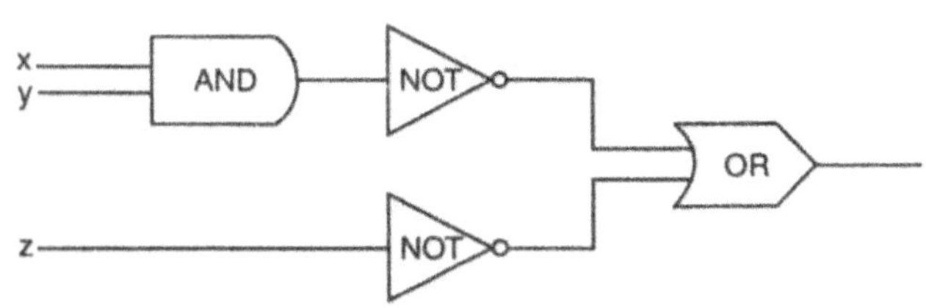

Sol. Given circuit is :

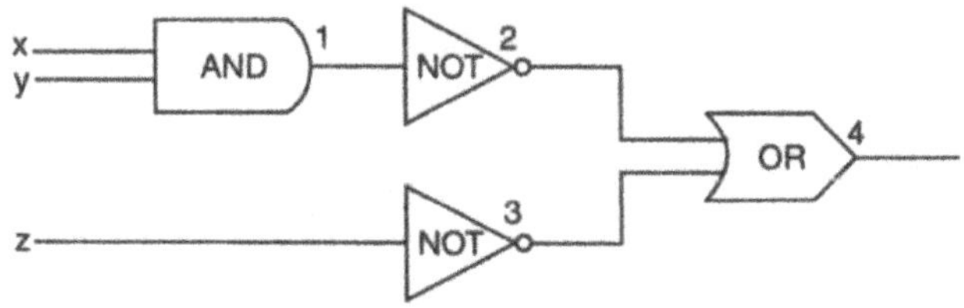

	Inputs			*Output at 1*	*Output at 2*	*Output at 3*	*Output of the circuit*
	x	*y*	*z*				
(*i*)	1	0	1	0	1	0	1
(*ii*)	0	1	1	0	1	0	1
(*iii*)	1	1	1	1	0	0	0
(*iv*)	0	0	1	0	1	0	1

Example 6. *Find the Boolean expression for the output of the given logic circuit :*

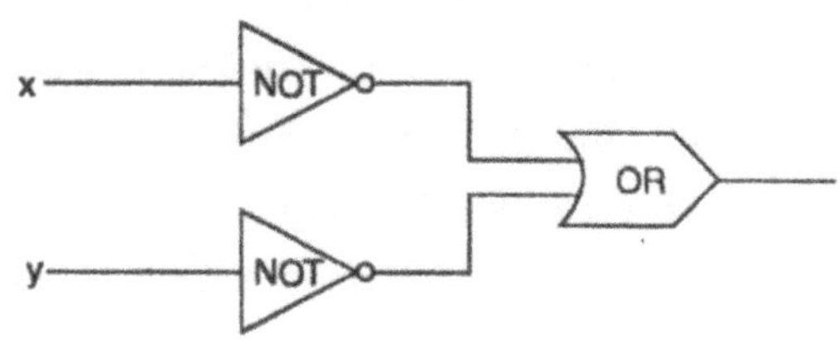

Also draw the truth table for the given logic circuit.

Sol. Given logic circuit is shown below.

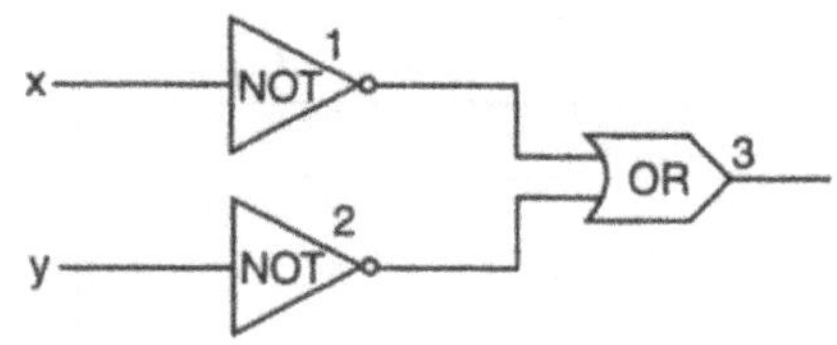

The inputs are x and y.

At point 1, the output of the NOT gate is x'.

At point 2, the output of the NOT gate is y'.

The inputs to the OR gate at point 3 are x' and y'.

∴ At point 3, the output of the OR gate is $\mathbf{x' + y'}$.

Truth table

Inputs				*Output*
x	y	x'	y'	$x' + y'$
0	0	1	1	1
1	0	0	1	1
0	1	1	0	1
1	1	0	0	0

Example 7. *Find the Boolean expression for the output of the given logic circuit :*

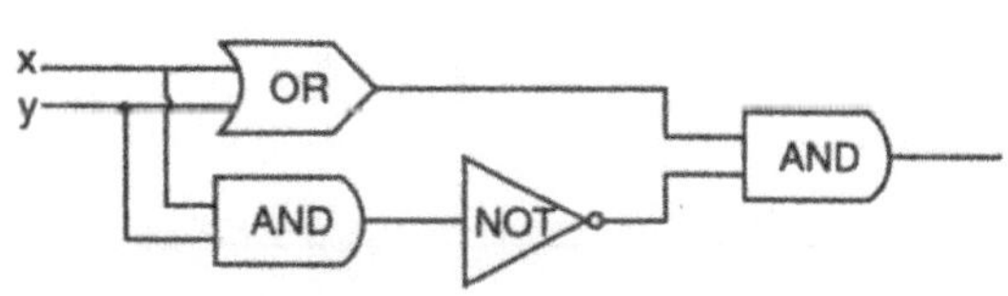

Also draw the truth table for the given logic circuit.

Sol. Given circuit is shown below.

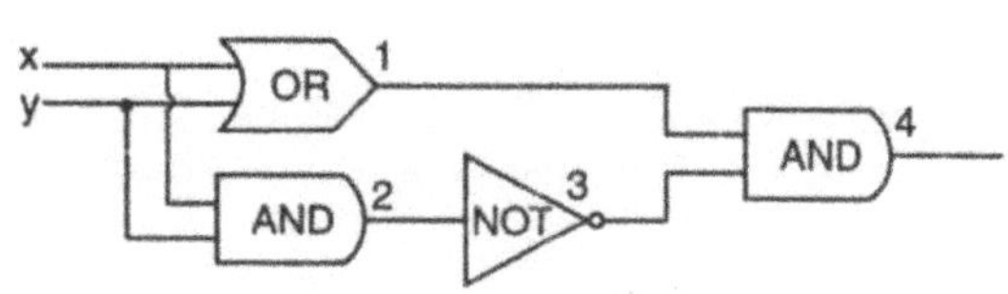

The inputs are x and y.

At point 1, the output of the OR gate is $x + y$

At point 2, the output of the AND gate is $x.y$

At point 3, the output of the NOT gate is $(x, y)'$
At point 4, the output of the AND gate is $\mathbf{(x + y).(x.y)'}$.

Truth Table

Inputs					*Output*
x	y	$x + y$	$x.y$	$(x.y)'$	$(x + y).(x.y)'$
0	0	0	0	1	0
1	0	1	0	1	1
0	1	1	0	1	1
1	1	1	1	0	0

Example 8. *Find the Boolean expression for the output of the given logic circuit. Also draw the truth table for the given logic circuit.*

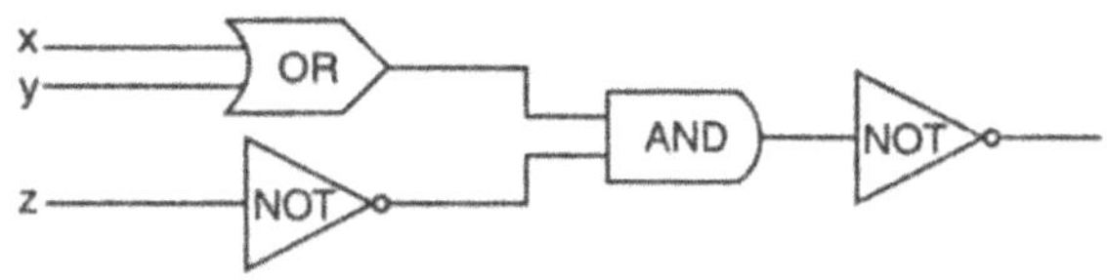

Sol. Given circuit is shown below.

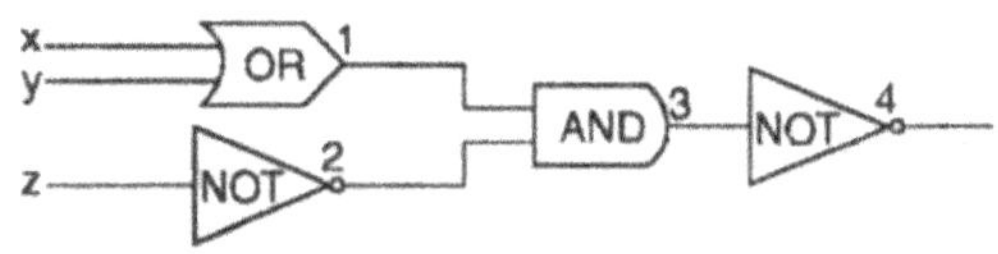

The inputs are x, y and z.

At point 1, the output of the OR gate is $x + y$

At point 2, the output of the NOT gate is z'

At point 3, the output of the AND gate is $(x + y).z'$

At point 4, the output of the NOT gate is $((x + y).z')'$

$\therefore$ Output of the circuit is $\mathbf{((x + y).z')'}$.

Truth Table

Inputs						*Output*
x	y	z	$x + y$	z'	$(x + y).z'$	$((x + y).z')'$
0	0	0	0	1	0	1
1	0	0	1	1	1	0
0	1	0	1	1	1	0
0	0	1	0	0	0	1
1	1	0	1	1	1	0
1	0	1	1	0	0	1
0	1	1	1	0	0	1
1	1	1	1	0	0	1

EXERCISE 10.8

1. Find the value of the Boolean expression $(x + y').\ y$ in the switching algebra if :
   (*i*) $x = 0, y = 1$ (*ii*) $x = 1, y = 1$.
2. Find the value of the Boolean expression $(x + y).x.z'$ in the switching algebra if :
   (*i*) $x = 0, y = 1, z = 1$ (*ii*) $x = 1, y = 0, z = 0$.
3. Find the value of the Boolean expression $(x.y + x'.z).\ (x + y.z)$ in the switching algebra if :
   (*i*) $x = 1, y = 1, z = 1$ (*ii*) $x = 0, y = 1, z = 0$.
4. Construct an input/output table for the Boolean function $f(x, y)$ in the switching algebra if :
   (*i*) $f(x, y) = (x + y).(x' + y).y$ (*ii*) $f(x, y) = (x.y' + x'.y).x$.
5. Construct an input/output table for the Boolean function $f(x, y, z)$ in the switching algebra if :
   (*i*) $f(x, y, z) = (x + y + z)'.(x.y' + z)$ (*ii*) $f(x, y, z) = x.y.z' + x'.y'.z + x'.y'.z'$.
6. Write the Boolean expression and the Boolean function $f(x, y)$ in the switching algebra by using its input/output table.

Inputs		*Output*
x	y	$f(x, y)$
0	0	0
1	0	1
0	1	0
1	1	1

7. Write the Boolean expression and the Boolean functions $f(x, y)$ in the switching algebra by using its input/output table.

Inputs		*Output*
x	y	$f(x, y)$
0	0	1
1	0	1
0	1	1
1	1	0

8. Write the Boolean expression and the Boolean function $f(x, y, z)$ in the switching algebra by using its input/output table.

Inputs			*Output*
x	y	z	$f(x, y, z)$
0	0	0	1
1	0	0	0
0	1	0	0
0	0	1	1
1	1	0	0
1	0	1	0
0	1	1	1
1	1	1	1

9. Write the Boolean expression and the Boolean function $f(x, y, z)$ in the switching algebra by using its input/output table :

Inputs			*Output*
x	y	z	$f(x, y, z)$
0	0	0	0
1	0	0	1
0	1	0	0
0	0	1	0
1	1	0	0
1	0	1	1
0	1	1	1
1	1	1	0

10. Find the output of the following inputs for the given circuit :

(i) $x = 0, y = 1$

(ii) $x = 1, y = 1$.

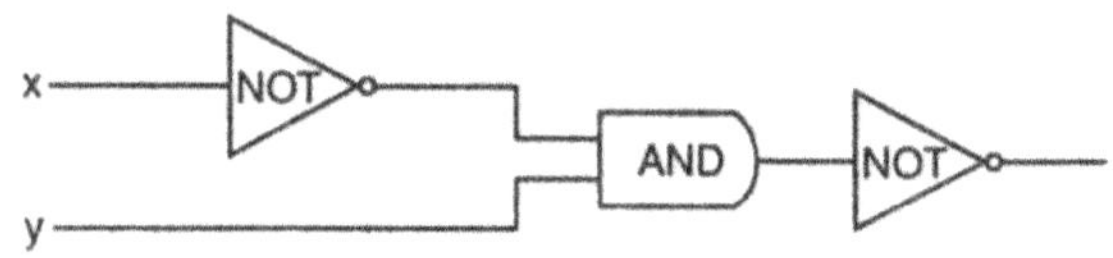

11. Find the output of the following inputs for the given circuit :

(i) $x = 0$, $y = 1, z = 0$

(ii) $x = 1, y = 0, z = 1$

(iii) $x = 1, y = 1, z = 0$

(iv) $x = 1, y = 1, z = 1$.

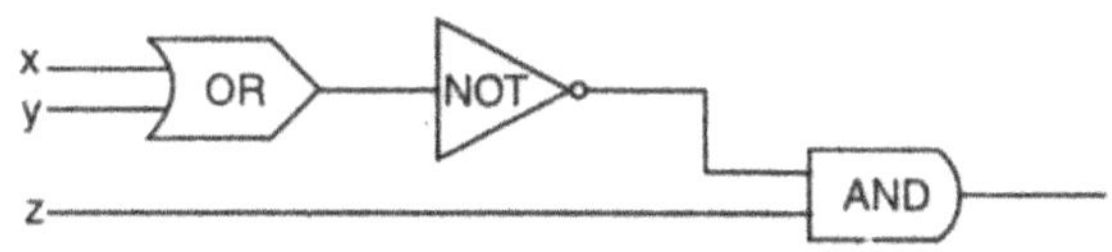

12. Find the Boolean expression for the output of the given logic circuit :

Also draw the truth table for the given logic circuit.

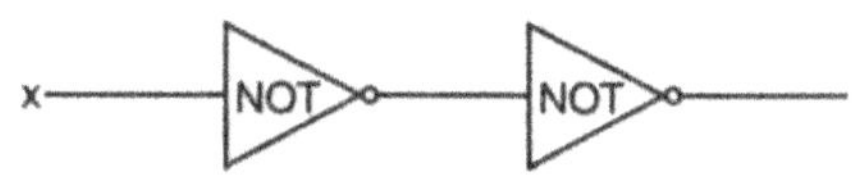

13. Find the Boolean expression for the output of the given logic circuit :

Also draw the truth table for the given logic circuit and indicate the output when $x = 0$, $y = 1$.

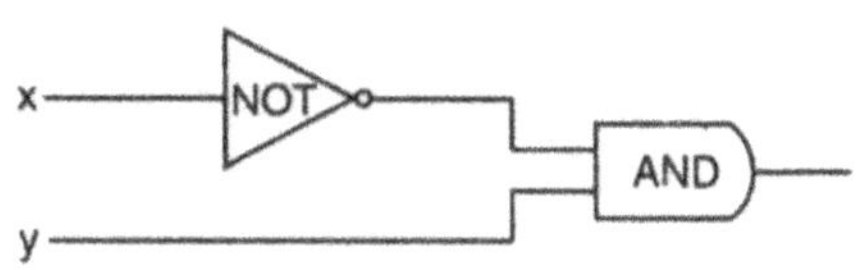

14. Find the Boolean expression for the output of the given logic circuit :

Also draw the truth table for the given logic circuit.

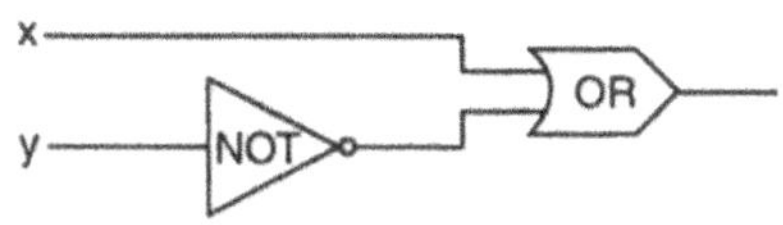

15. Find the Boolean expression for the output of the given logic circuit :

Also draw the truth table for the given logic circuit.

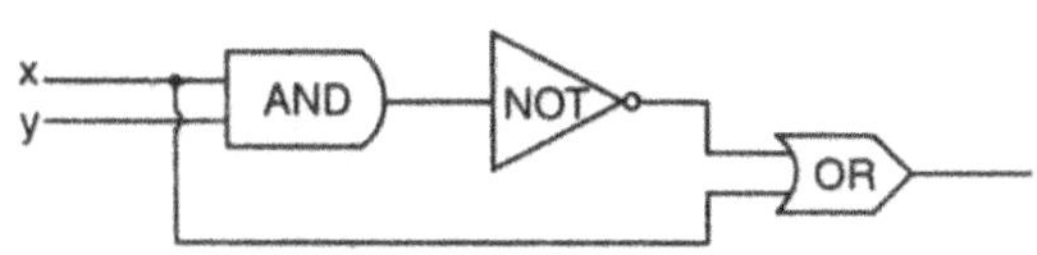

Answers

1. (i) 0 (ii) 1 2. (i) 0 (ii) 1

3. (i) 1 (ii) 0

4. (*i*)

Inputs						*Output*
x	y	x'	$x + y$	$x' + y$	$(x + y).(x' + y)$	$f(x, y) = (x + y).(x' + y).y$
0	0	1	0	1	0	0
1	0	0	1	0	0	0
0	1	1	1	1	1	1
1	1	0	1	1	1	1

(*ii*)

Inputs							*Output*
x	y	x'	y'	$x.y'$	$x'.y$	$x.y' + x'.y$	$f(x, y) = (x.y' + x'.y).x$
0	0	1	1	0	0	0	0
1	0	0	1	1	0	1	1
0	1	1	0	0	1	1	0
1	1	0	0	0	0	0	0

5. (*i*)

Inputs									*Output*
x	y	z	$x + y$	$x + y + z$	$(x + y + z)'$	y'	$x.y'$	$x.y' + z$	$f(x, y, z) = (x + y + z)'.(x.y' + z)$
0	0	0	0	0	1	1	0	0	0
1	0	0	1	1	0	1	1	1	0
0	1	0	1	1	0	0	0	0	0
0	0	1	0	1	0	1	0	1	0
1	1	0	1	1	0	0	0	0	0
1	0	1	1	1	0	1	1	1	0
0	1	1	1	1	0	0	0	1	0
1	1	1	1	1	0	0	0	1	0

(*ii*)

Inputs									*Output*
x	y	z	x'	y'	z'	$x.y.z'$	$x'.y'.z$	$x.'\,y'.z'.$	$f(x, y, z) = x.\,y.z' + x'.y'.z + x'.\,y'\,.z'$
0	0	0	1	1	1	0	0	1	1
1	0	0	0	1	1	0	0	0	0
0	1	0	1	0	1	0	0	0	0
0	0	1	1	1	0	0	1	0	1
1	1	0	0	0	1	1	0	0	1
1	0	1	0	1	0	0	0	0	0
0	1	1	1	0	0	0	0	0	0
1	1	1	0	0	0	0	0	0	0

6. x, $f(x, y) = x$.

7. $x' + y'$, $f(x, y) = x' + y'$

8. $x'.y' + y.z$, $f(x, y, z) = x'.y' + y.z$

9. $x.y' + x'.y.z$, $f(x, y, z) = x.y' + x'.y.z$

10. (*i*) 0 (*ii*) 1 (*iii*) 0 (*iv*) 0.

11. (*i*) 0 (*ii*) 0

12. x,

Input	*Output*
x	x
0	0
1	1

13. $x'.y$,

Input		*Output*
x	y	$x'.y$
0	0	0
1	0	0
0	1	1
1	1	0

14. $x + y'$,

Inputs		*Output*
x	y	$x + y'$
0	0	1
1	0	1
0	1	0
1	1	1

15. $(x.\ y)' + x$,

Inputs		*Output*
x	y	$(x.y)' + x$
0	0	1
1	0	1
0	1	1
1	1	1

10.27. CONVERSION OF BOOLEAN EXPRESSION TO LOGIC CIRCUIT

In this section, we shall consider the method of drawing logic circuit to a given Boolean expression. The three logic gates AND , OR and NOT are said to be logically complete because any Boolean expression may be realised using only these three gates. The method of constructing logic circuits for Boolean expression using only these three gates is given in the following working rules.

WORKING RULES FOR CONSTRUCTING LOGIC CIRCUITS

Step I. *Identify complements of variables in the given expression.*

Step II. *Draw one NOT gate for each variable whose complement appear in the expression.*

Step III. *Start with the innermost paranthesis and obtain it by using a logic gate OR or AND.*

Step IV. *Combine the output of the innermost paranthesis to the input of the next innermost paranthesis. This process is continued till we get the logic circuit of the given expression.*

Example 1. *Find a logic circuit corresponding to the Boolean expression $x' + y$. Also draw its input/output table.*

Sol. Boolean expression is $x' + y$.

Given expression contains complement of the variable x. In the first step, we draw NOT gate for x.

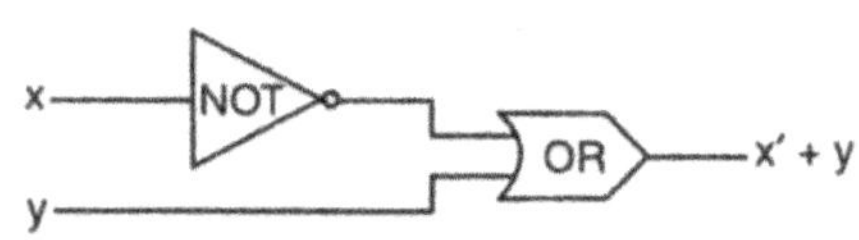

In the next step, we combine. x' and y by using an OR gate. This gives the logic circuit of the expression $x' + y$.

Input/output table

Inputs			*Output*
x	y	x'	$x' + y$
0	0	1	1
1	0	0	0
0	1	1	1
1	1	0	1

Example 2. *Find a logic circuit corresponding to the Boolean expression $x.(x' + y)$. Also draw its input/output table.*

Sol. Boolean expression is $x.(x' + y)$. Given expression contains complement of the variable x. In the first step we draw NOT gate for x. In the next step, we combine $x' + y$ by an OR gate. Next, we combine x and $x' + y$ by an AND gate. This gives the logic circuit of the expression $x.(x' + y)$.

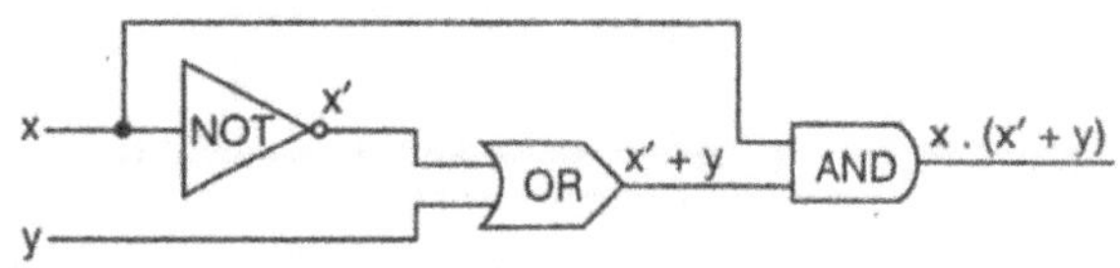

Input/output table

Inputs				*Output*
x	y	x'	$x' + y$	$x.(x' + y)$
0	0	1	1	0
1	0	0	0	0
0	1	1	1	0
1	1	0	1	1

Example 3. *Find a logic circuit corresponding to the Boolean expression $(x.(y' + z)) + y$. Also draw its input/output table.*

Sol. Boolean expression is $(x.(y' + z)) + y$. Given expression contains complement of only one variable y. In the first step we draw NOT gate for y. In the second step, we combine y' and z by an OR gate to get $y' + z$. In the third step, we combine x and $y' + z$ by an AND gate to get $x.(y' + z)$. In the fourth step, we combine $x.(y' + z)$ and y by an OR gate to get the required expression $x.(y' + z) + y$.

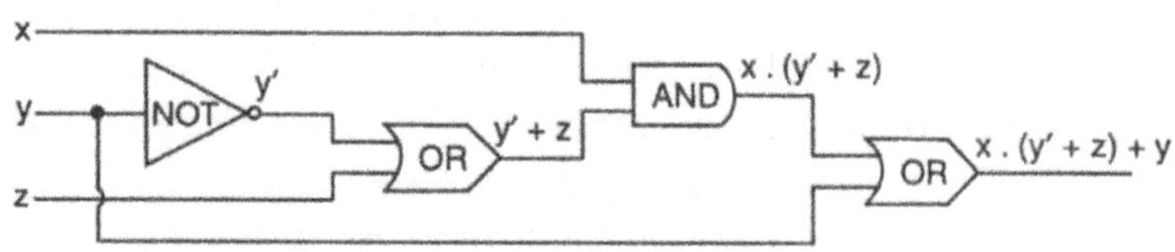

Input/output table

Inputs						*Output*
x	y	z	y'	$y' + z$	$x.(y' + z)$	$x.(y' + z) + y$
0	0	0	1	1	0	0
1	0	0	1	1	1	1
0	1	0	0	0	0	1
0	0	1	1	1	0	0
1	1	0	0	0	0	1
1	0	1	1	1	1	1
0	1	1	0	1	0	1
1	1	1	0	1	1	1

Example 4. *Find a logic circuit corresponding to the given input/output table.*

Inputs		*Output*
x	y	
0	*0*	*0*
1	*0*	*1*
0	*1*	*1*
1	*1*	*1*

Sol. Let $f(x, y)$ be the Boolean expression corresponding to the given input/output table. By **Boole's expansion theorem,** we have

$$f(x, y) = f(0, 0).x'.y' + f(1, 0).x.y' + f(0, 1).\ x'.y + f(1, 1).x.y.$$

$$\therefore\quad f(x, y) = 0.x'.y' + 1.x.y' + 1.x'.y + 1.x.y$$

$$= x.y' + x'.y + x.y$$

$f(x, y)$ can be simplified by using the properties of Boolean algebra.

We have $f(x, y) = x.y' + x'.y + x.y$

$$= x.y' + (x' + x).y$$

$$= x.y' + 1.y = x.y' + y.$$

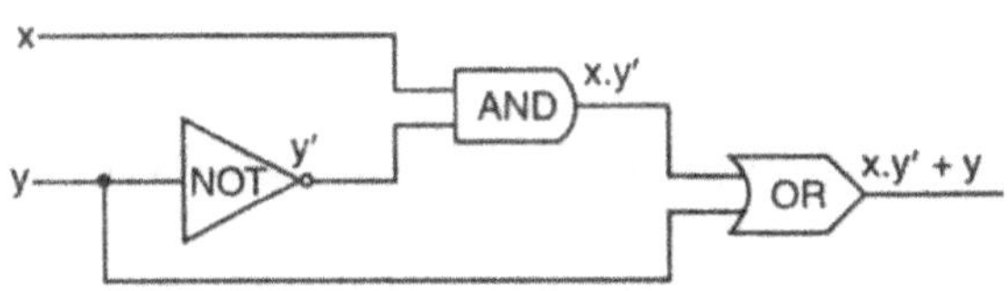

Boolean expression $x.y' + y$ contains complement of the variable y. In the first step, we draw NOT gate for y. In the second step, we combine x and y' by an AND gate to get $x.y'$. In the third step we combine $x.\ y'$ and y by an OR gate to get $x.y' + y$.

Example 5. *Find a logic circuit corresponding to the given input/output table :*

Inputs			*Output*
x	y	z	
1	*1*	*1*	*1*
1	*1*	*0*	*0*
1	*0*	*1*	*0*
1	*0*	*0*	*1*
0	*1*	*1*	*0*
0	*1*	*0*	*0*
0	*0*	*1*	*1*
0	*0*	*0*	*0*

Sol. Let $f(x, y, z)$ be the Boolean expression corresponding to the given input/output table. By **Boole's expansion theorem,** we have

$$f(x, y, z) = f(1, 1, 1).x.y.z + f(1, 1, 0).x.y.z'$$

$$+ f(1, 0, 1)\ x.y'.z + f(1, 0, 0).x\ y'\ z'$$

$$+ f(0, 1, 1).x'.y.z + f(0, 1, 0)\ x'.y.z'$$

$$+ f(0, 0, 1).x'.y'.z + f(0, 0, 0).x'.y'.z'$$

$$= 1.x.y.z + 0.x.y.z' + 0.x.y'.z$$

$$+ 1.x.y'.z' + 0.x'.y.z + 0.x'.y.z'$$

$$+ 1.x'.y'.z + 0.x'.y'.z'$$

$= x.y.z + 0 + 0 + x.y'.z' + 0 + 0 + x'.y'.z + 0$

$= x.y.z + x.y'.z' + x'.y'.z.$

Boolean expression for $f(x, y, z)$ contains complement of x, y and z variables.

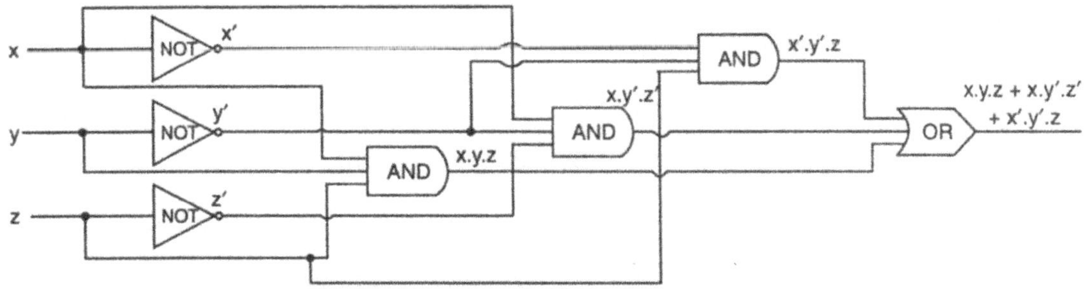

In the first step we draw NOT gates for x, y, z. In the second step we draw AND gates to get $x.y.z$, $x.y'.z'$ and $x'.y'.z$. In the third step we combine $x.y.z$, $x.y'.z'$ and $x'.y'.z$ by an OR gate to get the required Boolean expression $x.y.z + x.y'.z' + x'.y'.z$.

EXERCISE 10.9

1. Find a logic circuit corresponding to the Boolean expression $x.y'$. Also draw its input/output table.
2. Find a logic circuit corresponding to the Boolean expression $x' + y'$. Also draw its input/output table.
3. Find a logic circuit corresponding to the Boolean expression $x' + x.y'$. Also draw its input/output table.
4. Find a logic circuit corresponding to the Boolean expression $x.y + x'.y'$. Also draw its input/output table.
5. Find a logic circuit corresponding to the Boolean expression $(x + y'.z) + z$. Also draw its input/output table.
6. Find a logic circuit corresponding to the Boolean expression $(x' + y.z') + y$. Also draw its input/output table.
7. Find a logic circuit corresponding to the given input/output table.

Inputs		*Output*
x	y	
0	0	1
1	0	1
0	1	0
1	1	0

8. Find a logic circuit corresponding to the given input/output table.

Inputs		*Output*
x	y	
0	0	1
1	0	1
0	1	0
1	1	1

9. Find a logic circuit corresponding to the given input/output table.

Inputs			*Output*
x	y	z	
0	0	0	0
1	0	0	1
0	1	0	0
0	0	1	0
1	1	0	0
1	0	1	0
0	1	1	1
1	1	1	1

10. Find a logic circuit corresponding to the given input/output table.

Inputs			*Output*
x	y	z	
0	0	0	1
1	0	0	0
0	1	0	1
0	0	1	1
1	1	0	0
1	0	1	0
0	1	1	0
1	1	1	0

Answers

1.

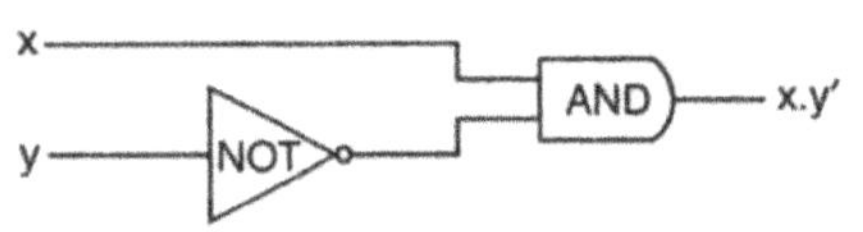

Inputs			*Output*
x	y	y'	$x.y'$
0	0	1	0
1	0	1	1
0	1	0	0
1	1	0	0

2.

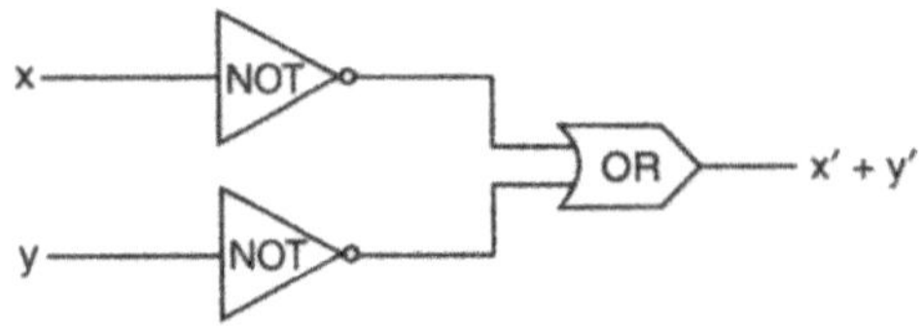

Inputs				*Output*
x	y	x'	y'	$x' + y'$
0	0	1	1	1
1	0	0	1	1
0	1	1	0	1
1	1	0	0	0

3.

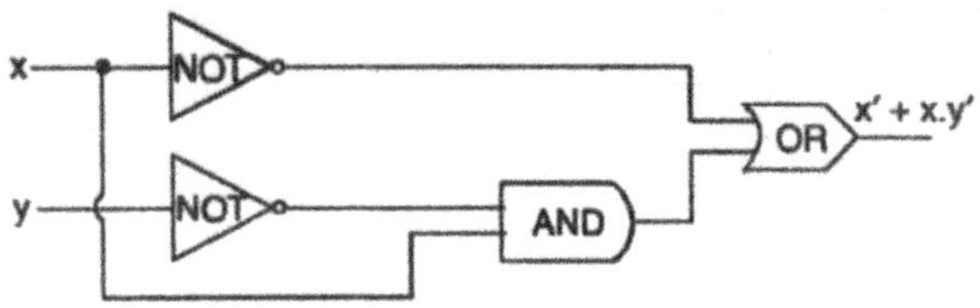

Inputs					*Output*
x	y	x'	y'	$x.y'$	$x' + x.y'$
0	0	1	1	0	1
1	0	0	1	1	1
0	1	1	0	0	1
1	1	0	0	0	0

4.

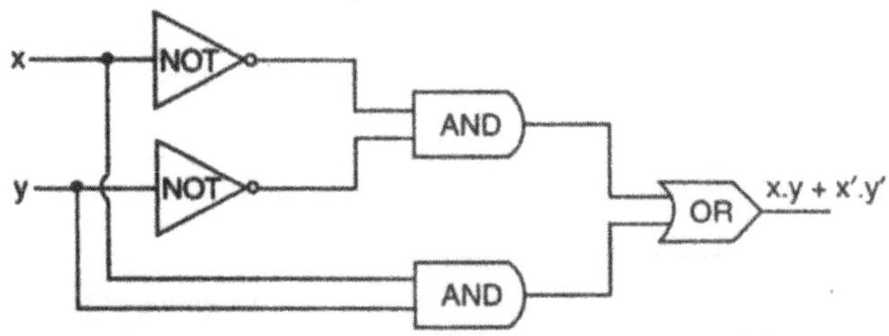

Inputs						*Output*
x	y	x'	y'	$x.y$	$x'.y'$	$x.y + x'.y'$
0	0	1	1	0	1	1
1	0	0	1	0	0	0
0	1	1	0	0	0	0
1	1	0	0	1	0	1

5.

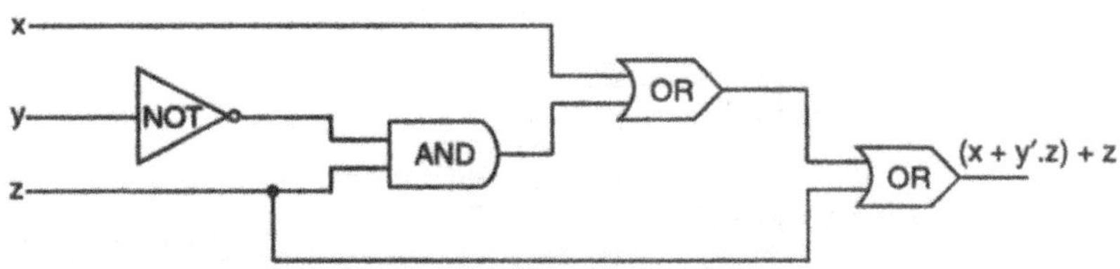

Inputs						*Output*
x	y	z	y'	$y'.z$	$x + y'.z$	$(x + y'.z) + z$
0	0	0	1	0	0	0
1	0	0	1	0	1	1
0	1	0	0	0	0	0
0	0	1	1	1	1	1
1	1	0	0	0	1	1
1	0	1	1	1	1	1
0	1	1	0	0	0	1
1	1	1	0	0	1	1

6.

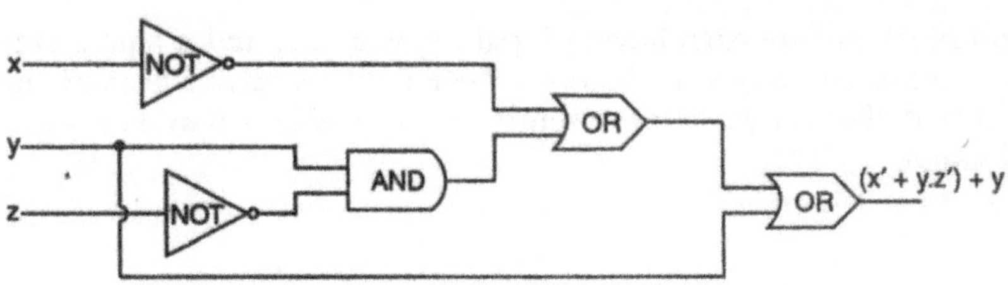

Inputs							*Output*
x	y	z	x'	z'	$y.z'$	$x' + y.z'$	$(x' + y.z') + y$
0	0	0	1	1	0	1	1
1	0	0	0	1	0	0	0
0	1	0	1	1	1	1	1
0	0	1	1	0	0	1	1
1	1	0	0	1	1	1	1
1	0	1	0	0	0	0	0
0	1	1	1	0	0	1	1
1	1	1	0	0	0	0	1

7.

8.

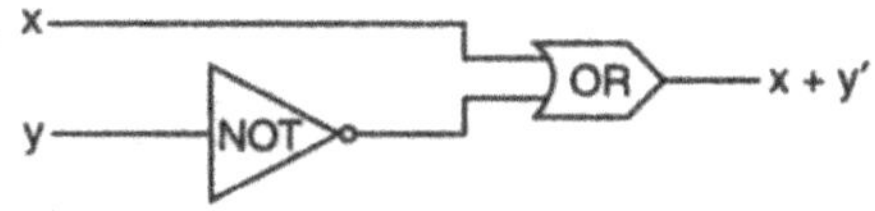

9.

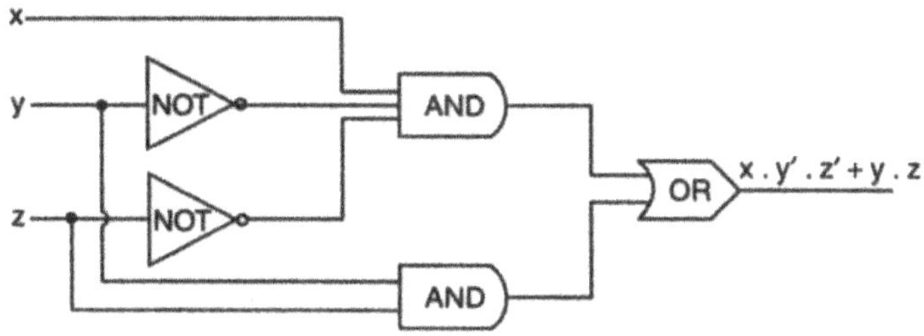

10.

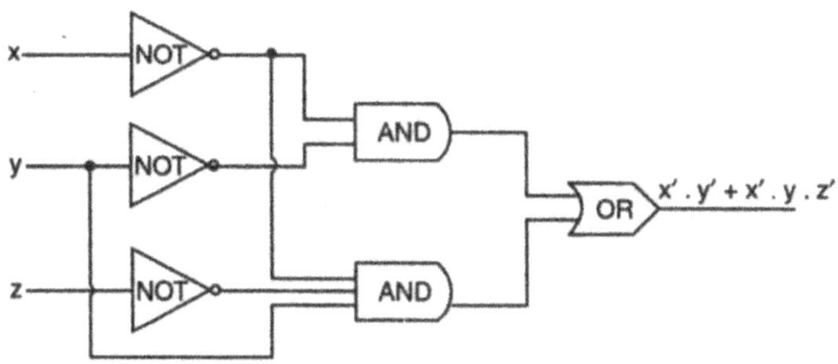

Hints

7. $x'.y' + x.y' = (x' + x).y' = 1.y' = y'.$

8. $x'.y' + x.y' + x.y = (x' + x).\ y' + x.y = 1.y' + x.y = y' + x.y = (y' + x).(y' + y)$

$= (y' + x).1 = y' + x = x + y'.$

10.28. EQUIVALENT LOGICAL CIRCUITS

Two logical circuits each having inputs $x, y, z,$ ...... and a single output are said to be **equivalent logical circuits** if whenever the circuits receive the same input, they produce the same output. Thus, two logical circuits are equivalents if and only if their input/output tables are same.

WORKING RULES TO CHECK EQUIVALENCE OF TWO LOGICAL CIRCUITS

Step I. *Find the Boolean expression of the first circuit and draw its input/output table.*

Step II. *Find the Boolean expression of the second circuit and draw its input/output table.*

Step III. *Compare the truth tables.*

Step IV. *If the tables are identical then say that the given circuits are equivalent.*

Example 1. *Show that the logical circuits given below are equivalent :*

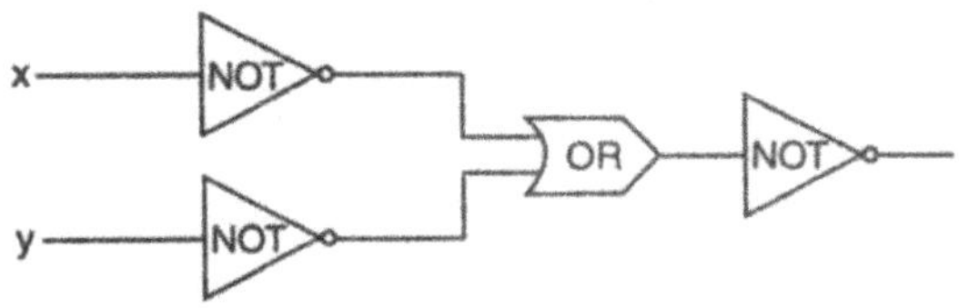

Sol. First Circuit.

The inputs are x and y.

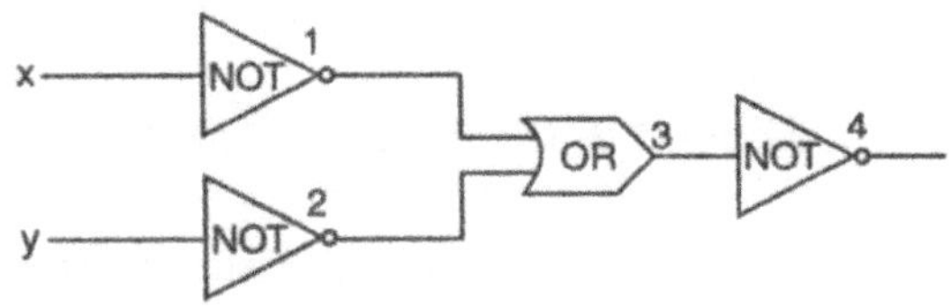

At point 1, the output of the NOT gate is x'.

At point 2, the output of the NOT gate is y'.

At point 3, the output of the OR gate is $x' + y'$.

At point 4, the output of the AND gate is $(x' + y')'$.

$\therefore$ Boolean expression of the given circuit is $(x' + y')'$.

Input/output table of $(x' + y')'$

Inputs					*Output*
x	y	x'	y'	$x' + y'$	$(x' + y')'$
0	0	1	1	1	0
1	0	0	1	1	0
0	1	1	0	1	0
1	1	0	0	0	1

Second Circuit.

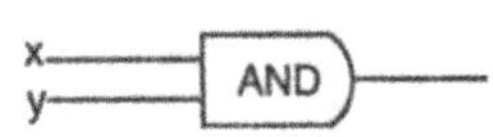

The inputs are x and y.

The output of the AND gate is $x.y$.

$\therefore$ Boolean expression of the given circuit is $x.y$.

Input/output table of x.y

Inputs		*Output*
x	y	$x.y$
0	0	0
1	0	0
0	1	0
1	1	1

∴ The input/output tables are same.

∴ Given circuits are equivalent.

Example 2. *Show that the logical circuits given below are equivalent :*

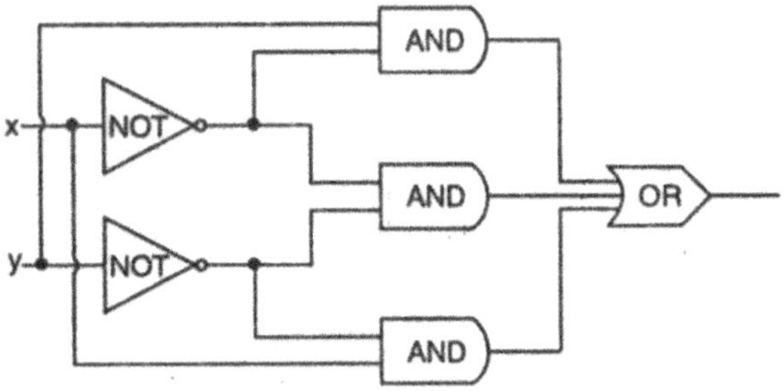

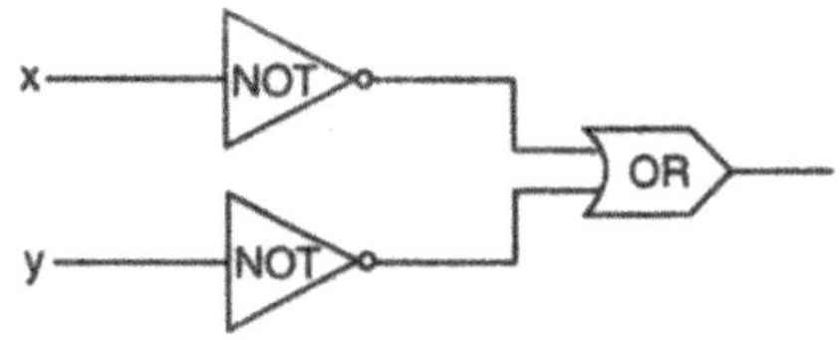

Sol. First Circuit

The inputs are x and y.

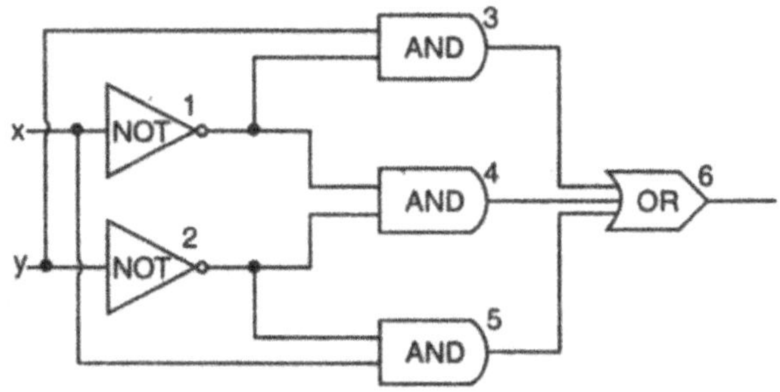

At point 1, the output of the NOT gate is x'.

At point 2, the output of the NOT gate is y'.

At point 3, the output of the AND gate is $x'.y$.

At point 4, the output of the AND gate is $x'.y'$.

At point 5, the output of the AND gate is $x.y'$.

At point 6, the output of the OR gate is $x'.y + x'.y' + x.y'$.

∴ Boolean expression of given circuit is $x'.y + x'\ y' + x.\ y'$.

Input/output table of x′.y + x′.y′ + x.y′

Inputs							*Output*
x	y	x'	y'	$x'.y$	$x'.y'$	$x.y'$	$x'.y + x'.y' + x.y'$
0	0	1	1	0	1	0	1
1	0	0	1	0	0	1	1
0	1	1	0	1	0	0	1
1	1	0	0	0	0	0	0

Second Circuit

The inputs are x and y.

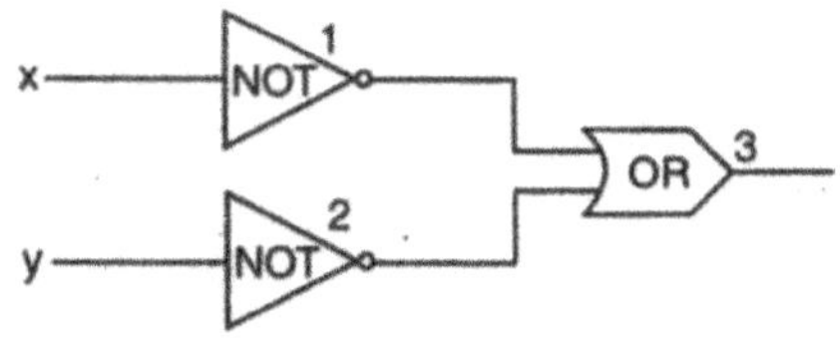

At point 1, the output of the NOT gate is x'.

At point 2, the output of the NOT gate is y'.

At point 3, the output of the OR gate is $x' + y'$.

∴ Boolean expression of given circuit is $x' + y'$.

Input/output table of x′ + y′

Inputs				*Output*
x	y	x'	y'	$x'+y'$
0	0	1	1	1
1	0	0	1	1
0	1	1	0	1
1	1	0	0	0

∴ The input/output tables are same.

∴ Given circuits are equivalent.

EXERCISE 10.10

1. Show that the logical circuits given below are equivalent :

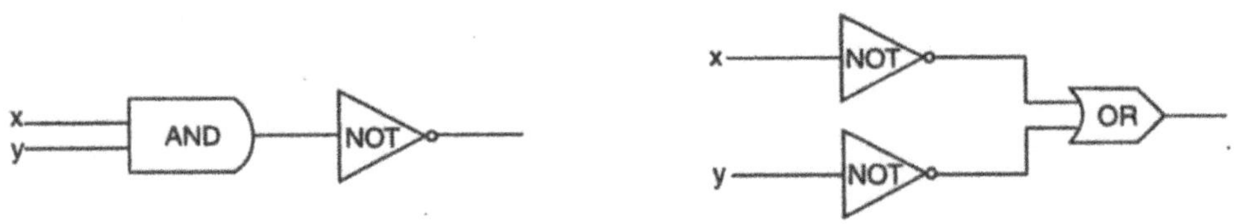

2. Show that the logical circuits given below are equivalent :

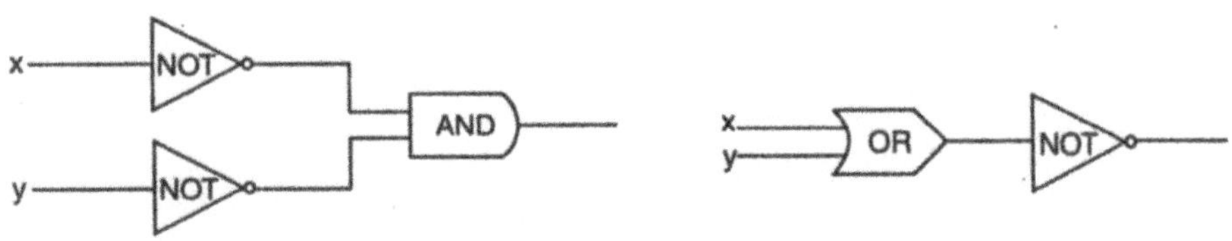

3. Show that the logical circuits given below are equivalent :

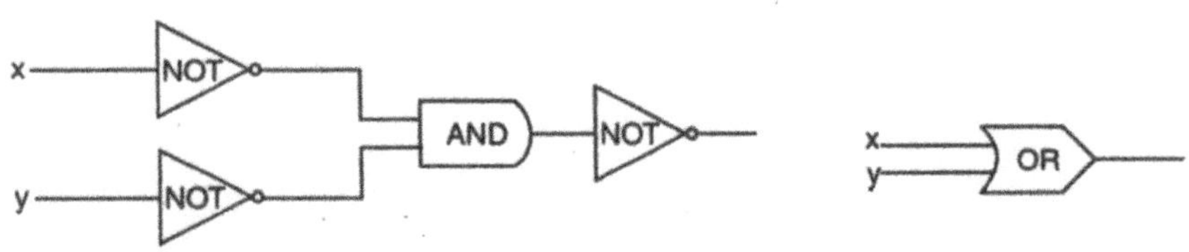

4. Show that the logical circuits given below are equivalent :

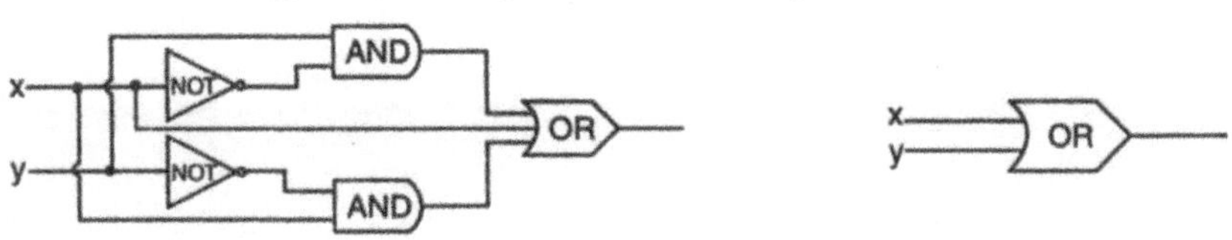

SECTION IV

11

Graphs

11.1. INTRODUCTION

Graphs are used to represent problems involving arrangement of discrete objects, where we are not interested in the internal properties of the objects rather we are concerned with the relationship among these discrete objects. Sets containing discrete objects and binary relations on them can be represented and studied very conveniently by graphs. Graph theory was born in 1736 with Euler's paper in which he solved the famous **Königsberg bridge problem.**

Graph theory begins with very simple geometric ideas and has many powerful applications. For the last two decades, the development in the field of graph theory has increased unprecedentedly.

In the modern times, graph theory is employed in many areas such as computers, communication, engineering, physical sciences etc. Applied mathematicians and engineers play a very vital role in the development and applicability of graph theory.

Graphs are of directed and undirected types. In this chapter, we shall be considering only undirected graphs. For us, a graph will mean an undirected graph.

I. TYPES OF GRAPHS

11.2. GRAPH

A **Graph** G is defined as a pair (V, E), where V is a set of elements called **vertices** and E is a set of unordered pairs of vertices called **edges.**

The vertices of a graph are denoted by $v_1, v_2, \ldots\ldots$ and the edges are denoted by $e_1, e_2, \ldots$.

A graph is represented by a diagram in which the vertices are represented by small dots (or small circles) and each edge $e = \{v_i, v_j\}$ is represented by a line, drawn straight or curved, long or short, connecting its endpoints v_i and v_j. The main consideration is the incidence between the edges and vertices.

For example, let $G = (V, E)$ be a graph, where

$$V = \{v_1, v_2, v_3, v_4, v_5\},$$
$$E = \{e_1, e_2, e_3, e_4, e_5, e_6\}$$

and

$$e_1 = \{v_1, v_2\}, \quad e_2 = \{v_2, v_3\},$$
$$e_3 = \{v_3, v_4\}, \quad e_4 = \{v_2, v_4\},$$
$$e_5 = \{v_4, v_4\}, \quad e_6 = \{v_2, v_3\}.$$

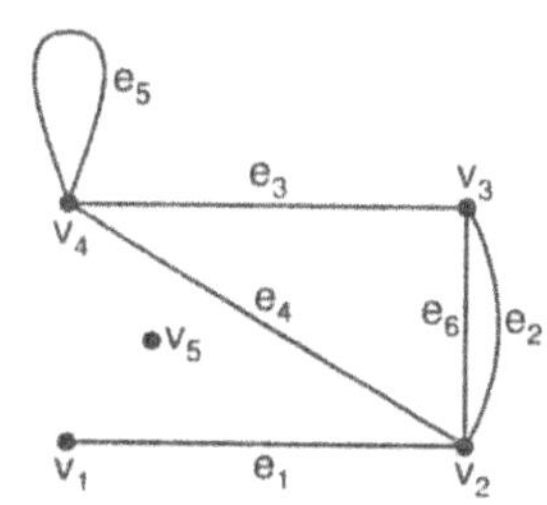

The graph G is shown in the adjoining diagram. This is a graph with five vertices and six edges. The edge $\{v_i, v_j\}$ can also be written as $\{v_j, v_i\}$.

If the edge e_k is equal to $\{v_i, v_j\}$ then the vertices v_i and v_j are called the **end points** or **end vertices** of the edge e_k and e_k is said to **connect** v_i and v_j. Also, the edge e_k is said to be **incident** on each of its end points v_i and v_j. In the above example, the vertices v_2 and v_3 are the end vertices of the edge e_2. Also, the edge e_2 is incident on the vertices v_2 and v_3.

Two vertices v_i and v_j are called **adjacent vertices** if there is an edge $\{v_i, v_j\}$ in the graph. In the above example, the vertices v_3 and v_4 are adjacent but the vertices v_1 and v_4 are not adjacent.

An edge having same vertex at its both end vertices is called a **loop.** In the above example, the edge $e_5 = \{v_4, v_4\}$ is a loop.

Two edges having same pair of distinct end vertices are called **parallel edges.** In the above example, the edges e_2 and e_6 are parallel because both have the same end-vertices v_2 and v_3.

The **order** of a graph is the number of its vertices. For example, the order of above graph is 5.

The **size** of a graph is the number of edges in the graph. For example, the size of the above graph is 6.

A graph is said to be a **null graph** if the set E of edges is empty. The diagram of the null graph $G = (\{v_1, v_2, v_3\}, \phi)$ is shown in the figure.

A null graph is also called a **discrete graph.**

A graph consisting of a single vertex and n loops is called a **bouquet** and is denoted by B_n. The bouquets B_2 and B_3 are shown in the diagram.

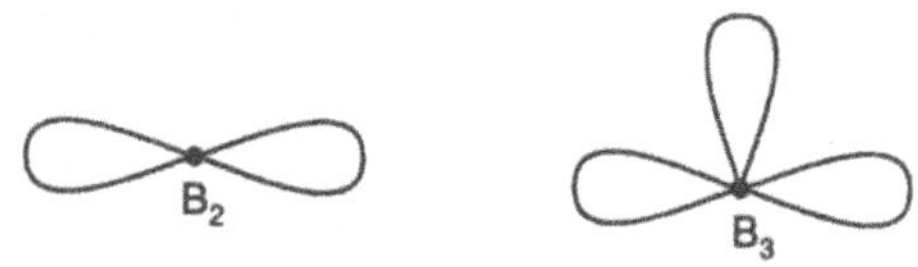

A graph is said to be a **finite graph** if it has a finite number of vertices and a finite number of edges. A finite graph with one vertex and no edge is called a **trivial graph.**

A graph which is not finite is called an **infinite graph.** A portion of an infinite graph is shown in the figure. In our further discussion we shall be considering only finite graphs. For us, the term **graph** shall mean a finite graph.

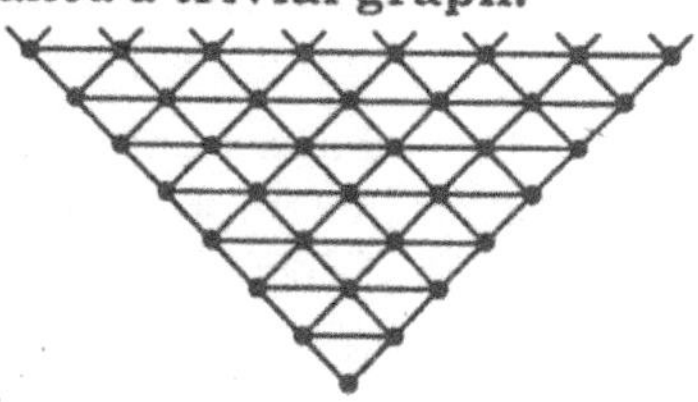

Example 1. *Draw the diagram of the graph $G = (V, E)$, where*
$V = \{1, 2, 3, 4, 5, 6\}$ and
$E = \{\{1, 2\}, \{1, 3\}, \{1, 4\}, \{2, 5\}, \{2, 6\}, \{3, 5\}, \{3, 6\}, \{4, 5\}\}$.

Sol. The vertices are 1, 2, 3, 4, 5 and 6. Let the edges $\{1, 2\}$, $\{1, 3\}$, $\{1, 4\}$, $\{2, 5\}$, $\{2, 6\}$, $\{3, 5\}$, $\{3, 6\}$, $\{4, 5\}$ be denoted by $e_1, e_2, e_3, e_4, e_5, e_6, e_7, e_8$ respectively.

The diagram of the graph is shown in the figure.

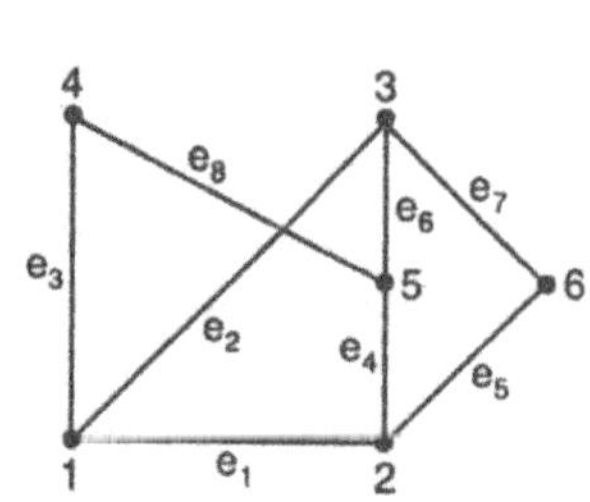

11.3. SHAPE OF GRAPH

Let $G(V, E)$ be a graph. In representing the graph G by a diagram, the length of the edges shown in the diagram is immaterial. The shape of a graph is also not fixed. The diagrams of the same graph may look quite different in shape but technically they are same.

Let $G = (V, E)$ be a graph, where

$$V = \{1, 2, 3, 4\} \quad \text{and} \quad E = \{e_1, e_2, e_3, e_4, e_5, e_6\}$$

and

$$e_1 = \{1, 2\}, e_2 = \{2, 3\}, e_3 = \{1, 2\}, e_4 = \{3, 4\}, e_5 = \{2, 4\}, e_6 = \{1, 4\}.$$

The following are the diagrams of the graph G drawn differently.

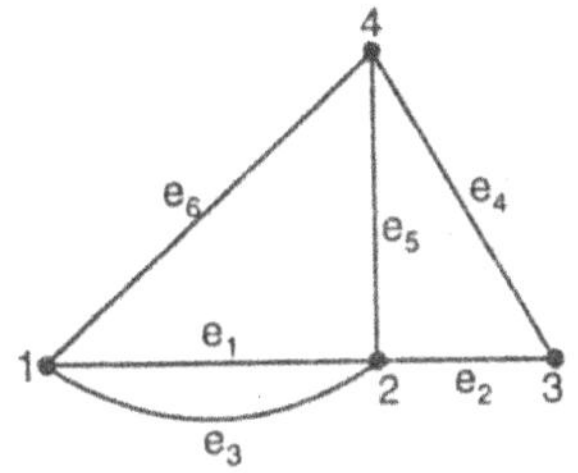

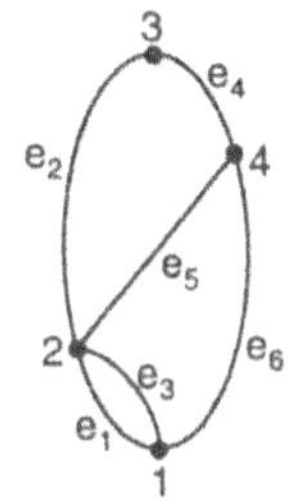

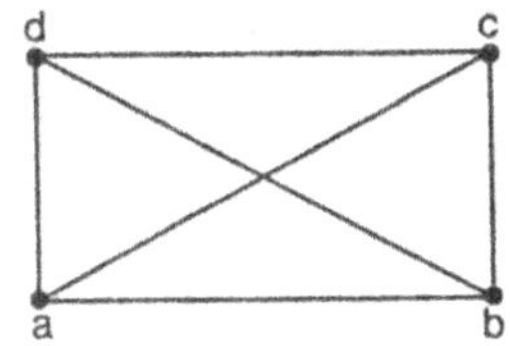

Remark. Consider the graph shown in the adjoining figure. In this graph, the edges $\{a, c\}$ and $\{b, d\}$ seem to be intersecting at a point which is not a vertex. Such edges should be thought of as being in different planes and thus having no common point.

11.4. DEGREE OF A VERTEX

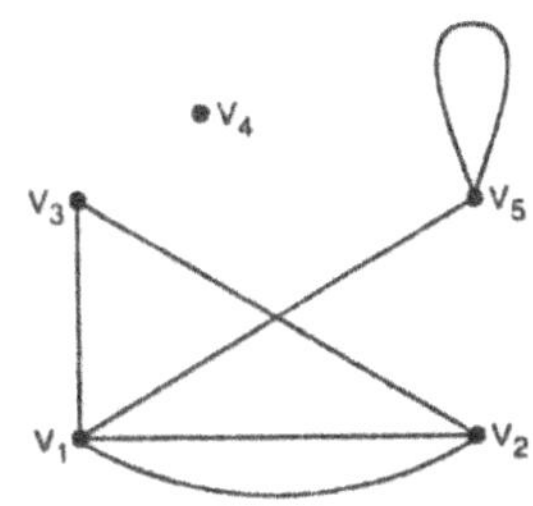

Let $G = (V, E)$ be a graph. The **degree** of a vertex in the graph is equal to the number of edges in G which are incident on this vertex *i.e.*, the number of edges having this vertex as an end vertex. The degree of a vertex v is denoted by $d(v)$. In the graph shown in the figure, we have

$$d(v_1) = 4, \quad d(v_2) = 3,$$
$$d(v_3) = 2, \quad d(v_4) = 0,$$
$$d(v_5) = 3.$$

It is interesting to note that

$$4 + 3 + 2 + 0 + 3 = 12 = 2 \times \text{no. of edges}.$$

Two adjacent edges are said to be in **series** if their common end vertex is of degree two. In the above example, the edges $\{v_1, v_3\}$ and $\{v_2, v_3\}$ are in series because their common end vertex v_3 is of degree 2.

If the vertices in a graph $G = (V, E)$ are $v_1, v_2, \ldots\ldots, v_n$ and $d(v_i) = d_i$, $1 \le i \le n$, then the sequence $d_1, d_2, \ldots\ldots, d_n$ is called the **degree sequence** of the graph G.

11.5. TYPES OF VERTICES

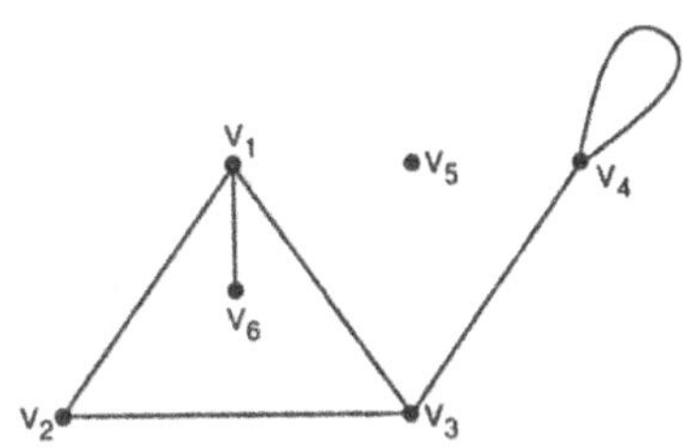

Let $G = (V, E)$ be a graph. A vertex of the graph G is called an **even vertex** or an **odd vertex** according as its degree is an even number or an odd number.

In the graph shown in the figure, we have

$$d(v_1) = 3, \quad d(v_2) = 2,$$
$$d(v_3) = 3, \quad d(v_4) = 3,$$
$$d(v_5) = 0, \quad d(v_6) = 1.$$

$\therefore$ The vertices v_2 and v_5 are even vertices and the vertices v_1, v_3, v_4 and v_6 are odd vertices.

A vertex is called an **isolated vertex** if it is not an end vertex of any edge in the graph. In the above example, the vertex v_5 is an isolated vertex.

A vertex is called a **pendant vertex** if it is of degree one *i.e.*, only one edge is incident on this vertex. In the above example, the vertex v_6 is a pendant vertex.

Remark. A pendant vertex of a graph is also called an **end vertex of the graph.**

11.6. EULER'S FIRST THEOREM OF GRAPH THEORY

Theorem I. Prove that the sum of the degrees of the vertices of a graph is equal to twice the number of edges in the graph. *(KU 2006 C)*

Proof. An edge of the graph may or may not be a loop.

If an edge is a loop at a vertex then by definition it contributes twice to the degree of that vertex. If an edge is not a loop then it contributes to the degrees of two distinct vertices.

$\therefore$ When the degrees of the vertices are added, each edge is counted exactly two times.

$\therefore$ Sum of degrees of all vertices is equal to twice the number of edges.

This theorem is called the *Euler's first theorem of graph theory.*

Theorem II. Prove that every graph has an even number of odd vertices.

Proof. Let x = sum of degrees of even vertices

and y = sum of degrees of odd vertices

$\therefore$ $x + y$ = sum of degrees of all vertices

By **Euler's theorem,** the sum of the degrees of all vertices is twice the number of edges.

Let $x + y = 2k$, where k is the number of edges.

Since sum of even numbers is also even, the sum x is an even number.

$\Rightarrow$ $2k - x$ is an even number. ($\because$ $2k$ is even)

$\Rightarrow$ y is an even number.

By definition, y is the sum of degrees of odd vertices.

$\therefore$ y is the sum of some odd numbers.

$\therefore$ Number of odd numbers must be an even number because the sum of odd number of odd numbers is always odd.

$\therefore$ The number of odd vertices must be even.

Example 2. *For the graph shown in the figure verify that :*

(*i*) *the sum of degrees of the vertices is twice the number of edges.*

(*ii*) *the number of odd vertices is even.*

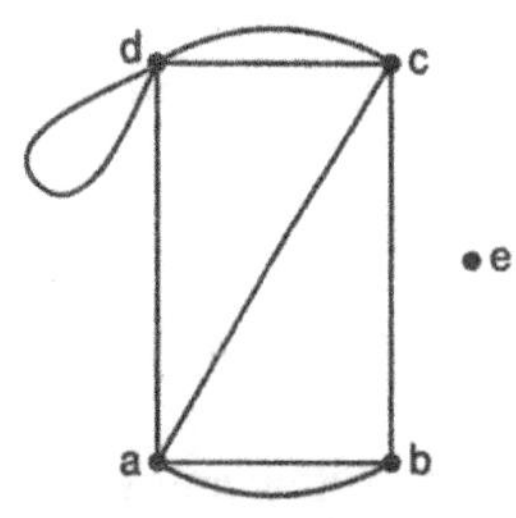

Sol. Here $V = \{a, b, c, d, e\}$

From the graph, we see that

$$d(a) = 4,\ d(b) = 3,\ d(c) = 4,\ d(d) = 5,\ d(e) = 0$$

and number of edges is 8.

(*i*) Sum of degrees of vertices

$$= 4 + 3 + 4 + 5 + 0 = 16$$
$$= 2 \times 8 = 2(\text{no. of edges}).$$

(*ii*) Odd vertices are b and d.

$\therefore$ There are 2 odd vertices, which is an even number.

Example 3. *It is known that there exists a graph with 12 vertices and 28 edges in which the degree of each vertex is either 3 or 5. Find the number of vertices of degree 5.*

Sol. Let the number of vertices of degree 5 be m.

$\therefore$ Number of vertices of degree $3 = 12 - m$

Number of edges = 28

∴ By **Euler's theorem,**

$$5m + 3(12 - m) = 2(28)$$

$\Rightarrow$ $5m + 36 - 3m = 56 \Rightarrow m = 10$

∴ There are **10 vertices** of degree 5 each.

Example 4. *Find whether it is possible for a graph to have 12 edges such that 2 of the vertices have degree 3 each and the remaining vertices have degree 4 each.*

Sol. Number of edges = 12

Let the number of vertices be n.

Since 2 vertices have degree 3 each, $n - 2$ vertices must have degree 4 each.

By **Euler's theorem,**

$$2(3) + (n - 2)(4) = 2(12)$$

$\Rightarrow$ $6 + 4n - 8 = 24 \Rightarrow 4n = 26 \Rightarrow n = 6.5$, which is not possible.

∴ Graph with given data is not possible.

EXERCISE 11.1

1. Draw the diagram of the graph $G = (V, E)$, where

$V = \{1, 2, 3, 4, 5\}$

and $E = \{\{1, 3\}, \{1, 5\}, \{4, 5\}, \{2, 4\}, \{4, 4\}\}$.

2. Find the order and size of the graph shown in the figure.

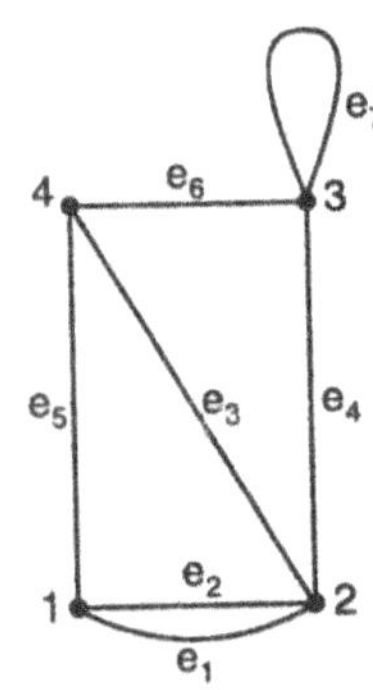

3. For the graph shown in the figure, verify that :
   (*i*) the sum of degrees of the vertices is twice the number of edges.
   (*ii*) the number of odd vertices is even.

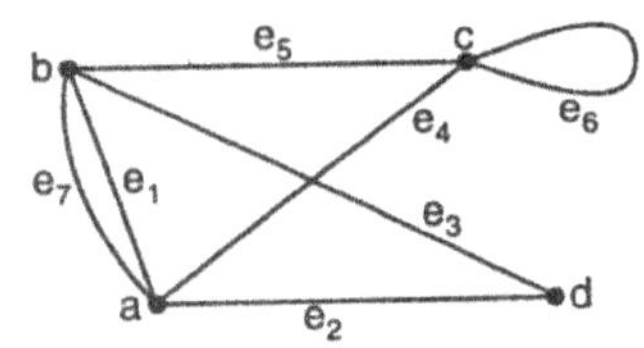

4. Find the number of edges in a graph having 5 vertices out of which 2 vertices are of degree 3 and 3 vertices are of degree 2.
5. How many vertices are necessary for a graph having 16 edges and each vertex is of degree 2 ?
6. Show that there does not exist a graph with 5 vertices with degrees equal to 2, 3, 1, 2, 3.
7. A graph has 21 edges, 3 vertices of degree 4 each and other vertices are of degree 3 each. Find the number of vertices in the graph.
8. Show that there is no graph with 12 vertices and 28 edges in which the degree of each vertex is either 3 or 6.

Answers

1. 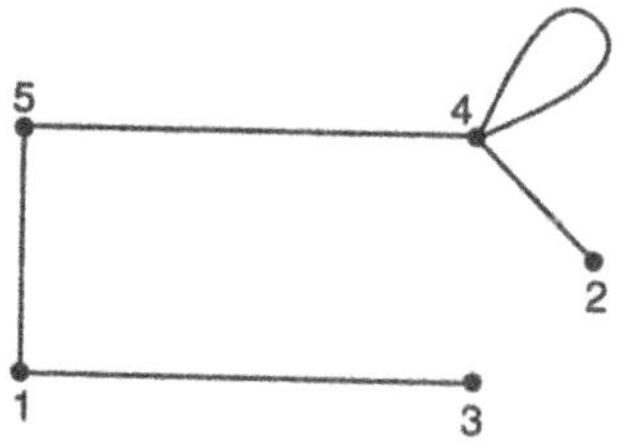

2. 4, 7

4. 6

5. 16

7. 13.

11.7. TYPES OF GRAPHS

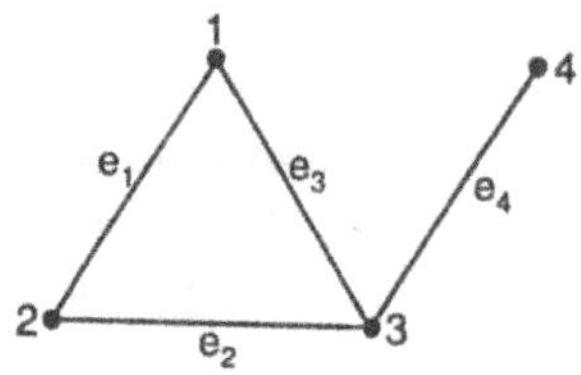

I. Simple Graph. A graph is called a **simple graph** if it has neither loops nor parallel edges. The graph shown in the figure is a simple graph.

II. Complement of a Simple Graph. The **complement of a simple graph** $G = (V, E)$ is defined as the simple graph $\overline{G} = (V, F)$ in which there is an edge between two vertices u and v if and only if there is no edge between u and v in G. The figure given below shows a graph G and its complement $\overline{G}$.

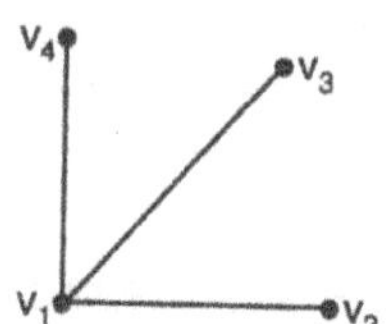

G

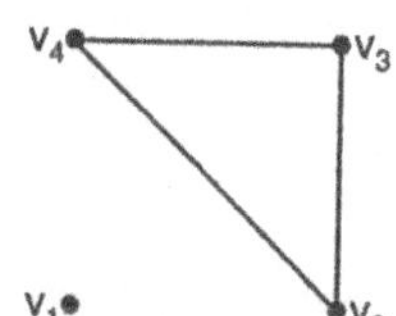

$\overline{G}$

It is easy to see that the complement of $\overline{G}$ is G.

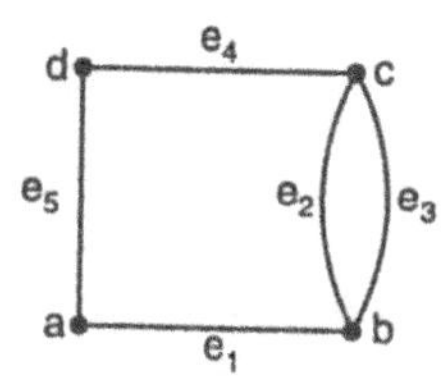

III. Multigraph. A graph is called a **multigraph** if it has no loops. The graph shown in the figure is a multigraph.

Example 1. *Draw all simple graphs with vertices a, b, c, d and two edges, one of which is* $\{a, b\}$.

Sol. Vertices are a, b, c, d.

Since ${}^4C_2 = 6$, edges which connect distinct vertices are $\{a, b\}, \{b, c\}, \{c, d\}, \{a, d\}, \{a, c\}, \{b, d\}$.

$\therefore$ There are 5 edges, other than $\{a, b\}$, connecting distinct vertices.

$\therefore$ There are 5 possible simple graphs. These graphs are given below :

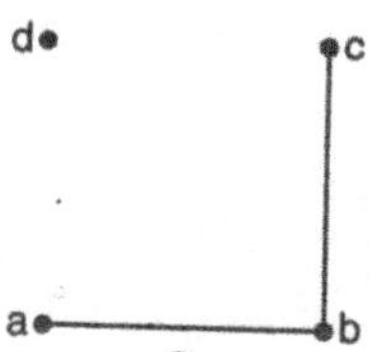

G_1

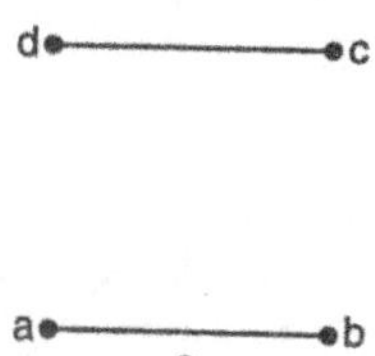

G_2

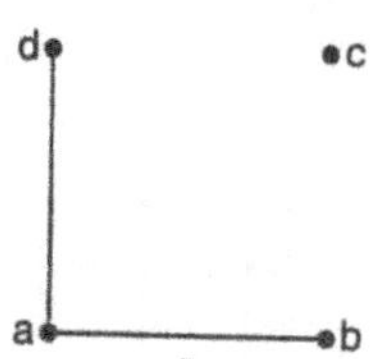

G_3

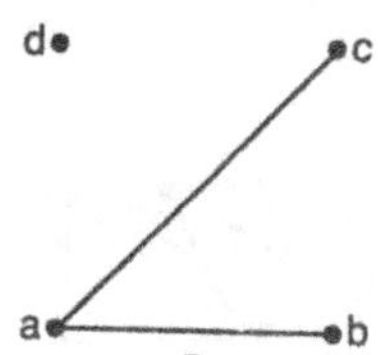

G_4

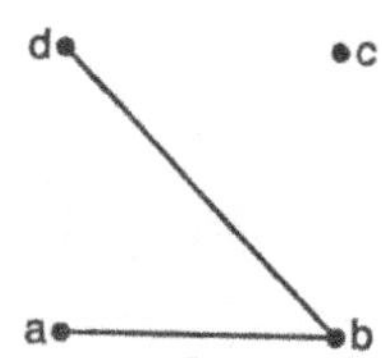

G_5

Example 2. *Show that the degree of any vertex in a simple graph of n vertices cannot exceed n – 1.*

Sol. Let $G(V, E)$ be a simple graph of n vertices.

If the graph is a trivial graph then the result holds obviously. So, let the graph be not a trivial graph.

Since the graph is simple, G cannot have loops and parallel edges. Let v be any arbitrary but fixed vertex of the graph.

Each vertex other than v, can be connected with v by an edge at most once, because there are no parallel edges. Thus, these $n - 1$ vertices can increase the degree of v by $n - 1$. Also there are no loops, so the vertex v itself cannot increase the degree of v.

$\therefore$ The degree of v is at most $n - 1$.

$\therefore$ In general, the degree of any vertex cannot exceed $n - 1$.

Example 3. *Is there is a simple graph with degree sequence 2, 3, 2, 5 ?*

Sol. The degree sequence of the graph is 2, 3, 2, 5.

$\therefore$ Number of vertices = 4

Since the graph is simple, the degree of any vertex cannot exceed $4 - 1$ *i.e.*, 3.

$\therefore$ There cannot be a simple graph with given degree sequence because a vertex of a simple graph of 4 vertices cannot have degree 5.

Example 4. *Prove that a non-trivial simple graph without any isolated vertex must have at least one pair of vertices whose degrees are equal.* (KU 2006)

Sol. Let $G = (V, E)$ be a non-trivial simple graph. Let G has $n(> 1)$. vertices. Let v be any arbitrary but fixed vertex of G. Since G is simple, the graph G cannot have loops and parallel edges.

$\therefore$ v cannot increase the degree of itself. Also, every vertex other than v can increase the degree of v by at most one.

$\therefore$ The vertices, other than v, can increase the degree of v by at most $n - 1$.

$\therefore$ The degree of v cannot exceed $n - 1$.

$\therefore$ Every vertex in G has degree less than n.

Since there is no isolated vertex in the graph, the degree of no vertex can be 0. So, the degree of each vertex can be any number from 1 to $n - 1$.

Numbers (pigeonholes) less than n are 1, 2, 3, ..., $n - 1$ and the vertices (pigeons) are n.

Since $n - 1 < n$, by **pigeonhole principle,** at least two vertices must have degree.

IV. Regular Graph. A graph is called a **regular graph** if the degree of all vertices in the graph are equal. If the degree of each vertex is k then the regular graph is called a **k-regular graph** or a **regular graph of degree k.**

The graphs (*i*), (*ii*) and (*iii*) given below are 2-regular graphs. The graph (*iv*) is a 3-regular graph.

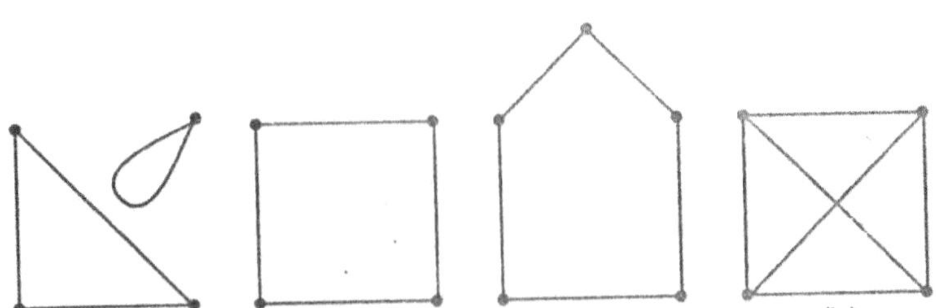

(i) (ii) (iii) (iv)

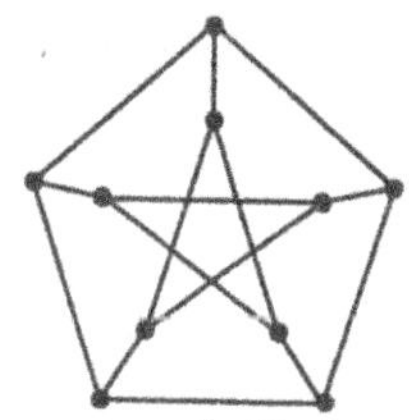

It may be observed that if k is an odd number then the number of vertices in a k-regular graph cannot be odd.

The 3-regular graph of 10 vertices and 15 edges as shown in the diagram is called the **peterson graph.**

Remark. By definition, a regular graph may or may not be a simple graph.

Example 5. *Show that it is not possible to have a group of seven persons such that each person in the group knows exactly three other persons in the group.*

Sol. Let V be the set of seven vertices $v_1, v_2, \ldots\ldots, v_7$ representing seven persons. We join two vertices only when the corresponding persons know each other. Since each person knows three other persons, each vertex is an end point of 3 edges.

$\therefore$ The graph is a 3-regular graph.

$\therefore$ Total number of degrees of all vertices $= 3 \times 7 = 21$

$\therefore$ By **Euler's theorem**, the number of edges in the graph is 21/2, which is impossible.

$\therefore$ It is not possible to have a group of seven persons such that each person in the group knows exactly three other persons in the group.

V. Complete Graph. A simple graph is called a **complete graph** if there is exactly one edge joining every pair of vertices in the graph. A complete graph having n vertices is denoted by K_n. In the complete graph K_n, the degree of each vertex of the graph is $n - 1$. Below are given some complete graphs :

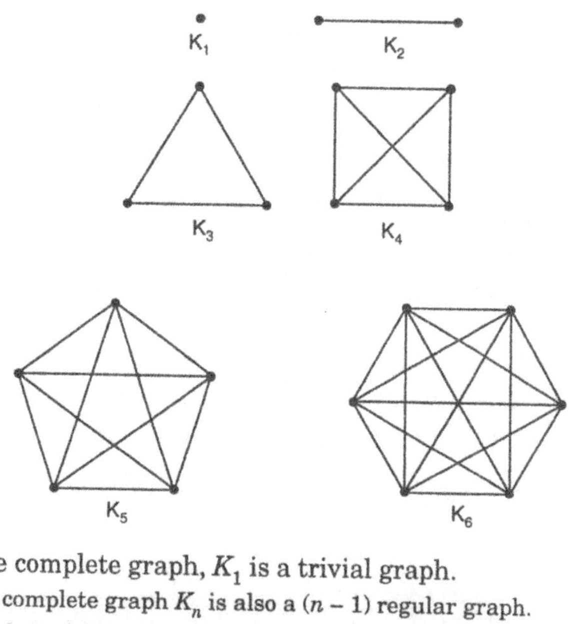

By definition the complete graph, K_1 is a trivial graph.

Remarks. 1. The complete graph K_n is also a $(n - 1)$ regular graph.

2. A complete graph is always a regular graph but a regular graph may not be a complete graph.

Example 6. *Find the number of edges in a complete graph with n vertices.*

Sol. Let K_n be a complete graph with $n(> 1)$vertices. By definition there is exactly one edge joining every pair of distinct vertices in K_n.

$\therefore$ No. of edges in K_n

= no. of ways of selecting 2 vertices from n vertices

= no. of combinations of n vertices taking 2 at a time.

$$= {}^nC_2 = \frac{n(n-1)}{2}.$$

The result is also true for the complete graph K_1 because $\frac{1(1-1)}{2} = 0$.

VI. Bipartite Graph. A simple graph $G = (V, E)$ is called a **bipartite graph** if the vertices of the graph can be partitioned into two sets X and Y such that every edge in the graph is between a vertex in X and a vertex in Y and this bipartite graph is written as $G = (X, Y, E)$. In the bipartite graph $G = (X, Y, E)$, a vertex in X(resp. Y) may not be connected with a vertex in Y(resp. X) by an edge of the graph.

A bipartite graph $G(X, Y, E)$ is called a **complete bipartite graph** if every vertex in X is connected with every vertex in Y by an edge of the graph. If $n(X) = m$ and $n(Y) = n$, then the complete bipartite graph is denoted by $K_{m,n}$. The complete bipartite graph $K_{m,n}$ has mn edges. The following are some examples of complete bipartite graphs :

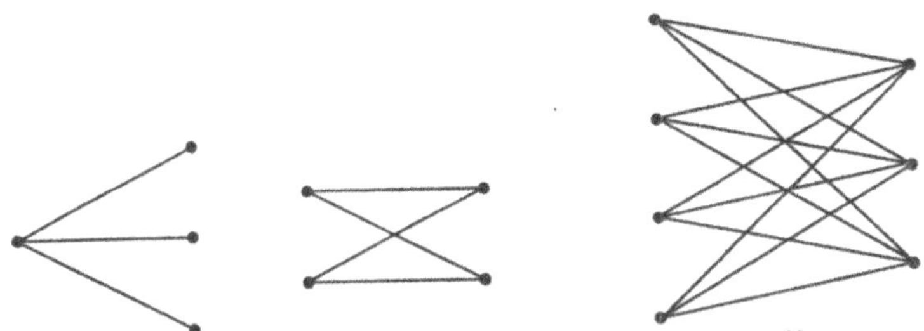

$K_{1,3}$ $K_{2,2}$ $K_{4,3}$

Example 7. *Show that the complement of a bipartite graph may not be a bipartite graph.*

Sol. Let $G = (X, Y, E)$ be a bipartite graph where, $X = \{1, 2, 3\}$, $Y = \{4, 5, 6\}$ and $E = \{\{1, 4\}, \{1, 5\}, \{1, 6\}, \{2, 4\}, \{2, 5\}, \{2, 6\}, \{3, 4\}, \{3, 5\}, \{3, 6\}\}$.

The graph G is shown in the figure (*i*). The complement of G is the graph $\overline{G} = (X \cup Y, F)$, where $F = \{\{1, 2\}, \{2, 3\}, \{1, 3\}, \{4, 5\}, \{5, 6\}, \{4, 6\}\}$. The graph $\overline{G}$ is shown in the figure (*ii*). The diagram of $\overline{G}$ shows that this graph is not a bipartite graph.

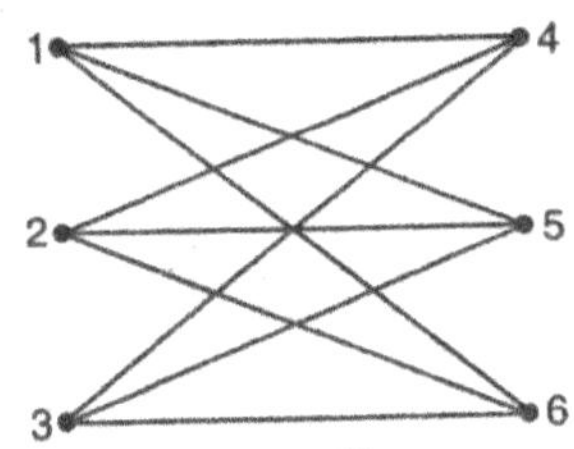

(*i*)

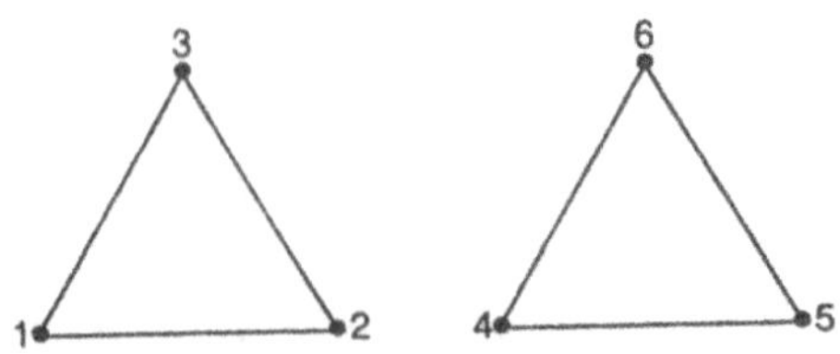

(*ii*)

$\therefore$ The complement of a bipartite graph may not be a bipartite graph.

Example 8. *Find the number of edges in a complete bipartite graph* $K_{m,n}$.

Sol. Let $K_{m,n} = G(X, Y, E)$.

$\therefore$ X and Y have m and n vertices respectively.

Let v be any arbitrary but fixed vertex in X. Since the graph $K_{m,n}$ is simple, there are neither loops nor parallel edges in the graph.

$\therefore$ There are exactly n edges which are adjacent to v. The set X has m vertices.

$\therefore$ By the *fundamental principle of counting*, the total number of edges in the graph $= m \times n =$ **mn.**

Example 9. *Using Euler's theorem, find the size of the complete bipartite graph* $K_{m,n}$.

Sol. Let $K_{m,n} = (X, Y, E)$.

$\therefore$ X and Y have m and n vertices respectively.

Since the graph $K_{m,n}$ is simple, there are neither loops nor parallel edges in the graph.

$\therefore$ The degree of each vertex in X is n.

$\therefore$ Sum of degrees of all vertices in $X = mn$

Similarly, the sum of degrees of all vertices in Y is nm *i.e.*, mn

$\therefore$ Sum of degrees of all vertices in $K_{m,n} = mn + mn = 2mn$

$\therefore$ By **Euler's theorem**, the number of edges in $K_{m,n} = \frac{2mn}{2} =$ **mn.**

VII. Weighted Graph. A graph G is called a **weighted graph** if each edge e of G is assigned a non-negative real number $w(e)$ as its weight.

For example, the graph shown in the figure is a weighted graph.

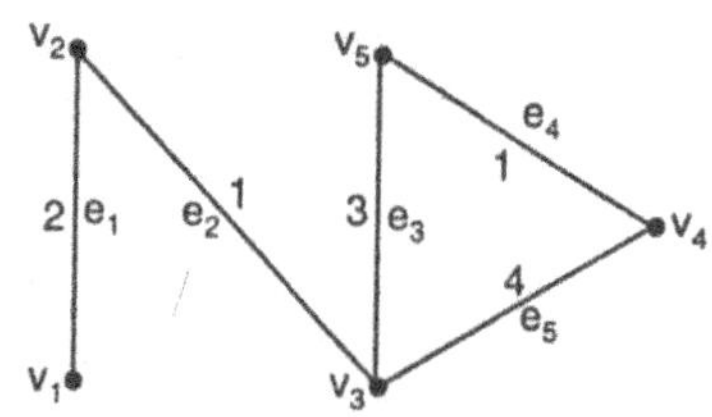

In this graph, we have

$$w(e_1) = 2,\ w(e_2) = 1,\ w(e_3) = 3,\ w(e_4) = 1,\ w(e_5) = 4.$$

If H is a subgraph* of a weighted graph G then the sum of the weights of all the edges in H is called the **weight of the subgraph** H and is denoted by $w(H)$. The graph H shown in the figure is a subgraph of the graph given in the above example.

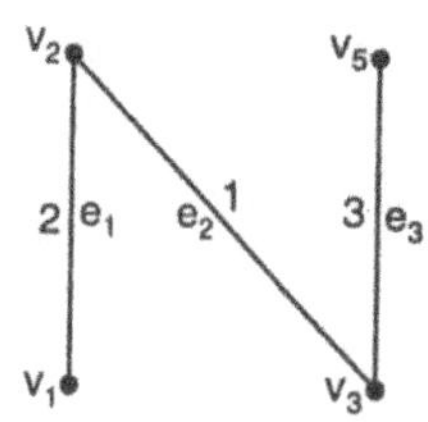

Here, weight of subgraph H of G

$$= w(H) = 2 + 1 + 3 = 6.$$

VIII. Planar Graph. A graph is said to be a **planar graph** if it can be drawn on a plane in such a way that its edges do not cross one another except at common vertices.

A graph which is not a planar graph is called a **non planar graph.**

It may be noticed that a subgraph of a planar graph is a planar graph and any graph having at least one non planar subgraph is also non planar.

A simple planar graph is called a **maximal planar graph** if the graph becomes a non-planar graph if any two non-adjacent vertices in it are joined by an edge.

Illustrations 1. Let G be a graph shown in the figure. The graph G is a planar graph because its edges are not crossing one another.

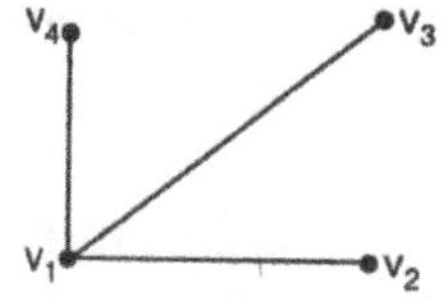

2. The graph K_4, the complete graph of four vertices is shown in the figure (*i*) In this graph the edges $\{v_1, v_3\}$ and $\{v_2, v_4\}$ are intersecting.

From this figure, it appears that the graph K_4 is not a planar graph.

* A graph (V', E') is called a **subgraph** of a graph (V, E) if $V' \subseteq V$ and $E' \subseteq E$.

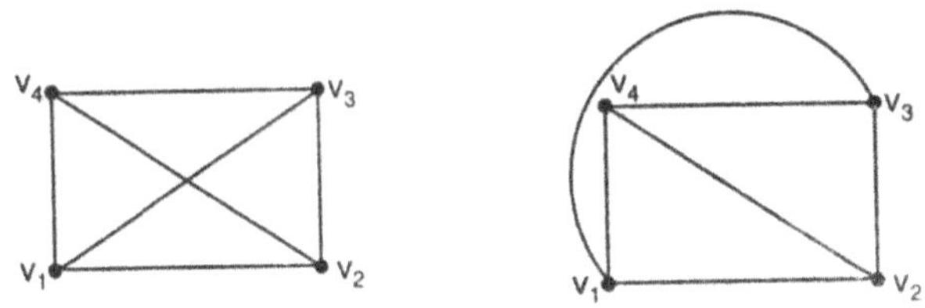

(i) (ii)

But this is not true. The graph K_4 can also be drawn on a plane as given in the figure (ii). In this representation of K_4, no two edges are intersecting. Thus, the graph K_4 is a planar graph.

3. The graph G shown in the figure (i) is not appearing to be a planar graph because edges $\{v_1, v_4\}$ and $\{v_5, v_6\}$ are crossing each other. The edges $\{v_3, v_4\}$ and $\{v_6, v_7\}$ are also intersecting.

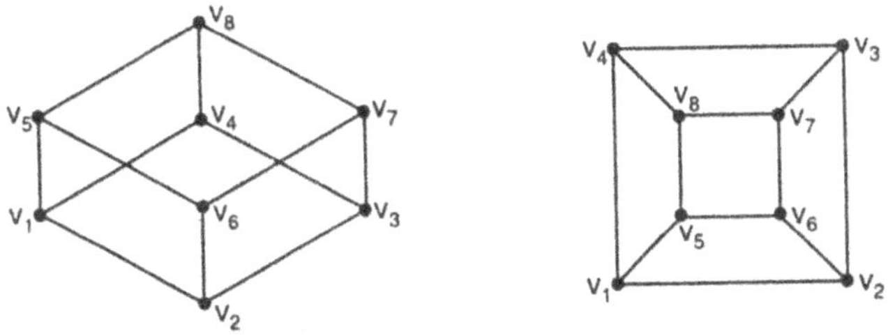

(i) (ii)

This graph G can also be drawn on a plane as shown in the figure (ii). In this representation of the graph G, no two edges are crossing one another. The representation (ii) of the graph G forces us to declare the graph G as a planar graph.

Example 10. *Show that the graph shown in the figure is a planar graph.*

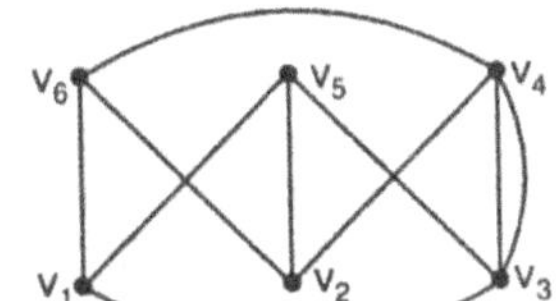

Sol. The graph given in the figure is not appearing to be a planar graph because the edges $\{v_1, v_5\}$ and $\{v_2, v_6\}$ are intersecting. The edges $\{v_2, v_4\}$ and $\{v_3, v_5\}$ are also intersecting.

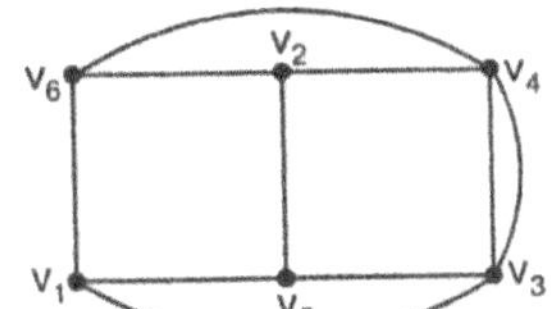

We interchange the positions of v_2 and v_5 and the graph takes the shape as shown in the figure. In this representation, no two edges are intersecting.

$\therefore$ The given graph is a planar graph.

Example 11. *Only by using geometrical considerations, show that the complete graph K_5 is not a planar graph.*

Sol. Let 1, 2, 3, 4 and 5 represent the vertices of the complete graph K_5.

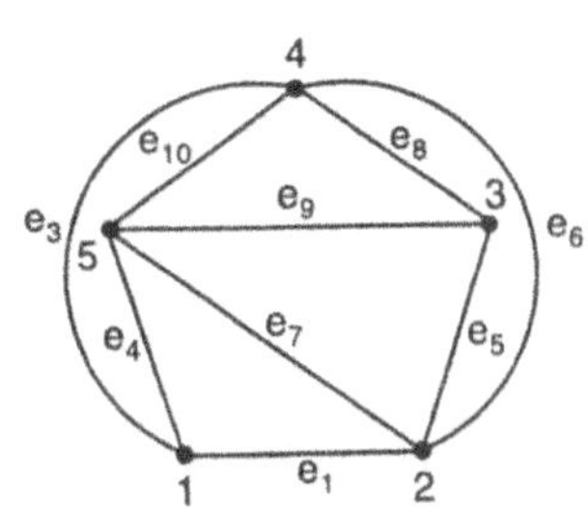

Since ${}^5C_2 = 10$, there are 10 edges in K_5. Let these edges be $e_1 = \{1, 2\}$, $e_2 = \{1, 3\}$, $e_3 = \{1, 4\}$, $e_4 = \{1, 5\}$, $e_5 = \{2, 3\}$, $e_6 = \{2, 4\}$, $e_7 = \{2, 5\}$, $e_8 = \{3, 4\}$, $e_9 = \{3, 5\}$ and $e_{10} = \{4, 5\}$. We draw the edges e_1, e_5, e_8, e_{10} and e_4. The pentagon formed divides the plane of the graph into two regions one inside and one outside. We draw the edges e_7 and e_9 inside the pentagon.

Till now, no two edges have intersect themselves. We draw edges e_3 and e_6, outside the pentagon so that there are no intersections of edges. Now we left with only one edge e_2. From the figure, it is clear that the edge e_2 joining v_1 and v_3 cannot be drawn either inside the pentagon or outside the pentagon without intersecting any edge.

$\therefore$ The graph K_5 is not a planar graph.

Remark. It should be noticed that the graph shown in the diagram is a *maximal planar graph.*

EXERCISE 11.2

1. Is there a simple graph G with degree sequence 2, 1, 5, 3, 2, 7 ?
2. If k is an odd number, then show that the number of vertices in a k-regular graph cannot be odd.
3. Using Euler's theorem, find the size of the complete graph K_n.
4. Find the fewest vertices needed to construct a complete graph with at least 200 edges.
5. Show that if a bipartite graph $G = (X, Y, E)$ is regular, both X and Y have the same number of elements.
6. Show that the number of edges in the bipartite graph $G(X, Y, E)$ is maximum if X and Y have the same number of elements.
7. Show that the graph shown in the figure is a planar graph.

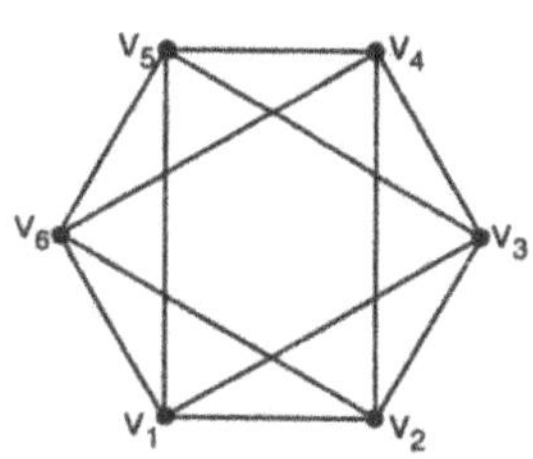

Q. 7

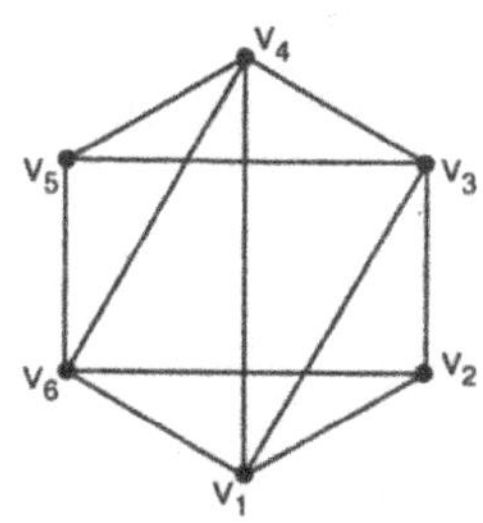

Q. 8

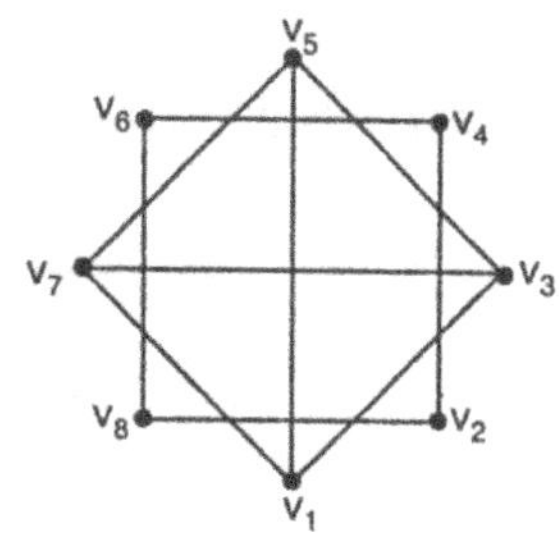

Q. 9

8. Show that the graph shown in the figure is a planar graph.
9. Show that the graph shown in the figure is a planar graph.

Answers

1. No.

3. $\frac{n(n-1)}{2}$

4. 21

7.

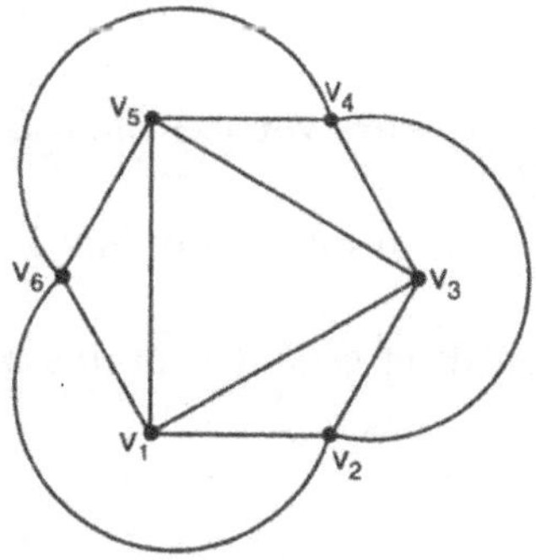

8.

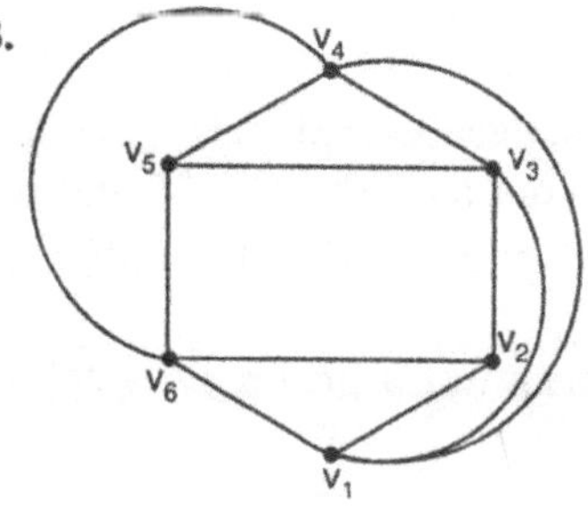

9.

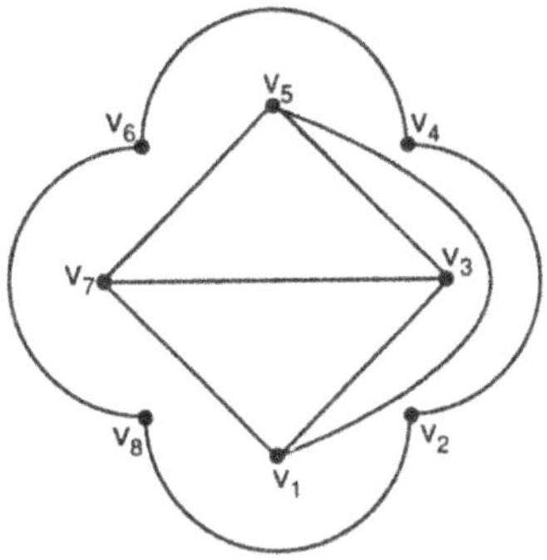

Hints

3. Each vertex increases the sum of degrees of all other vertices by $n - 1$.

4. $\frac{n(n-1)}{2} \geq 200 \quad \Rightarrow \quad n \geq 21.$

11.8. SUBGRAPH

Let $G = (V, E)$ be a graph. A graph $H = (V', E')$ is called a **subgraph** of G if $V' \subseteq V$ and $E' \subseteq E$.

In the given figure, the graphs H and K are subgraphs of the graph G :

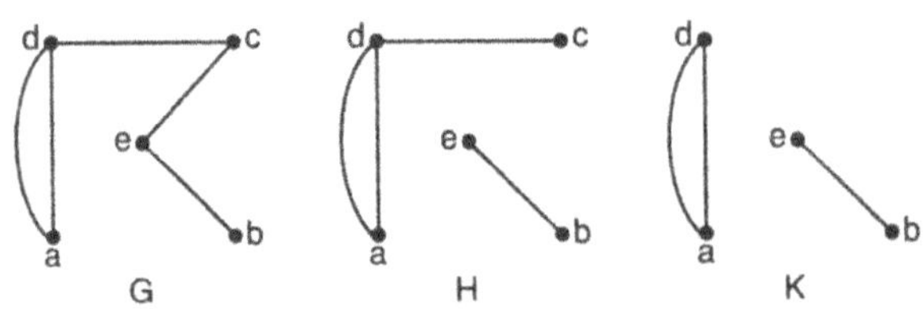

From the definition of subgraph, we observe that :

(*i*) Every graph is a subgraph of itself.

(*ii*) Every subgraph of a subgraph of a graph G is a subgraph of G.

(*iii*) A graph containing single vertex of G and no edge is a subgraph of G.

(*iv*) A graph containing single edge of G together with its end vertices is a subgraph of G.

A subgraph $H(V', E')$ of a graph $G(V, E)$ is called the subgraph **induced** by the vertices of H if its edge set E' contains all edges in G whose end points belong to V'.

In the above example, the edge $\{c, e\}$ of G is not in H and c, e are vertices of H. So, the subgraph H is not a subgraph of G induced by the vertices of H.

It can be checked that the subgraph K of the graph G is a subgraph of G induced by the vertices of K.

A subgraph of a graph G is called a **spanning subgraph** of G if it contains all the vertices of G.

Let G be a graph and v, a vertex of G. The subgraph of G obtained by deleting v from G and deleting all edges in G containing v as an end vertex is denoted by $G - v$.

Let G be a graph and e, an edge in G. The subgraph of G obtained by deleting the edge e from G is denoted by $G - e$. The following graphs illustrate the above two definitions :

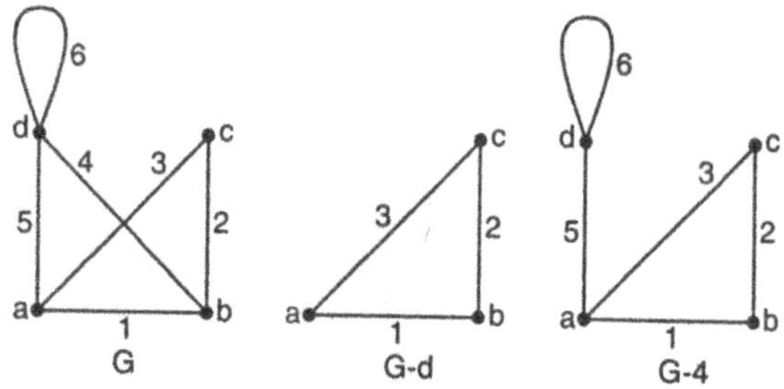

11.9. GRAPH ISOMORPHISM

Let $G = (V, E)$ and $G' = (V', E')$ be two graphs. A mapping $f : V \to V'$ is called a **graph isomorphism** if f is one-one, onto and there is an edge between $f(u)$ and $f(v)$ in G' if and only if there is an edge between u and v in G. In this case, we say that the graphs G and G' are **isomorphic**.

Consider the graphs G and G' shown in the figure :

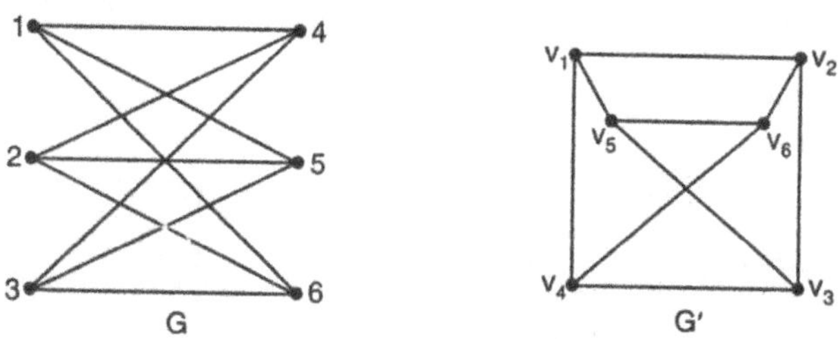

Define $f : \{1, 2, 3, 4, 5, 6\} \to \{v_1, v_2, v_3, v_4, v_5, v_6\}$ by

$$f(1) = v_1, f(2) = v_6, f(3) = v_3, f(4) = v_2, f(5) = v_5, f(6) = v_4.$$

Clearly f is one-one and onto. It can be verified that $\{i, j\}$ is an edge in G if and only if $\{f(i), f(j)\}$ is an edge in G'.

A simple graph is called **self complementary** if it is isomorphic to its complement.

For example, let G be a simple graph and $\overline{G}$ be its complement.

Let $V = \{v_1, v_2, v_3, v_4\}$ be the set of vertices of the graph G.

Define $f : V \to V$ by

$$f(v_1) = v_2, \; f(v_2) = v_4, \; f(v_3) = v_1, \; f(v_4) = v_3.$$

$\therefore$ f is one-one and onto.

Also $\{v_i, v_j\}$ is an edge in G if and only if $\{f(v_i), f(v_j)\}$ is an edge in $\overline{G}$.

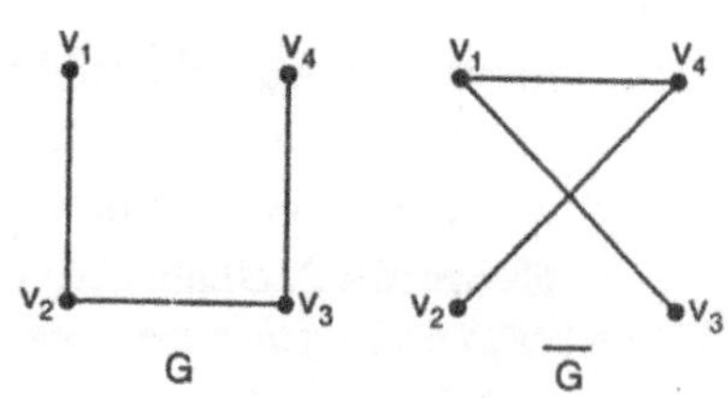

$\therefore$ f is a graph isomorphism.

$\therefore$ G and $\overline{G}$ are isomorphic.

$\therefore$ The graph G is self complementary.

Remarks 1. If $f: G \to G'$ is a graph isomorphism, then there is an edge between $f(u)$ and $f(v)$ in G' if and only if there is an edge between u and v in G. This property of graph isomorphism f is also expressed as '*f preserves adjacency and non-adjacency*'.

2. If $f: G \to G'$ is a graph isomorphism then G and G' have :

(*i*) same number of vertices

(*ii*) same number of edges

(*iii*) equal number of vertices with a given degree.

Example 1. *Show that the given graphs G_1 and G_2 are isomorphic graphs :*

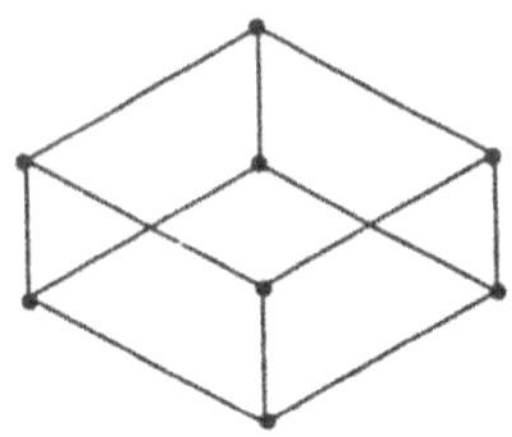

G_1

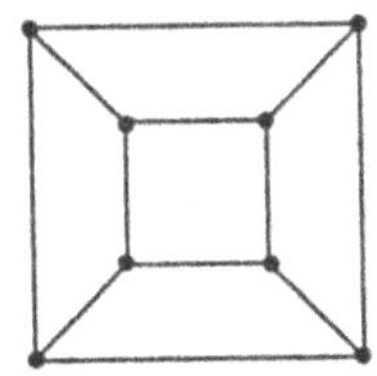

G_2

Sol. Let $u_1, u_2, ..., u_8$ and $v_1, v_2, ..., v_8$ be the vertices of the graphs G_1 and G_2 respectively as marked in the given graphs.

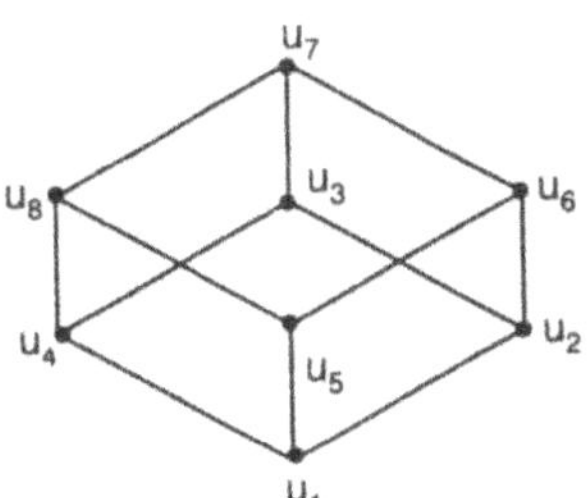

G_1

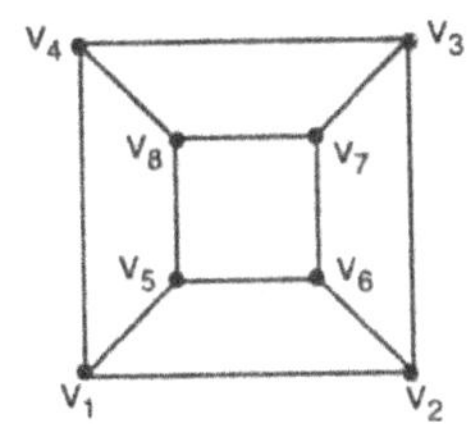

G_2

Define a mapping f from the set $V_1 = \{u_1, u_2, ..., u_8\}$ to the set $V_2 = \{v_1, v_2, ..., v_8\}$ by

$$f(u_i) = v_i,\ 1 \le i \le 8.$$

$\therefore$ f is a one-one and onto mapping.

Also $\{u_i, u_j\}$ is an edge in the graph G_1 if and only if $\{f(u_i), f(u_j)\}$ *i.e.*, $\{v_i, v_j\}$ is an edge in the graph G_2.

In other words, the mapping f preserves adjacency and non-adjacency.

$\therefore$ $f: G_1 \to G_2$ is a graph isomorphism.

$\therefore$ The given graphs G_1 and G_2 are isomorphic.

Example 2. *Show that every simple graph of order n is isomorphic to a subgraph of the complete graph with n vertices.*

Sol. Let $U = \{u_1, u_2, ..., u_n\}$ be the vertex set of a simple graph G of order n.

Let $V = \{v_1, v_2, ..., v_n\}$ be the vertex set of the complete graph K_n.

Let $H = (V, E)$ be the subgraph of K_n, where $\{v_i, v_j\} \in E$ if and only if $\{u_i, u_j\}$ is an edge in G. Define $f : G \to H$ by $f(u_i) = v_i$, $1 \le i \le n$.

$\therefore$ f is one-one and onto.

Also $\{f(u_i), f(u_j)\}$ *i.e.*, $\{v_i, v_j\}$ is an edge in H if and only if $\{u_i, u_j\}$ is an edge in G.

$\therefore$ f is a graph isomorphism.

$\therefore$ The simple group G is isomorphic to the subgraph H of the complete graph K_n.

EXERCISE 11.3

1. The graph G is shown in the figure. Draw the following subgraph of G :

   (*i*) $G - a$, $G - b$, $G - e$

   (*ii*) $G - 1$, $G - 6$, $G - 3$.

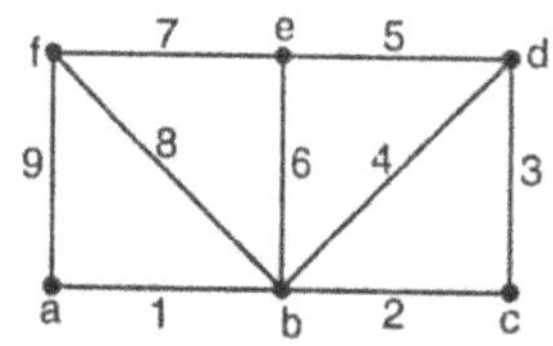

2. If the graphs G and G' are isomorphic then show that :

   (*i*) order of G and G' are equal

   (*ii*) size of G and G' are equal.

3. Show that two simple graphs are isomorphic if and only if their complements are also isomorphic.

Answers

1. (*i*)

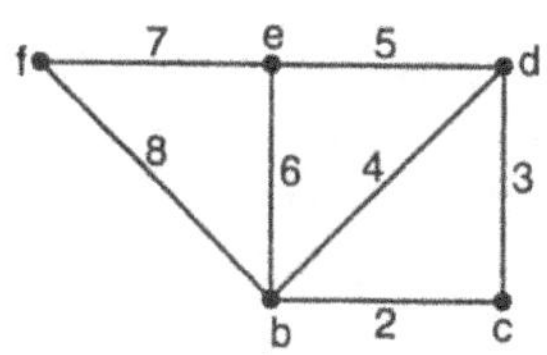

G-a

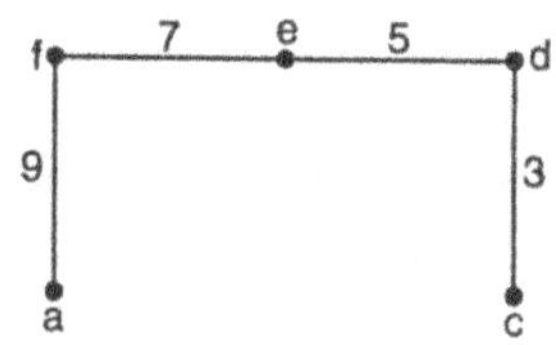

G-b

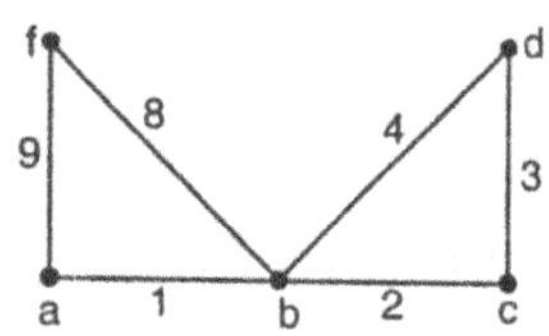

G-e

(*ii*)

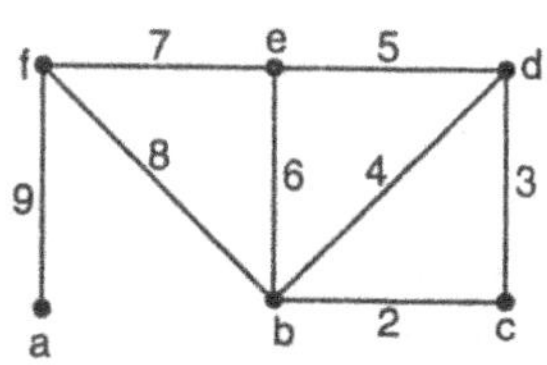

G-1

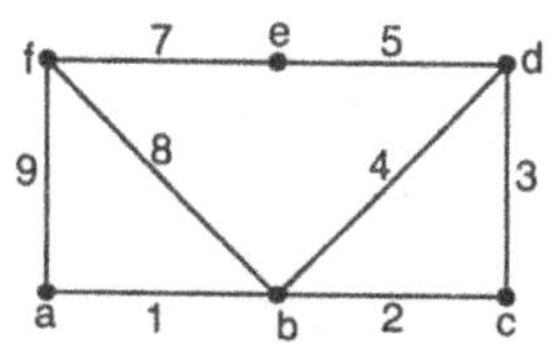

G-6

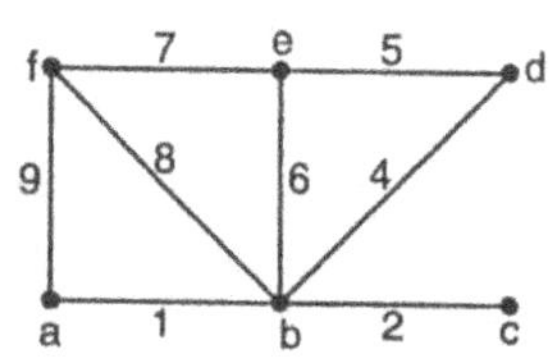

G-3

Hint

2. Let $f : G \to G'$ be a graph isomorphism.

   $\therefore$ Order of G and G' are equal because f is one-one and onto. Also, size of G and G' are equal because f preserves adjacency and non-adjacency.

II. MATRIX REPRESENTATION OF GRAPHS

11.10. ADJACENCY MATRIX OF A GRAPH

Let G be a graph with n vertices $v_1, v_2, \ldots\ldots, v_n$. The **adjacency matrix** of the graph G is a $n \times n$ matrix $A = [a_{ij}]$, where :

$$a_{ij} = \begin{cases} \text{no. of edges joining } v_i \text{ and } v_j & \text{if } i \neq j \\ 2(\text{no. of loops at } v_i) & \text{if } i = j \end{cases}$$

Since, the edge $\{v_i, v_j\}$ is same as the edge $\{v_j, v_i\}$, we have $a_{ij} = a_{ji}$.

$\therefore$ The adjacency matrix A is always *symmetric.*

If there are no loops in the graph then every diagonal entry a_{ii} is '0'.

If there are no parallel edges in the graph then every non-diagonal entry is either 0 or 1.

$\therefore$ In the adjacency matrix of a simple graph, every diagonal entry is 0 and every nondiagonal entry is either 0 or 1.

Also, in the adjacency matrix of the complete graph K_n, every diagonal entry is 0 and every non-diagonal entry is 1.

The adjacency matrix of a graph depends upon the ordering of its vertices. If the order of some vertices is changed, the adjacency matrix would also be changed and the original adjacency matrix can be obtained from this changed matrix just by interchanging some rows and columns.

In the adjacency matrix A of a graph G, the sum of entries in the ith row (or column) is always equal to its degree *i.e.,* the number of edges having ith vertex as an end-vertex. Also, the sum of all the entries in the matrix A is equal to twice the number of edges in its graph G. These rules may be used to verify the correctness of the adjacency matrix of the given graph.

Example 1. *Find the adjacency matrix of the graph G given in the figure.*

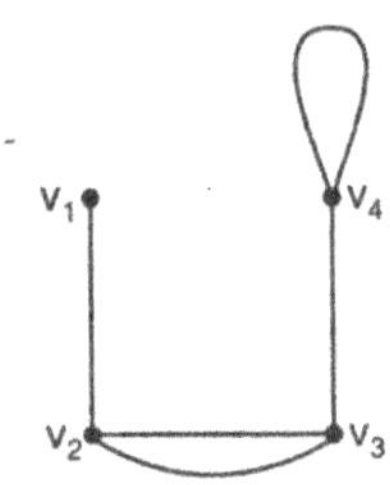

Sol. The vertices of the graph are v_1, v_2, v_3 and v_4.

$\therefore$ The adjacency matrix of G is a 4×4 matrix.

Let $A = [a_{ij}]$ be the adjacency matrix of the graph G.

The edges in the graph are $\{v_1, v_2\}$, $\{v_2, v_3\}$, $\{v_2, v_3\}$, $\{v_3, v_4\}$ and $\{v_4, v_4\}$.

Edge $\{v_1, v_2\}$ $\Rightarrow$ $a_{12} = a_{21} = 1$

Parallel edges $\{v_2, v_3\}$, $\{v_2, v_3\}$ $\Rightarrow$ $a_{23} = a_{32} = 2$

Edge $\{v_3, v_4\}$ $\Rightarrow$ $a_{34} = a_{43} = 1$

Loop $\{v_4, v_4\}$ $\Rightarrow$ $a_{44} = 2(1) = 2$

All other entries of A are zero.

$$\therefore \text{ Adjacency matrix of } G = A = \begin{bmatrix} a_{11} & a_{12} & a_{13} & a_{14} \\ a_{21} & a_{22} & a_{23} & a_{24} \\ a_{31} & a_{32} & a_{33} & a_{34} \\ a_{41} & a_{42} & a_{43} & a_{44} \end{bmatrix} = \begin{bmatrix} \mathbf{0} & \mathbf{1} & \mathbf{0} & \mathbf{0} \\ \mathbf{1} & \mathbf{0} & \mathbf{2} & \mathbf{0} \\ \mathbf{0} & \mathbf{2} & \mathbf{0} & \mathbf{1} \\ \mathbf{0} & \mathbf{0} & \mathbf{1} & \mathbf{2} \end{bmatrix}.$$

Example 2. *Find the graph whose adjacency matrix is* $\begin{bmatrix} 2 & 2 & 1 & 2 \\ 2 & 0 & 2 & 1 \\ 1 & 2 & 2 & 0 \\ 2 & 1 & 0 & 0 \end{bmatrix}$.

Sol. Let $A = \begin{bmatrix} 2 & 2 & 1 & 2 \\ 2 & 0 & 2 & 1 \\ 1 & 2 & 2 & 0 \\ 2 & 1 & 0 & 0 \end{bmatrix}$.

The order of A is 4×4.

Let $G = (V, E)$ be the graph of A.

$\therefore$ G has 4 vertices. Let $V = \{v_1, v_2, v_3, v_4\}$.

$a_{11} = 2 = 2(1) \Rightarrow$ There is a loop at v_1

$a_{12} = 2 \Rightarrow \{v_1, v_2\}, \{v_1, v_2\}$ are parallel edges

$a_{13} = 1 \Rightarrow \{v_1, v_3\}$ is an edge

$a_{14} = 2 \Rightarrow \{v_1, v_4\}, \{v_1, v_4\}$ are parallel edges

$a_{22} = 0 \Rightarrow$ There is no loop at v_2

$a_{23} = 2 \Rightarrow \{v_2, v_3\}, \{v_2, v_3\}$ are parallel edges

$a_{24} = 1 \Rightarrow \{v_2, v_4\}$ is an edge

$a_{33} = 2 \Rightarrow$ There is a loop at v_3

$a_{34} = 0 \Rightarrow \{v_3, v_4\}$ is not an edge

$a_{44} = 0 \Rightarrow$ There is no loop at V_4

$\therefore \quad E = \{\{v_1, v_1\}, \{v_1, v_2\}, \{v_1, v_2\}, \{v_1, v_3\}, \{v_1, v_4\}, \{v_1, v_4\}, \{v_2, v_3\}, \{v_2, v_3\}, \{v_2, v_4\}, \{v_3, v_3\}\}$

$\therefore$ The graph G is known. The diagram of the graph G is shown in the adjoining figure.

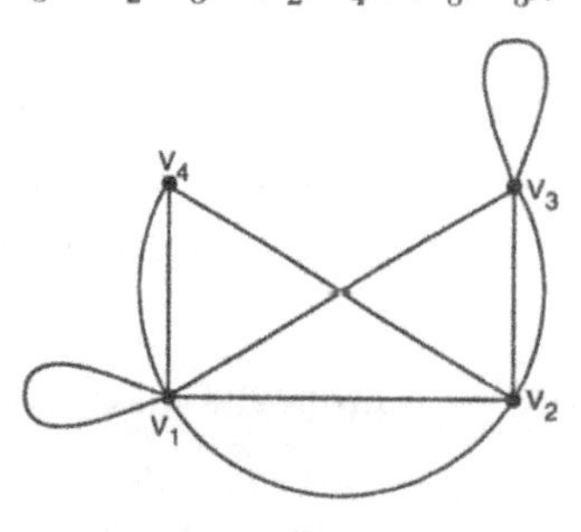

Remark. For finding the edges in the graph of A, we need consider the entries of A which are on the main diagonal and above it *i.e.*, we are required to consider only those a_{ij} for which $i \le j$.

11.11. INCIDENCE MATRIX OF A GRAPH

The concept of **incidence matrix** is defined only for simple graphs. Let G be a simple graph with n vertices $v_1, v_2, \ldots\ldots, v_n$ and m edges $e_1, e_2, \ldots\ldots, e_m$. The **incidence matrix** of the graph G is a $n \times m$ matrix $B = [b_{ij}]$ defined as follows :

For $1 \le i \le n$, ith row of B corresponds to the vertex v_i.

For $1 \le j \le m$, jth column of B corresponds to the edge e_j. If the edge e_j of G joins the vertices v_i and v_k, then the entries b_{ij} and b_{kj} are taken as 1 and all other entries in the jth column are 0.

In an incidence matrix, each entry is either 0 or 1.

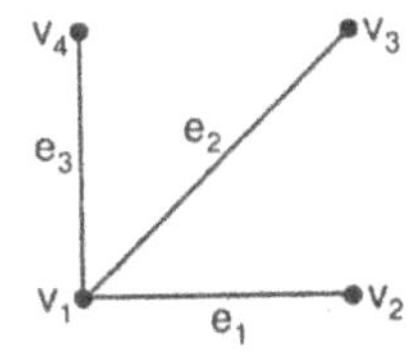

Example 3. *Find the incidence matrix of the simple graph G given in the figure.*

Sol. The vertices of the graph G are v_1, v_2, v_3, v_4 and the edges are e_1, e_2, e_3.

$\therefore$ The incidence matrix of G is of order 4×3.

Let $B = [b_{ij}]$ be the incidence matrix of the graph G.

$$e_1 = \{v_1, v_2\} \Rightarrow b_{11} = 1, b_{21} = 1$$
$$e_2 = \{v_1, v_3\} \Rightarrow b_{12} = 1, b_{32} = 1$$
$$e_3 = \{v_1, v_4\} \Rightarrow b_{13} = 1, b_{43} = 1$$

All other entries of B are zero.

$\therefore$ Incidence matrix of G

$$= B = \begin{bmatrix} b_{11} & b_{12} & b_{13} \\ b_{21} & b_{22} & b_{23} \\ b_{31} & b_{32} & b_{33} \\ b_{41} & b_{42} & b_{43} \end{bmatrix} = \begin{bmatrix} \mathbf{1} & \mathbf{1} & \mathbf{1} \\ \mathbf{1} & \mathbf{0} & \mathbf{0} \\ \mathbf{0} & \mathbf{1} & \mathbf{0} \\ \mathbf{0} & \mathbf{0} & \mathbf{1} \end{bmatrix}.$$

Example 4. *Find the simple graph whose incidence matrix is*

$$\begin{bmatrix} 1 & 0 & 1 & 1 & 0 \\ 0 & 1 & 1 & 0 & 1 \\ 1 & 0 & 0 & 0 & 1 \\ 0 & 1 & 0 & 1 & 0 \end{bmatrix}.$$

Sol. Let
$$B = \begin{bmatrix} 1 & 0 & 1 & 1 & 0 \\ 0 & 1 & 1 & 0 & 1 \\ 1 & 0 & 0 & 0 & 1 \\ 0 & 1 & 0 & 1 & 0 \end{bmatrix}$$

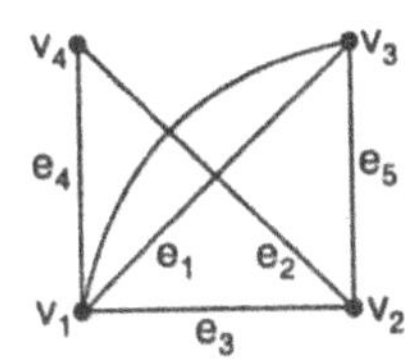

The order of B is 4×5.

Let $G(V, E)$ be the graph of B.

$\therefore$ G has 4 vertices and 5 edges.

Let $V = \{v_1, v_2, v_3, v_4\}$ and $E = \{e_1, e_2, e_3, e_4, e_5\}$.

From the matrix B, we have

$$e_1 = \{v_1, v_3\},\ e_2 = \{v_2, v_4\},\ e_3 = \{v_1, v_2\},\ e_4 = \{v_1, v_4\},\ e_5 = \{v_2, v_3\}$$

$\therefore$ The simple graph G is known. The diagram of the graph G is shown in the figure.

EXERCISE 11.4

1. Find the adjacency matrices of the following simple graphs :

(*i*)

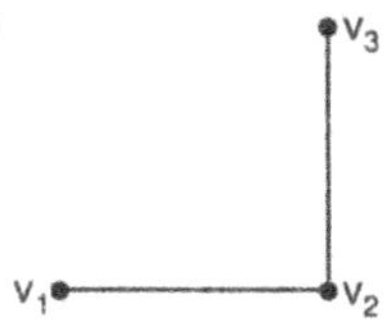

(*ii*)

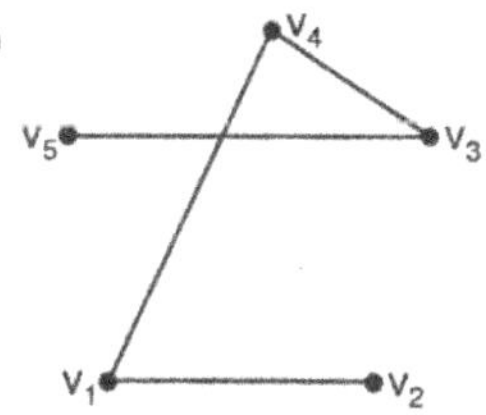

2. Find the adjacency matrices of following complete graphs :

(*i*)

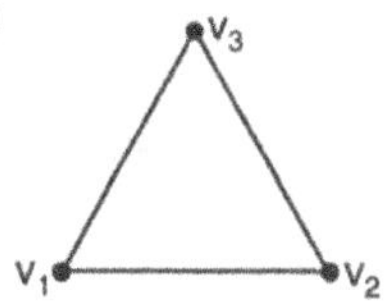

(*ii*)

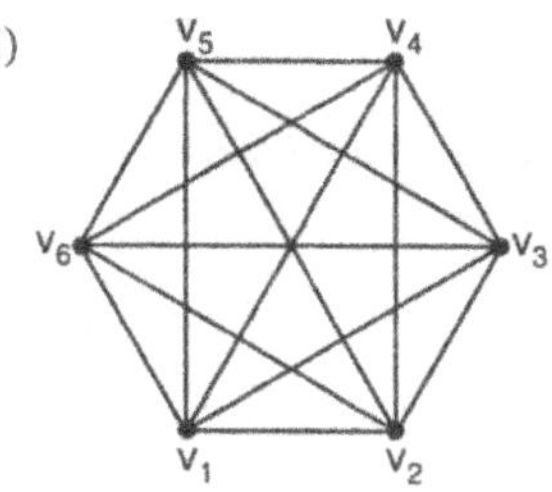

3. Find the adjacency matrices of the following graphs :

(*i*)

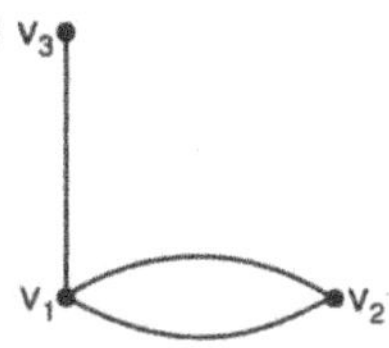

(*ii*)

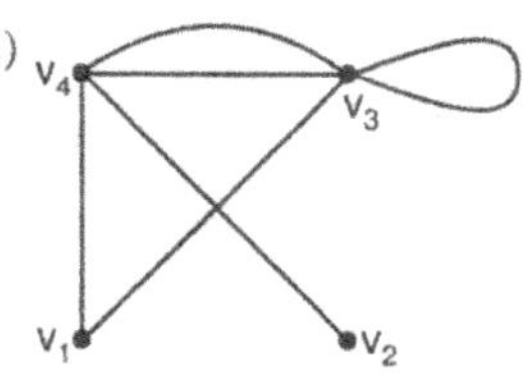

(*iii*)

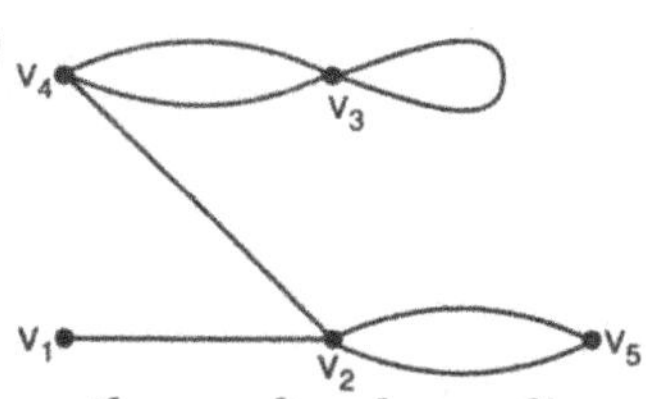

(*iv*)

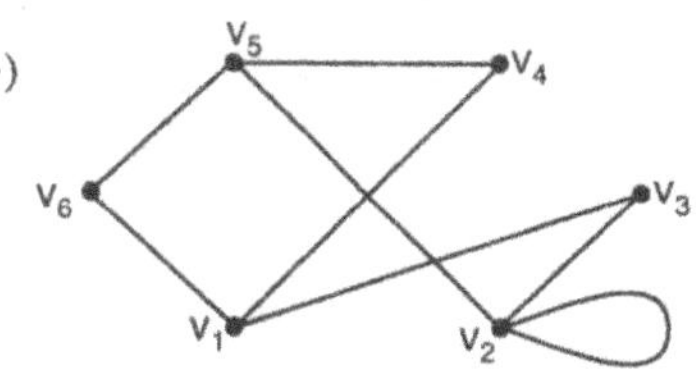

4. Draw the graphs whose adjacency matrices are given below :

(*i*) $\begin{bmatrix} 0 & 1 \\ 1 & 0 \end{bmatrix}$

(*ii*) $\begin{bmatrix} 0 & 1 & 1 \\ 1 & 0 & 1 \\ 1 & 1 & 0 \end{bmatrix}$

(*iii*) $\begin{bmatrix} 2 & 1 \\ 1 & 2 \end{bmatrix}$

(*iv*) $\begin{bmatrix} 2 & 2 & 1 \\ 2 & 2 & 0 \\ 1 & 0 & 0 \end{bmatrix}$

(*v*) $\begin{bmatrix} 0 & 1 & 1 & 2 \\ 1 & 0 & 2 & 1 \\ 1 & 2 & 2 & 0 \\ 2 & 1 & 0 & 0 \end{bmatrix}$

(*vi*) $\begin{bmatrix} 2 & 1 & 1 & 1 & 1 \\ 1 & 0 & 1 & 0 & 1 \\ 1 & 1 & 2 & 0 & 0 \\ 1 & 0 & 0 & 0 & 2 \\ 1 & 1 & 0 & 2 & 0 \end{bmatrix}$.

5. Find the incidence matrices of the following simple graphs :

(*i*)

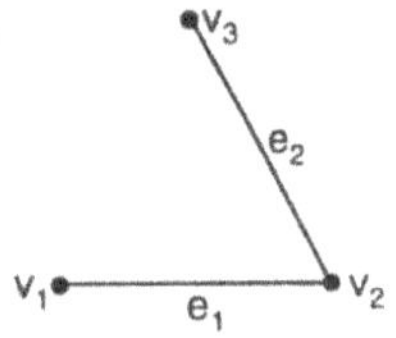

(*ii*)

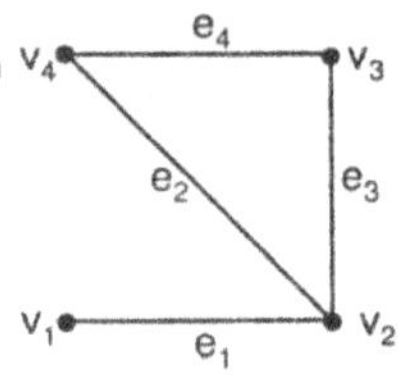

(*iii*)

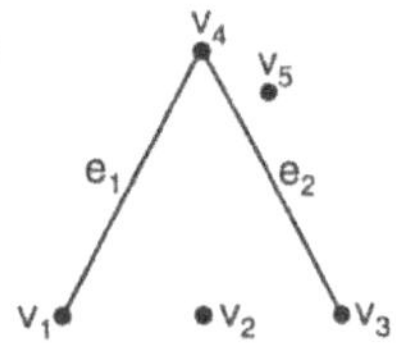

(*iv*)

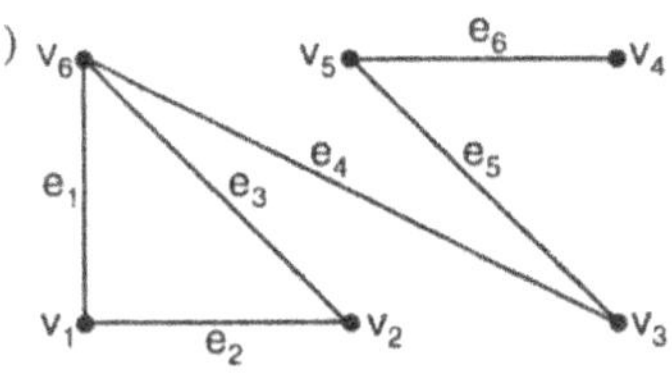

6. Draw the simple graphs whose incidence matrices are given below :

(*i*) $\begin{bmatrix} 1 & 0 & 1 \\ 0 & 1 & 0 \\ 1 & 0 & 0 \\ 0 & 1 & 1 \end{bmatrix}$ (*ii*) $\begin{bmatrix} 1 & 1 & 0 & 0 & 1 \\ 1 & 0 & 1 & 1 & 0 \\ 0 & 1 & 0 & 1 & 0 \\ 0 & 0 & 1 & 0 & 1 \end{bmatrix}$

(*iii*) $\begin{bmatrix} 1 & 0 & 1 & 1 & 0 & 0 \\ 1 & 1 & 0 & 0 & 0 & 0 \\ 0 & 0 & 0 & 0 & 0 & 1 \\ 0 & 0 & 1 & 0 & 1 & 1 \\ 0 & 1 & 0 & 1 & 1 & 0 \end{bmatrix}$ (*iv*) $\begin{bmatrix} 1 & 0 & 0 & 0 \\ 0 & 0 & 1 & 1 \\ 0 & 0 & 0 & 1 \\ 0 & 0 & 1 & 0 \\ 0 & 1 & 0 & 0 \\ 1 & 1 & 0 & 0 \end{bmatrix}$.

Answers

1. (*i*) $\begin{bmatrix} 0 & 1 & 0 \\ 1 & 0 & 1 \\ 0 & 1 & 0 \end{bmatrix}$ (*ii*) $\begin{bmatrix} 0 & 1 & 0 & 1 & 0 \\ 1 & 0 & 0 & 0 & 0 \\ 0 & 0 & 0 & 1 & 1 \\ 1 & 0 & 1 & 0 & 0 \\ 0 & 0 & 1 & 0 & 0 \end{bmatrix}$

2. (*i*) $\begin{bmatrix} 0 & 1 & 1 \\ 1 & 0 & 1 \\ 1 & 1 & 0 \end{bmatrix}$ (*ii*) $\begin{bmatrix} 0 & 1 & 1 & 1 & 1 & 1 \\ 1 & 0 & 1 & 1 & 1 & 1 \\ 1 & 1 & 0 & 1 & 1 & 1 \\ 1 & 1 & 1 & 0 & 1 & 1 \\ 1 & 1 & 1 & 1 & 0 & 1 \\ 1 & 1 & 1 & 1 & 1 & 0 \end{bmatrix}$

3. (*i*) $\begin{bmatrix} 0 & 2 & 1 \\ 2 & 0 & 0 \\ 1 & 0 & 0 \end{bmatrix}$ (*ii*) $\begin{bmatrix} 0 & 0 & 1 & 1 \\ 0 & 0 & 0 & 1 \\ 1 & 0 & 2 & 2 \\ 1 & 1 & 2 & 0 \end{bmatrix}$

(*iii*) $\begin{bmatrix} 0 & 1 & 0 & 0 & 0 \\ 1 & 0 & 0 & 1 & 2 \\ 0 & 0 & 2 & 2 & 0 \\ 0 & 1 & 2 & 0 & 0 \\ 0 & 2 & 0 & 0 & 0 \end{bmatrix}$ (*iv*) $\begin{bmatrix} 0 & 0 & 1 & 1 & 0 & 1 \\ 0 & 2 & 1 & 0 & 1 & 0 \\ 1 & 1 & 0 & 0 & 0 & 0 \\ 1 & 0 & 0 & 0 & 1 & 0 \\ 0 & 1 & 0 & 1 & 0 & 1 \\ 1 & 0 & 0 & 0 & 1 & 0 \end{bmatrix}$

4. (*i*) 

(*ii*) 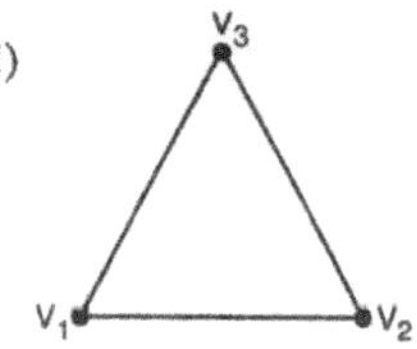

(*iii*) 

(*iv*) 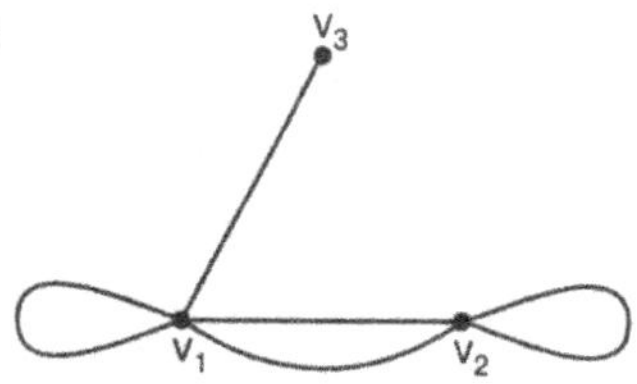

(*v*) 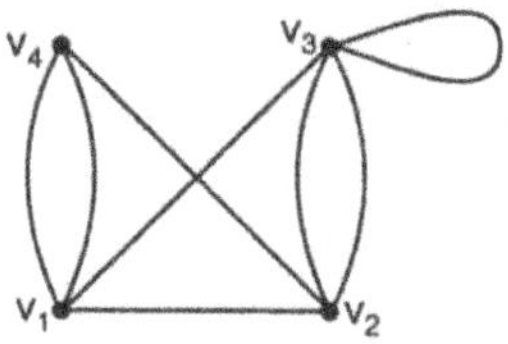

(*vi*) 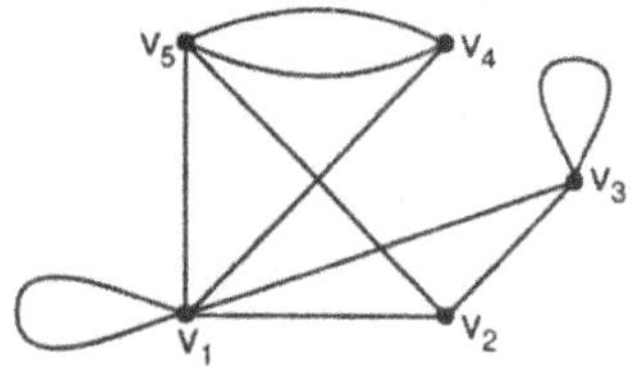

5. (*i*) $\begin{bmatrix} 1 & 0 \\ 1 & 1 \\ 0 & 1 \end{bmatrix}$

(*ii*) $\begin{bmatrix} 1 & 0 & 0 & 0 \\ 1 & 1 & 1 & 0 \\ 0 & 0 & 1 & 1 \\ 0 & 1 & 0 & 1 \end{bmatrix}$

(*iii*) $\begin{bmatrix} 1 & 0 \\ 0 & 0 \\ 0 & 1 \\ 1 & 1 \\ 0 & 0 \end{bmatrix}$

(*iv*) $\begin{bmatrix} 1 & 1 & 0 & 0 & 0 & 0 \\ 0 & 1 & 1 & 0 & 0 & 0 \\ 0 & 0 & 0 & 1 & 1 & 0 \\ 0 & 0 & 0 & 0 & 0 & 1 \\ 0 & 0 & 0 & 0 & 1 & 1 \\ 1 & 0 & 1 & 1 & 0 & 0 \end{bmatrix}$

6. (*i*) 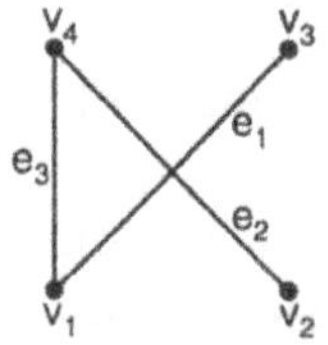

(*ii*) 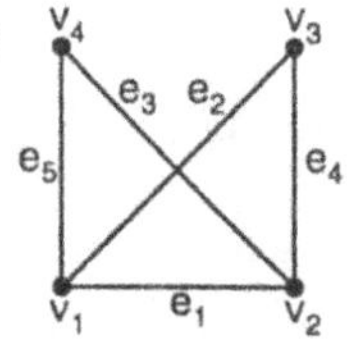

(*iii*) 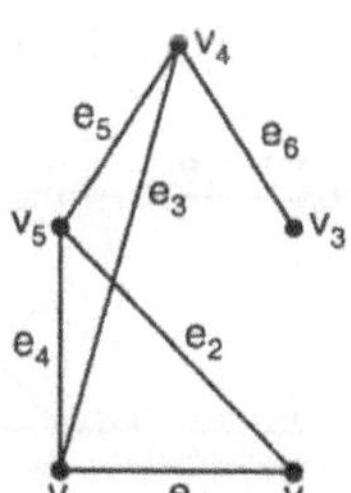

(*iv*) 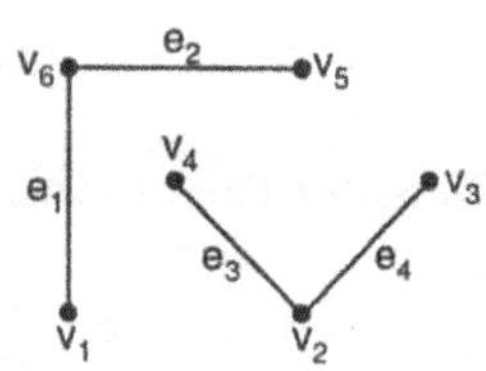

III. PATHS AND CIRCUITS

11.12. PATH

Let $G = (V, E)$ be a graph. An alternating sequence $v_0, e_1, v_1, e_2, v_2, \ldots\ldots, v_{n-1}, e_n, v_n$ of vertices and edges, beginning and ending with vertices is called a **path** in the graph G if :

(*i*) each edge contains the vertices preceding and following it. Thus, e_i contains the vertices v_{i-1} and v_i for $1 \le i \le n$

(*ii*) a vertex may be repeated in the above sequence

(*iii*) no edge is repeated in the above sequence.

Vertices with which a path begins and ends are called the **terminal vertices** of the path.

The number of edges in a path is called the **length** of the path. Thus an edge is a path of length one.

A path $v_0, e_1, v_1, e_2, v_2, \ldots\ldots, v_{n-1}, e_n, v_n$ is called a **simple path** if no vertex is repeated in the sequence of the path, except for the possibility, $v_1 = v_n$.

It may be observed that a loop cannot be included in a simple path.

11.13. SHORTEST PATH

A path between two vertices u and v of a graph G is called a **shortest path** between the vertices u and v if the number of edges in the path is minimum *i.e.*, the length of the path is minimum.

For example, consider the graph given in the figure.

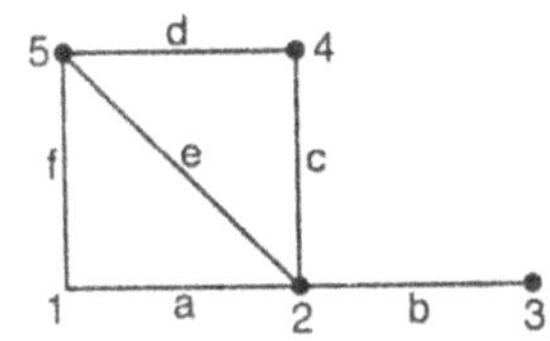

5, e, 2, b, 3 and 5, d, 4, c, 2, b, 3 are two paths between the vertices 5 and 3 of the graph. The lengths of these paths are 2 and 3 respectively.

The shortest path between the vertices 5 and 3 is 5, e, 2, b, 3, because there does not exist any path between 5 and 3 whose length is less than 2.

11.14. CIRCUIT

Let $G = (V, E)$ be a graph. A path $v_0, e_1, v_1, e_2, v_2, \ldots\ldots, v_{n-1}, e_n, v_n$ is called a **circuit** in the graph G if $v_0 = v_n$.

In other words, a circuit is a path which begins and ends at the same vertex.

A circuit $v_0, e_1, v_1, e_2, v_2, \ldots\ldots, v_{n-1}, e_n, v_n$ with $v_0 = v_n$ is called a **simple circuit** if no vertex is repeated in the sequence of the circuit except for $v_0 = v_n$.

Equivalently, we can say that a simple circuit is a simple path with coincident terminal vertices.

Illustration 1. Consider the graph given in the diagram. This graph has 6 vertices and 7 edges.

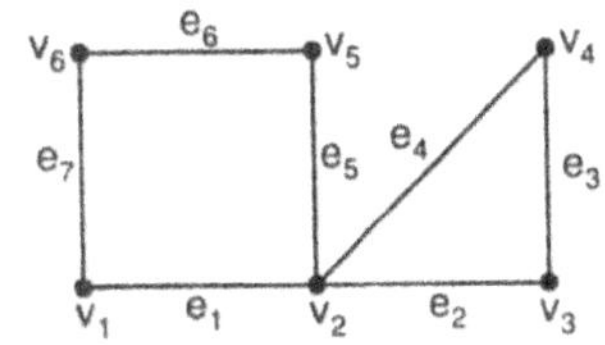

(*i*) The alternating sequence $v_1, e_1, v_2, e_4, v_4, e_4, v_2$ is not a path in the graph because the edge e_4 is repeated in the sequence.

(*ii*) The alternating sequence $v_1, e_1, v_2, e_4, v_4, e_3, v_3, e_2, v_2, e_5, v_5$ is a path in the given graph because no edge is repeated.

In this path the vertex v_2 is repeated.

∴ This path is not a simple path.

In the above path the terminal vertices are v_1 and v_5.

(*iii*) The alternating sequence $v_1, e_1, v_2, e_2, v_3, e_3, v_4$ is a path in the given graph because no edge is repeated. In this path, the vertices are also not repeated.

∴ This path is a simple path.

(*iv*) The alternating sequence $v_1, e_1, v_2, e_2, v_3, e_3, v_4, e_4, v_2, e_5, v_5, e_6, v_6, e_7, v_1$ is a path in the given graph because no edge is repeated.

In this sequence the terminal vertices are same.

∴ The given sequence is a circuit.

In this circuit, the vertex v_2 is repeated. Thus, this circuit is not a simple circuit.

(*v*) The alternating sequence $v_2, e_2, v_3, e_3, v_4, e_4, v_2$ is a path in the given graph because no edge is repeated. In this sequence the terminal vertices are same.

∴ The given sequence is a circuit.

In this circuit, no vertex is repeated except for the terminal vertices.

∴ This circuit is a simple circuit.

Illustration 2. Consider the graph given in the diagram. This graph has 5 vertices and 6 edges with parallel edges e_2, e_3 with end vertices v_1, v_5 and a loop e_6 at v_3.

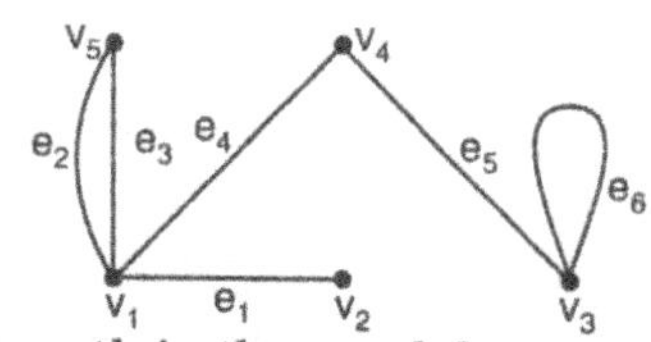

(*i*) The alternating sequence $v_5, e_3, v_1, e_4, v_4, e_5, v_3, e_6, v_3$ is a path in the graph because no edge is repeated. This path is not a simple path because the vertex v_3 is repeated.

(*ii*) The alternating sequence $v_3, e_5, v_4, e_4, v_1, e_3, v_5$ is a simple path in the graph because no edge is repeated and also no vertex is repeated.

(*iii*) The alternating sequence v_1, e_3, v_5, e_2, v_1 is a simple circuit in the graph because no edge is repeated and no vertex is repeated except for the terminal vertices.

Example 1. *For the given graph, which of the following alternating sequences of vertices and edges are path in the graph :*

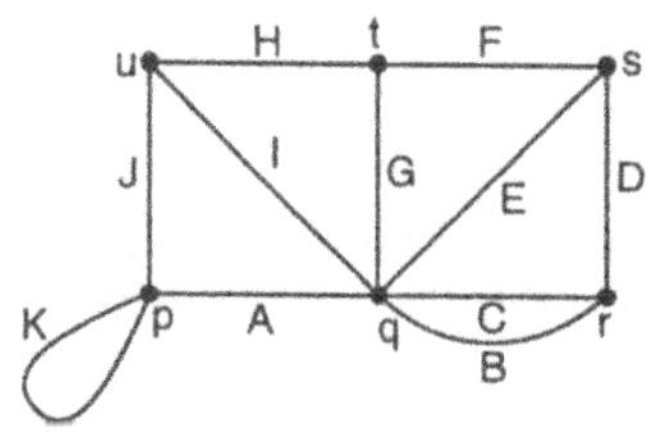

(*i*) *p, A, q, B, r, C, q* (*ii*) *p, K, p*

(*iii*) *q, B, r, C, q*

(*iv*) *p, K, p, A, q, B, r, C, q, A, p*

(*v*) *u, I, q, G, t, F, s, D, r, C, q, A, p, K, p.*

Sol. (*i*) The given sequence is *p, A, q, B, r, C, q*.

In this sequence, no edge is repeated.

∴ Given sequence is a path.

(*ii*) The given sequence is *p, K, p*.

In this sequence, no edge is repeated.

∴ Given sequence is a path.

(*iii*) Given sequence is *q, B, r, C, q*.

In this sequence, no edge is repeated.

∴ Given sequence is a path.

(*iv*) Given sequence is *p, K, p, A, q, B, r, C, q, A, p*.

In this sequence, the edge *A* is repeated.

∴ Given sequence is not a path.

(*v*) Given sequence is *u, I, q, G, t, F, s, D, r, C, q, A, p, K, p*.

In this sequence, no edge is repeated.

∴ Given sequence is a path.

Example 2. *For the given graph, classify the following alternating sequences of vertices and edges as path, simple path, circuit, simple circuit :*

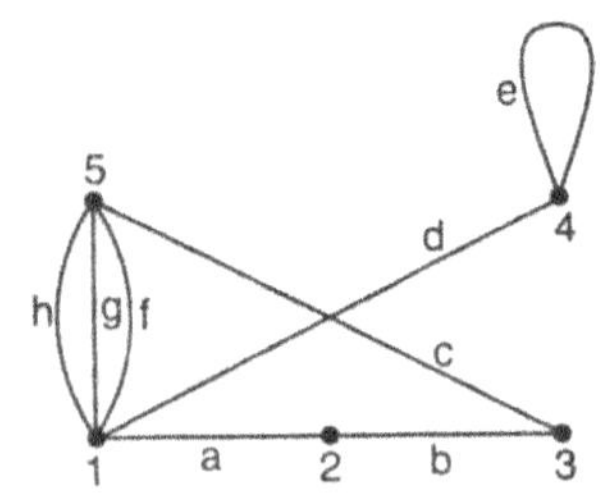

(*i*) *1, a, 2, b, 3, c, 5, g, 1, f, 5*

(*ii*) *1, f, 5, g, 1, h, 5*

(*iii*) *5, h, 1, a, 2, b, 3*

(*iv*) *3, b, 2, a, 1, f, 5, c, 3*

(*v*) *3, b, 2, a, 1, f, 5, g, 1, h, 5, c, 3.*

Sol. (*i*) Given sequence is 1, *a*, 2, *b*, 3, *c*, 5, *g*, 1, *f*, 5.

In this sequence, no edge is repeated. The vertices are not distinct. The terminal vertices are not same.

∴ The given sequence is a path.

(*ii*) Given sequence is 1, *f*, 5, *g*, 1, *h*, 5.

In this sequence, no edge is repeated.

The vertices are not distinct. The terminal vertices are not same.

∴ The given sequence is a path.

(*iii*) Given sequence is 5, *h*, 1, *a*, 2, *b*, 3.

In this sequence, no edge is repeated. The vertices are distinct. The terminal vertices are distinct.

∴ The given sequence is a simple path.

(*iv*) Given sequence is 3, *b*, 2, *a*, 1, *f*, 5, *c*, 3.

In this sequence, no edge is repeated. The vertices are distinct except for the terminal vertices, which are same.

∴ The given sequence is a simple circuit.

(*v*) The given sequence is 3, *b*, 2, *a*, 1, *f*, 5, *g*, 1, *h*, 5, *c*, 3.

In this sequence, no edge is repeated. The vertices, other than the terminal vertices, are not distinct. The terminal vertices are same.

∴ The given sequence is a circuit.

IV. EULERIAN PATHS AND CIRCUITS

11.15. EULER PATH

A path in a graph is called an **Euler path** if it includes every edge of the graph exactly once*.

---

*For an *Euler path,* it is sufficient to say that the path must include every edge of the graph, because no edge can repeat itself in a path.

In other words, an Euler path in a graph G is an alternating sequence $v_0, e_1, v_1, e_2, v_2, \ldots\ldots, v_{n-1}, e_n, v_n$ of vertices and edges, beginning and ending with vertices such that :

(*i*) each edge contains the vertices preceding and following it

(*ii*) a vertex may be repeated in the above sequence

(*iii*) no edge is repeated in the above sequence

(*iv*) every edge of the graph is included in the above sequence.

Remark. Condition (*iv*) is making a path, an Eulerian path.

11.16. EULER CIRCUIT

A circuit in a graph is called an **Euler circuit** if it includes every edge of the graph exactly once.

In other words, an Euler circuit in a graph G is an alternating sequence $v_0, e_1, v_1, e_2, v_2, \ldots\ldots, v_{n-1}, e_n, v_n$ of vertices and edges beginning and ending with vertices such that :

(*i*) each edge contains the vertices preceding and following it

(*ii*) a vertex may be repeated in the above sequence

(*iii*) no edge is repeated in the above sequence

(*iv*) every edge of the graph is included in the above sequence

(*v*) the terminal vertices of the above sequence are same *i.e.,* $v_0 = v_n$.

Remarks. 1. Conditions (*i*) – (*iii*) ensures that the sequence is a path

2. Conditions (*i*) – (*iv*) ensures that the sequence is an Euler path.

3. Condition (*v*) is making an Euler path, an Euler circuit.

Thus an Euler circuit is an Euler path with identical terminal vertices.

11.17. EULER GRAPH

A graph is called an **Euler graph** if there exists an Euler circuit in the graph.

Illustration 1. Consider the graph given in the diagram. This graph has 5 vertices and 5 edges.

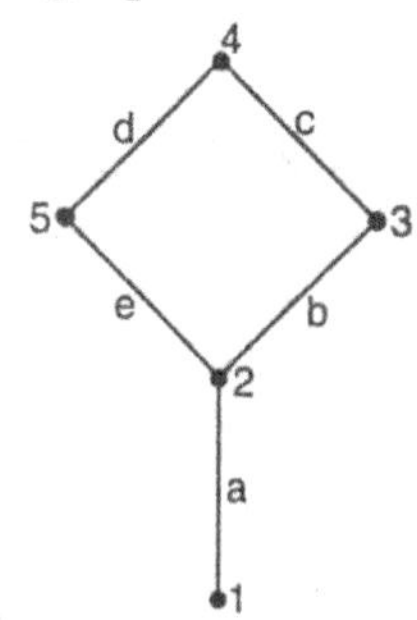

(*i*) The alternating sequence 1, *a*, 2, *e*, 5, *d*, 4, *c*, 3, *b*, 2, *a*, 1 is not an Euler path because this sequence is not even a path.

(*ii*) The alternating sequence 1, *a*, 2, *e*, 5, *d*, 4, *c*, 3 is a path. This path is not an Euler path because the edge *b* of the graph is not included in the sequence.

(*iii*) The alternating sequence 1, *a*, 2, *e*, 5, *d*, 4, *c*, 3, *b*, 2 is a path. This path is an Euler path because this path is covering every edge of the given graph.

(*iv*) The alternating sequence 1, *a*, 2, *b*, 3, *c*, 4, *d*, 5, *e*, 2 is also an Euler path in the given graph.

It is interseting to note that this graph has no Euler circuit. The reader is advised to put in little labour to see if there is an Euler circuit in this graph.

Illustration. 2. Consider the graph given in the diagram. This graph has 4 vertices and 8 edges.

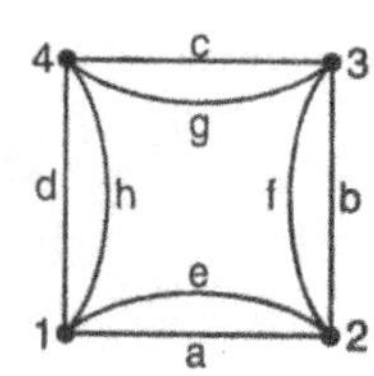

(*i*) The alternating sequence 1, *a*, 2, *b*, 3, *c*, 4, *d*, 1 is a circuit in the graph. But this circuit is not an Euler circuit because the edges *e, f, g, h* occurring in the graph are not covered in the sequence.

(*ii*) The alternating sequence 1, a, 2, b, 3, c, 4, d, 1, e, 2, f, 3, g, 4, h, 1 is an Euler path which is also an Euler circuit because the given sequence is a circuit in G and all edges of the graph are covered in the sequence. Thus, the given graph is an Euler graph.

(*iii*) The alternating sequence 1, a, 2, f, 3, c, 4, h, 1, e, 2, b, 3, g, 4, d, 1 is also an Euler path which is also an Euler circuit in the given graph.

11.18. CONNECTED GRAPH

A graph is called a **connected graph** if there is a path from any vertex to any other vertex in the graph.

A graph which is not connected is called a **disconnected graph.**

If a graph is disconnected then the various connected pieces of the graph are called the **components** of the graph.

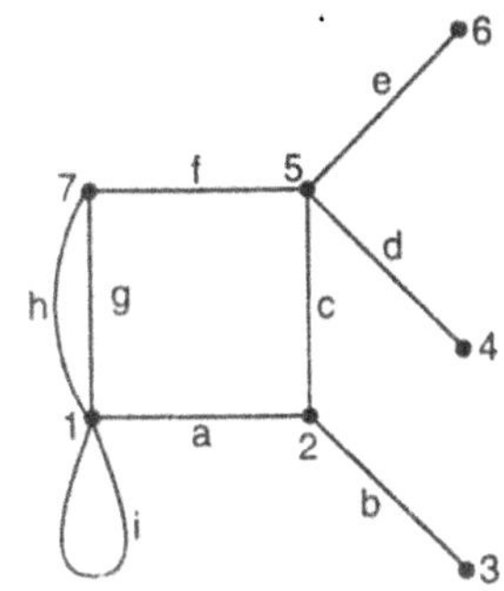

Illustration 1. The graph shown in the figure is a connected graph because there is a path from any vertex to any other vertex in the graph.

6, e, 5, d, 4 is a path from the vertex 6 to the vertex 4.

1, a, 2, c, 5, d, 4 is a path from the vertex 1 to the vertex 4.

Illustration 2. The graph shown in the figure is not a connected graph because there is no path from the vertex 2 to the vertex 4. This disconnected graph has two connected pieces. Thus, this graph has two components.

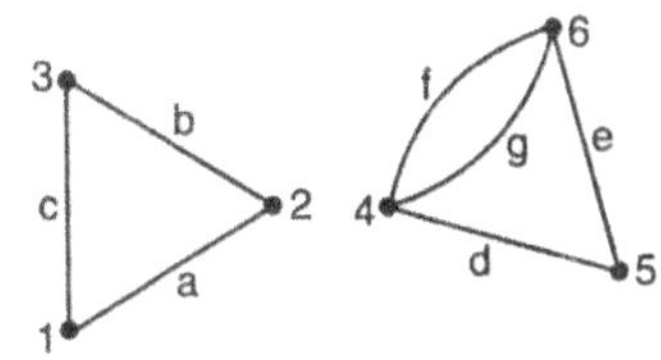

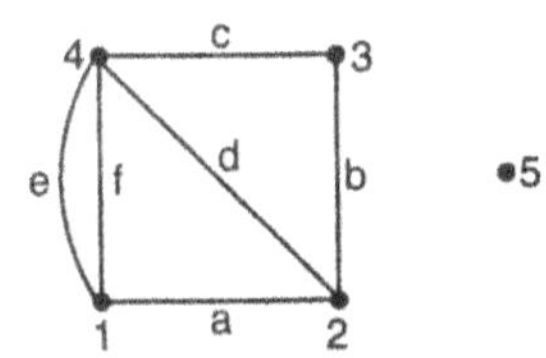

Illustration 3. A graph having at least one isolated vertex is always a disconnected graph. In the given graph, there are two components. The graph containing only the isolated vertex 5 is one of the components.

Let $G = (V, E)$ be a graph and $e \in E$. The edge e is called a **bridge** in the graph G if the graph $G - e$ is disconnected.

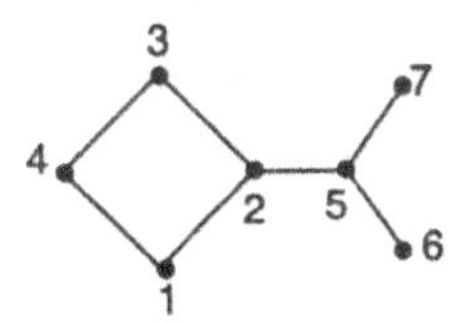

For example, the edge {2, 5} is a bridge in the graph G given in the figure. It can be checked that the graph G – {2, 5} is not a connected graph.

Remarks 1. Every complete graph is a connected graph.

2. Every discrete graph of n vertices is a disconnected graph having n components.

Theorem. Let G be a connected graph and u, v be two distinct vertices of G. Prove that there exists a simple path between the vertices u and v.

Proof. The given graph G is connected.

$\therefore$ By definition, there exists a path between u and v. Let $v_0, e_1, v_1, e_2, v_2, \ldots\ldots, v_{n-1}, e_n, v_n$ be a path between u and v, where $v_0 = u$ and $v_n = v$.

If $v_0, v_1, v_2, \ldots\ldots, v_{n-1}, v_n$ are all distinct then this path is a simple path.

So, let $v_0, v_1, v_2, \ldots\ldots, v_{n-1}, v_n$ be not all distinct.

$\therefore$ There exist at least one vertex that appears at v_i and v_j both in the path with $i < j$.

We rewrite the above path as $v_0, e_1, v_1, e_2, v_2, \ldots\ldots, e_i, v_i, e_{i+1}, v_{i+1}, \ldots\ldots, v_j, e_{j+1}, v_{j+1}, \ldots\ldots, v_{n-1}, e_n, v_n$. We remove the terms $e_{i+1}, v_{i+1}, \ldots\ldots, v_j$ from the path. We still have a path in the graph G between u and v and containing fewer edges. We continue this process until each repeated vertex appears only once in the path. This gives us a simple path between u and v.

This proves the result.

11.19. IMPORTANT THEOREMS

The following are some of the very important theorems concerning *Euler paths* and *Euler circuits* in a graph.

Theorem I. If a graph has a vertex of odd degree then there cannot be any Euler circuit in the graph.

Proof. Let G be a graph having a vertex of odd degree. If possible let there be a Euler circuit C in the graph. Therefore the circuit C includes each and every edge of G exactly once. Let v be any arbitrary vertex of the graph G. If deg$(v) = 0$, then it is not an odd number. Let deg $(v) > 0$. Thus there are edges with end point v and so the vertex v is a part of the circuit C. Everytime the circuit C meets v it goes through two 'new' edges with end point v using one edge to arrive at v and another to leave v. So the degree of v is a multiple of 2 *i.e.*, it is an even number.

$\therefore$ There is no vertex of odd degree. This is impossible.

$\therefore$ There cannot be any Euler circuit in the graph G.

Theorem II. If a graph is connected and every vertex has even degree then there exists an Euler circuit in the graph. *(KU 2007)*

Proof. Let G be a connected graph and let each vertex of G be of even degree. Let v be any arbitrary vertex of G. We construct a path starting with v and going through the edges of G so that no edge is traced more than once. Since every vertex is of even degree, we can leave every vertex we enter. Also, the degree of v is even, so we shall eventually reach v, when the construction of the path comes to an end. If this closed path *i.e.*, the circuit, say C, includes all the edges of G, this is the required Euler circuit.

Let C be not a Euler circuit. We remove from G all the edges used in C and obtain a subgraph G' of G formed by all the vertex and the remaining edges. The degrees of all vertices of G' are also even because the difference of two even numbers is also an even number. Since the graph G is connected, the subgraph G' must touch C at least at one vertex, say v. Starting from v, we construct a path in the graph G'. Since the vertices of G' are of even degrees, this path in G' must terminate at v. This closed path, say C_1, can be combined with C to form a new path which starts and ends at v and has more edges than. (This process can be repeated until we get a circuit that covers all the edges of G. This gives the required Euler circuit of the graph G.

Theorem III. If a graph has more than two vertices of odd degrees then there cannot be any Euler path in the graph.

Proof. Let G be a graph having vertices v_1, v_2, v_3 of odd degrees. If possible, let there be a Euler path P in the graph. Therefore the path P includes each and every edge of G exactly once. Let v be any non-terminal vertex of the path P. Every time the path P meets v it goes through two 'new' edges with end point v using one edge to arrive at v and another to leave v. So the degree of v is a multiple of 2 *i.e.*, it is an even number.

∴ The vertices v_1, v_2, v_3 of odd degrees can only be the terminal vertices of the path P. One, out of v_1, v_2, v_3, is at the beginning of P and another at the end of P and the remaining one vertex at the end of an untravelled edge.

∴ This 'untravelled edge' is not included in the Euler path P. This is impossible.

∴ There cannot be any Euler path in the graph G.

Theorem IV. If a graph is connected and has exactly two vertices of odd degrees then there exists an Euler path in the graph. Also, any Euler path in G must begin at one vertex of odd degree and end at the other vertex of odd degree.

Proof. Let G be a connected graph and let the vertices v_1, and v_2 be of odd degrees. Let G' be the graph having all vertices and edges of G and one additional edge $\{v_1, v_2\}$.

∴ The graph G' is connected and every vertex of G' has even degree.

∴ There exists a Euler circuit C in the graph G'. (By **Theorem II**)

Removing the edge $\{v_1, v_2\}$ from the circuit C, we get a path with terminal vertices v_1 and v_2. This path includes each and every edge of G exactly once because the circuit C of G' includes each and every edge of G' exactly once.

∴ This path is a Euler path of G and has terminal vertices v_1 and v_2.

Example 3. *Show that the graph given in the diagram is Eulerian.*

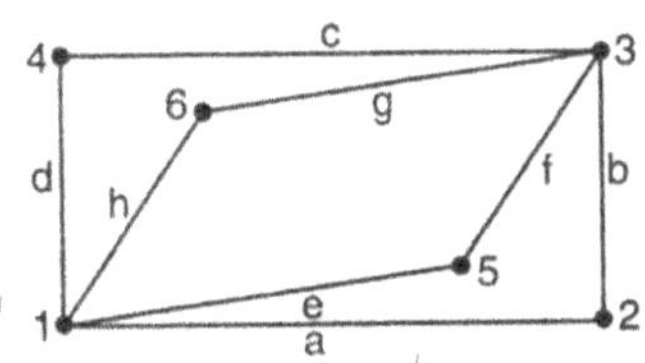

Sol. Given graph has 6 vertices and 8 edges.

Consider the path 1, a, 2, b, 3, c, 4, d, 1, e, 5, f, 3, g, 6, h, 1.

This is a path with identical terminal vertices and every edge is included in this path.

∴ This path is an Euler circuit.

∴ The given graph is Eulerian.

Example 4. (Königsberg Bridge Problem) *Two islands A and B formed by the Pregel river in Königsberg were connected to each other and to the banks C and D with seven bridges as shown in the figure. Can a person start at any of the four land areas A, B, C or D walk over each of the seven bridges exactly once and return to the starting point.*

Sol. Consider the graph G shown in the figure. The vertices a, b, c and d of the graph represent the land areas A, B, C and D respectively. An edge in the graph represents the bridge between the corresponding land areas.

∴ There are seven edges in the graph G.

In the graph G, we have

$d(a) = 5$, $d(b) = 3$, $d(c) = 3$, $d(d) = 3$.

∴ The graph has vertices of odd degree.

∴ G cannot have any Euler circuit in the graph. (By **Theorem I**)

∴ There cannot exist a path in the graph G including every edge exactly once and identical terminal vertices.

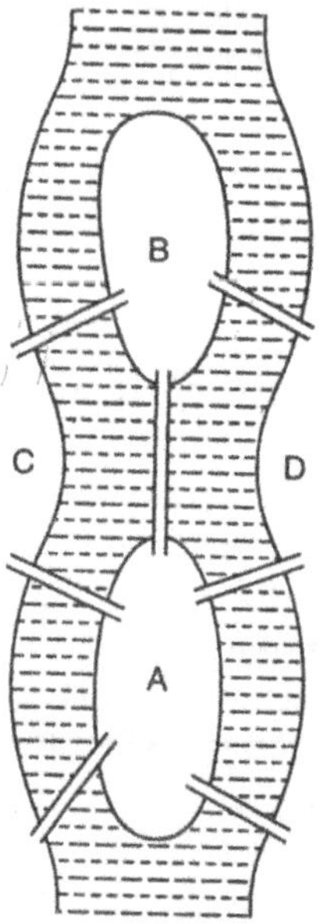

Königsberg Bridge Problem

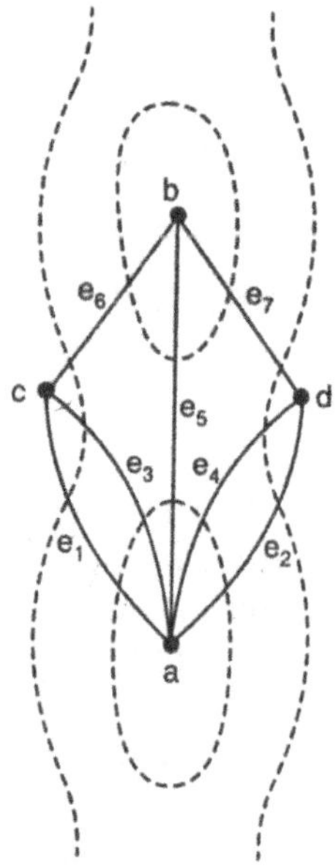

Graph G

∴ There exists no path in the Königsberg including every bridge exactly once and identical starting and ending point for the person.

∴ It is impossible for a person to start at any of the four land areas *A, B, C* or *D* walkover each of the seven bridges exactly once and return to the starting point.

V. HAMILTONIAN PATHS AND CIRCUITS

11.20. HAMILTONIAN PATH

A path in a graph is called a **Hamiltonian path** if it includes every vertex in the graph exactly once except for the possibility of identical terminal vertices of the path.

In other words, a Hamiltonian path in a graph *G* is an alternating sequence $v_0, e_1, v_1, e_2, v_2, \ldots\ldots, v_{n-1}, e_n, v_n$ of vertices and edges, beginning and ending with vertices such that :

(*i*) each edge contains the vertices preceding and following it

(*ii*) no vertex is repeated in the above sequence except for the possibility, $v_0 = v_n$

(*iii*) no edge is repeated in the above sequence

(*iv*) every vertex of the graph is included in the above sequence.

Remark. Conditions (*ii*) and (*iv*) are making a path, a Hamiltonian path.

11.21. HAMILTIONIAN CIRCUIT

A circuit in a graph is called a **Hamiltonian circuit** if it includes every vertex in the graph exactly once, except for identical terminal vertices of the path.

In other words, a Hamiltonian circuit in a graph is an alternating sequence $v_0, e_1, v_1, e_2, v_2, \ldots\ldots, v_{n-1}, e_n, v_n$ of vertices and edges beginning and ending with vertices such that :

(*i*) each edge contains the vertices preceding and following it

(*ii*) no vertex is repeated in the above sequence except for $v_0 = v_n$

(*iii*) no edge is repeated in the above sequence

(*iv*) every vertex of the graph is included in the above sequence.

Remark. Every Hamilton circuit is a Hamilton path with identical terminal vertices.

11.22. HAMILTONIAN GRAPH

A graph is called a **Hamiltonian graph** if there exists a Hamiltonian circuit in the graph.

Important observations. If a graph is an Euler graph then there exists an Euler circuit in the graph *i.e.*, the graph has a path with identical terminal vertices such that every edge is included exactly once and may not include every vertex of the graph. On the other hand, if a graph is a Hamiltonian graph then there exists a Hamiltonian circuit in the graph *i.e.*, the graph has a path with identical terminal vertices such that every vertex is included exactly once except for the terminal vertices and may not include every edge of the graph.

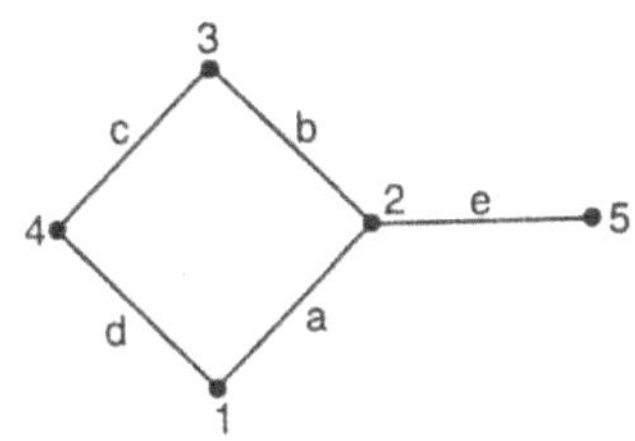

Illustration 1. Consider the graph given in the diagram. This graph has 5 vertices and 5 edges.

The path 5, *e*, 2, *b*, 3, *c*, 4, *d*, 1 is a Hamiltonian path because each vertex is included exactly once in the path. This path is not a Hamiltonian circuit because the terminal vertices are not identical.

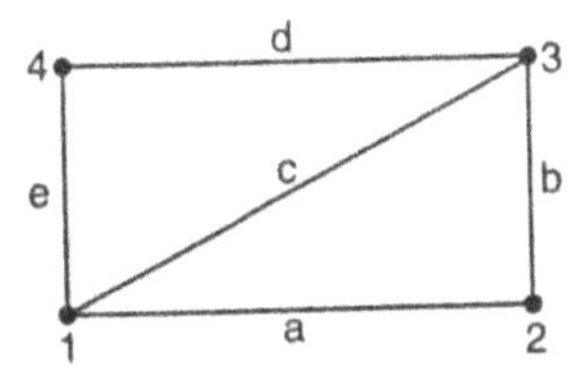

Illustration 2. Consider the graph given in the diagram. This graph has 4 vertices and 5 edges.

The path 1, *a*, 2, *b*, 3, *c*, 1, *e*, 4 is not a Hamiltonian path because the vertex 1 is repeated.

The path 1, *a*, 2, *b*, 3, *d*, 4, *e*, 1 is a Hamiltonian circuit because the path includes every vertex and no vertex is repeated except for the terminal vertices.

Thus, the given graph is a Hamitonian graph.

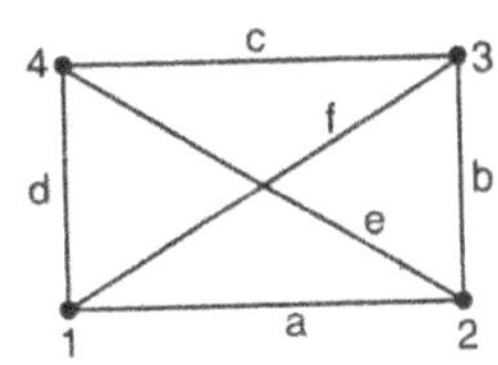

Illustration 3. Every complete graph has a Hamiltonian circuit.

Consider the graph K_4 :

The path 1, *a*, 2, *b*, 3, *c*, 4, *d*, 1 is a Hamiltonian circuit.

∴ Every complete graph is a Hamitonian graph.

Example 5. *Show that the graph given in the diagram is non-Eulerian and Hamiltonian.*

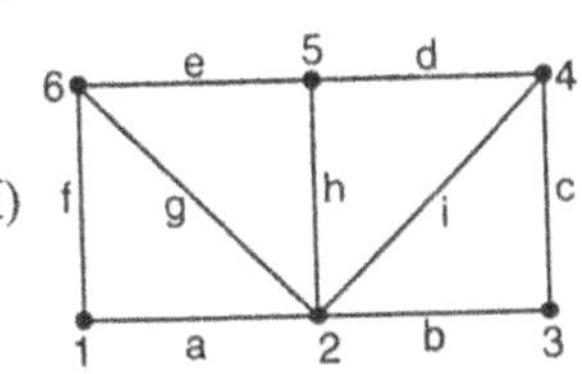

Sol. Given graph has 6 vertices and 9 edges.

The vertex 6 is of degree 3, which is an odd number.

∴ This graph cannot have an Euler circuit. (Using **Result III**)

∴ This graph is non-Eulerian.

Consider the path 1, *a*, 2, *b*, 3, *c*, 4, *d*, 5, *e*, 6, *f*, 1

This is a path with identical terminal vertices and every vertex is included in this path. Also each vertex occur only once except for the terminal vertices.

∴ This path is a Hamiltonian circuit.

∴ The given graph is Hamiltonian.

Remark. Note that the edges *g*, *h* and *i* are not included in the Hamiltonian circuit.

11.23. DISTANCE AND DIAMETER

Let G be a connected graph. Let u and v be any two vertices in the graph. By definition, there exists a path in the graph from the vertex u to the vertex v. The length of this path is the number of edges in the path. The length of a shortest path from the vertex u to the vertex v is called the **distance** between the vertices u and v and written $d(u, v)$.

The maximum distance between any two vertices in G is called the **diameter** of the graph and written diam (G).

For example, consider the graph given in the figure.

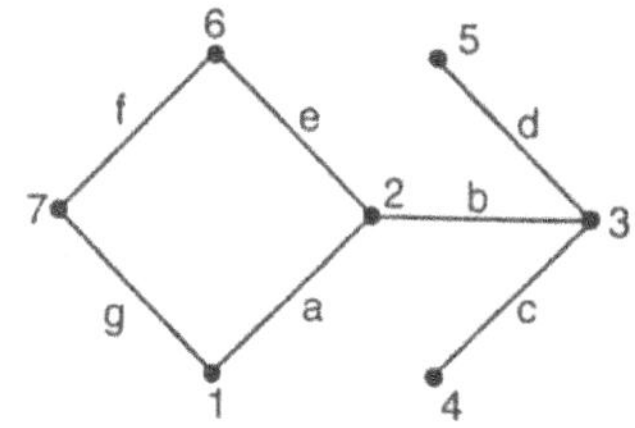

We have $d(2, 3) = 1, d(1, 3) = 2, d(7, 3) = 3, d(7, 4) = 4$

Here the maximum value of $d(u, v)$ for any vertices u and v of G is 4.

$\therefore$ diam $(G) = 4$.

11.24. TRAVELLING SALESMAN PROBLEM

In this section, we shall discuss a problem which is closely related to **Hamiltonian circuits.** This problem is popularly known as **Travelling Salesman Problem.** This problem is stated as follows :

A salesman is required to visit a number of cities during a trip. Given the distances between the cities, in what order should he travel so as to visit every city precisely once and return home, with the minimum distance travelled.

Let the cities be represented by $v_0, v_1, ..., v_n$ and the road joining the cities v_i and v_j be represented by $\{v_i, v_j\}$.

$\therefore$ We get a graph with vertices $v_0, v_1, ..., v_n$ and edges $\{v_i, v_j\}$, where $0 \le i < j \le n$. Let this graph be denoted by G. Let the edges of this graph be denoted by $e_1, e_2, e_3, ...$

Let w_i be the weight of the edge e_i, where w_i is the length of the road represented by the edge e_i. Also, there is an edge (*i.e.*, road) between any two vertices (*i.e.*, cities), the graph G is a weighted complete graph.

We want to find a path for the salesman so that each vertex (*i.e.*, city) in the graph is included and at the same time, no vertex (*i.e.*, city) is repeated except for the terminal vertices and also the total distance covered is minimum.

In other words, we to want to find a *Hamiltonian circuit* in the graph with the additional condition that the sum of weights of the edges in the circuit is minimum.

Theoretically, this problem can be solved by finding all the Hamiltonian circuits in the graph and identifying that circuit for which the sum of weights of all edges is minimum. But this procedure is not feasible if n is large. This is a very technical problem and requires deep study of the subject. Keeping in view our limitations, we shall discuss only one method called the **nearest-neighbour method** to solve this problem. This method give reasonably good results.

11.25. NEAREST–NEIGHBOUR METHOD FOR SOLVING TRAVELLING SALESMAN PROBLEM

Let $G = (V, E, w)$ be a weighted complete graph. Here w is the weight function from the set E of edges to the set of positive real numbers. The following steps are followed to identify the Hamiltonian circuit in the graph G such that the sum of the weights of edges in the circuit is minimum.

Step I. *Choose any vertex and call it v_0.*

Step II. *Find the vertex (other than v_0) which is nearest to v_0 i.e., the weight of the concerned edge is minimum. Let us call this vertex as v_1. Let the edge between v_0 and v_1 be e_1, say.*

Step III. *Find the vertex (other than v_0 and v_1) which is nearest to v_1. Let this vertex be denoted by v_2. Let the edge between v_1 and v_2 be e_2, say.*

$\therefore$ *We get the path v_0, e_1, v_1, e_2, v_2.*

Step IV. *Repeat the process of* **step III** *till all the vertices are included in the path. If there are $n + 1$ vertices in the graph, then the final form of the path looks like*

$v_0, e_1, v_1, e_2, v_2, e_3, v_3, \ldots\ldots, v_{n-1}, e_n, v_n$.

Step V. *We include the edge joining the vertices v_n and v_0, say e_{n+1}, to this path so that this path is extended to the circuit*

$v_0, e_1, v_1, e_2, v_2, e_3, v_3, \ldots\ldots, v_{n-1}, e_n, v_n, e_{n+1}, v_0$.

This is a Hamiltonian circuit in the graph G and the sum of the weights of the edges in this circuit is approximately minimum. A Hamiltonian circuit with this property is called a **shortest Hamiltonian circuit.**

The circuit $v_0, e_1, v_1, e_2, v_2, e_3, v_3, \ldots\ldots, v_{n-1}, e_n, v_n, e_{n+1}, v_0$ is the required shortest Hamiltonian circuit by the nearest-neighbour method.

Remark. The *nearest-neighbour method* is also applicable to the weighted graphs which are not complete, provided the process of finding edges in the above method is not hampered.

Example 6. *Find a shortest Hamiltonian circuit for the given weighted graph by the nearest-neighbour method.*

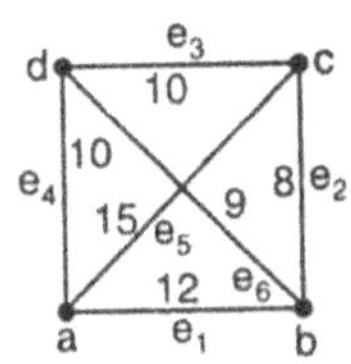

Sol. The given graph has 4 vertices and 6 edges $e_1, e_2, e_3, e_4, e_5, e_6$ with respective weights 12, 8, 10, 10, 15, 9.

Let us start the nearest-neighbour method with the vertex a.

The distances of the vertex a from the vertices b, c, d (other than a) are 12, 15, 10 respectively. The vertex d is nearest to the vertex a. The edge between a and d is e_4.

The distances of the vertex d from the vertices b, c (other than a and d) are 9, 10 respectively. The vertex b is nearest to the vertex d. The edge between d and b is e_6.

Now there is only one uncovered vertex *i.e.*, c. The edge between b and c is e_2. The edge between c and a is e_5.

$\therefore$ The sequence of vertices in the circuit is a, d, b, c, a.

$\therefore$ The required shortest Hamiltonian circuit is $a, e_4, d, e_6, b, e_2, c, e_5, a$.

EXERCISE 11.5

1. For the given graph, which of the following alternating sequences of vertices and edges are paths in the graph :

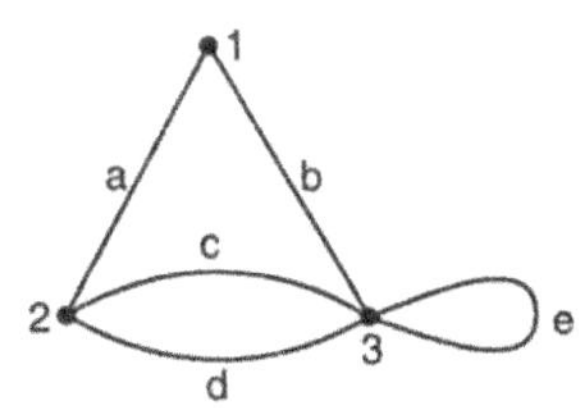

(*i*) 2, c, 3, c, 2, a, 1 (*ii*) 2, c, 3, d, 2, a, 1

(*iii*) 3, e, 3, d, 2 (*iv*) 1, a, 2, c, 3, b, 1, a, 2

(*v*) 1, b, 3, e, 3, d, 2.

2. For the given graph, which of the following alternating sequences of vertices and edges are simple paths in the graph :

(*i*) 2, *b*, 3, *c*, 4, *d*, 3

(*ii*) 2, *b*, 3, *c*, 4, *e*, 1

(*iii*) 1, *a*, 2, *b*, 3, *f*, 5, *g*, 1

(*iv*) 3, *c*, 4, *d*, 3, *b*, 2

(*v*) 5, *f*, 3, *c*, 4, *c*, 3.

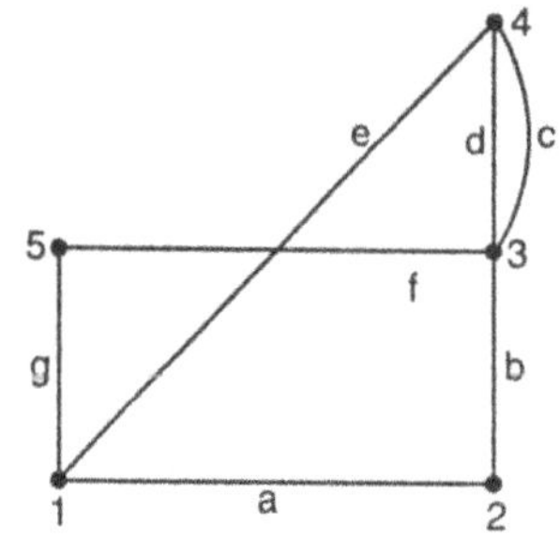

3. For the given graph, which of the following alternating sequences of vertices and edges are circuits in the graph :

(*i*) 1, *a*, 2, *b*, 3, *g*, 4, *f*, 5, *e*, 1

(*ii*) 1, *a*, 2, *b*, 3, *g*, 4, *h*, 4, *f*, 5, *e*, 1

(*iii*) 3, *g*, 4, *c*, 2, *a*, 1, *d*, 4, *g*, 3

(*iv*) 2, *a*, 1, *d*, 4, *c*, 2

(*v*) 4, *h*, 4, *f*, 5, *e*, 1.

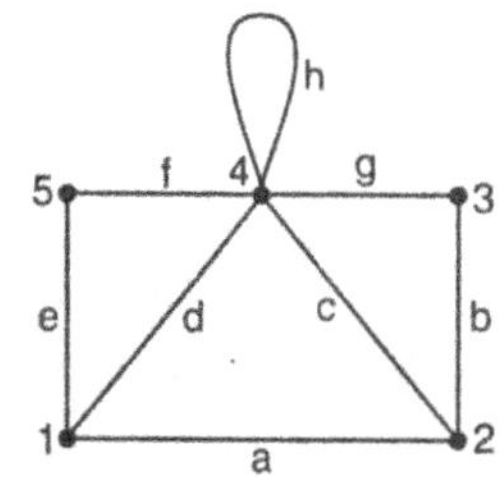

4. For the given graph, which of the following sequences of vertices and edges are simple circuits in the graph :

(*i*) 1, *a*, 2, *h*, 2, *b*, 3, *e*, 4, *c*, 1

(*ii*) 1, *a*, 2, *b*, 3, *d*, 4, *c*, 1

(*iii*) 2, *f*, 5, *g*, 5, *f*, 2

(*iv*) 4, *e*, 3, *b*, 2, *a*, 1, *c*, 4

(*v*) 3, *e*, 4, *d*, 3.

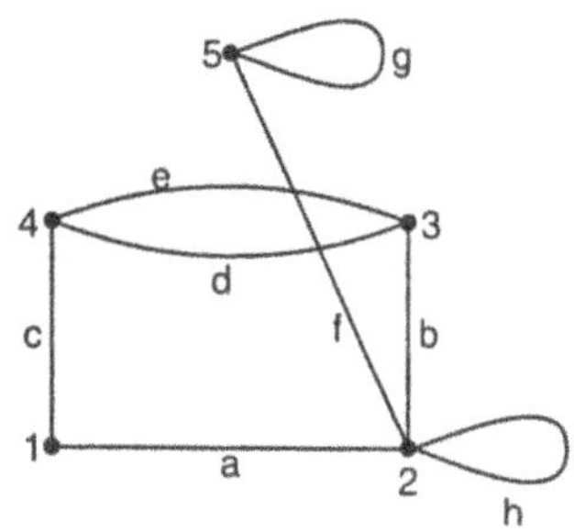

5. For the given graph, which of the following alternating sequences of vertices and edges are Euler paths in the graph :

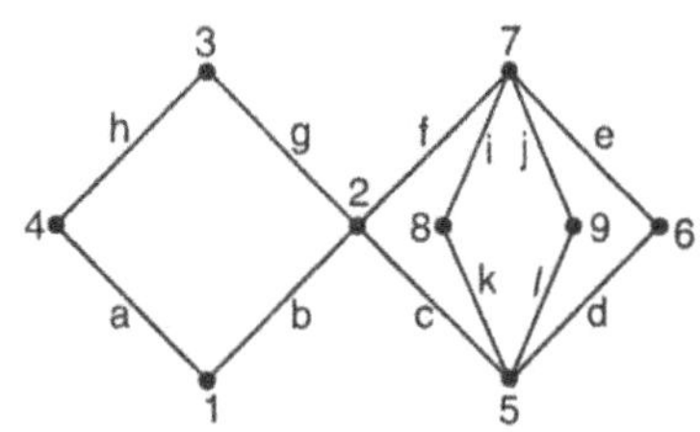

(*i*) 1, *a*, 4, *h*, 3, *g*, 2, *f*, 7, *e*, 6, *d*, 5, *c*, 2, *b*, 1

(*ii*) 5, *d*, 6, *e*, 7, *j*, 9, *l*, 5, *k*, 8, *i*, 7, *f*, 2, *g*, 3, *h*, 4, *a*, 1, *b*, 2, *c*, 5

(*iii*) 4, *h*, 3, *g*, 2, *f*, 7, *e*, 6, *d*, 5, *l*, 9, *j*, 7, *i*, 8, *k*, 5, *c*, 2, *b*, 1, *a*, 4

(*iv*) 2, *f*, 7, *i*, 8, *k*, 5, *l*, 9, *j*, 7, *e*, 6, *d*, 5, *c*, 2, *b*, 1, *a*, 4, *h*, 3, *g*, 2.

6. For the given graph, which of the following alternating sequences of vertices and edges are Euler circuits in the graph :

(*i*) 1, *a*, 2, *c*, 3, *d*, 4, *e*, 5, *g*, 1

(*ii*) 2, *b*, 2, *a*, 1, *f*, 5, *g*, 1, *f*, 5, *e*, 4, *d*, 3, *c*, 2

(*iii*) 5, *f*, 1, *g*, 5, *e*, 4, *d*, 3, *c*, 2, *b*, 2, *a*, 1, *g*, 5

(*iv*) 5, *f*, 1, *a*, 2, *b*, 2, *c*, 3, *d*, 4, *e*, 5.

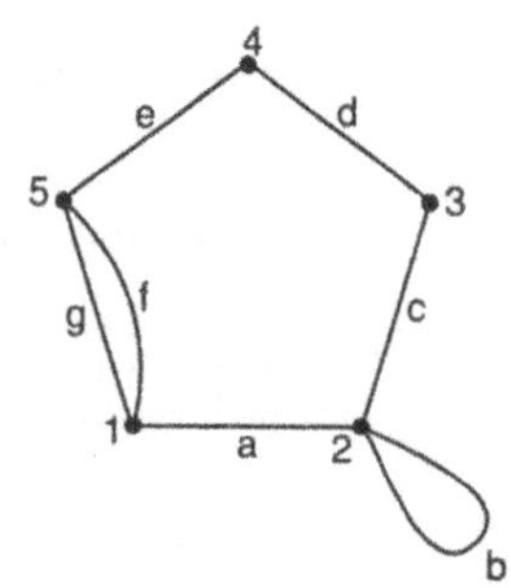

7. Find two Euler paths in the given graph.

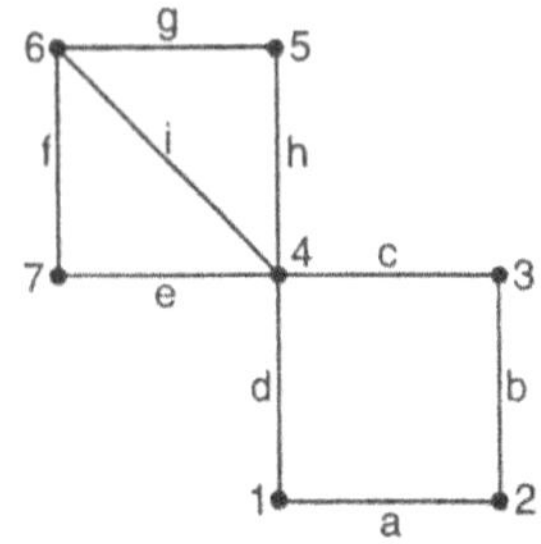

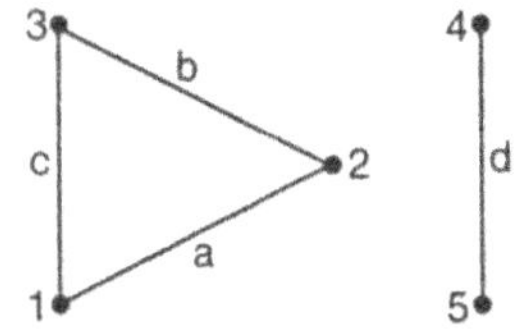

Q. 7 Q. 8

8. Show that there cannot be an Euler path in the given graph.
9. Identify an Euler circuit in the given graph.

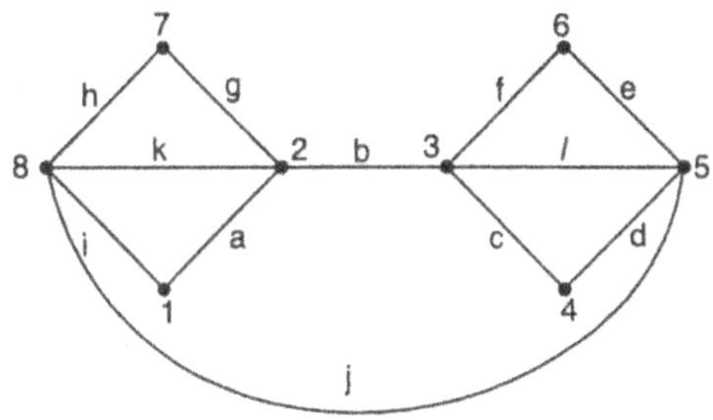

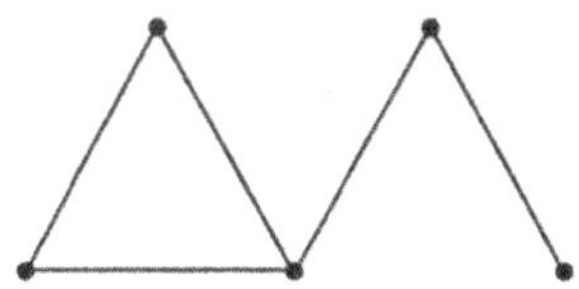

Q. 9 Q. 10

10. Find the diameter of the given connected graph.
11. Find the diameter of the given connected graph.

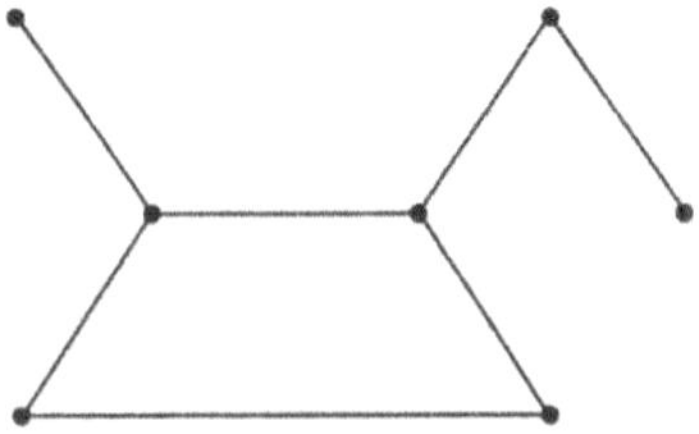

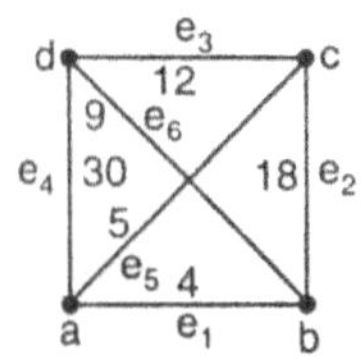

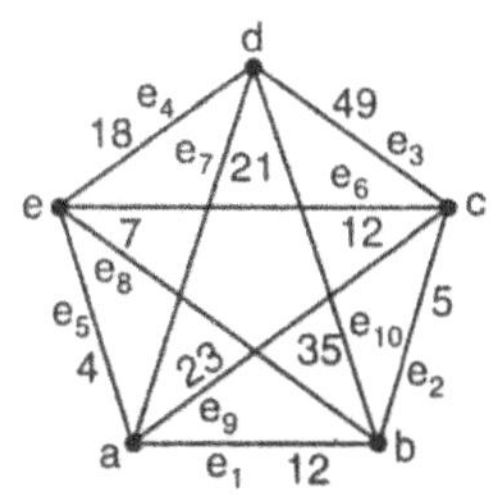

Q. 11 Q. 12 Q. 13

12. Find a shortest Hamiltonian circuit in the given weighted graph by the nearest-neighbour method. Also find the sum of the weights of the edges in the circuit.

13. Find the shortest Hamiltonian circuit in the given weighted graph by the nearest-neighbour method. Also find the sum of weights of the edges in the circuit.

14. Find the shortest Hamiltonian circuits in the given weighted graph by the nearest-neighbour method starting with vertex a and vertex d. Also find the sum of weights of the edges in the circuits.

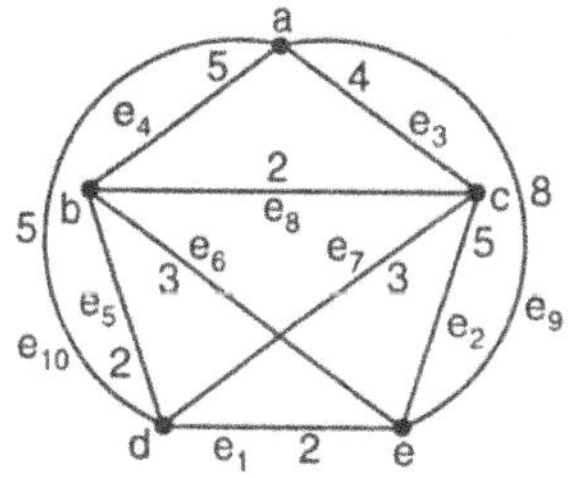

Answers

1. (ii), (iii), (v) 2. (ii), (iii) 3. (i), (ii), (iv) 4. (ii), (iv), (v)
5. (ii), (iii), (iv) 6. No
7. 4, d, 1, a, 2, b, 3, c, 4, h, 5, g, 6, f, 7, e, 4, i, 6 ; 6, g, 5, h, 4, e, 7, f, 6, i, 4, d, 1, a, 2, b, 3, c, 4
9. 8, h, 7, g, 2, k, 8, i, 1, a, 2, b, 3, l, 5, e, 6, f, 3, c, 4, d, 5, j, 8
10. 3 11. 4
12. $a, e_1, b, e_6, d, e_3, c, e_5, a$; 30
13. $a, e_5, e, e_8, b, e_2, c, e_3, d, e_7, a$; 86
14. $a, e_3, c, e_8, b, e_5, d, e_1, e, e_9, a$; 18
$d, e_5, b, e_8, c, e_3, a, e_9, e, e_1, d$; 18

Hint

8. Edge d cannot be connected with any of a, b, or c.

12

Trees

12.1. INTRODUCTION

A particular type of graphs are called *trees*. Thus, every tree is a graph. The study of trees is considered to be very important in graph theory. Many types of games and puzzles are solved by using trees. The sorting of mail according to pincode is done according to a tree.

12.2. TREE

A graph is called a **tree** if it is connected and without simple circuits.

A tree has at least one vertex.

Illustrations I. Consider the graph given in the figure. This graph is not connected.

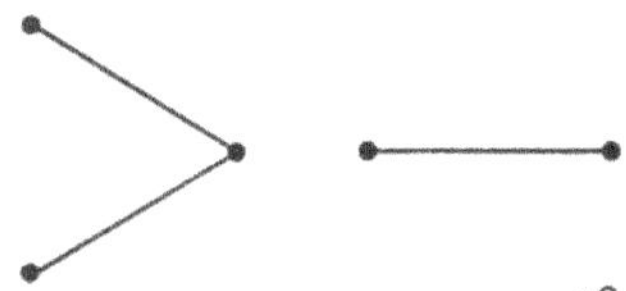

$\therefore$ It is not a tree.

II. Consider the graph given in the figure. This graph is connected.

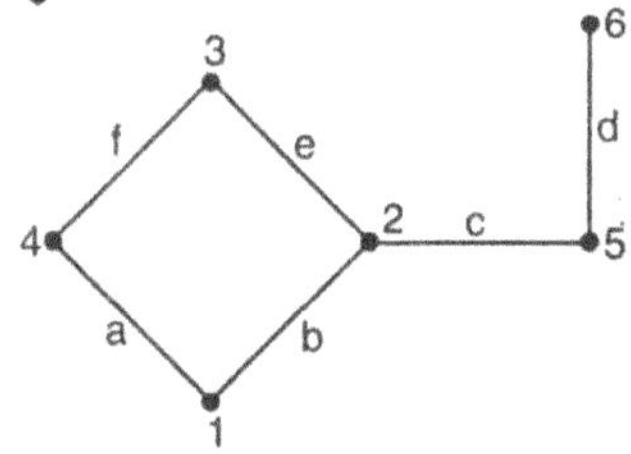

1, b, 2, e, 3, f, 4, a, 1 is a simple circuit in the graph.

$\therefore$ The given graph is not a tree.

III. Consider the graphs given below.

All the graphs are trees because every graph is connected and also free from simple circuits.

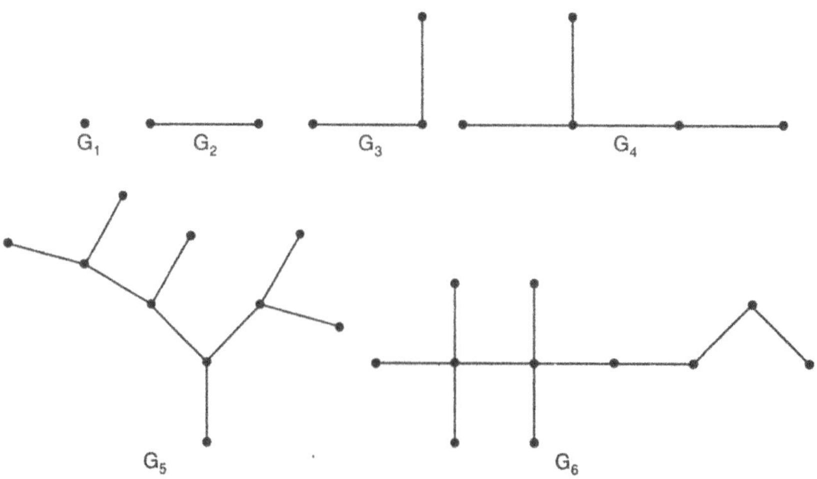

It may be observed that if there are n vertices in a tree then there are exactly $n - 1$ edges.

Since loops and parallel edges form simple circuits, no tree can have loops or parallel edges. Thus, every tree is a simple graph.

A graph having no simple circuit is also called a **acyclic graph.** Thus, a tree is a connected acyclic graph.

A vertex of degree one in a tree is called a **terminal node** or a **leaf.**

A vertex of degree greater than one is called an **internal node** or a **branch node.**

A collection of disjoint trees is called a **forest.**

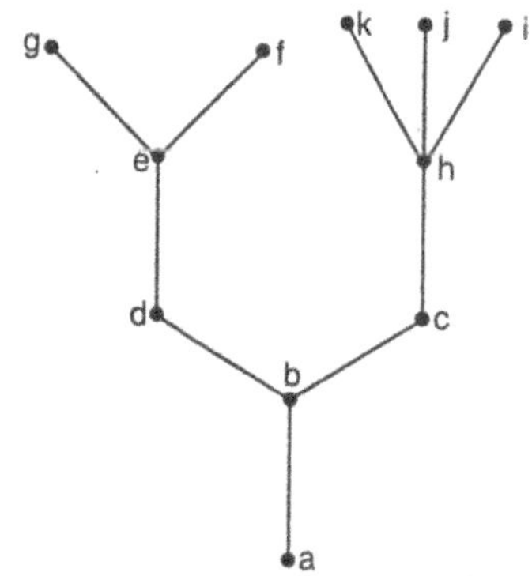

Consider the tree shown in the figure. This tree has 11 vertices and 10 edges. The vertices a, g, f, k, j and i are terminal nodes of the tree. The vertices b, d, e, c and h are internal nodes of the tree.

12.3. IMPORTANT THEOREMS

Theorem I. Prove that there is exactly one path between every pair of vertices in a tree.

Proof. Let u and v be any two vertices in a tree T. Since the tree T is a connected graph, there exists a path between u and v. If possible, let there exists two distinct paths between the vertices u and v. These two paths taken together form a circuit in T. We shall show that this circuit contains a simple circuit. Let this circuit be denoted by

$$v_0, e_1, v_1, e_2, v_2, \ldots\ldots, v_{n-1}, e_n, v_n (= v_0)$$

If $v_0, v_1, v_2, \ldots\ldots, v_{n-1}$ are all distinct then this circuit is a simple circuit.

Let $v_0, v_1, v_2, \ldots\ldots, v_{n-1}$ be not all distinct.

$\therefore$ There exist at least one vertex that appears at v_i and v_j in the circuit with $i < j$.

The circuit can also be expressed as

$$v_0, e_1, v_1, e_2, v_2, \ldots\ldots, e_i, v_i, e_{i+1}, v_{i+1}, \ldots\ldots, v_j, e_{j+1}, v_{j+1}, \ldots\ldots, v_{n-1}, e_n, v_n (= v_0).$$

We remove the terms $e_{i+1}, v_{i+1}, \ldots\ldots, v_j$ from the circuit. We still have a circuit in T containing fewer edges. We continue this process until each repeated vertex except v_0, v_n appears only once in the circuit. This gives us a simple circuit in the tree, which is impossible.

$\therefore$ There exists exactly one path between u and v.

Theorem II. Let G be a graph such that there is exactly one path between every pair of vertices. Prove that the graph G is a tree.

Proof. Since there exists a path between every pair of vertices of G, the graph G is connected.

We shall show that there is no simple circuit in the graph G. If possible, let there be a simple circuit in the graph G, say C. The circuit C may or may not be a loop.

Case I. C is a loop.

Let C be the loop u, e, u. Let v be any other vertex of G.

$\therefore$ There exists a path, say, $u, e_1, u_1, e_2, u_2, \ldots\ldots, u_{n-1}, e_n, v$ between u and v.

$\therefore$ $u, e, u, e_1, u_1, e_2, u_2, \ldots\ldots, u_{n-1}, e_n, v$ is also a path between u and v.

$\therefore$ There are two distinct paths between u and v, which is impossible.

$\therefore$ The simple circuit C cannot be a loop.

Case II. C is not a loop.

In this case, C has at least two distinct vertices, say, w_1 and w_2. The circuit C gives two distinct paths between the vertices w_1 and w_2, which is also impossible.

$\therefore$ The graph G cannot have any simple circuit.

$\therefore$ The graph G is a tree.

Theorem III. Prove that a tree with n vertices has n–1 edges. *(KU 2006)*

Proof. We shall prove this theorem by induction on the number of vertices.

Let $n = 1$. We know that a tree of one vertex has no edge.

$\therefore$ The theorem is true for $n = 1$.

Let the result be true for all trees with less than n vertices.

Let T be a tree with n vertices.

Let e be any arbitrary but fixed edge in the tree T and e connects the vertices u and v of T.

$\therefore$ u, e, v is a path between u and v. Since there is exactly one path between every pair of vertices in a tree, there is no other path between u and v other than u, e, v.

$\therefore$ The graph $T - e$ obtained by deleting e from T is a disconnected graph. The graph $T - e$ has two components, say, T_1 and T_2, one containing u and the other containing v. Since there are no simple circuits in T, the connected graphs T_1 and T_2 cannot have simple circuits.

$\therefore$ By definitions, the graphs T_1 and T_2 are both trees.

Let the number of vertices in the trees T_1 and T_2 be k and $n–k$ respectively.

$$\therefore \qquad k < n \quad \text{and} \quad n - k < n$$

$\therefore$ By the induction hypothesis, the number of edges in T_1 and T_2 are respectively $k–1$ and $n - k - 1$.

$\therefore$ Number of edges in the graph $T - e$

$$= (k - 1) + (n - k - 1) = n - 2$$

$\therefore$ Number of edges in the tree T

$$= (n - 2) + 1 = n - 1.$$

$\therefore$ By induction, a tree with n vertices has $n - 1$ edges.

Theorem IV. *Prove that a tree that has more than one vertex has at least one vertex of degree one.* (KU 2006 C)

Proof. Let T be a tree having n vertices.

$\therefore$ The number of edges in the tree is $n - 1$.

$\therefore$ By **Euler's theorem**

sum of degrees of all vertices $= 2(n - 1)$ ...(1)

Let v be any arbitrary but fixed vertex of T.

Since T has more than one vertex and is also connected, v cannot be an isolated vertex of T.

$\therefore$ $\deg(v) \neq 0$ *i.e.*, $\deg(v) \geq 1$

If possible, let there be no vertex of T having degree one.

$\therefore$ $\deg(v) \geq 2$

$\therefore$ Sum of degrees of all vertices $\geq 2n$

$\therefore$ (1) $\Rightarrow$ $2(n - 1) \geq 2n$ $\Rightarrow$ $-2 \geq 0$, which is impossible.

$\therefore$ Our supposition is wrong.

$\therefore$ The tree T has at least one vertex of degree one.

Example 1. *Let u and v be any two vertices in a tree T and the edge e (= {u, v}) is not in T. Show that the graph G containing T and the edge e is not a tree.*

Sol. u and v are vertices in the tree T. The tree T is connected.

∴ There exists a path between u and v. Since every path between two vertices contains a simple path between those vertices, there exists a simple path, not containing e, between u and v. This simple path between u and v together with the edge e forms a simple circuit in the graph G.

∴ The graph G cannot be a tree.

Example 2. *Let u and v be any two vertices in a tree T and the edge e (= {u, v}) is in T. Show that the graph T–e is not a tree.*

Sol. u and v are vertices in the tree T.

∴ There exists exactly one path between u and v. Since $e \in T$ and u, e, v is a path between u and v, there is no other path between u and v.

∴ The graph $T - e$ has no path between u and v.

∴ The graph $T - e$ is disconnected.

∴ The graph $T - e$ is not a tree.

Remark. Let T be a tree containing the edge $\{u, v\}$. The diagrams of T and $T - \{u, v\}$ are given below :

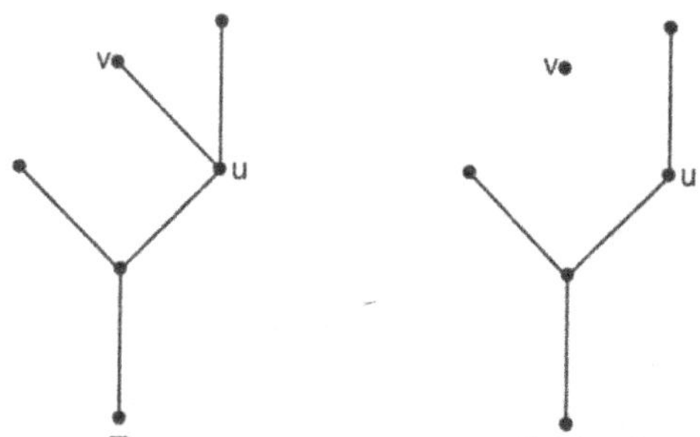

It is clear that $T - \{u, v\}$ is not connected.

Example 3. *Show that a connected graph with n vertices and n–1 edges is a tree.*

Sol. Let G be a connected graph with n vertices and n–1 edges. If possible, let G be not a tree. Since G is connected, the graph G must have at least one simple circuit, say, C.

Let e be any arbitrary but fixed edge in the circuit C. The graph G–e obtained by deleting e from G is still connected and having n vertices and n–2 (= $n - 1 - 1$) edges.

Since $G - e$ is connected, the edge e is not a bridge. We continue this process of locating edges that are not bridges and deleting them until, we get a connected subgraph G_1 with n vertices and $n - \lambda$ edges, where $\lambda > 1$, in which every edge is a bridge.

The connected graph G_1 is free from simple circuit, for otherwise we would have located another bridge in the connected graph G_1.

∴ The graph G_1 is a tree.

Since G_1 has n vertices, it must have n–1 edges.

$\Rightarrow \qquad n - \lambda = n - 1 \quad \Rightarrow \quad \lambda = 1$, which is impossible.

$\therefore$ Our supposition is wrong.

$\therefore$ The graph G is a tree.

Example 4. *Show that a graph with n vertices and n – 1 edges and having no simple circuit is a tree.* (KU 2007)

Sol. Let G be a graph with n vertices and $n - 1$ edges and having no simple circuit. If possible, let G be not a tree. Since G is having no simple circuit, the graph G must not be connected.

Let the components of G be $G_i (i = 1, 2, ..., k)$ such that G_i has n_i vertices, where

$$n_1 + n_2 + ... + n_k = n.$$

Thus each component G_i is connected and without simple circuits.

$\therefore$ Each G_i is a tree.

$\therefore$ Number of edges in tree $G_i = n_i - 1$

$\therefore$ Number of edges of G = sum of edges of components $G_1, G_2, ..., G_k$

$$= (n_1 - 1) + (n_2 - 1) + ... + (n_k - 1)$$
$$= (n_1 + n_2 + ... + n_k) - k = n - k$$

$\Rightarrow \qquad n - k = n - 1 \quad \Rightarrow \quad k = 1$, which is impossible, because $k > 1$.

$\therefore$ Our supposition is wrong.

$\therefore$ The graph G is a tree.

12.4. SPANNING TREE

Let G be a connected graph. A subgraph T of graph G is called a **spanning tree** of G if T is a tree and T includes all the vertices of G.

For example, let G be a connected graph as shown in the figure :

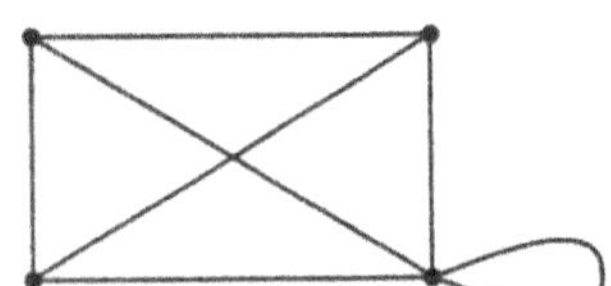

G

Below are given six spanning trees $T_1, T_2, T_3, T_4, T_5, T_6$ of the graph G.

Each of the trees given below has $3 (= 4 - 1)$ edges.

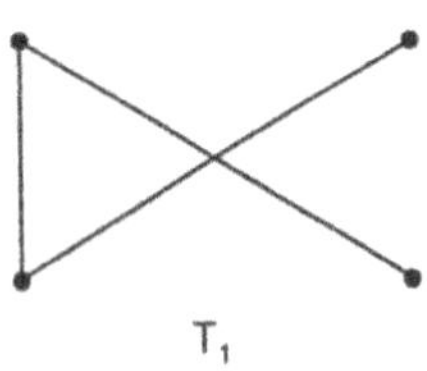

T_1

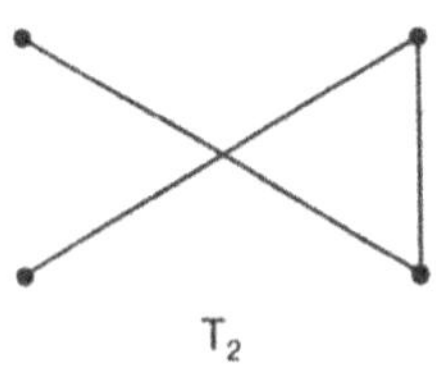

T_2

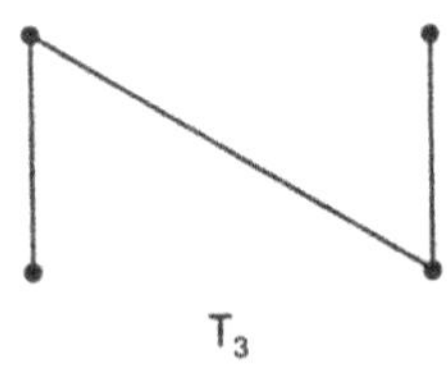

T_3

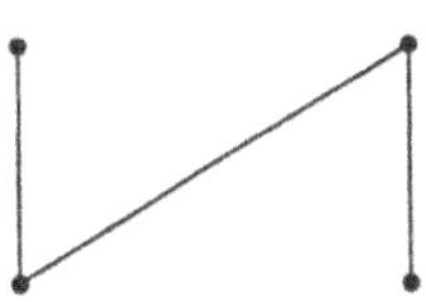

T_4

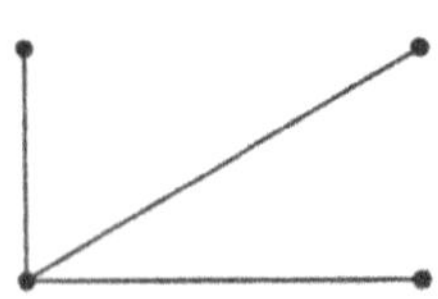

T_5

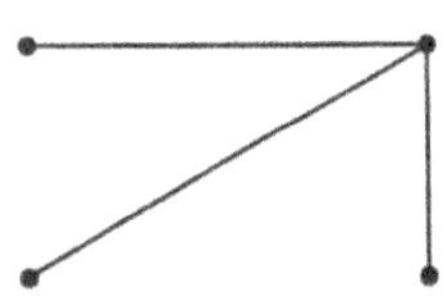

T_6

12.5. MINIMAL SPANNING TREE

Let G be a connected weighted graph. A spanning tree T of G is called a **minimal spanning tree** of G if the weight of the tree T is minimum *i.e.*, the sum of the weights of the edges in the tree T is minimum.

The minimal spanning tree of a connected weighted graph is not unique. Different methods may give different minimal spanning tree. The weight of a minimal spanning tree of a connected weighted graph is always unique.

We shall discuss two methods of finding a minimal spanning tree of a given connected weighted graph.

12.6. FIRST METHOD OF FINDING MINIMAL SPANNING TREE

Let G be a connected weighted graph of n vertices. We arrange the edges of G in the *descending order* of weights. The minimal spanning tree of the graph G is to contain all the vertices of the graph G. Our job is to delete some of the edges so that the remaining graph may have only $n - 1$ edges and it is a tree.

We proceed sequentially, deleting each edge that does not disconnect the remaining graph until $(n - 1)$* edges remains. This gives the required minimal spanning tree of the given connected weighted graph G of n vertices.

Example 5. *Find the minimal spanning tree of the given connected weighted graph G of 3 vertices.*

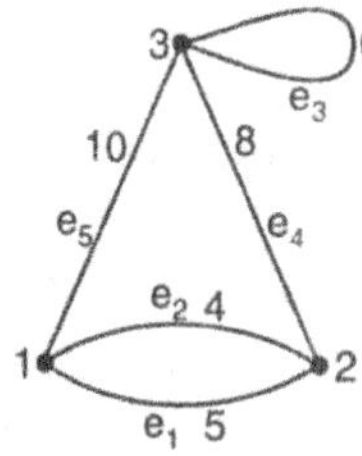

Sol. The given graph G has 3 vertices and 5 edges.

The required tree would have 3 vertices and 2 (= 3 – 1) edges.

The edges with weights are :

e_1	e_2	e_3	e_4	e_5
5	4	6	8	10

The edges with descending weights are :

e_5	e_4	e_3	e_1	e_2
10	8	6	5	4

We delete the edge e_5 from G because the remaining graph, shown in figure (*i*) is not disconnected.

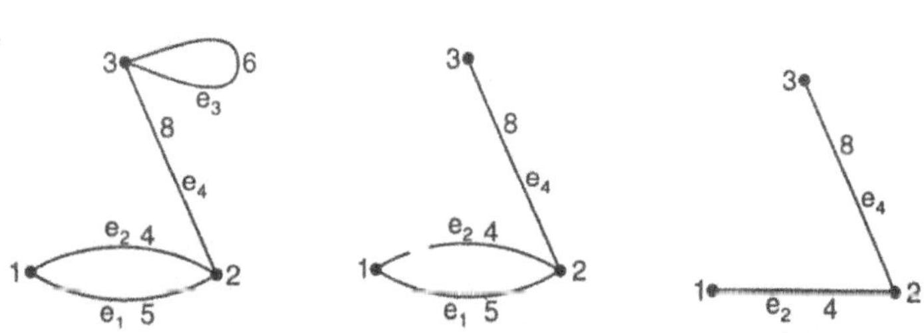

(*i*) (*ii*) (*iii*)

We cannot delete the edge e_4 from the graph in figure (*i*) because the remaining graph is bound to be disconnected.

We delete the edge e_3 from the graph in figure (*i*) because the remaining graph, shown in figure (*ii*) is not disconnected.

***Recall** : *A connected graph of n vertices is a tree if and only if it has n – 1 edges.*

We delete the edge e_1 from the graph in figure (*ii*) because the remaining graph, shown in figure (*iii*) is not disconnected. We need not consider other remaining edges because the graph (*iii*) has 2 edges.

∴ This is the required minimal spanning tree of the given connected weighted graph.

The weight of this minimal spanning tree is **12** (= 4 + 8).

Example 6. *Find the minimal spanning tree of the given connected weighted graph G of 4 vertices.*

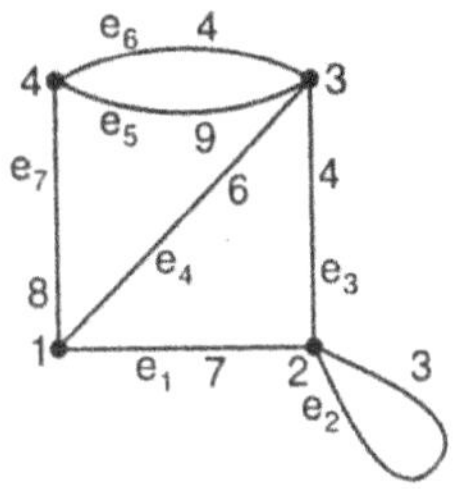

Sol. The given graph G has 4 vertices and 7 edges. The required tree would have 4 vertices and 3 (= 4 – 1) edges.

The edges with weights are :

e_1	e_2	e_3	e_4	e_5	e_6	e_7
7	3	4	6	9	4	8

The edges with descending weights are :

e_5	e_7	e_1	e_4	e_3	e_6	e_2
9	8	7	6	4	4	3

We delete the edge e_5 from G because the remaining graph, shown in figure (*i*) is not disconneted.

We delete the edge e_7 from the graph in figure (*i*) because the remaining graph, shown in figure (*ii*) is not disconnected.

We delete the edge e_1 from the graph in figure (*ii*) because the remaining graph shown in figure (*iii*) is not disconnected.

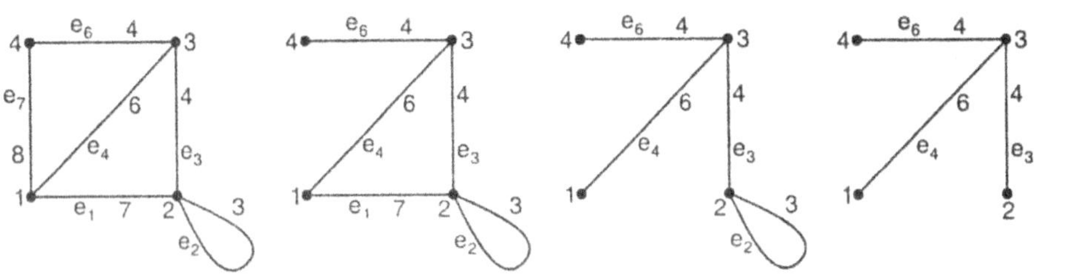

(*i*) (*ii*) (*iii*) (*iv*)

We cannot delete the edge e_4 from the graph in figure (*iii*) because the remaining graph is bound to be disconnected.

Similarly, the edges e_3 and e_6 cannot be deleted from the graph in figure (*iii*).

We delete the edge e_2 from the graph in figure (*iii*) because the remaining graph, shown in figure (*iv*) is not disconnected.

We need not consider other remaining edges because the graph (*iv*) has 3 edges.

∴ This is the required minimal spanning tree of the given connected weighted graph. The weight of this minimal spanning tree is **14** (= 4 + 6 + 4).

12.7. SECOND METHOD (KRUSKAL) OF FINDING THE MINIMAL SPANNING TREE

Let G be a connected weighted graph of n vertices. We arrange the edges of G in the *ascending order* of weights. The minimal spanning tree of the graph G is to contain all the vertices of the graph G. Our job is to delete some of the edges so that the remaining graph may have only $n-1$ edges and it is a tree.

We proceed sequentially, adding each edge that does not result in a simple circuit in the new graph until $(n-1)$* edges are added. This gives the required minimal spanning tree of the given connected weighted graph G of n vertices.

Example 7. *Find the minimal spanning tree of the given connected weighted graph G of 3 vertices by using Kruskal method.*

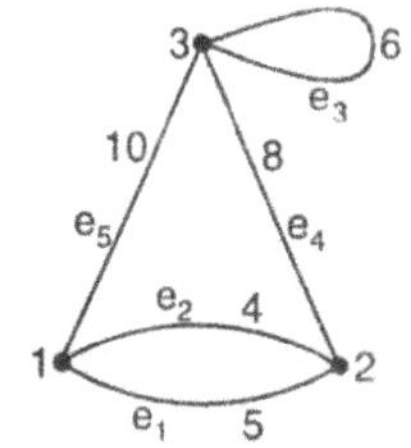

Sol. The given graph G has 3 vertices and 5 edges. The required tree would have 3 vertices and 2 (= 3 – 1) edges.

The edges with weights are :

e_1	e_2	e_3	e_4	e_5
5	4	6	8	10

The edges with ascending weights are :

e_2	e_1	e_3	e_4	e_5
4	5	6	8	10

We add the edge e_2 because the addition of e_2 to the graph in figure (*i*) does not make any simple circuit in the new graph as shown in figure (*ii*).

We cannot add the edge e_1 to the graph in figure (*ii*) because 1, e_2, 2, e_1, 1 would form a simple circuit in the new graph.

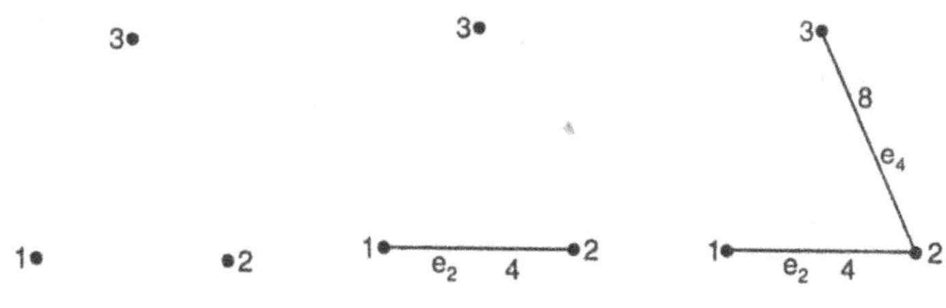

(*i*) (*ii*) (*iii*)

Similarly, we cannot add the edge e_3 to the graph in figure (*ii*) because 3, e_3, 3 would form a simple circuit in the new graph.

We add the edge e_4 because the addition of e_4 to the graph in figure (*ii*) does not make any simple circuit in the new graph as shown in figure (*iii*).

We need not consider other remaining edges because the graph (*iii*) has 2 edges.

∴ This is the required minimal spanning tree of the given connected weighted graph. The weight of this minimal spanning tree is **12** (= 4 + 8).

***Recall :** *A graph of n vertices having no simple circuits is a tree if and only if it has n – 1 edges.*

Example 8. *Find the minimal spanning tree of the given connected weighted graph G of 4 vertices by using Kruskal method.*

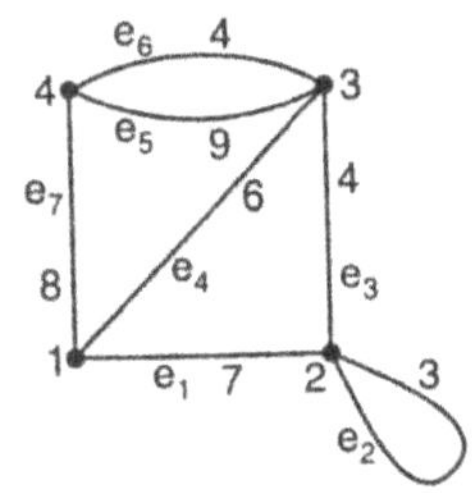

Sol. The given graph G has 4 vertices and 7 edges. The required tree would have 4 vertices and 3 (= 4 – 1) edges.

The edges with weights are :

e_1	e_2	e_3	e_4	e_5	e_6	e_7
7	3	4	6	9	4	8

The edges with ascending weights are :

e_2	e_6	e_3	e_4	e_1	e_7	e_5
3	4	4	6	7	8	9

We cannot add the edge e_2 to the graph in figure (*i*) because 2, e_2, 2 would form a simple circuit in the new graph.

We add the edge e_6 because the addition of e_6 to the graph (*i*) does not make any simple circuit in the new graph as shown in the figure (*ii*). We add the edge e_3 because the addition of e_3 to the graph (*ii*) does not make any simple circuit in the new graph as shown in the figure (*iii*).

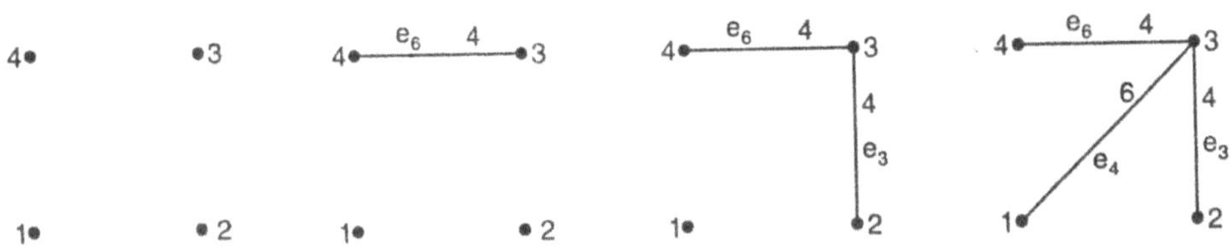

(*i*) (*ii*) (*iii*) (*iv*)

We add the edge e_4 because the addition of e_4 to the graph (*iii*) does not make any simple circuit in the new graph as shown in the figure (*iv*). We need not consider other remaining edges because the graph (*iv*) has 3 edges.

∴ This is the required minimal spanning tree of the given connected weighted graph. The weight of this minimal spanning tree is **14** (= 4 + 6 + 4).

Remark. We have discussed the same problems of finding *minimal spanning tree* by using both methods. This would help the reader to compare the different procedures followed in these methods.

EXERCISE 12.1

1. Which of the following graphs are trees ?

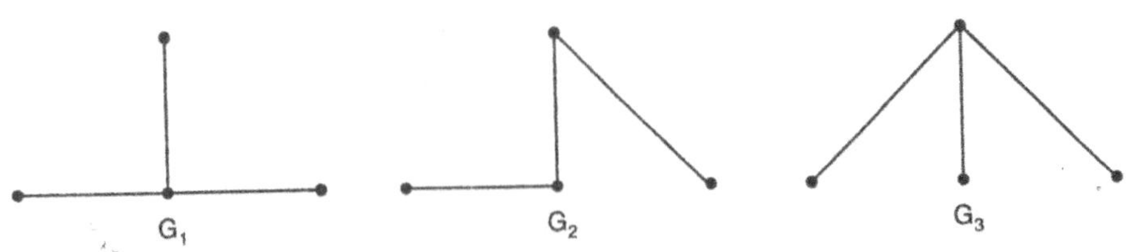

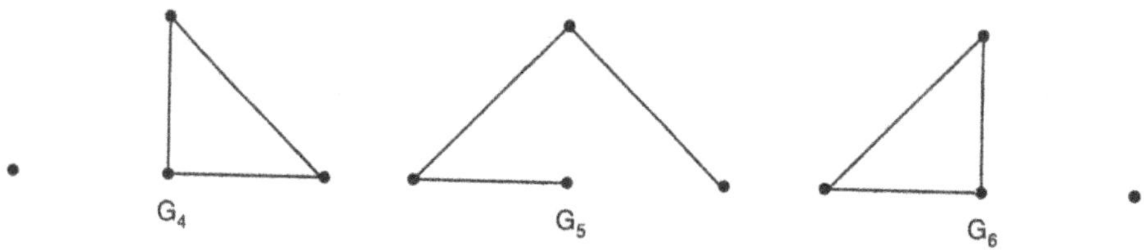

2. Which of the following graphs are trees ?

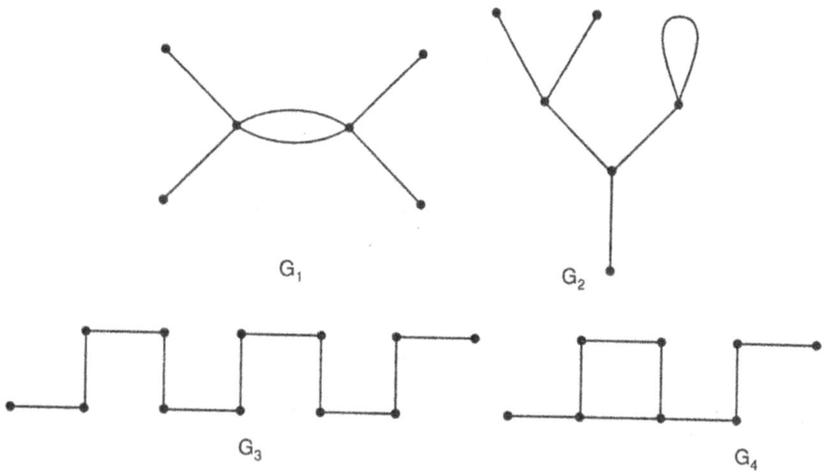

3. Find the minimal spanning trees of the following connected weighted graph by the first method :

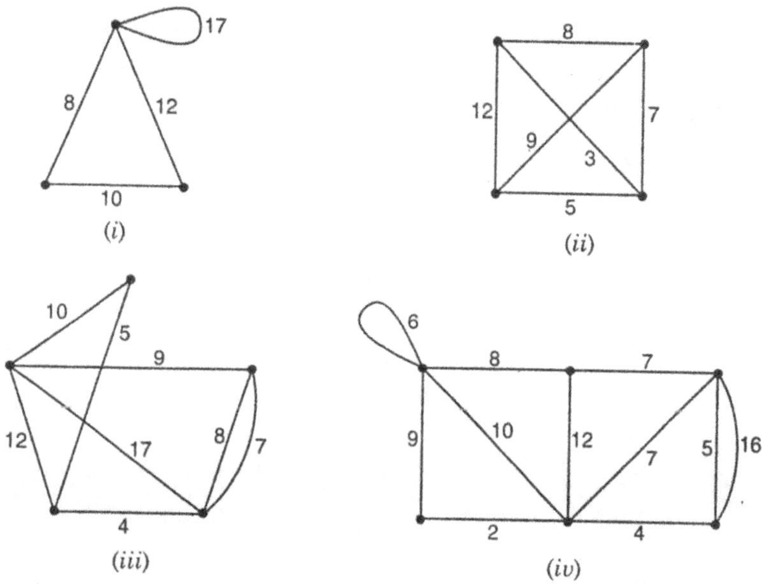

4. Find the minimal spanning trees of the graphs given in problem 3. by the Kruskal method.

Answers

1. G_1, G_2, G_3, G_5

2. G_3

3. (*i*)

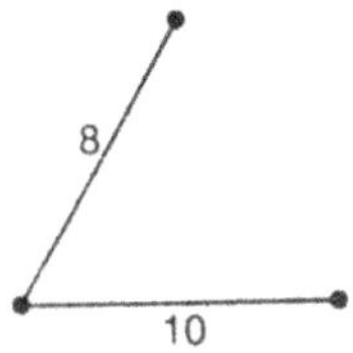

(*ii*)

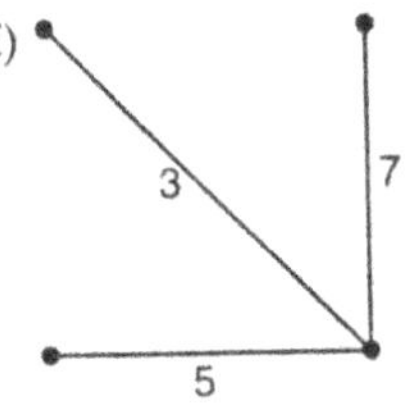

(*iii*)

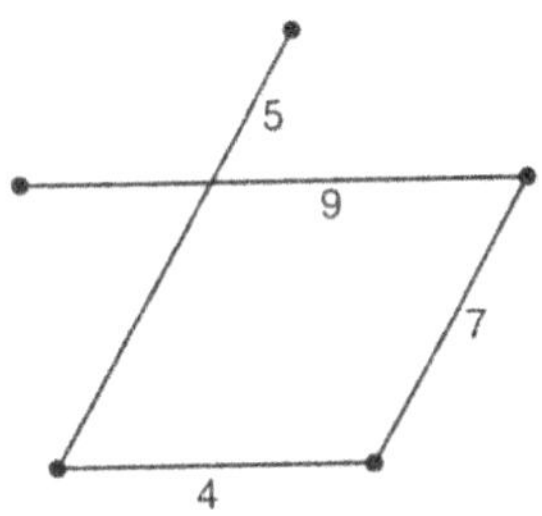

(*iv*)

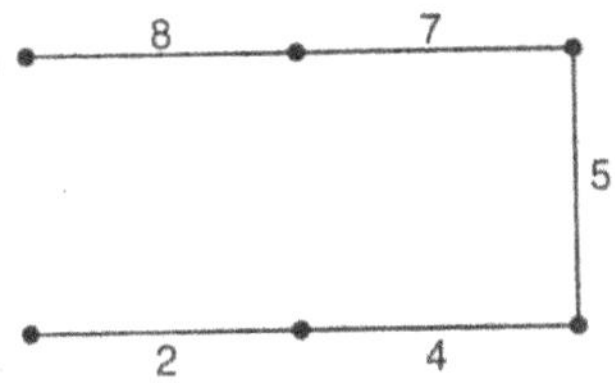

4. (*i*)

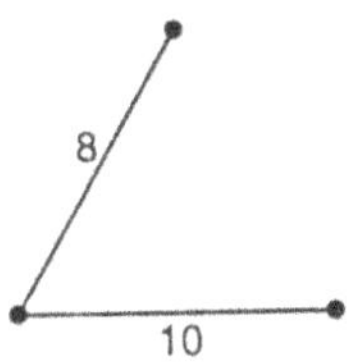

(*ii*)

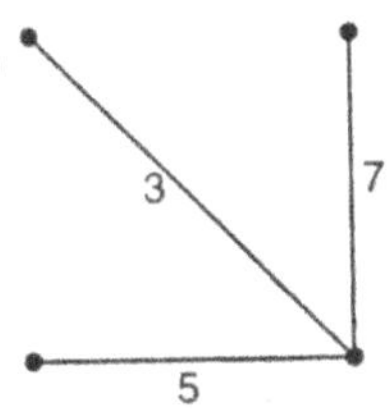

(*iii*)

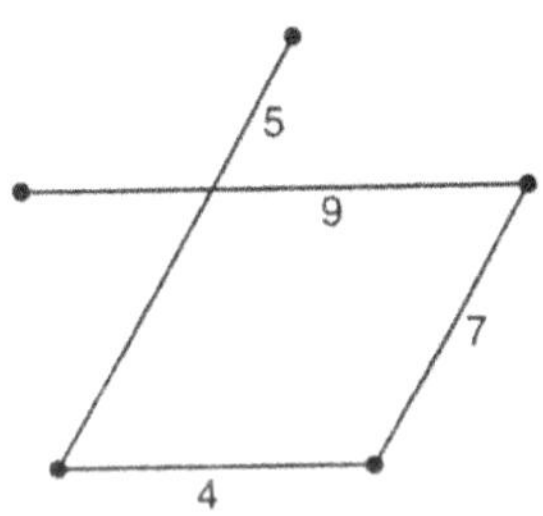

(*iv*)

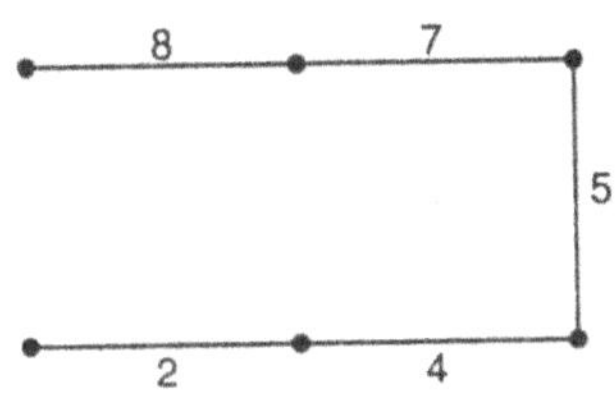

13

Finite State Machines

13.1. INTRODUCTION

A device which receives a set of input signals and produces a corresponding set of output signals is called an *information processing machine* or simply a *machine*. For example, a car is an information processing machine with the input signals being the depression of the accelerator and the angular position of the steering wheel and with the output signals being the speed and direction of the car. A digital computer is also an information processing machine with the user's program and data being the input signals and the result of the computation on the print-out being the output signals. A machine may or may not have memory. If a machine is without memory then its output at any instant depends only on the input at that instant. But if a machine has memory then its output at any instant depends on the input at that instant and also on the inputs at previous instants because the machine can recall the happenings of the past. A calculator is an example of a machine with memory. The summary of the past history of a machine at any particular instant is called its *state*. A machine may have any number of states. A machine with finite number of states is called a *finite state machine*.

In this chapter, we shall study mathematical models of finite state machines.

13.2. DIRECTED GRAPH

A **directed graph** G is defined as a pair (V, E), where V is a set of elements called **vertices** and E is a set of ordered pairs of vertices called **directed edges** or **arcs.**

The vertices of a directed graph are denoted by $v_1, v_2, ...$ and the directed edges are denoted by $e_1, e_2, ...$.

A directed graph is represented by a diagram in which the vertices are represented by small dots (or small circles) and each directed edge $e = (v_i, v_j)$ is represented by a line, drawn straight or curved, long or short, connecting the vertices v_i and v_j and an arrow is marked to show that this edge is from v_i to v_j. We say that this directed edge begins at v_i and ends at v_j.

For example, $G = (V, E)$ is a directed graph, where

$$V = \{v_1, v_2, v_3\} \text{ and } E = \{e_1, e_2, e_3, e_4, e_5\}$$

and $\quad e_1 = (v_2, v_3), e_2 = (v_1, v_2), e_3 = (v_3, v_1), e_4 = (v_3, v_1), e_5 = (v_3, v_3).$

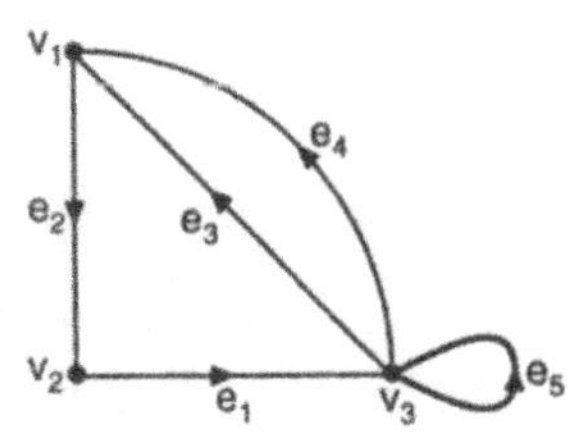

A directed graph is also called a **digraph.**

Let G be a directed graph. The **outdegree** of a vertex v of G is the number of edges beginning at v and the **indegree** of v is the number of edges ending at v. A vertex with zero indegree is called a **source** and a vertex with zero outdegree is called a **sink.**

In the above example,

$$\text{outdeg}(v_1) = 1, \text{outdeg}(v_2) = 1, \text{outdeg}(v_3) = 3$$
$$\text{indeg}(v_1) = 2, \quad \text{indeg}(v_2) = 1, \quad \text{indeg}(v_3) = 2.$$

13.3. ADJACENCY MATRIX OF A DIRECTED GRAPH

Let G be a directed graph with n vertices $v_1, v_2, \ldots\ldots, v_n$. The **adjacency matrix** of the directed graph G is a $n \times n$ matrix $A = [a_{ij}]$, where :

$$a_{ij} = \text{no. of directed edges from } v_i \text{ to } v_j.$$

The entries of the adjacency matrix A are all non-negative integers. Conversely, every $n \times n$ matrix A with non-negative integer entries uniquely defines a directed graph with n vertrices.

Example 1. *Find the adjacency matrix of the directed graph G given in the figure.*

Sol. The vertices of the directed graph are v_1, v_2, v_3 and v_4.

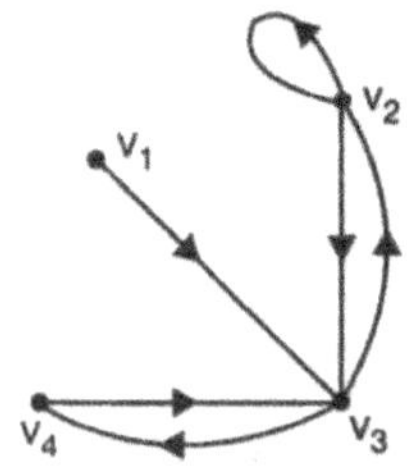

∴ The adjacancy matrix of G is a 4 × 4 matrix.

Let $A = [a_{ij}]$ be the adjacency matrix of the directed graph G.

The directed edges in the graph are (v_1, v_3), (v_2, v_2), (v_2, v_3), (v_3, v_2), (v_3, v_4) and (v_4, v_3).

There are no parallel directed edges in the graph.

∴ $a_{13} = 1, a_{22} = 1, a_{23} = 1, a_{32} = 1, a_{34} = 1, a_{43} = 0$ and all other entries of A are zero.

∴ Adjacency matrix of G

$$= A = \begin{bmatrix} a_{11} & a_{12} & a_{13} & a_{14} \\ a_{21} & a_{22} & a_{23} & a_{24} \\ a_{31} & a_{32} & a_{33} & a_{34} \\ a_{41} & a_{42} & a_{43} & a_{44} \end{bmatrix} = \begin{bmatrix} \mathbf{0} & \mathbf{0} & \mathbf{1} & \mathbf{0} \\ \mathbf{0} & \mathbf{1} & \mathbf{1} & \mathbf{0} \\ \mathbf{0} & \mathbf{1} & \mathbf{0} & \mathbf{1} \\ \mathbf{0} & \mathbf{0} & \mathbf{1} & \mathbf{0} \end{bmatrix}.$$

Example 2. *Find the directed graph whose adjacency matrix is*

$$\begin{bmatrix} 1 & 0 & 0 & 1 \\ 2 & 1 & 0 & 1 \\ 1 & 1 & 1 & 0 \\ 0 & 0 & 1 & 1 \end{bmatrix}.$$ (K.U. 2006 S)

Sol. Let
$$A = \begin{bmatrix} 1 & 0 & 0 & 1 \\ 2 & 1 & 0 & 1 \\ 1 & 1 & 1 & 0 \\ 0 & 0 & 1 & 1 \end{bmatrix}$$

The order of A is 4 × 4.

Let $G = (V, E)$ be the directed graph of A.

∴ G has four vertices. Let $V = \{v_1, v_2, v_3, v_4\}$

$a_{11} = 1 \Rightarrow$ There is a directed edge from v_1 to v_1 *i.e.*, a loop at v_1

$a_{14} = 1 \Rightarrow$ There is a directed edge from v_1 to v_4

$a_{21} = 2 \Rightarrow$ There are two directed edges from v_2 to v_1
$a_{22} = 1 \Rightarrow$ There is a loop at v_2
$a_{24} = 1 \Rightarrow$ There is a directed edge from v_2 to v_4
$a_{31} = 1 \Rightarrow$ There is a directed edge from v_3 to v_1
$a_{32} = 1 \Rightarrow$ There is a directed edge from v_3 to v_2
$a_{33} = 1 \Rightarrow$ There is a loop at v_3
$a_{43} = 1 \Rightarrow$ There is a directed edge from v_4 to v_3
$a_{44} = 1 \Rightarrow$ There is a loop at v_4

$\therefore \quad E = \{(v_1, v_1)\ (v_1, v_4), (v_2, v_1), (v_2, v_1), (v_2, v_2), (v_2, v_4), (v_3, v_1), (v_3, v_2), (v_3, v_3), (v_4, v_3), (v_4, v_4)]$

$\therefore$ The directed graph G is known. The diagram of the directed graph G is shown in the adjoining figure.

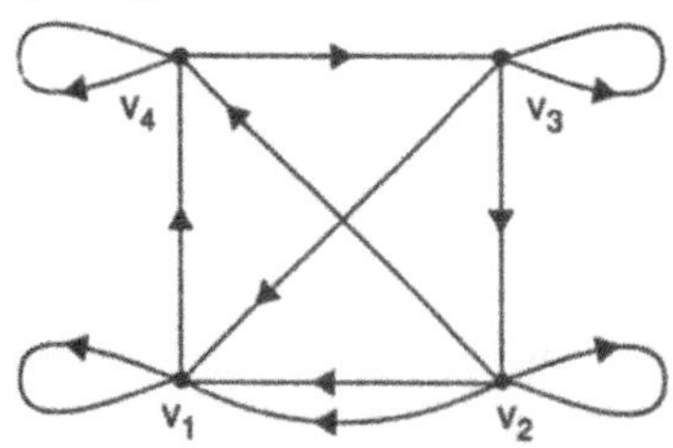

Example 3. *Find the directed graph whose adjacency matrix is* $\begin{bmatrix} 1 & 0 & 2 & 1 \\ 2 & 1 & 0 & 2 \\ 1 & 1 & 0 & 1 \\ 1 & 2 & 1 & 0 \end{bmatrix}$.

(KU 2007)

Sol. Let $\quad A = \begin{bmatrix} 1 & 0 & 2 & 1 \\ 2 & 1 & 0 & 2 \\ 1 & 1 & 0 & 1 \\ 1 & 2 & 1 & 0 \end{bmatrix}$

The order of A is 4×4.
Let $G = (V, E)$ be the directed graph of A.
$\therefore \quad G$ has four vertices. Let $V = \{v_1, v_2, v_3, v_4\}$.

$a_{11} = 1 \Rightarrow$ There is a directed edge from v_1 to v_1 *i.e.*, a loop at v_1
$a_{13} = 2 \Rightarrow$ There are two directed edges from v_1 to v_3
$a_{14} = 1 \Rightarrow$ There is a directed edge from v_1 to v_4
$a_{21} = 2 \Rightarrow$ There are two directed edges from v_2 to v_1
$a_{22} = 1 \Rightarrow$ There is a loop at v_2
$a_{24} = 2 \Rightarrow$ There are two directed edges from v_2 to v_4
$a_{31} = 1 \Rightarrow$ There is a directed edge from v_3 to v_1
$a_{32} = 1 \Rightarrow$ There is a directed edge from v_3 to v_2
$a_{34} = 1 \Rightarrow$ There is a directed edge from v_3 to v_4
$a_{41} = 1 \Rightarrow$ There is a directed edge from v_4 to v_1
$a_{42} = 2 \Rightarrow$ There are two directed edges from v_4 to v_2
$a_{43} = 1 \Rightarrow$ There is a directed edge from v_4 to v_3

$\therefore\quad E = \{(v_1, v_1), (v_1, v_3), (v_1, v_3), (v_1, v_4), (v_2, v_1), (v_2, v_1), (v_2, v_2), (v_2, v_4), (v_2, v_4), (v_3, v_1), (v_3, v_2), (v_3, v_4), (v_4, v_1), (v_4, v_2), (v_4, v_2), (v_4, v_3)\}$

$\therefore$ The directed graph G is known. The diagram of the directed graph G is shown in the adjoining figure.

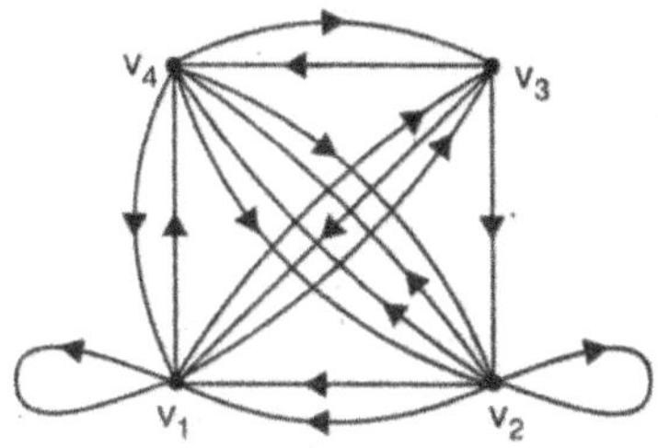

EXERCISE 13.1

1. Find the adjacency matrices of the following directed graphs :

(i)

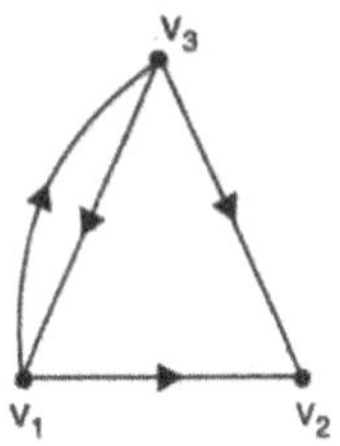

(ii)

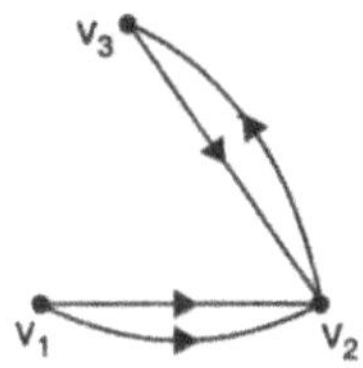

(iii)

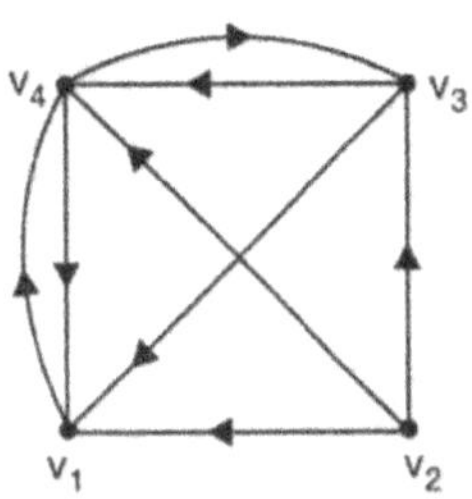

(iv)

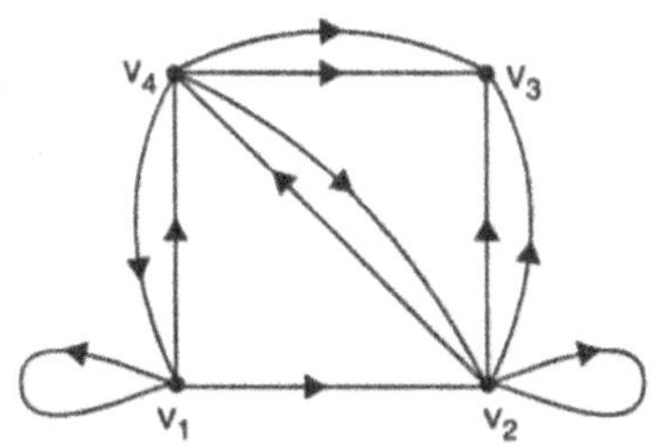

2. Find the directed graph whose adjacency matrix is :

(i) $\begin{bmatrix} 1 & 2 & 0 \\ 2 & 1 & 1 \\ 0 & 0 & 1 \end{bmatrix}$ (ii) $\begin{bmatrix} 2 & 1 & 1 \\ 0 & 2 & 0 \\ 1 & 1 & 1 \end{bmatrix}$

(iii) $\begin{bmatrix} 1 & 2 & 1 & 0 \\ 2 & 1 & 0 & 1 \\ 1 & 1 & 1 & 1 \\ 2 & 1 & 0 & 1 \end{bmatrix}$ (iv) $\begin{bmatrix} 1 & 2 & 1 & 2 \\ 2 & 0 & 2 & 1 \\ 1 & 2 & 1 & 0 \\ 2 & 1 & 0 & 0 \end{bmatrix}$. (KU 2006)

Answers

1. (i) $\begin{bmatrix} 0 & 1 & 1 \\ 0 & 0 & 0 \\ 1 & 1 & 0 \end{bmatrix}$ (ii) $\begin{bmatrix} 0 & 2 & 0 \\ 0 & 0 & 1 \\ 0 & 1 & 0 \end{bmatrix}$

(iii) $\begin{bmatrix} 0 & 0 & 0 & 1 \\ 1 & 0 & 1 & 1 \\ 1 & 0 & 0 & 1 \\ 1 & 0 & 1 & 0 \end{bmatrix}$ (iv) $\begin{bmatrix} 1 & 1 & 0 & 1 \\ 0 & 1 & 2 & 1 \\ 0 & 0 & 0 & 0 \\ 1 & 1 & 2 & 0 \end{bmatrix}$.

2. (*i*)

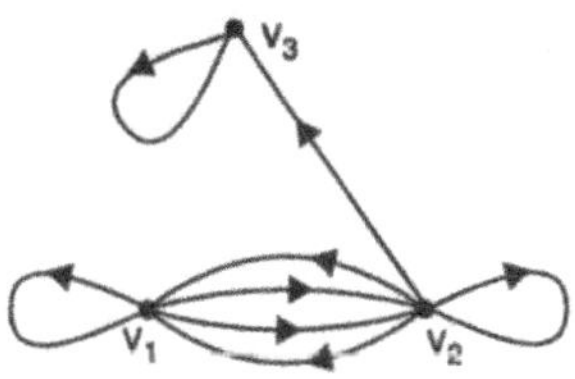

(*ii*)

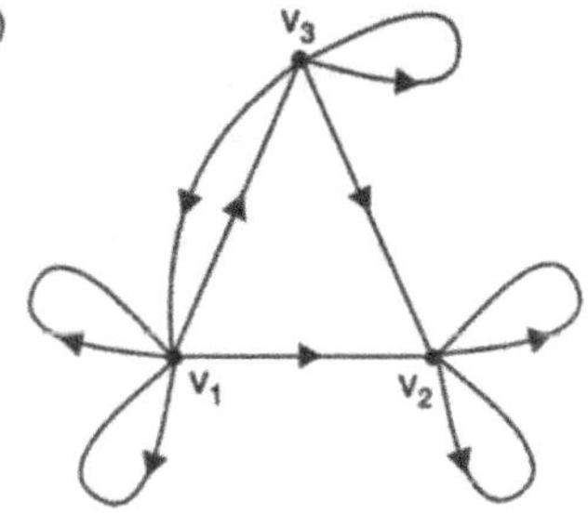

(*iii*)

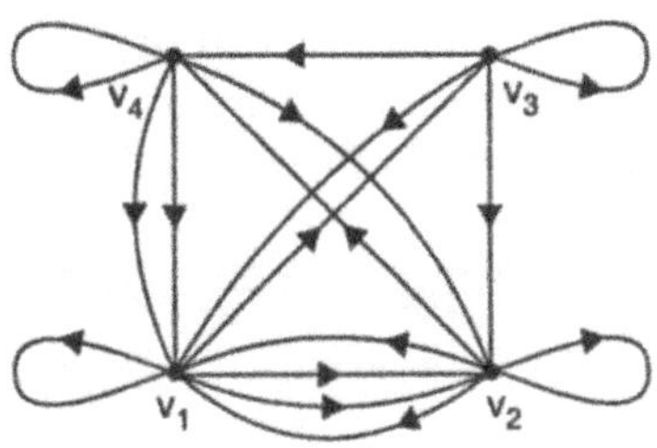

(*iv*)

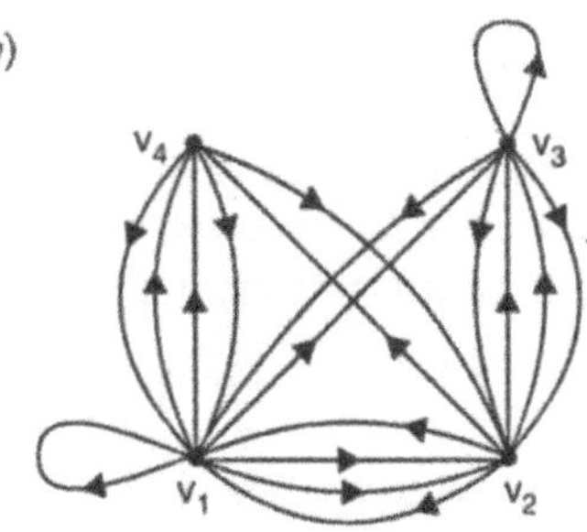

13.4. ALPHABET, WORDS, LANGUAGE

Any nonempty set of symbols is called an **alphabet.**

For example, $\{a, b, c, ..., y, z\}$, $\{a, b, c\}$, $\{0, 1\}$, $\{b, c, d\}$, $\{\alpha, \beta, \gamma\}$ are alphabets.

The elements of an alphabet are called its **letters.**

If A is an alphabet then a finite sequence of elements of A, without using commas, is called a **word** or a **string** on the alphabet A.

For example, let $A = \{a, b, c\}$ be an alphabet. The sequences *aabac, bbbac, cba* are words on the alphabet A.

We write *aa* as a^2, *aaa* as a^3 etc. Thus, the words *aabac* and *bbbacc* can be written as a^2bac and b^3ac^2 respectively.

The empty sequence of letters, denoted by λ, is also considered to be a word on A, called the **empty word.**

The set of all words on A is denoted by A^*.

A collection of words on an alphabet A is called a **language** over the alphabet A.

$\therefore$ A language over an aphabet A is a subset A^*.

For example, let $A = \{a, b, c\}$ be an alphabet. The set $\{ab^m c^n : m, n \in \mathbf{N}\}$ *i.e.*,

$\{abc, abc^2, ab^2c, ab^2c^2, ...\}$ is a language over the alphabet A.

13.5. FINITE STATE AUTOMATON (FSA) (MOORE MACHINE) (*KU 2007*)

A finite state automaton* is a finite state machine which does not 'print' an output using an output alphabet distinct from the input alphabet.

A finite state automaton M consists of following five parts :

1. A finite set I of inputs, called *input alphabet.*
2. A finite set S of states of M.
3. A subset Y of S, of 'accepting' or 'yes' states.
4. An initial state s_0 in S.

5. A next-state function F from $S \times I$ into S.

This finite state automaton is also denoted as $M(I, S, Y, s_0, F)$.

A finite state automaton is also called a **Moore machine.**

13.6. STATE DIAGRAM OF A FINITE STATE AUTOMATON

Let $M(I, S, Y, s_0, F)$ be a finite state automaton. The state diagram $D(M)$ of the given finite state automaton M is a labeled** directed graph drawn as follows :

The vertices of $D(M)$ are the elements of S *i.e.*, the states of the machine M. The vertices of $D(M)$ are shown by circles. The accepting states are shown by double circles.

If $s_i \in S$, $a_j \in I$ and $F(s_i, a_j)$*** $= s_k$ then we draw a directed edge in $D(M)$ beginning at s_i and ending at s_k and label this edge by the input a_j. The initial state s_0 is indicated by means of a special directed edge which terminates at s_0 but having no initial vertex. Also, if two or more inputs cause the same change of states then we label a single directed edge by all such inputs instead of having a directed edge for each of such inputs.

The diagram thus formed is called the **state diagram** of the given finite state automaton.

Remark. For each vertex s_i and each letter a_j in I, there is a directed edge leaving s_i and labeled by a_j.

$\therefore$ The outdegree of each vertex is equal to the number of letters in the alphabet I.

Example 1. *Let $M(I, S, Y, s_0, F)$ be a finite state automaton, where $I = \{a, b\}$, $S = \{s_0, s_1, s_2\}$, $Y = \{s_1\}$ and the function $F : S \times I \to S$ is given by the adjoining table.*

F	a	b
s_0	s_0	s_1
s_1	s_1	s_2
s_2	s_2	s_2

Draw a state diagram of this automaton.

Sol. $S = \{s_0, s_1, s_2\}$ implies that there are three vertices in the state diagram of M. $Y = \{s_1\}$ implies that the vertex s_1 is to be shown by a double circle. From the table of F, we find the directed edges.

$F(s_0, a) = s_0 \Rightarrow$ There is a loop at s_0 labeled by a.

$F(s_0, b) = s_1 \Rightarrow$ There is a directed edge from s_0 to s_1 labeled by b.

$F(s_1, a) = s_1 \Rightarrow$ There is a loop at s_1 labeled by a.

$F(s_1, b) = s_2 \Rightarrow$ There is a directed edge from s_1 to s_2 labeled by b.

$F(s_2, a) = s_2 \Rightarrow$ There is a loop at s_2 labeled by a.

$F(s_2, b) = s_2 \Rightarrow$ There is a loop at s_2 labeled by b.

The state diagram of the given automaton is shown in the figure.

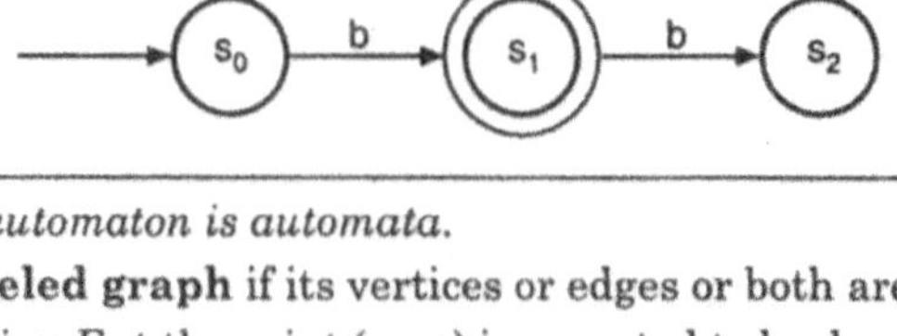

*The plural of the word *automaton is automata.*

A graph is called a **labeled graph if its vertices or edges or both are assigned some specific data.

***The value of the function F at the point (s_i, a_j) is expected to be denoted by $F((s_i, a_j))$. For the sake of simplicity, we are omitting one pair of parentheses and write this value as $F(s_i, a_j)$.

Remark. The above state diagram can also be drawn as shown in the diagram.

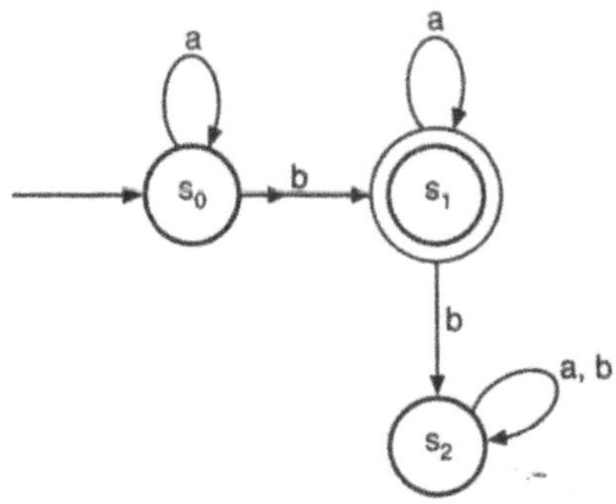

Example 2. *Let $M(I, S, Y, s_0, F)$ be a finite state automaton where $I = \{a, b\}$, $S = \{s_0, s_1, s_2, s_3\}$, $Y = \{s_3\}$ and the function $F : S \times I \to S$ is given by the adjoining table.*

F	a	b
s_0	s_0	s_1
s_1	s_1	s_2
s_2	s_2	s_3
s_3	s_3	s_0

Draw a state diagram of this automaton.

Sol. $S = \{s_0, s_1, s_2, s_3\}$ implies that there are four vertices in the state diagram of M. $Y = \{s_3\}$ implies that the vertex s_3 is to be shown by a double circle. From the table of F, we find the directed edges.

$F(s_0, a) = s_0 \Rightarrow$ There is a loop at s_0 labeled by a.
$F(s_0, b) = s_1 \Rightarrow$ There is a directed edge from s_0 to s_1 labeled by b.
$F(s_1, a) = s_1 \Rightarrow$ There is a loop at s_1 labeled by a.
$F(s_1, b) = s_2 \Rightarrow$ There is a directed edge from s_1 to s_2 labeled by b.
$F(s_2, a) = s_2 \Rightarrow$ There is a loop at s_2 labeled by a.
$F(s_2, b) = s_3 \Rightarrow$ There is a directed edge from s_2 to s_3 labeled by b.
$F(s_3, a) = s_3 \Rightarrow$ There is a loop at s_3 labeled by a.
$F(s_3, b) = s_0 \Rightarrow$ There is a directed edge from s_3 to s_0 labeled by b.

The state diagram of the given automaton is shown in the figure.

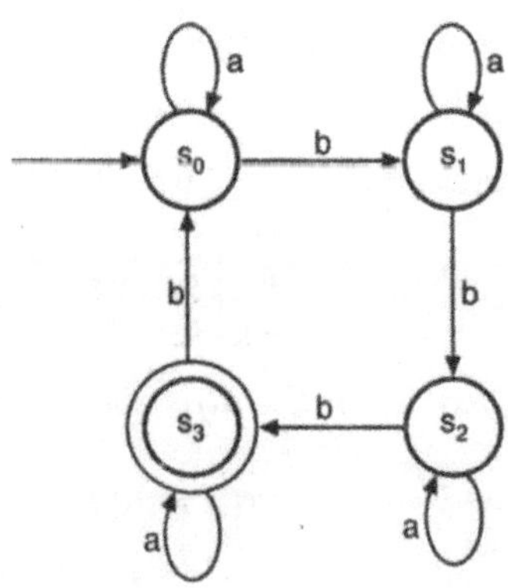

Example 3. *Find the finite state machine whose state diagram is given in the figure :*

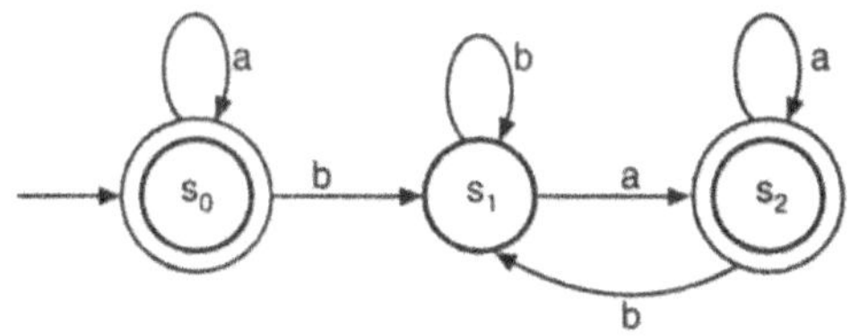

Sol. The vertices in the diagram are s_0, s_1 and s_2.

$\therefore$ $S = \{s_0, s_1, s_2\}$

Vertices marked by double circles are s_0 and s_2.

$\therefore$ $Y = \{s_0, s_2\}$

Also, the initial state in S is s_0. The directed edges are labeled by the symbols a, b.

$\therefore$ $I = \{a, b\}$

From the diagram we have

$$F(s_0, a) = s_0,\ F(s_0, b) = s_1,\ F(s_1, a) = s_2,$$
$$F(s_1, b) = s_1,\ F(s_2, a) = s_2,\ F(s_2, b) = s_1.$$

$\therefore$ The required finite state automaton is $M(I, S, Y, s_0, F)$, where $I = \{a, b\}$, $S = \{s_0, s_1, s_2\}$, $Y = \{s_0, s_2\}$ and the function F is given by the adjoining table.

F	a	b
s_0	s_0	s_1
s_1	s_2	s_1
s_2	s_2	s_1

13.7. LANGUAGE DETERMINED BY A FINITE STATE AUTOMATON

Let $M(I, S, Y, s_0, F)$ be a finite state automaton.

$\therefore$ I is an alphabet.

Let $w = a_1a_2 \ldots\ldots a_m$ be a word on the alphabet I.

The initial state is s_0.

Let $F(s_{i-1}, a_i)$ be denoted by s_i, where $1 \le i \le m$.

$\therefore$ We get states $s_1, s_2, \ldots\ldots, s_m$. Some of these s_i's may be identical.

If $s_m \in Y$ *i.e.*, if s_m is an accepting state in S then we say that the automaton M recognizes or accepts the word w. The set of all words of the alphabet I which are accepted by the automaton M is called the **language determined by the automaton** M and is denoted by $L(M)$.

Example 4. *Let $M(I, S, Y, s_0, F)$ be a finite state automaton, where $I = \{a, b\}$, $S = \{s_0, s_1, s_2\}$, $Y = \{s_2\}$ and F is given be the adjoining table.*

F	a	b
s_0	s_0	s_1
s_1	s_1	s_2
s_2	s_0	s_1

Find whether the following words on the alphabet I are accepted by the automaton M or not :

(*i*) *abaa* (*ii*) *babbb*

(*iii*) *abaab* (*iv*) *babababab.*

Sol. (*i*) The word is *abaa*.

We have $F(s_0, a) = s_0,\ F(s_0, b) = s_1,\ F(s_1, a) = s_1, F(s_1, a) = s_1.$

$\therefore$ The sequence of states for this word is

$$s_0 \xrightarrow{a} s_0 \xrightarrow{b} s_1 \xrightarrow{a} s_1 \xrightarrow{a} s_1$$

Also, $s_1 \notin Y$.

$\therefore$ The word *abaa* is not accepted by the automaton *M*.

(*ii*) The word is *babbb*.

The sequence of states for this word is

$$s_0 \xrightarrow{b} s_1 \xrightarrow{a} s_1 \xrightarrow{b} s_2 \xrightarrow{b} s_1 \xrightarrow{b} s_2$$

Also, $s_2 \in Y$.

$\therefore$ The word *babbb* is accepted by the automaton *M*.

(*iii*) The word is *abaab*.

The sequence of states for this word is

$$s_0 \xrightarrow{a} s_0 \xrightarrow{b} s_1 \xrightarrow{a} s_1 \xrightarrow{a} s_1 \xrightarrow{b} s_2$$

Also, $s_2 \in Y$. $\therefore$ The word *abaab* is accepted by the automaton *M*.

(*iv*) The word is *babababab*.

The sequence of states for this word is

$$s_0 \xrightarrow{b} s_1 \xrightarrow{a} s_1 \xrightarrow{b} s_2 \xrightarrow{a} s_0 \xrightarrow{b} s_1 \xrightarrow{a} s_1 \xrightarrow{b} s_2 \xrightarrow{a} s_0 \xrightarrow{b} s_1$$

Also, $s_1 \notin Y$. $\therefore$ The word *babababab* is not accepted by the automaton *M*.

Example 5. *Let the state diagram of a finite state automaton be as shown in the figure. Find whether the following words on the alphabet {a, b} are accepted by this automaton or not :*

(*i*) *aaba*

(*ii*) *aaab*

(*iii*) *aabbb*

(*iv*) *aaaab*.

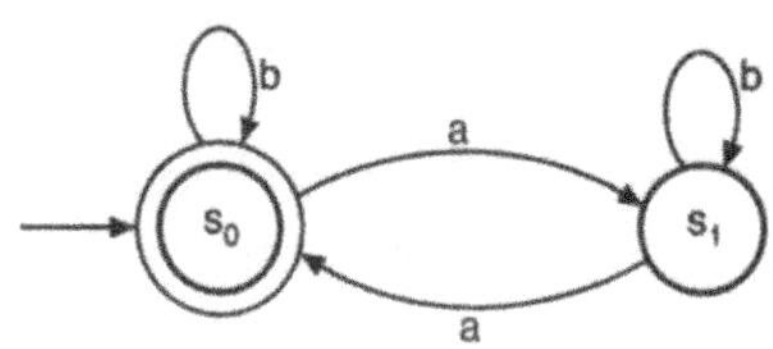

Sol. (*i*) The word is *aaba*.

The sequence of states for this word is

$$s_0 \xrightarrow{a} s_1 \xrightarrow{a} s_0 \xrightarrow{b} s_0 \xrightarrow{a} s_1$$

Since s_1 is not an accepting state, the word *aaba* is not accepted by the given automaton.

(*ii*) The word is *aaab*.

The sequence of states for this word is

$$s_0 \xrightarrow{a} s_1 \xrightarrow{a} s_0 \xrightarrow{a} s_1 \xrightarrow{b} s_1$$

Since s_1 is not an accepting state, the word *aaab* is not accepted by the given automaton.

(*iii*) The word is *aabbb*.

The sequence of states for this word is

$$s_0 \xrightarrow{a} s_1 \xrightarrow{a} s_0 \xrightarrow{b} s_0 \xrightarrow{b} s_0 \xrightarrow{b} s_0$$

Since s_0 is an accepting state, the word *aabbb* is accepted by the given automaton.

(*iv*) The given word is *aaaab*.

The sequence of states for this word is

$$s_0 \xrightarrow{a} s_1 \xrightarrow{a} s_0 \xrightarrow{a} s_1 \xrightarrow{a} s_0 \xrightarrow{b} s_0$$

Since s_0 is an accepting state, the word *aaaab* is accepted by the given automaton.

Example 6. *Design a finite state automaton that accepts only those strings over {0, 1}, that contain an even number of 1's.* (KU 2006)

Sol. Let $M(I, S, Y, s_0, F)$ be the required finite state automaton.

$\therefore$ $I = \{0, 1\}$ and s_0 is the initial state.

The automaton M is to accept only those words in which the number of 1 is any one of the numbers 0, 2, 4, 6, ...

$\therefore$ The word 0 is acceptable.

$\therefore$ We take s_0 as an accepting state and a loop at s_0 labeled by 0.

The word 1 is not acceptable.

$\therefore$ We introduce a non-accepting state s_1 and a directed edge from s_0 to s_1 labeled 1. We mark a loop labeled by 0 at s_1.

$\therefore$ The letter 0 will not cause a change in any state.

We mark a directed edge from s_1 to s_0 and label it 1.

Now, if the number of 1 in the word is even, the automaton is in state s_0 and if the number of 1 in the word is odd, the automaton is in non-accepting state s_1.

The state diagram of the required automaton is shown in the figure. In this automaton, we have $S = \{s_0, s_1\}$,

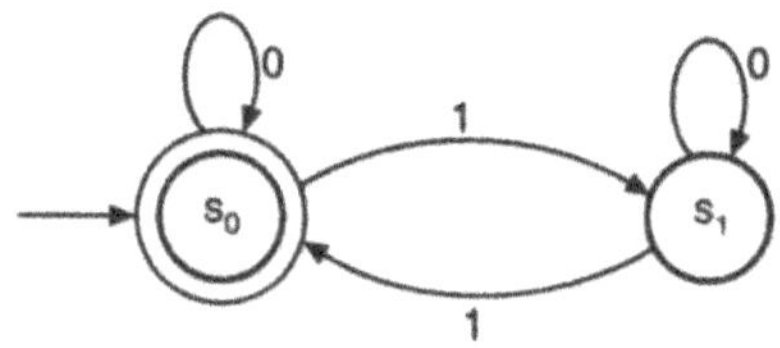

$Y = \{s_0\}$ and the function $F : S \times I \to S$ is given by the table in the figure.

F	0	1
s_0	s_0	s_1
s_1	s_1	s_0

Example 7. *Construct a finite state automaton that accepts only those strings over {0, 1} for which the last two input symbols are 1.* (KU 2006 C)

Sol. Let $M(I, s, Y, s_0, F)$ be the required finite state automaton.

$\therefore$ $I = \{0, 1\}$ and s_0 is the initial state. The automaton M is to accept only those strings which end in two 1's.

$\therefore$ The empty string (λ) and 1 are not acceptable and 11 is acceptable.

$\therefore$ We require three states s_0, s_1 and s_2.

We introduce directed edge from s_0 to s_1 labeled 1, directed edge from s_1 to s_2 labeled 1. We take only s_2 as an accepting state. We introduce a loop at s_2 labeled by 1.

$\therefore$ Before the last two 1's, if we are in any state, the last two 1's will make us to stay in the accepting state s_2. Also if there is any '0' in the string we shall want to go back to s_0. For this we introduce the following loop and directed edges :

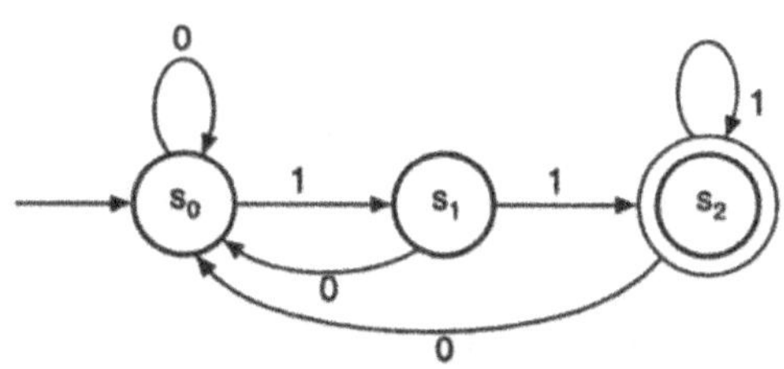

Loop at s_0 labeled 0, directed edge from s_1 to s_0 labeled 0, directed edge from s_2 to s_0 labeled 0.

The state diagram of the required automaton is shown in the figure. In this automaton, we have

$S = \{s_0, s_1, s_2\}$, $Y = \{s_2\}$ and the function $F : S \times I \to S$ is given by the table in the figure.

F	0	1
s_0	s_0	s_1
s_1	s_0	s_2
s_2	s_0	s_2

EXERCISE 13.2

1. Define a Moore machine and give an example of the same. (*KU 2007*)
2. Let $M(I, S, Y, s_0, F)$ be a finite state automaton, where $I = \{a, b\}$, $S = \{s_0, s_1, s_2\}$, $Y = \{s_2\}$ and the function $F : S \times I \to S$ is given by the adjoining table.

F	a	b
s_0	s_0	s_1
s_1	s_2	s_1
s_2	s_2	s_2

Draw a state diagram of this automaton.

3. Let $M(I, S, Y, s_0, F)$ be a finite state automaton, where $I = \{a, b\}$, $S = \{s_0, s_1, s_2\}$, $Y = \{s_1\}$ and the function $F : S \times I \to S$ is given by the adjoining table.

F	a	b
s_0	s_1	s_2
s_1	s_2	s_1
s_2	s_2	s_2

Draw a state diagram of this automaton.

4. Find the finite state automaton, whose state diagram is given in the figure.

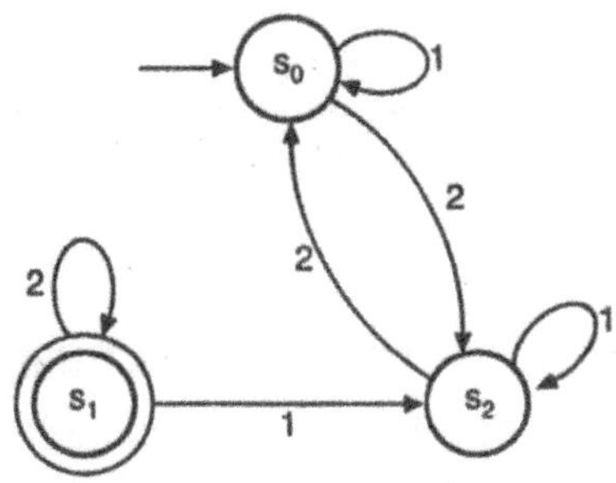

5. Let $M(I, S, Y, s_0, F)$, be a finite state automaton, where $I = \{0, 1\}$, $S = \{s_0, s_1, s_2\}$, $Y = \{s_1\}$ and the function $F : S \times I \to S$ is given by the adjoining table.

F	0	1
s_0	s_0	s_1
s_1	s_2	s_0
s_2	s_1	s_2

Find which of the following words on the alphabet $I = \{0, 1\}$ are accepted by the automaton :

(*i*) 011 (*ii*) 0010

(*iii*) 11100 (*iv*) 0001101101.

6. Let the state diagram of a finite state automaton be shown in the figure.
Find which of the following words on the alphabet I = {0, 1} are accepted by the automaton :

(*i*) *abab*

(*ii*) *abbb*

(*iii*) *aaabbb*

(*iv*) *bababaabbb*.

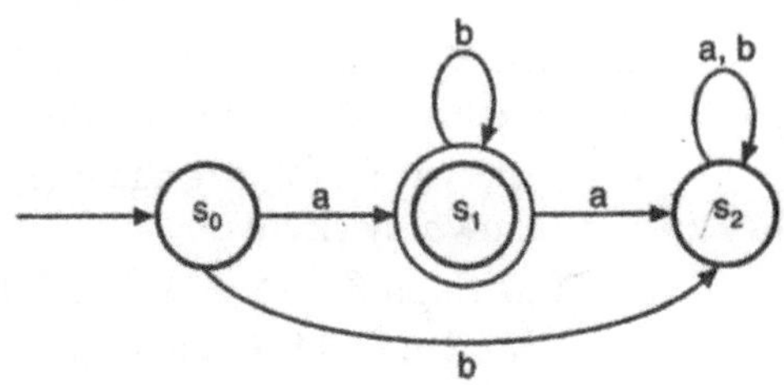

7. Let $I = \{l, m\}$. Construct a finite machine automaton which will accept only those words from I which end in two l's.
8. Find the state diagram of a finite state automaton that accepts only those words over $\{a, b\}$ that contain an even number of a.

Answers

2. 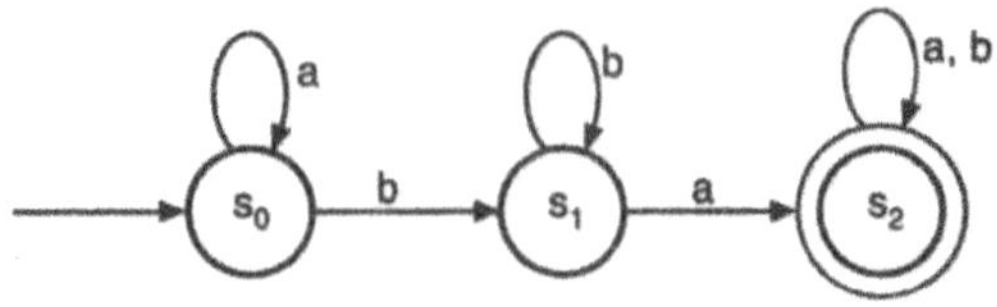

3. 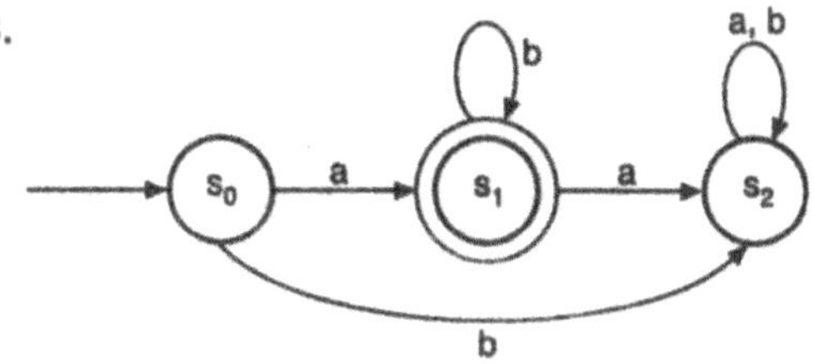

4. $M(I, S, Y, s_0, F)$, where $I = \{1, 2\}$, $S = \{s_0, s_1, s_2\}$, $Y = \{s_1\}$,

F	1	2
s_0	s_0	s_2
s_1	s_2	s_1
s_2	s_2	s_0

5. (*iii*), (*iv*)

6. (*ii*)

7. 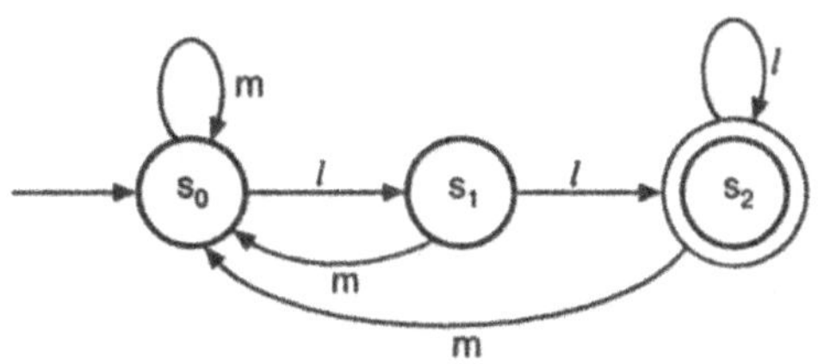

8. 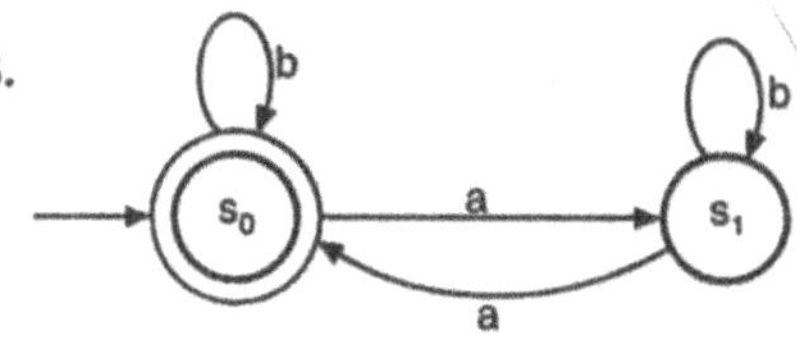

13.8. FINITE STATE MACHINE (FSM)

In this section, we shall study finite state machines which also 'print' an output using an output alphabet distinct from input alphabet.

A general finite state machine M consists of following six parts :

1. A finite set I of inputs, called *input alphabet.*
2. A finite set S of states of M.
3. A finite set O of outputs, called *output alphabet.*
4. An initial state s_0 in S.
5. A next-state function F from $S \times I$ into S.
6. An output function G from $S \times I$ into O.

This finite state machine is also denoted as $M(I, S, O, s_0, F, G)$.

13.9. STATE DIAGRAM OF A FINITE STATE MACHINE

Let $M(I, S, O, s_0, F, G)$ be a finite state machine. The state diagram $D(M)$ of the given finite state machine M is a labeled directed graph drawn as follows :

The vertices of $D(M)$ are the elements of S *i.e.,* the states of the machine M. The vertices of $D(M)$ are shown by circles.

The functions $F : S \times I \to S$ and $G : S \times I \to O$ define a function

$H : S \times I \to S \times O$ by

$$H(s_i, a_j) = (F(s_i, a_j), G(s_i, a_j)), \text{ where } (s_i, a_j) \in S \times I.$$

If $s_i \in S$, $a_j \in I$ and $H(s_i, a_j) = (F(s_i, a_j), G(s_i, a_j)) = (s_k, x_r)$, say, then we draw a directed edge in $D(M)$ beginning at s_i and ending at s_k and label this edge by the pair of input a_j and output x_r. The input a_j is placed near the base of edge and the output x_r is placed in the middle of the edge. The initial state s_0 is indicated by means of a special directed edge which terminates at s_0 but having no initial vertex. The diagram thus formed is called the *state diagram* of the given finite state machine.

Remark. Each directed edge in the state diagram represents a transition from the present state to the next state. $H(s_i, a_j) = (s_k, x_r)$ implies that there is a directed edge (s_i, s_k) and labeled by the pair a_j, x_r. This directed edge signifies that if the present state of the machine is s_i and the present input is a_j then the result is output x_r and the next state will be s_k.

13.10. EQUIVALENCE OF FINITE STATE MACHINES

Two finite state machines are said to be **equivalent** if starting from their respective initial states, they will produce the same output sequence when they are given the same input sequence.

Example 1. *Let $M(I, S, O, s_0, F, G)$ be a finite state machine, where $I = \{a, b\}$, $S = \{s_0, s_1, s_2, s_3\}$, $O = \{x, y, z\}$ and the functions F and G are defined by :*

$$F(s_0, a) = s_1, \quad F(s_0, b) = s_0, \quad F(s_1, a) = s_2, \quad F(s_1, b) = s_3$$
$$F(s_2, a) = s_3, \quad F(s_2, b) = s_1, \quad F(s_3, a) = s_1, \quad F(s_3, b) = s_0,$$
$$G(s_0, a) = x, \quad G(s_0, b) = y, \quad G(s_1, a) = z, \quad G(s, b) = z,$$
$$G(s_2, a) = z, \quad G(s_2, b) = y, \quad G(s_3, a) = x, \quad G(s_3, b) = z.$$

Draw the state diagram of the machine M.

Sol. $S = \{s_0, s_1, s_2, s_3\}$ implies that there are four vertices in the state diagram of M.

Let $H : S \times I \to S \times O$ be defined by $H(s_i, a_j) = (F(s_i, a_j), G(s_i, a_j))$

$$\therefore \quad H(s_0, a) = (s_1, x), \quad H(s_0, b) = (s_0, y),$$
$$H(s_1, a) = (s_2, z), \quad H(s_1, b) = (s_3, z),$$
$$H(s_2, a) = (s_3, z), \quad H(s_2, b) = (s_1, y),$$
$$H(s_3, a) = (s_1, x), \quad H(s_3, b) = (s_0, z).$$

The function H can also be shown in a tabular form as given in the figure.

From the function H, we find the directed edges.

H	a	b
s_0	s_1, x	s_0, y
s_1	s_2, z	s_3, z
s_2	s_3, z	s_1, y
s_3	s_1, x	s_0, z

$H(s_0, a) = (s_1, x) \Rightarrow$ There is a directed edge from s_0 to s_1 labeled by the pair a, x

$H(s_0, b) = (s_0, y) \Rightarrow$ There is a loop at s_0 labeled by the pair b, y

$H(s_1, a) = (s_2, z) \Rightarrow$ There is a directed edge from s_1 to s_2 labeled by the pair a, z

$H(s_1, b) = (s_3, z) \Rightarrow$ There is a directed edge from s_1 to s_3 labeled by the pair b, z

$H(s_2, a) = (s_3, z) \Rightarrow$ There is a directed edge from s_2 to s_3 labeled by the pair a, z

$H(s_2, b) = (s_1, y) \Rightarrow$ There is a directed edge from s_2 to s_1 labeled by the pair b, y

$H(s_3, a) = (s_1, x) \Rightarrow$ There is a directed edge from s_3 to s_1 labeled by the pair a, x

$H(s_3, b) = (s_0, z) \Rightarrow$ There is a directed edge from s_3 to s_0 labeled by the pair b, z

The state diagram of the given finite state machine is shown in the figure.

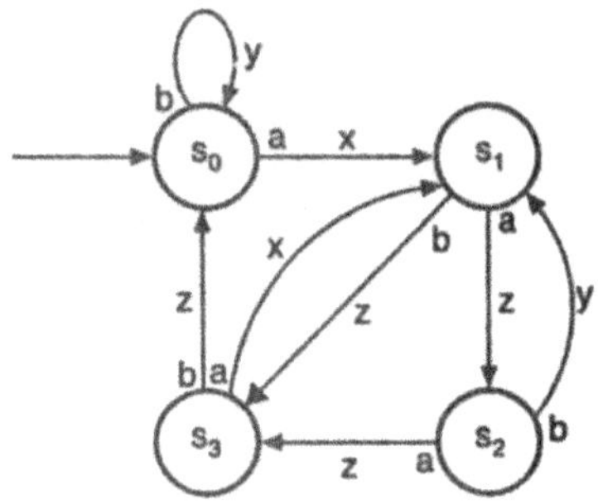

Remark. Table of the function $H : S \times I \to S \times O$ is called the **state table** of the given finite state mechine.

Example 2. *Let $M(I, S, O, s_0, F, G)$ be a finite state machine, where $I = \{a, b\}$, $S = \{s_0, s_1, s_2, s_3\}$, $O = \{x, y, z\}$. Let $H : S \times I \to S \times O$ be the functions defined by $H(s_i, a_j) = (F(s_i, a_j)), G(s_i, a_j))$, where $s_i \in S$, $a_j \in I$. The values of the function H are given in the adjoining table.*

H	a	b
s_0	s_2, y	s_1, z
s_1	s_2, x	s_3, y
s_2	s_2, y	s_1, z
s_3	s_3, z	s_0, x

(*i*) *Draw the state diagram of M* :

(*ii*) *Find the output word if the input word is $a^2b^2ab^2a^2b$.*

Sol. (*i*) $S = \{s_0, s_1, s_2, s_3\}$ implies that there are four vertices in the state diagram of ***M***.

$H(s_0, a) = (s_2, y) \Rightarrow$ There is a directed edge from s_0 to s_2 labeled by the pair a, y
$H(s_0, b) = (s_1, z) \Rightarrow$ There is a directed edge from s_0 to s_1 labeled by the pair b, z
$H(s_1, a) = (s_2, x) \Rightarrow$ There is a directed edge from s_1 to s_2 labeled by the pair a, x
$H(s_1, b) = (s_3, y) \Rightarrow$ There is a directed edge from s_1 to s_3 labeled by the pair b, y
$H(s_2, a) = (s_2, y) \Rightarrow$ There is a loop at s_2 labeled by the pair a, y
$H(s_2, b) = (s_1, z) \Rightarrow$ There is a directed edge from s_2 to s_1 labeled by the pair b, z
$H(s_3, a) = (s_3, z) \Rightarrow$ There is a loop at s_3 labeled by the pair a, z
$H(s_3, b) = (s_0, x) \Rightarrow$ There is a directed edge from s_3 to s_0 labeled by the pair b, x

The state diagram of the given finite state machine is shown in the figure.

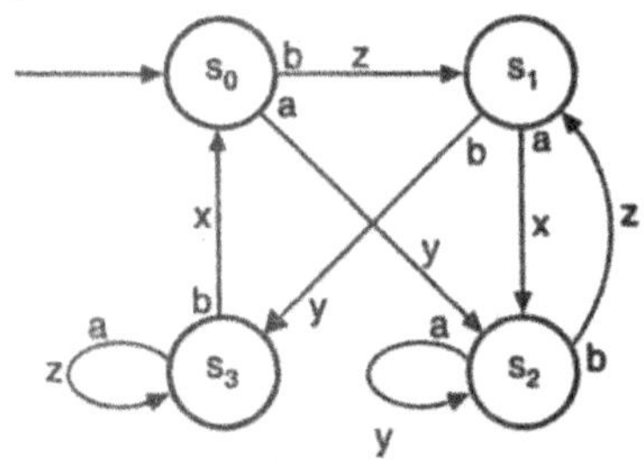

(*ii*) The given input word is $a^2b^2ab^2a^2b$, *i.e.*, *aabbabbaab*. Starting at the initial state s_0, we move from state to state by the edges which are labeled respectively by the given input symbols.

$\therefore$ We have

$$s_0 \xrightarrow{a} s_2 \xrightarrow{a} s_2 \xrightarrow{b} s_1 \xrightarrow{b} s_3 \xrightarrow{a} s_3 \xrightarrow{b} s_0 \xrightarrow{b} s_1 \xrightarrow{a} s_2 \xrightarrow{a} s_2 \xrightarrow{b} s_1$$

Using the output letters on the above states, in order, we get the output word *yyzyzxzxyz* or $\mathbf{y^2zyzxzxyz}$.

EXERCISE 13.3

1. Let $M(I, S, O, s_0, F, G)$ be a finite state machine, where $I = \{\alpha, \beta\}$, $S = \{s_0, s_1, s_2\}$, $O = \{x, y, z\}$ and the functions $F : S \times I \to S$ and $G : S \times I \to O$ as defined as follows

$$F(s_0, \alpha) = s_1, \quad F(s_0, \beta) = s_2, \quad F(s_1, \alpha) = s_2,$$
$$F(s_1, \beta) = s_1, \quad F(s_2, \alpha) = s_0, \quad F(s_2, \beta) = s_1,$$
$$G(s_0, \alpha) = x, \quad G(s_0, \beta) = y, \quad G(s_1, \alpha) = x,$$
$$G(s_1, \beta) = z, \quad G(s_2, \alpha) = z, \quad G(s_2, \beta) = y.$$

Draw the state diagram of the machine M.

2. Let $M(I, S, O, s_0, F, G)$ be a finite state machine, where $I = \{a, b\}$, $S = \{s_0, s_1, s_2\}$, $O = \{x, y, z\}$. Let $H : S \times I \to S \times O$ be the function defined by $H(s_i, a_j) = (F(s_i, a_j), G(s_i, a_j))$, where $s_i \in S$, $a_j \in I$. The values of the function H are given in the adjoining table.

H	a	b
s_0	s_0, z	s_2, y
s_1	s_0, y	s_2, x
s_2	s_1, x	s_2, z

(i) Draw the state diagram of M.

(ii) Find the output word if the input word is $aba^2b^2ab^2a^2ba^2$.

Answers

1.

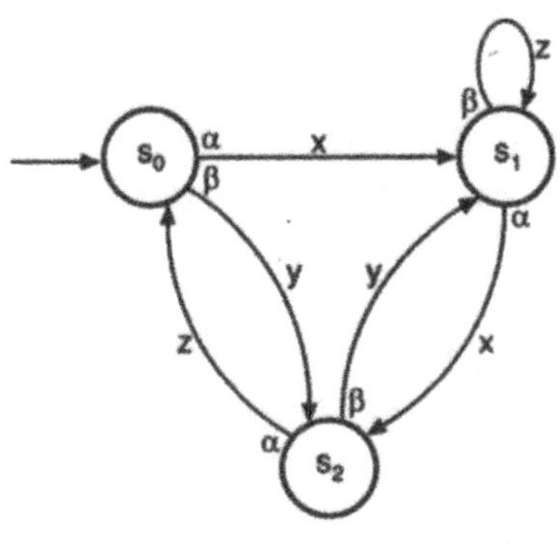

2. (i)

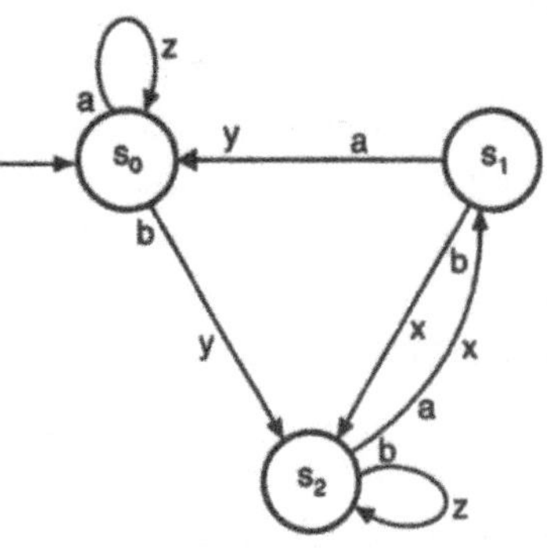

(ii) $zyxy^2zx^2zxy^2xy$.

EXAMINATION PAPER

K.U., KURUKSHETRA

BA./B.Sc., Part III, Examination, April 2007

Paper III (BM 304)

Discrete Mathematics

Time : 3 Hours] [*Maximum Marks : B.A—30*
B.Sc.—45

Note : Attempt any **Five** questions, selecting at least **One** question from each section. All questions carry equal marks.

SECTION A

1. (*a*) Find the number of equivalence relations that can be defined on a set having three elements.

(*b*) Prove that the interval [0, 1] is uncountable.

2. (*a*) How many people among 4,00,000 people are born at the same time ?

(*b*) A bag contains 5 white, 7 black and 6 red balls. A ball is drawn at random. Find the probability that it is a red or a white ball.

3. (*a*) State and prove Baye's Theorem.

(*b*) A coin is tossed 6 times. What is the probability of getting at least 4 heads ?

SECTION B

4. (*a*) Solve the recurrence relation

$$a_n = 2a_{n-1} - a_{n-2},\ n \geq 2, \text{ where } a_0 = 1,\ a_1 = 4.$$

(*b*) Find the total solutions of the difference equation

$$a_n - 5a_{n-1} + 6a_{n-2} = 1.$$

5. (*a*) Prove that the generating function $(1 - 4z)^{-2}$ can be expressed as $a_n = (n + 1)4^n$.

(*b*) Construct a truth table for the statement $(p \vee q) \wedge \sim r$.

SECTION C

6. (*a*) Prove that every chain is a distributive lattice.

(*b*) Let L be a complemented and distributive lattice. Prove that the join-irreducible elements of L other than 0 are its atoms.

7. (*a*) Give an example of a lattice which is not distributive.

(*b*) Let B be a Boolean algebra. Prove that if $a, b \in$ B then $(a \vee b)' = a' \wedge b'$ and $(a \wedge b)' = a' \vee b'$.

8. (*a*) Express $E(x, y, z) = x(y'z')'$ in its complete sum-of-products form.
   (*b*) Construct the truth-table for the Boolean polynomial
   $$p : B_3 \to B = \{0, 1\}, \quad \text{where } p(x, y, z) = (x \wedge y) \vee (x \vee (y' \wedge z)).$$

SECTION D

9. (*a*) Prove that if every vertex of a connected graph has even degree then G has a Euler circuit.
   (*b*) Prove that a graph G with $e = v - 1$ that has no circuit must be a tree.
10. (*a*) Find the graph where adjacency matrix is
   $$\begin{bmatrix} 1 & 0 & 2 & 1 \\ 2 & 1 & 0 & 2 \\ 1 & 1 & 0 & 1 \\ 1 & 2 & 1 & 0 \end{bmatrix}$$
   (*b*) Dэfine a Moore Machine and given an example of the same.

Solution of Question Not Already Covered in the Book

1. (*a*) Let $A = \{a, b, c\}$ be a set containing three elements. We know that every equivalence relation on A give rise to a partition of A and every partition of A give rise to an equivalence relation on A. Thus, the number of equivalence relations on A is exactly same as the number of partitions of A. The possible partitions of A are $\{a\}, \{b\}, \{c\}$; $\{a\}, \{b, c\}$; $\{b\}, \{a, c\}$; $\{c\}, \{a, b\}$ and $\{a, b, c\}$.
   These are 5 in number.
   $\therefore$ Exactly 5 equivalence relations can be defined on the set A.